AF357325

Process Design and Sustainable Development

Process Design and Sustainable Development

Editor

Peter Glavič

Basel • Beijing • Wuhan • Barcelona • Belgrade • Novi Sad • Cluj • Manchester

Editor
Peter Glavič
University of Maribor
Maribor
Slovenia

Editorial Office
MDPI
St. Alban-Anlage 66
4052 Basel, Switzerland

This is a reprint of articles from the Special Issue published online in the open access journal *Processes* (ISSN 2227-9717) (available at: https://www.mdpi.com/journal/processes/special_issues/Process_Design_Development).

For citation purposes, cite each article independently as indicated on the article page online and as indicated below:

Lastname, A.A.; Lastname, B.B. Article Title. *Journal Name* **Year**, *Volume Number*, Page Range.

ISBN 978-3-7258-0707-9 (Hbk)
ISBN 978-3-7258-0708-6 (PDF)
doi.org/10.3390/books978-3-7258-0708-6

Contents

About the Editor

Peter Glavič

Peter Glavič graduated in Chemical Technology and, later, in Economics and Business; he then earned his Master and Doctoral degrees in Chemistry. Prof Glavič held managerial positions in the paper, chemical, ceramics, and metallurgical industries for 9 years. He served as a professor of Chemical Engineering for 30 years. His research focused on process systems' engineering, environmental engineering, education, and sustainable development. He participated in many international research projects financed by the EU, the NATO, national agencies, companies, etc. He published more than 120 scientific articles, 110 papers at international conferences, 85 scientific reports, 140 professional reports, and 20 textbooks. Prof Glavič has more than 3200 citations and an h-index of 24. He was a Member of the Slovenian Parliament for 8 years, the Vice-Rector and President of the Slovenian Academy of Engineering, and the Chairman of several professional bodies in the Slovenian Chemical Society, the Society of Economists in Maribor, and the University Centre for professors emeriti and retired professors. He is currently the Editor-in-Chief of MDPI's Standards journal, a Guest Editor of Processes, and a member of the Editorial Board in some other journals. He has previously been a member of many international scientific committees and professional bodies.

Editorial

Special Issue on "Process Design and Sustainable Development"

Peter Glavič

Department of Chemistry and Chemical Engineering, University of Maribor, Smetanova 17, 2000 Maribor, Slovenia; peter.glavic@um.si; Tel.: +386-414-09611

Citation: Glavič, P. Special Issue on "Process Design and Sustainable Development". *Processes* **2023**, *11*, 117. https://doi.org/10.3390/pr11010117

Received: 28 December 2022
Revised: 29 December 2022
Accepted: 29 December 2022
Published: 31 December 2022

Thirty years ago, at the United Nations' (UN) Earth Summit in Rio de Janeiro, Brazil, 178 countries adopted Agenda 21, a global partnership for sustainable development to improve human lives and protect the environment [1]. Twenty years ago, The Johannesburg Declaration on Sustainable Development was accepted to eradicate poverty and ensure sustainable development. Ten years ago, the UN Conference on Sustainable Development in Rio de Janeiro (Rio+20) adopted the outcome document "The Future We Want", in which they decided to start a process to develop a set of Sustainable Development Goals (SDGs). The UN 2030 Agenda for Sustainable Development, with its 17 SDGs, was adopted in September 2015, and in December 2015 the Paris Agreement on Climate Change was accepted by the Parties to the United Nations Framework Convention on Climate Change, its goal being to limit global warming to well below 2 °C, preferably to 1.5 °C, compared to pre-industrial levels [2]. To achieve this long-term temperature goal, countries aim to reach a global peak of greenhouse gas emissions as soon as possible to achieve a climate-neutral world by mid-century.

Currently, the Earth is already about 1.1 °C warmer than it was in the late 1800s, and emissions continue to rise. Global GHG emissions have increased by 53%, from 32.52 Gt/a (billion tons per year) in 1990 to 49.76 Gt/a in 2019 [3]. To limit global warming to no more than 1.5 °C, emissions need to be reduced by 45% by 2030 and reach net zero by 2050 [4]. The European Union (EU) and its 27 member states, accepted the European Green Deal plan to transform the EU into a modern, resource-efficient, and competitive economy with no net GHG emissions by 2050 [5], and at least 55% less net greenhouse gas emissions by 2030, compared to 1990 levels. GHG emissions in the EU decreased by 32% between 1990 and 2020 [6]. Preliminary estimates indicate that emissions rebounded in 2022, exceeding the pre-COVID-19 levels [7]. At COP27 (Conference of the Parties, Egypt, 2022), the EU's climate policy chief, Frans Timmermans, even promised to exceed the EU's 2030 reduction target by 2%, to 57% [8].

In 2019, approximately 34% (20 Gt) of total net anthropogenic GHG emissions came from the energy supply sector; 24% (14 Gt) from industry; 22% (13 Gt) from agriculture, forestry and other land use; 15% (8.7 Gt) from transport; and 6% (3.3 Gt) from buildings [9]. Average annual GHG emissions growth between 2010 and 2019 slowed compared to the previous decade in energy supply (from 2.3% to 1.0%) and industry (from 3.4% to 1.4%) but remained roughly constant at about 2% in the transport sector.

Industry emitted 29.4% of all the GHGs: 24.2% originated from energy use, while industrial processes emitted an additional 5.2% of GHGs [10]. Global process industries are among the most intensive GHG emissions activities in industry, e.g., iron and steel—7.2%, chemical and petrochemical—3.6%, food and tobacco—1.0%, non-ferrous metals—0.7%, and paper and pulp—0.6%. Aside the emissions from energy production, process industries emit GHGs in their production processes, e.g., chemicals—2.2% and cement—3%. Further, transport, energy use in buildings, and fugitive emissions from energy production must also be added.

Process design and sustainable development are connected in their aim to reduce the negative effects of GHG emissions on human development. Process design is detrimental to energy and material usage during production and usage, as well as for GHG emissions. By designing for circular economy and efficiency improvements, and by applying process optimization, we can reduce GHG emissions and enlighten energy and material recycling. We can also extend the lifetime of products.

The number of papers dealing with the topic (keyword: process design and sustainable development, Web of Science) is increasing—from 821 to 3 586 per year in the period 2012–2021 (Figure 1). This follows the two most evident megatrends in the process industries: climate change and resource scarcity. This Special Issue on Process Design and Sustainable Development includes 16 papers, which can be grouped into process design, sustainable development, and other topics.

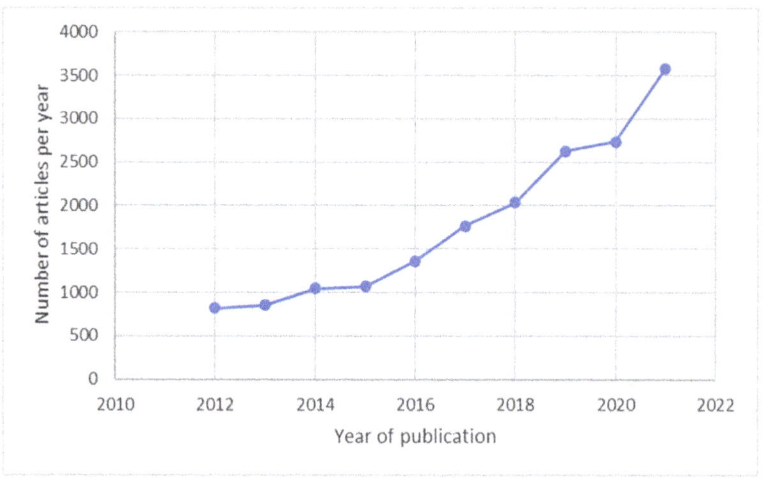

Figure 1. Number of papers on process design and sustainable development, Web of Science.

Design for sustainable development

The definitions of process design and sustainable development are described separately and together, as is the process design for sustainability, including environmental, economic, and social dimensions [11]. A case study of the EU27 chemical industry in the period 1990–2019, with a 47% increase in production and 54% reduction in GHG emissions, indicates that the decoupling of production and emissions is viable. The most urgent future development areas in process industries are (a) carbon capture with utilization or storage; (b) process analysis, simulation, synthesis, and optimization tools; and (c) zero waste, circular economy, and resource efficiency.

Design as a discipline has changed to address the complexity of the current problems [12]. Sustainable design, inclusive design, codesign, and social design have emerged as examples of such issues. Social, environmental, and cultural trends have affected design, but the design process itself remains almost unchanged. An alternative design process method and new criteria are needed to develop innovations in response to these trends. An extended design process includes distributed agencies, transparency, and pertinence, but it is also important to keep in mind the principle of proportionality.

The guiding principles of sustainability, sustainable design, green engineering, and sustainable engineering need to be updated to include the present state of human knowledge [13]. The updated principles include traditional and more recent items: a holistic approach, sustainability hierarchies, sustainable consumption, resource scarcity, equalities within and between generations, all stakeholders' engagement, and internalizing externalities. Equal importance is given to the principles of environmental, social, economic, and tridimensional systems' approach.

Project management is crucial for achieving business excellence in a new industrial paradigm [14]. Therefore, the impact of different levels of project management maturity on business excellence has been determined in the context of Industry 4.0. A sample of 124 organizations of different sizes from 27 countries, covering services, education, industry, and finance, was selected. A significant connection was found between project management maturity and business excellence. However, there was no evidence of a mediating effect of Industry 4.0 readiness on the relationship between project management maturity and business excellence.

Design for increased resource efficiency

Process industries are regular users of fossil fuels for energy production and processing. Energy usage relates to GHG emissions. In 2010, the European Commission proposed a 10-year strategy, Europe 2020, with five targets for sustainable development. One of them was the 20-20-20 target—to reduce greenhouse gas emissions by at least 20% compared to 1990 levels, increase the share of renewable energy in final energy consumption to 20%, and achieve a 20% increase in energy efficiency. The analysis of sustainability indicators, which were transformed into a synthetic measure for comparability of the resulting values, identified the best countries to fulfil the sustainable growth aims [15].

The automotive sector accounts for 16.2% of global energy consumption and 25% of CO_2 emissions from energy-related sectors [16]. If present trends persist, the worldwide transportation demand for energy, and CO_2 emissions from energy, are expected to double by 2050. Electric vehicles (EVs—hybrid, plug-in hybrid, and battery) emit only half of the conventional emissions [17]. Aside from the social preferences for environmental protection and energy policies, rising fuel prices are promoting EV adoption. Barriers to this adoption include the high purchase prices of EVs, limited driving distances, and long charging times. The multicriteria decision-making methods show that battery capacity and lifespan, and high costs are the most important barriers to EV transition, while government support promotes it.

A hybrid vehicle (HV) is one that draws its power from two or more different energy sources; the internal combustion engine (ICE) performs better at maintaining the high speed of the vehicle than a standard electric motor, while the electric motor is more effective at providing torque or turning power [18]. Integration of the ICE with the thermoelectric generator (TEG) takes advantage of ICE waste heat at the exhaust gate and predicts their reliability. Researchers seek to recycle wasted energy to produce electricity by integrating TEG with ICE, which rely on the electrical conductivity of the thermal conductor strips. The proposed metaheuristic verification improves the TECs' efficiency by 32.21% and reduces fuel consumption by up to 19.63%.

Spent grains from microbreweries are mostly formed by malting barley. Transforming spent grains from waste to raw materials in the production of nontraditional flour requires a drying process [19]. A natural convection solar dryer (NCSD) was evaluated as an alternative to a conventional electric convective dryer (CECD) for the dehydration process of local microbrewers' spent grains. Suitable models (empirical, neural networks, and computational fluid dynamics, CFD) were used to simulate both types of drying processes under different conditions. NCSD produced a dried product with a comparatively better quality; overall operating costs were greatly reduced, and the NCSD was about 40–45% cheaper than the CECD.

The relationship among carbon dioxide emissions and linkage effects using the input–output data of the information and communications technology (ICT) industry between South Korea and the USA was examined [20]. The linkage effects were analyzed to determine the impact of the ICT industry on the national economy and CO_2 emissions. The results indicated that the ICT manufacturing industry in Korea has high backward and forward linkage effects, while the US has a small influence on both effects. The ICT service industries of the two countries have small backward linkage effects, so their influence on other industries is small. CO_2 emissions from ICT service are higher than from ICT manufacturing in both countries, South Korea and the US.

During watermelon fruit production, about half of the fruit—namely the rind—is usually discarded as waste. To transform such waste into a useful product such as flour, thermal treatment is needed. The drying temperature of the rind is most important to produce flour with the best characteristics. In [21], a multi-objective optimization (MOO, also known as multicriteria decision making) procedure was applied to define the optimum drying temperature for the rind flour fabrication to be used in bakery products. The group of process indicators comprised acidity, pH, water-holding capacity, oil-holding capacity, and batch time; they represented conflicting objectives that were balanced by the MOO procedure using the weighted distance method to find their optimal values.

Design for optimal operation

An efficient and flexible production system can contribute to production solutions. A detailed mathematical model of the system under investigation was presented in [22]. The evolutionary algorithm was the most efficient solution to the problem associated with the model. A computer application for this model was created, and it was run multiple times. The runtime results are also presented in this study. The sample task was taken from a real company, but it was simplified for reasons of transparency. Order-driven production can be mathematically described and assigned to a well-managed model. The problem was solved with a variation of the genetic algorithm.

The optimal cleaning scheduling of a heat exchanger unit undergoing fouling was analyzed and optimized for uncertain operating conditions in [23]. The fouling process, the flexibility assessment, and the optimization algorithm were presented. The fouling kinetic model was described, uncertain operating conditions were handled by the flexibility index, and the scheduling optimization and the corresponding decisional algorithm were discussed. The scheduling algorithm was applied to a case study, a sensitivity analysis with respect to two possible uncertain variables followed, and the results obtained with the process parameters accounted for the mechanical cleaning methodology, and for both, chemical and mechanical ones.

Electrostatic desalting is one of the most popular methods to remove saline water from crude oil by applying an external electrostatic field. Complex phenomena, such as the frequency of collisions and coalescence, were included in a CFD model for better understanding and optimization of the desalting process from the perspectives of both process safety and improvement [24]. The main process parameters are electric field strength, water content, temperature (through oil viscosity), and droplet size on the collision time or frequency of collision between a pair of droplets. First, a simplified collision model between two droplets in water-in-oil emulsions was developed. Then, this model was employed to perform a complete process analysis determining the effect of the main process variables on the collision frequency.

In [25], the cooperative optimization of two-stage desulfurization processes in the slime-fluidized bed boiler (FBB) was studied, and a model-based optimization strategy was proposed to minimize the operational cost of the desulfurization system. First, a mathematical model for the FBB with a two-stage desulfurization process was established. Then, several parameters affecting the SO_2 concentration at the outlet of the slime-fluidized bed boiler were simulated and deeply analyzed. Finally, the optimization operation problems under different sulfur contents were studied with the goal of minimizing the total desulfurization cost. The optimized operation reduced the total desulfurization cost by 9%.

In Fischer–Tropsch (FT) synthesis, synthesis gas (or syngas, i.e., a mixture of H_2 and CO) is converted into a variety of hydrocarbon products, including paraffin, olefins, and value-added chemicals [26]. A two-dimensional CFD scale-up model of the FT reactor was developed to thermally compare the Microfibrous-Entrapped-Cobalt-Catalyst (MFECC) and the conventional Packed Bed Reactor (PBR). The model implemented an advanced predictive detailed kinetic model to study the effect of a thermal runaway on C5+ hydrocarbon product selectivity. The results demonstrated the superior capability of the MFECC bed in mitigating hotspot formation due to its ultra-high thermal conductivity. A significant improvement in C5+ hydrocarbon selectivity was observed in the MFECC bed in contrast to a significantly low number for the PBR.

Rheology studies the stress and deformation of matter that may flow; not only does it include liquids, but also soft solids and substances such as sludge, suspensions, and human body fluids [27]. The rheological characterization of fluids using a rheometer is an essential task in food processing, materials, healthcare, or even industrial engineering; in some cases, the high cost of a rheometer and the issues related to the possibility of developing both electrorheological and magnetorheological tests in the same instrument must be overcome.

Conclusions

The Special Issue on Process Design and Sustainable Development covers a highly important and rapidly expanding area of research and development. The articles included in the Special Issue cover a selection of important topics from the area. These topics will be interesting for many scientists, researchers, and engineers. As of December 2022, the articles have received approximately 20,000 views and 50 citations.

Conflicts of Interest: The author declares no conflict of interest.

References

1. United Nations, Sustainable Development Goals. Available online: https://sdgs.un.org/goals (accessed on 27 December 2022).
2. United Nations Climate Change, The Paris Agreement. Available online: https://unfccc.int/process-and-meetings/the-paris-agreement/the-paris-agreement (accessed on 27 December 2022).
3. Our World in Data, Greenhouse Gas Emissions. Available online: https://ourworldindata.org/greenhouse-gas-emissions (accessed on 27 December 2022).
4. United Nations Climate Action, Net Zero Coalition. Available online: https://www.un.org/en/climatechange/net-zero-coalition (accessed on 27 December 2022).
5. European Commission, A European Green Deal. Available online: https://commission.europa.eu/strategy-and-policy/priorities-2019-2024/european-green-deal_en (accessed on 27 December 2022).
6. European Environment Agency, Total Greenhouse Gas Emission Trends and Projections in Europe. Available online: https://www.eea.europa.eu/ims/total-greenhouse-gas-emission-trends (accessed on 27 December 2022).
7. Eurostat, Quarterly Greenhouse Gas Emissions in the EU. Available online: https://ec.europa.eu/eurostat/statistics-explained/index.php?title=Quarterly_greenhouse_gas_emissions_in_the_EU (accessed on 27 December 2022).
8. Euronews, EU Climate Chief Announces Updated Emissions Goal at COP27. Available online: https://www.euronews.com/green/2022/11/15/not-backtracking-eu-climate-chief-announces-updated-emissions-goal-at-cop27 (accessed on 27 December 2022).
9. Intergovernmental Panel on Climate Change (IPCC), Climate Change 2022, Mitigation of Climate Change. Available online: https://www.ipcc.ch/report/ar6/wg3/downloads/report/IPCC_AR6_WGIII_SPM.pdf (accessed on 27 December 2022).
10. Our World in Data, Emissions by Sector. Available online: https://ourworldindata.org/emissions-by-sector (accessed on 27 December 2022).
11. Glavič, P.; Novak Pintarič, Z.; Bogataj, M. Process Design and Sustainable Development—A European Perspective. *Processes* **2021**, *9*, 148. [CrossRef]
12. Hernandez, R.J.; Goñi, J. Responsible Design for Sustainable Innovation: Towards an Extended Design Process. *Processes* **2020**, *8*, 1574. [CrossRef]
13. Glavič, P. Updated Principles of Sustainable Engineering. *Processes* **2022**, *10*, 870. [CrossRef]
14. Fajsi, A.; Morača, S.; Milosavljević, M.; Medić, N. Project Management Maturity and Business Excellence in the Context of Industry 4.0. *Processes* **2022**, *10*, 1155. [CrossRef]
15. Širá, E.; Kotulič, R.; Kravčáková Vozárová, I.; Daňová, M. Sustainable Development in EU Countries in the Framework of the Europe 2020 Strategy. *Processes* **2021**, *9*, 443. [CrossRef]
16. Ritchie, H.; Roser, M. CO2 and Greenhouse Gas Emissions. Available online: https://ourworldindata.org/co2-and-other-greenhouse-gas-emissions (accessed on 24 March 2021).
17. Kuo, T.-C.; Shen, Y.-S.; Sriwattana, N.; Yeh, R.-H. Toward Net-Zero: The Barrier Analysis of Electric Vehicle Adoption and Transition Using ANP and DEMATEL. *Processes* **2022**, *10*, 2334. [CrossRef]
18. Abed, A.M.; Seddek, L.F.; Elattar, S. Building a Digital Twin Simulator Checking the Effectiveness of TEG-ICE Integration in Reducing Fuel Consumption Using Spatiotemporal Thermal Filming Handled by Neural Network Technique. *Processes* **2022**, *10*, 2701. [CrossRef]
19. Capossio, J.P.; Fabani, M.P.; Reyes-Urrutia, A.; Torres-Sciancalepore, R.; Deng, Y.; Baeyens, J.; Rodriguez, R.; Mazza, G. Sustainable Solar Drying of Brewer's Spent Grains: A Comparison with Conventional Electric Convective Drying. *Processes* **2022**, *10*, 339. [CrossRef]

20. Mun, J.; Yun, E.; Choi, H. A Study of Linkage Effects and Environmental Impacts on Information and Communications Technology Industry between South Korea and USA: 2006–2015. *Processes* **2021**, *9*, 1043. [CrossRef]
21. Capossio, J.P.; Fabani, M.P.; Román, M.C.; Zhang, X.; Baeyens, J.; Rodriguez, R.; Mazza, G. Zero-Waste Watermelon Production through Nontraditional Rind Flour: Multiobjective Optimization of the Fabrication Process. *Processes* **2022**, *10*, 1984. [CrossRef]
22. Fülöp, M.T.; Gubán, M.; Gubán, A.; Avornicului, M. Application Research of Soft Computing Based on Machine Learning Production Scheduling. *Processes* **2022**, *10*, 520. [CrossRef]
23. Di Pretoro, A.; D'Iglio, F.; Manenti, F. Optimal Cleaning Cycle Scheduling under Uncertain Conditions: A Flexibility Analysis on Heat Exchanger Fouling. *Processes* **2021**, *9*, 93. [CrossRef]
24. Ramirez-Argaez, M.A.; Abreú-López, D.; Gracia-Fadrique, J.; Dutta, A. Numerical Study of Electrostatic Desalting Process Based on Droplet Collision Time. *Processes* **2021**, *9*, 1226. [CrossRef]
25. Xiao, Y.; Xia, Y.; Jiang, A.; Lv, X.; Lin, Y.; Zhang, H. Research on Optimization of Coal Slime Fluidized Bed Boiler Desulfurization Cooperative Operation. *Processes* **2021**, *9*, 75. [CrossRef]
26. Abusrafa, A.E.; Challiwala, M.S.; Wilhite, B.A.; Elbashir, N.O. Thermal Assessment of a Micro Fibrous Fischer Tropsch Fixed Bed Reactor Using Computational Fluid Dynamics. *Processes* **2020**, *8*, 1213. [CrossRef]
27. Hernández-Rangel, F.J.; Saavedra-Leos, M.Z.; Morales-Morales, J.; Bautista-Santos, H.; Reyes-Herrera, V.A.; Rodríguez-Lelis, J.M.; Cruz-Alcantar, P. Continuous Improvement Process in the Development of a Low-Cost Rotational Rheometer. *Processes* **2020**, *8*, 935. [CrossRef]

 *processes*

Article

Building a Digital Twin Simulator Checking the Effectiveness of TEG-ICE Integration in Reducing Fuel Consumption Using Spatiotemporal Thermal Filming Handled by Neural Network Technique

Ahmed M. Abed [1,2,*], Laila F. Seddek [3] and Samia Elattar [4,5]

[1] Department of Industrial Engineering, College of Engineering, Prince Sattam Bin Abdulaziz University, Alkharj P.O. Box 16273, Saudi Arabia
[2] Industrial Engineering Department, Zagazig University, Zagazig P.O. Box 44519, Egypt
[3] Department of Mathematics, College of Science and Humanities in Al-Kharj, Prince Sattam Bin Abdulaziz University, Alkharj P.O. Box 11942, Saudi Arabia
[4] Department of Industrial and Systems Engineering, College of Engineering, Princess Nourah Bint Abdulrahman University, Riyadh P.O. Box 11564, Saudi Arabia
[5] Department of Industrial Engineering, Alexandria Higher Institute of Engineering and Technology (AIET), Alexandria P.O. Box 21311, Egypt
* Correspondence: a.abed@psau.edu.sa; Tel.: +966-509506811

Abstract: Scholars seek to recycle wasted energy to produce electricity by integrating thermoelectric generators (TEGs) with internal combustion engines (ICE), which rely on the electrical conductivity, β, of the thermal conductor strips. The TEG legs are alloyed from iron, aluminum and copper in a strip shape with specific characteristics that guarantee maximum thermo-electric transformation, which has fluctuated between a uniform, Gaussian, and exponential distribution according to the structure of the alloy. The ICE exhaust and intake gates were chosen as the TEG sides. The digital simulator twin model checks the integration efficiency through two sequential stages, beginning with recording the causes of thermal conductivity failure via filming and extracting their data by neural network procedures in the feed of the second stage, which reveal that the cracks are a major obstacle in reducing the TEG-generated power. Therefore, the interest of the second stage is predicting the cracks' positions, $P_{i,j}$, and their intensity, Q_P, based on the ant colony algorithm which recruits imaging data (STTF-NN-ACO) to install the thermal conductors far away from the cracks' positions. The proposed metaheuristic (STTF-NN-ACO) verification shows superiority in the prediction over [Mat-ACO] by 8.2% and boosts the TEGs' efficiency by 32.21%. Moreover, increasing the total generated power by 12.15% and working hours of TEG by 20.39%, reflects reduced fuel consumption by up to 19.63%.

Keywords: waste heat recovery; damage detection; non-destructive testing; thermal filming; digital twin; optimization; influencing factors

Citation: Abed, A.M.; Seddek, L.F.; Elattar, S. Building a Digital Twin Simulator Checking the Effectiveness of TEG-ICE Integration In Reducing Fuel Consumption Using Spatiotemporal Thermal Filming Handled by Neural Network Technique. *Processes* **2022**, *10*, 2701. https://doi.org/10.3390/pr10122701

Academic Editors: Davide Papurello and Jie Zhang

Received: 1 November 2022
Accepted: 12 December 2022
Published: 14 December 2022

1. Introduction

A hybrid vehicle (HV) is one that draws its power from two or more different energy sources. The fundamental idea behind hybrid cars is that the various motors perform better at various speeds; the internal combustion engine (ICE) performs better at maintaining high speed of the vehicle than a standard electric motor while the electric motor is more effective at providing torque, or turning power. Speeding up and switching from one to the other at the right moment results in a win-win situation for energy efficiency, which increases fuel economy. Therefore, the main objective is to guarantee the integration of the ICE with the thermoelectric generator (TEG) to take advantage of ICE waste heat at the exhaust gate and predicting their reliability. The value of accurate time series significance does not need to be emphasized. There have been decades of studies on this issue, which classified the solutions approach into two most prevalent methods: the first is statistical (e.g., S-ARIMA, Holt-Winter, Box-Jenkins, etc.) and the other is the Mat-heuristic training approach based on analyzing the time series (e.g., LSTM, CNN, TLNN, and recurrent network models). If there were not two options, each would have advantages over the other. If statistics is followed, the why and what of every result must be explained; while if Mat-heuristic training is followed, findings may be superior but are unwarranted [1]. Therefore, validation is an important stage. This paper intends to design a digital simulator twin to check the usability of waste heat of an internal combustion engine (ICE) for transforming it into electricity efficiently via a thermoelectric generator (TEG). The digital twin relies on TEG and has two plates (i.e., engine gasket), one of both is a hot gasket plate connected to the exhaust gate, while the cold gasket plate is connected to the intake gate through thermal conductor strips. The digital twin was used to test three different alloy textures of gaskets and their legs. The main purpose is to determine the most efficient thermal convection for different gaskets. Therefore, this paper reviews the main causes of heat conduction failure, which are cracks in the hot gasket plate that prevent conductivity by predicting these causes using the metaheuristic approach that helps track these causes and reduce transformation efficiency for electric power. Figure 1 illustrates the cause-and-effect diagram for the failure of thermal conductivity, which cancels the TEG integration.

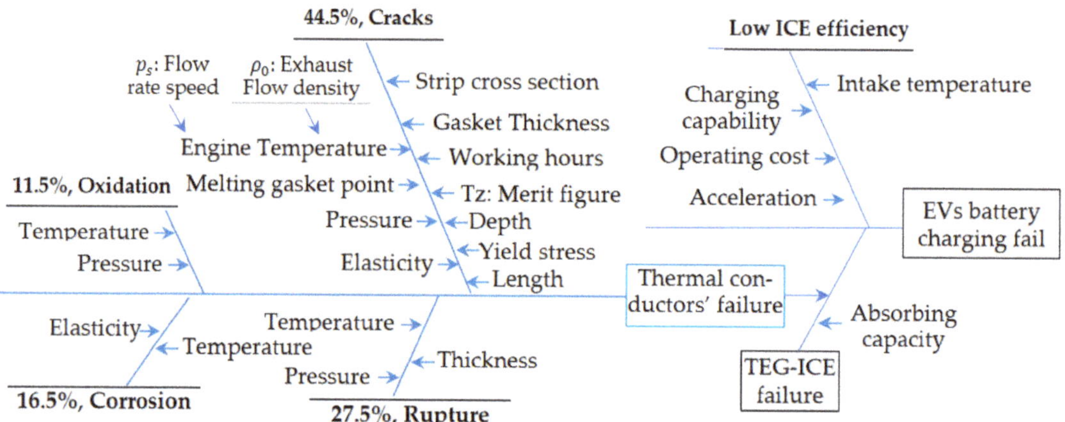

Figure 1. The cause and effect diagram for HVs charging failure due to heat conduction failure.

The authors try to predict the cracks' positions via a Mat-heuristic network model to sketch the installation of the thermal conductor's path more securely far away from these cracks. The experimental observations record the causes of cracks using spatiotemporal thermal filming for seeking a crack-free path above the gasket surface for installing the thermal conductors to study the working span life of the integration of TEG and ICE. The digital twin studies the efficiency of integration, the amount of transferred heat and the electric power generated. This paper provides a comprehensive vision of the main technical development of HVs and emerging technologies for their future application in sustaining the battery for a long time. Key technologies regarding batteries, such as charging technology, electric motors, control, and charging infrastructure of HVs are summarized and considered a keystone for our motivation [2–4]. This paper also highlights the technical challenges for the improvement of operating performance via efficiency, reliability, programming of electronic components, heat waste management, and safety of HVs in the coming stages as deduced from Winslow et al. [5]. Therefore, the interest is in using IoT in tracking the HV battery, which is fed by TEG instead of a native generator and studying this approach through the digital twin model. The development of electric vehicles HVs is an urgent necessity, especially after it was proven that an average of 25% of carbon emissions (CE) in the UK and USA are due to internal combustion engines (ICE), which represent 80% of the change in air characteristics, where two-thirds of the air is saturated with CO_2, nitrogen oxide represents a third, and 50% is hydrocarbons, which are known as Greenhouse Gases (GHG) [6,7]. There are ten common components in HHV (auxiliary battery, transformer, generator, electric traction motor, exhaust management system, fuel tank, ICE, electronic unit of power control, thermal system management, transmission), and experts advise not to discharge the traction battery, which leads to spoiling and reducing its lifespan [8,9]. Therefore, this paper suggests studying the replacement of the native generator with the thermoelectric generator, TEG, to take advantage of the heat generated during movement. The TEG is fed from the internal combustion engine and/or from the electric traction engine, to support the power level of the battery, which is at its maximum when also using the brakes repeatedly. The auxiliary battery is used to operate the vehicle before the traction battery is used, while the inverter converts direct current to low voltage alternating current to power the vehicle's accessories [10,11]. It is a major problem for the entire globe to benefit from energy waste in many applications. One of the famous frequent types of energy loss is low-grade thermal energy. Traditional methods for turning this thermal energy into electrical energy are ineffective, difficult, and expensive; however, the approved Seebeck effect is exploited by thermoelectric generators (TEGs), which may convert heat energy directly into electrical energy [12]. Peltier and Thomson's effects are also thermoelectric phenomena used in thermal measurements to achieve reversible thermal and electrical energy transformations and vice versa. The thermoelectric generators (TEG) work on the heat waste extracted from ICE generators or other sources that can absorb thermal energy [13,14] and track their behavior through the digital-twin simulator, which is fed by spatiotemporal thermal analysis films to guarantee steady-state electricity power generation. The main component in the TEG is the thermal conductors' legs, which must be better because of the impact on the generated electricity. Therefore, the authors present a smart prediction methodology for a time of failure to prevent the sudden stop that is considered costly [15]. Some of the heat waste sources and their sudden failure causes were discussed through 2020–2021 [16,17]. Therefore, the authors follow the published examples [18,19] for the importance of tracking and predicting the optimal operating conditions using one or more meta-heuristic optimization algorithms to measure the utilization of TEG accurately. The authors tend to use the STTF measurements fed into the digital twin simulator cloud to predict the efficiency of conductors working by a meta-heuristic algorithm, which integrates the STTF technique and a Cuckoo search to enhance the prediction model (STTF-NN-ACO) and is managed through IoT code inserted in the appendix. The rising source temperature, electrical conductivity, merit figure, melting point/volume, thermal conductor surface area, energy per unit area, gasket texture perpendicular slots', exhaust flow rate, exhaust temperature, intake temperature flow, operating period, and cost per W achieve higher power output and other factors or variables used in the proposed digital-twin Simulator model as cracks' span, depth, and length. The thermoelectric conductors' material was developed by improving the T_Z merit figure for enhancing the generated power, such as a ß-Zn4sp3 semiconductor compound that records a one-way 2.6 T_Z. The merit figure improved via seeking new materials formed in specific shape and produced an increase in the electric power at the lowest costs, such as Fe-Al-Cu alloys mixed with Lithium as shown in Table 1.

Table 1. The main components of three candidate alloys of TEG texture.

Alloy Sample	Cu%	Zn	Sn	Ni	Fe	Mn	Al %	Cr %	As	Hv	k W/m K	σ_T (MPa)	σ_v (MPa)
(LiFePO$_4$-Al) $\approx$ (Fe $-$ Al)	—	Rem.	1.2	—	0.006	—	4.65	20	—	85	13.5	310	105
(Cu-LiFePO$_4$) $\approx$ (Fe $-$ Cu)	15	Rem.	0.001	0.004	0.7	0.001	1.9	—	0.012	102	16.07	340	180
(LiAlH$_4$-Cu) $\approx$ (Al $-$ Cu)	77	Rem.	—	0.027	0.019	0.003	15	—	0.037	120	23.4	360	230

For instance, the cost per W is exclusively determined by energy per unit area and running duration when the fuel cost is minimal or virtually free, such as in waste heat recovery (energy recycling). The scientists thus focused on materials with better power output as opposed to conversion efficiency, such as a rare earth compound, $YbAl_3$ (Density: $\rho = 5.68$ Mg·m^{-3}), and other alloys that have a low value but yield energy of at least two times as much as any other material and can function across the temperature range of trash [20]. Therefore, a famous generator company expressed the desire to test the efficiency of the candidate three alloys, which are $(LiFePO_4\text{-}Al) \approx (Fe - Al)$, $(Cu\text{-}LiFePO_4) \approx (Cu - Fe)$ and $(LiAlH_4\text{-}Cu) \approx (Al - Cu)$ to form the thermal conductors' strips and texture of ICE gaskets (Hot TEG plates). The authors selected the ICE generator exhaust gasket and intake gasket textures to be woven as ducts of thermal conductors that perpendicularly end with legs (i.e., the two TEG's plates' thicknesses are 1.63 ±0.1 mm, and 2.15 mm, respectively) as illustrated in Figure 2. The machines based on steady current are subjected to failure when the electricity current rate decreases and seek a compensating source, which suggests the use TEG because of its ability to integrate with ICE generators and is deployed in many companies [21,22]. The authors tested three different types of alloys to determine their merit figure and predict their average malfunction time (i.e., reliability). For the first alloy (Fe-Al) the thermal properties of the terminals' alloy iron- chromium-aluminum fitted to thermal conduct propensity for cold brittleness if working at temperatures exceeding 1000 °C with chemical composition is 4.65% Al, 20% Cr, and balanced Fe, whose thermal conductivity is 13.5 W/m K. The second is iron-copper (i.e., Fe-Cu, called 3003, with 0.15% Cu and 0.7% Fe). The third is aluminum-copper (i.e., Al-Cu, with 15% Al with a grain size of ~47 μm) [23–25]. The digital-twin simulator model accurately represents the contact plates (gaskets' texture) and their joined legs with exhaust and intake flow points [26]. The challenge is to monitor the gasket's efficiency without stopping the generator [27–30]. Therefore, the authors suggest predicting the functional change of the gasket texture caused by cracks appearing by spatiotemporal thermal filming, which can feed the meta-heuristic optimization technique supported by mathematical equations to gain accurate prediction results. The tailored digital-twin simulator model [31] consists of an object (TEG and gaskets), a measuring tool (STTF), and a predicting meta-heuristic optimization algorithm, which uses some of the physical parameters related to cracks (e.g., the span, depth, and length in $m \times 10^{-4}$) that is measured by the cracks' center propagation digitally on the gasket surface texture, as discussed by [32,33]. Research is now being performed to overcome the fundamental drawback of thermoelectric generators, which is their relatively low efficiency, and to efficiently use them in a variety of applications for recovering waste heat. The automobile industry produces a significant amount of waste heat as a result of the low braking thermal efficiency of reciprocating engines, which is less than 30% for gasoline engines, according to Hotta et al. [21]. One of the important precautionary measures is not to allow the HVs battery to discharge more than 80% of the capacity so that the state of charge (SOC) does not fall below 20%. This is to protect the battery from over-discharging. This contribution is achieved using thermal conductors of copper, aluminum, and iron alloys between the thermal potential difference points that help generate an electric current that works within a discharge rate within 300 amps to try increasing the lifespan of the battery to an average of 3500 cycles for the Li-ion cells. The cells are 120 W/kg and must be compatible with traction motors that are divided into two types (AC and DC motors) and which are directly connected with the ECM unit to manage the battery in the so-called TEG, BMS, the amount of heat transferred, the voltage generated and refer to the system with voltage temperature monitoring (VTM) [34–36]. The digital twin is specially designed to be a core to improve HVs in reducing fuel consumption.

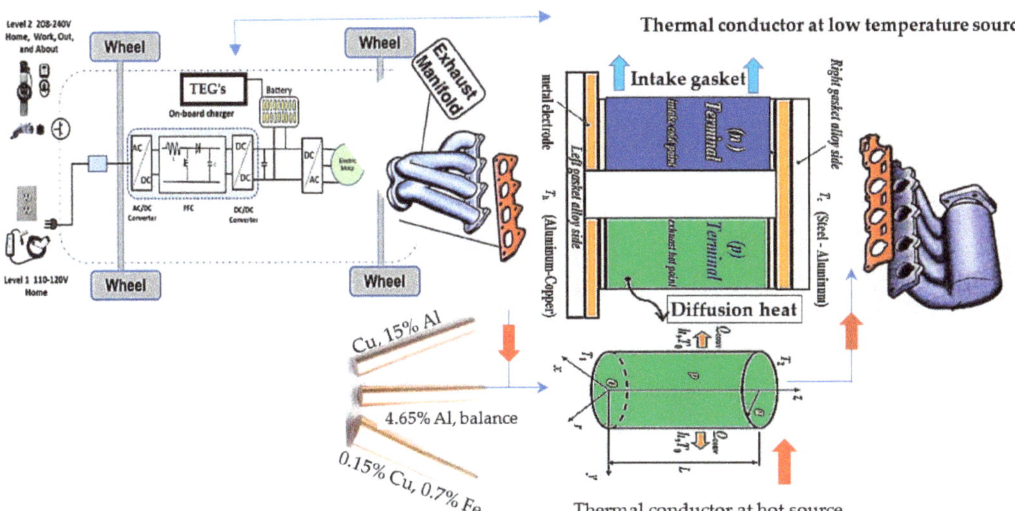

Figure 2. The components of the engine under study and TEG according to Dandan Pang, et al., (2022) [37].

2. The Architecture of the Digital-Twin Simulators' Paradigm

The exhaust temperature is the hot source useful for TEGs and allows them to convert waste heat into usable electric energy according to [38,39]. Figure 3 illustrates digital twin components and discusses the sequential two stages based on integrating the source [1] with TEG plates. The first stage aims to aggregate the physical measurement parameters [4], which is measured by specific means [3] to test the legs' behavior toward resist failures through crack generation and is named figuratively (reliability stage), to be fed for the prediction methodology [5] which tests the effectiveness of the thermo-conductor's leg material alloy [2] through the second stage. The two stages discussed in Algorithm 1.

Algorithm 1: Main Pseudocode of STTF-NN-ACO mechanism

//The components of the digital simulator's twin (ICE, TEG, and thermal conductor's strips)
//The mechanical and composite of tested alloys discussed by Admiral et al. [40], and Richard & Hertzberg [41]
//Stage (1): Check the reliability of the thermal conductors by imaging
Determine the TEG gasket plate's install (hot at the exhaust gate, cold at the intake gate)
Determine the gasket texture $((Fe - Al),(Fe - Cu),(Al - Cu))$
for alloy 1:3
    if (gasket texture and its strips from $((Fe - Al))$
        for 1 : causes
Select the high failure cause (Figure 1: Cause-and-effect diagram);
Determine the gasket surface coordinates (i denotes the x-axis and the j denotes y-axis);
Using Spatiotemporal Thermal Filming STTF to:
    Record the working time (hr.);
    Locate thermal conduction failure positions, $P_{(i,j)}^n$, the position of the starting of the cracks' point;
        for 1 : cracks
            Count the number of each crack's ramifications b;
            Determine the two longest ramifications lengths for each crack;
            Determine the tilt angle of each ramification to the x-axis;
            Determine the $P_{(i,j)}$ of the end terminal for each tracing ramification;
            Determine the intensity Q_P of each crack at specific position (the number of their ramifications);
//Stage (2): Predicting generated electricity STTF-NN-ACO
Predict the virtual curve line direction of cracks direction as illustrated in (Figure 4) by hybridizing meta-heuristic with the neural network to reduce the error of sketching the secure path of installing the thermal conductors' strips;
        Measure the efficiency, f (amount of heat conductivity, electrical power generation);
    Else if (gasket texture and its strips from next alloy)
    End
End

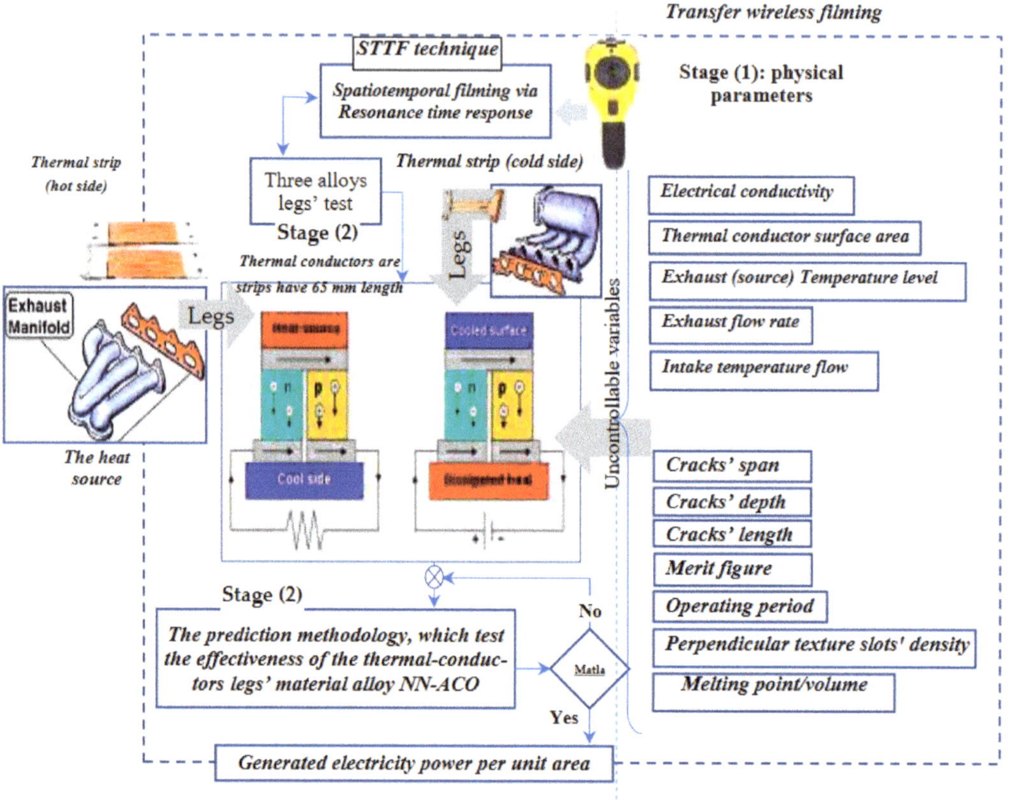

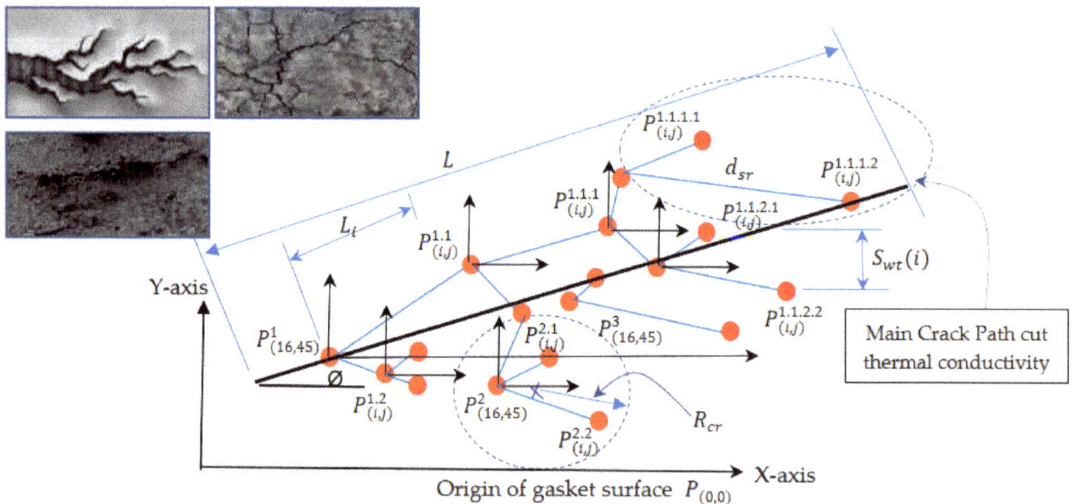

Figure 3. The digital twin simulator architecture.

Figure 4. A hypothetical visualization of the appearance and growth of cracks and how to track them. ● The crack position; ⟶ represent the coordinates; ⟶ dimension arrows.

The parameters suggested in the proposed mathematical equations are required to form the digital twin simulator for testing the effectiveness of three different alloys (iron, aluminum, and copper) to thermos-conductivity are shown in Table 2. Figure 4 illustrates the sketch of the recording and tracking the crack growth. The pseudocode of the proposed model describes the sequencing of tackling the data input and output to track the efficiency of the TEG-ICE integration.

Table 2. The ideal parameters and objective for proposed digital twin model.

Parameter	Description	Parameter	Description
D_x, D_y, D_v	The coefficient of terminals' cracks diffusion layer type 0.005 mm^2 day^{-1}	β	The electrical conductivity in the alloy [Sm^{-1}]
ρ_0	The exhaust flow density [s. mm^{-3}]	α	The Seebeck coefficient [V. K^{-1}]
p	The flow rate speed, mm^3 s^{-1}	∇	The Hamiltonian operator,
q	The dipole source, (0,1)	T	The temperature [K], 0.273 $\times$ 10^3 slits
k	The thermal conductivity [W/m. K]	b	The # of ramifications appeared on the gasket slots 0: 1000 (212)
Spatiotemporal Thermal Filming Parameters			
d_{sr}	The distance between two cracks' core on the leg surface [mm]	L_i	The crack segment length among red spots on spatiotemporal images
R_0	The radial distance from the gasket surface to picks imaging	R_{cr}	The radius of surrounding circle of terminals' cracks form closed shape.
r	The cracks growth rate per week (timespan between two sequential points), 3 weeks	h_r	The radius of the hotspot on the gasket surface,
d	The red crack diameter on the spatiotemporal image	$P_{i,j}$	The cracks' location on the gasket legs' surface
$\emptyset$	The angle of the cracks line slope line with the x-axis	$S_{wt}(i)$	The cracks' center deviation regression to thermal conductor path according to alloy
Q_p	Intensity of the crack is the area of a circle and surrounds all ramifications for specific crack and has position $P_{i,j}$ picked by spatiotemporal imaging, (Dark blue color, Green—Red)	ω	The relative area of closed perimeter of shared source for cracks by whole gasket area
The digital simulator twin parameters			
c_{fd}:	The cost of ICE fuel consumption ($0.808 l h^{-1})	P_w:	The power generated by ICE (KW)
c_σ:	Underutilization (dereliction) cost ($)	Q_{df}:	The ICE consumption, l. h^{-1}
c_d:	The electricity cost of kW·h^{-1} by a ICE, $/kW·h^{-1}	f:	Corrosion rate of alloy layer (mm.day^{-1})
A	Cross-section area	V:	Crack path length speed (mm/day)
t_m:	Generator uptime running (wk.)	$K_{i \in (1,2,3)}$:	Coefficients with constant values based on alloy type
t_s:	The red hotspot area imaging by STTF	t_{tc}^h:	The temperature losses from the hot gasket
$d_{p,q}1$	Diameter of the hotspot crack leg (mm)	t_{tc}^c:	The temperature losses from the cold gasket
g_s:	Slenderness ratio of gasket layer thickness tolerance ±0.14 mm	$Z\overline{T}$:	The thermoelectric material's performance index $= \alpha_p^2 \overline{T}/\upsilon_p\lambda_p$ And $\overline{T}$ and the average temperature of thermoelectric terminal conductors.
P_{min}:	The minimum absorption, kW	T_u:	The downtime due to replacement (hr.)
L_i:	Crack length mm $\rightarrow$ (oil spot)	$E(\xi)$:	The exponential distribution with rate ξ,
λ_p, σ_p:	The heat and electrical conductivity for the leg has type p, respectively.	α_h:	The Seebeck coefficient of the hot terminal
$T(r, z)$:	The temperature distribution in (r, z) plane	δ:	The standard deviation from average power, kW.
V_{opt}	Optimum value of crack path speed (μm/wk.)	m:	Total TEG-ICE generator running hours
I:	The electric current	z_n:	Number of red spots on the gasket layer
q_{pconv} (a, z):	The heat flow density at the gasket surface of p-type leg $= h[T(a, z) - T_0]$	c_t:	The gasket damage cost ($) (head or intake and exhaust) before analysis expected
Self ACO recruiting parameters			
$P_{i,j}^{nl}(t)$	The position where ants found food	η_{ij}	The visibility provides valuable information
N_i^{ml}	represents the area that has not been assigned	α, β, ρ	The pheromone trail evaporation rate, 4,1,0.8917
$\Delta\tau^{nl}$	Refers to the pheromone increment ≈ 0.0501	EAC_{ij}, EAC_{jk}	The electrothermal transformation failure cost
$nl_s(t)$	The ants that are planned to move to collect food from different places	ϑ	A binary parameter (0,1) illustrates the importance of the period of cycle time
Responses			
P_{watt}	Generated electrical power of the generator by K-W/hr.		
Fuel	The fuel consumption per working week		
η	The efficiency of the proposed integration based on heat transfer		
Di_r	The diffusion rate over time per month for each micrometer length		

The crack intensity Q_p in our calculation $\leq$ ($Q_{min} = 0$, $Q_{max} = 3$) is determined by the spatiotemporal filming for terminals' cracks. The thermal's pitch strength follows three distribution types as discussed in "Source characteristics of thermal filming." The thermal conductors must be installed far away at these positions, ($P_{i,j}$). The following approximations are proposed in order to streamline the process and guarantee correct simulation results. Ignore the crack paths' and treat it as a continuous straight line projection parallel to the nearest axis by angle $\emptyset$, $\mu = 0.005$ and, $\sigma^2 = 0.001$ [41,42]. The failure is achieved according to Equation (1):

$$\sum_{P_{1,1}}^{P_{i,j}} Q_P \geq S_{wt(i)} \; \forall i \tag{1}$$

The tracking also reveals that the cracks segment line discussed in Figure 4 above has length L, N (μ, σ^2, ξ), and M_i is ranged as $0 < M_i < L$. When the spatial image could not expect the behavior of the positions path or its intensity distribution is not considered, Equation (2) can be followed:

$$S_{wt} = 10 \, log \left[\sum 10^{0.1(L_i)} \right] \dots \tag{2}$$

The digital twin simulator architecture is sketched in the 2nd section through Figures 2–4 and tests the thermal conductors for electricity via two sequential stages. Tracking the failure causes in three different phases begins with hot gasket, strips' TEG legs, and cold gasket alloys to determine the significant factors to save them from failure in Section 3 (stage-1), while the methodology that relies on the neural network model, which is used to extract the data from STTF to support the ACO metaheuristic technique used to predict the functional change caused by crack diffusion to predict the cracks' position $P_{i,j}$ and their intensity Q_P as discussed in Section 3.1. The mathematical equations are used to enhance the digital twin behavior and NN output data parameters in checking the validity of TEG-ICE integration. Therefore, the data extracted from STTF is considered an initial value of searching using the meta-heuristic optimization algorithm to obtain the optimal results as discussed in Section 3.2, and to determine the battery used to receive generated electricity to serve the HVs sector. Finally, in the list validation in the conclusion discussed in Section 5, the authors suggest using other alloys that have a YbAl$_3$ compound in manufacturing the thermal conductors in the future work section.

3. Stage (1): Tracking Thermal Conductors' Failure Causes

This work was a consultant mission for one of the Egyptian generator companies, through the USCC office of Zagazig University, which provides the spatiotemporal thermal filming for the TEG legs through extensive experimental observations, and accumulating data are described as a controlled tracking dashboard. In view of the heat generated by the ICE at the exhaust duct and the movement of the rotating wheels during the use of the brakes at the air duct to generate a difference in the thermal effort and take advantage of the heat and convert it into an electric current by an electric generator and feed the traction batteries to operate the electric traction engine through an electronic control unit (ECM) to regulate the flow of electric power to control engine speed and torque by the transmission via IoT. This is considered very useful in the battery without discharge, as they depend on the TEG in a reciprocal manner by absorbing heat and converting it into electricity to maintain the efficiency of the movement generated when relying on the battery to transition to the mechanical transmission when speeds are less than 40 km/h. The dependence of HVs to begin working on the battery and the electric motor is a reason to save spent gasoline or ICE and reduce CE by an average of 15%. Traction batteries can be evaluated by five influencing factors: specific energy (Wh/kg), acceleration (W/kg), operating cost (cost/km/passenger), fast recharging capability (80% in 10 min), and capacity to absorb high currents during repeated braking [42,43]. The authors have used the thermal filming measurement that tracks the defective place on the studied surface as illustrated in Figures 5–8, which tracks each failure behavior for a specific alloy of strips of TEG legs and hot gasket texture. Figure 5 illustrates the behavior of (Fe-Al) in resisting the crack creation, $P_{i,j}$, which creates ramifications that are discrete around the crack centers uniformly, over t_m: 720 working hours, then coalesces and reduces the efficiency toward failure after t_m: 1440 working hours.

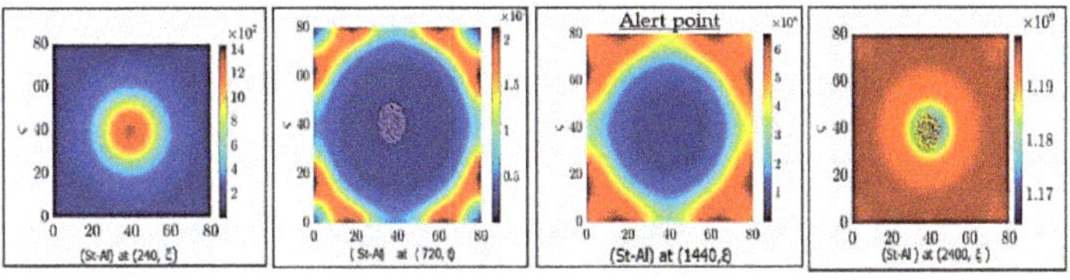

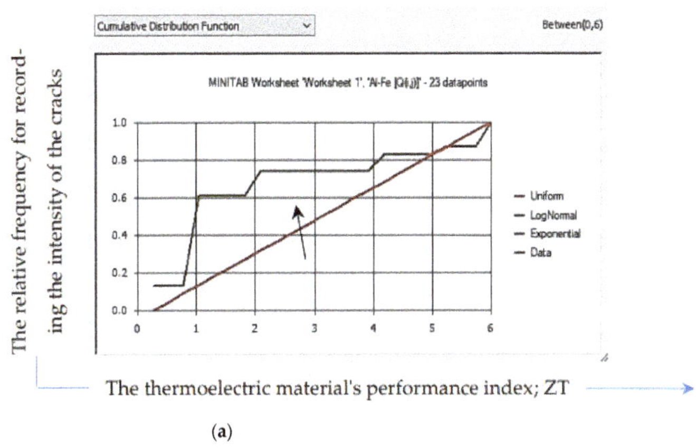

(**a**)

Figure 5. The change in the (Al-Fe) gasket alloy resistance working conditions for thermal transfer. (**a**) The behavior of the cracks' growth through 23 weeks for the Fe-Al alloy is uniform.

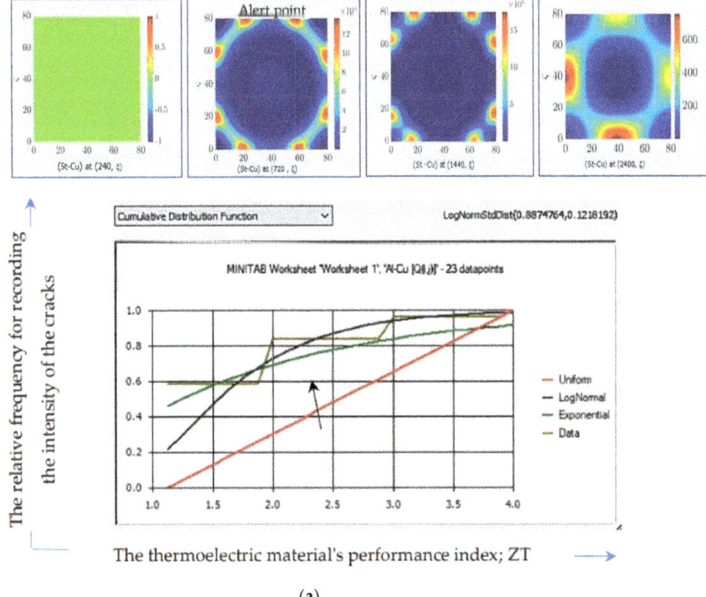

(**a**)

Figure 6. The change in the (Cu–Fe) gasket alloy resistance working condition for thermal transfer. (**a**) The behavior of the cracks' growth through 23 weeks for Al–Cu alloy is Gaussian. The arrow points to both lognormal and exponential which have low error than uniform.

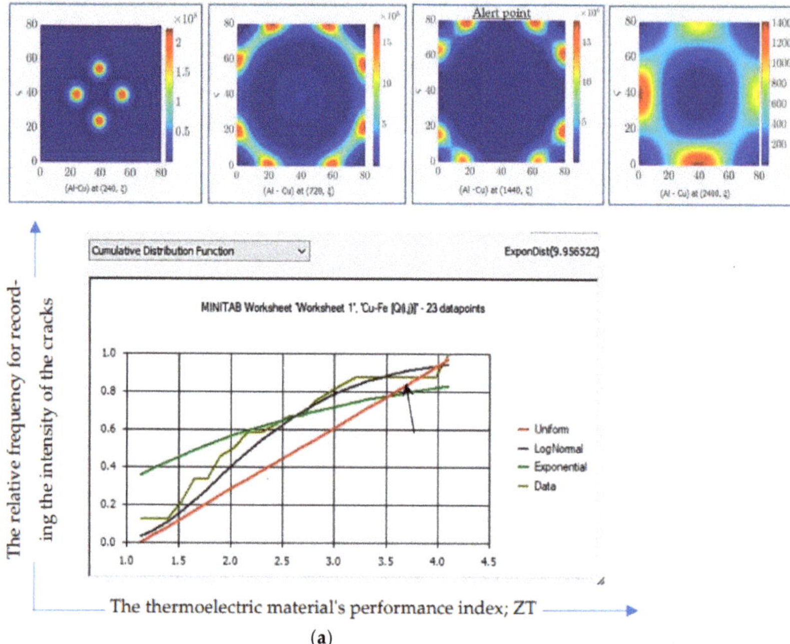

(a)

Figure 7. The change in the gasket (Al–Cu) alloy resistance working conditions for thermal transfer. (a) The behavior of the cracks' growth through 23 weeks for Cu–Fe alloy is exponential.

The cracks and their ramification behavior for the strengthening (Fe-Al) alloy for the gasket and TEG legs to save the thermal conductors via P_i- and Q_P-like uniform distributions as illustrated in Figure 5a and formulated as follows: $\begin{cases} P_{i,j} \sim U(0,L) \\ Q_P \sim U(Q_{min}, Q_{max}) \end{cases} \forall i,j = 1,2,3, \ldots, n \left(\text{mm} \times 10^{-1}\right)$ The conductor's failure begins in the middle and growth [between $(1, 4.1)$].

Figure 6 illustrates the (Cu-Fe) alloy after t_m: 720 working hours, which resists the terminals' cracks more than (Al-Fe), where the surface damaged slowly in Gaussian behavior than for the (Al-Cu) surface over t_m: 1440 working hours, which behaves exponentially as illustrated in Figure 7.

The cracks and their ramification behavior for strengthening the (Cu-Fe) alloy for the gasket to save the thermal conductors via $P_{i,j}$-like uniform distributions, and Q_P is a Gaussian distribution, as illustrated in Figure 6a and formulated as follows: $\begin{cases} P_{i,j} \sim U(0,L) \\ Q_P \sim U(\mu, \sigma^2) \end{cases} \forall i,j = 1,2,3, \ldots, n \left(\text{mm} \times 10^{-1}\right)$. The conductor's failure forming the surrounding circle is indicated in the following expression: [LogNormStdDist (0.7969939, 0.3705614)]. While the crack and their ramification behavior for strengthening the (Al-Cu) alloy for the gasket and TEG legs to save the thermal conductors is fixed for $P_{i,j}$, while the Q_P is exponentially distributed, as illustrated in Figure 7a and formulated as follows: $\begin{cases} P_{i,j} \sim M_i \\ Q_P \sim E(\lambda) \end{cases} \forall i,j = 1,2,3, \ldots, n \left(\text{mm} \times 10^{-1}\right)$. The conductor's failure creates four neighboring spots with following expression: [ExponDist (2.36975)].

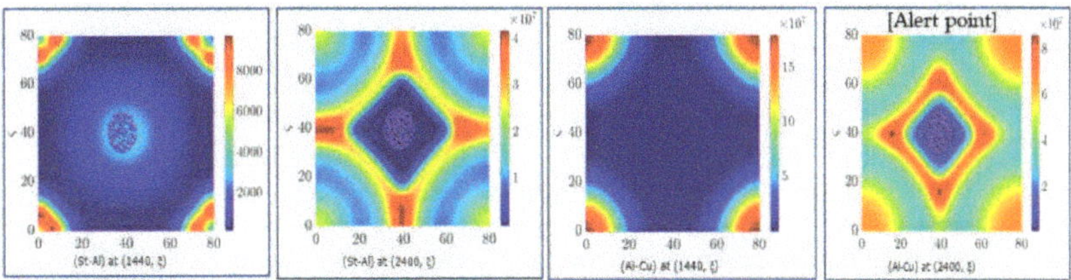

Figure 8. The change in the (Fe-Al) and (Al-Cu) alloys resistance working condition for thermal transfer.

Therefore, the authors' experiment uses a gasket lined on both sides, (Al-Fe) on the upper face and (Al-Cu) on the lower face. The authors tend to increase the thickness of the gasket alloy layer to 1.74 mm. The spatiotemporal evolution tracks the gasket terminals' cracks for the (Al-Fe) and the (Al-Cu) as illustrated in Figure 8, during t = (1440; 2400) working hours when $R_0 > 1$ (51 < b < 100), for which the ramifications increased to 100. Although the thickness increased, the films show the cracks exacerbate according to the ramification regression as illustrated in Figure 8, but resists for more than 2400 working hours, with $Dx = Dy = Dv = 0.2$.

The authors recommend fixing the thermal conductors away from colored positions (e.g., yellow and orange) that appear via STTF, which avoids the terminals' cracks over *tm*: 4392 working hours for (Al-Cu) and tm: 5184 working hours for the (Al-Fe) alloy.

3.1. Setting the Significant Parameters of Thermal Conductivity

The previous step in stage (1) was executed in the laboratory by tracking all significant causes that affect heat transmission effectively and reflect positively on generated power electricity over 23 weeks by analyzing the gasket case every 3 days (i.e., 45 gaskets' filming) for designing the suitable digital twin which can check the success of the integration of TEG-ICE. The second step in stage (1) will interest in extracting images' data by neural network supported by mathematical procedures such as the cracks' positions and their intensity precisely to study the generated electricity to charge HVs' batteries for a long time, which reflects positively in reducing fuel consumption and carbon emissions. The neural network will feed metaheuristic techniques such as ACO. The neural network was supplied with STTF images of the reasons for thermal conductivity failure after conversion to grayscale, with each pixel being divided by 256 as binary 0 to 1. After the original picture was reduced to 50-50 pixels, the number of input neurons was substantially decreased to just around two, as shown in Table 3. Analytical recording for the temperature distribution via the digital-twin simulator model using a STTF to heat effect is captured as illustrated in the merit Figures 9–11 and Figures 12–16 to seek efficiency about suggested cross sections of TEG legs, which also discusses the TEG-ICE system sketch and illustrates the coordinates of the heat stream. This study aims to attain the highest utilization rate for the electric generator operations (generated power, working days, and profit), in Equations (6)–(11).

Table 3. The neural network tunes.

	Parameters	Down	Up
L_1	Neuron number	2	17
L_2	Learning rate	0.012	0.39
L_3	Training epoch	210	2550
L_4	Momentum constant	0.11	0.95
L_5	Number of training runs	3	7

The digital twin is designed to test the legs' behavior toward resist failures as cracks' generation and track the cracks' span, depth, and length and is named figuratively (reliability stage). The crack positions are the points that failed in heat transfer and must be tracked to increase the working period time efficiently by installing the TEG legs far away from the cracks' positions. Analysis of the STTF images illustrated in Figures 5–8 cleared us of the TEG-IEC integration failure caused by crack intensity, Q_P, which is proportional to merit figure, temperature, melting point, their cross section, gasket texture plate thickness and their flow rate speed and exhaust flow density as expressed in Equation (3), which affect the thermal conductors which affect electrical conductivity (i.e., TEG legs) within their working time as illustrated in Figures 9 and 10. If these cracks are independently created at a constant average rate, the analogous continuous equation states that the intensity at a particular position meets a negative exponential distribution per unit of time as expressed in Equation (4).

$$\nabla \cdot \left(\frac{\omega}{\rho_0(\nabla p - q)} \right) - p\frac{\rho_0}{\omega} = Q_P \tag{3}$$

$$\beta = \frac{Q_p e^{-kd_{sr}}}{4\pi r} \; \forall i, j \in gasket\ coordinate \quad \Delta\ 0.01\text{m} \times 10^{-4} \tag{4}$$

Figure 9 illustrates that the intake, exhaust temperature, and the merit figure have a significant impact on generated power that can be gained from the conductors molded from (Fe-Al, Al-Cu), while Figure 10 illustrates that the cross section area of the conductor and the melting point have a direct impact on controlling the cracks' span. Figure 11 illustrates the main reason for the diffusions' crack per mm $\times 10^{-2}$ and affect intake temperature and melting point of the candidate alloys.

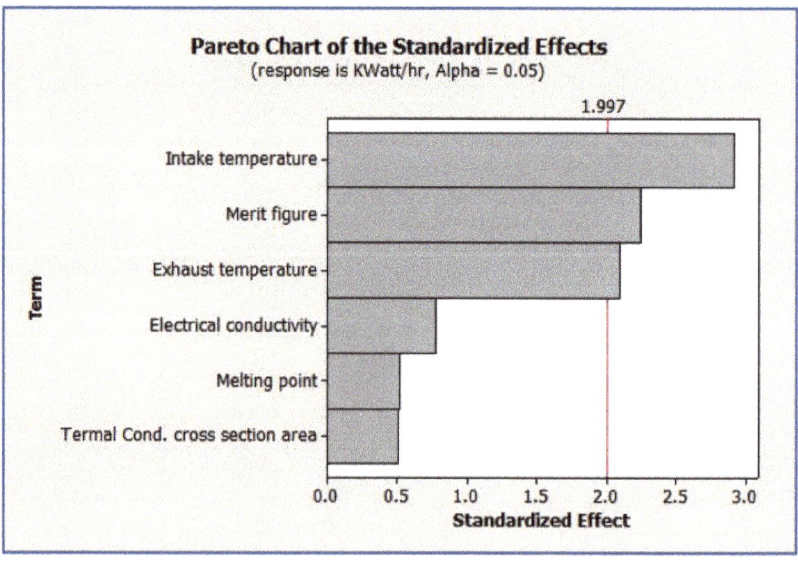

Figure 9. The significant parameters affected by power generation.

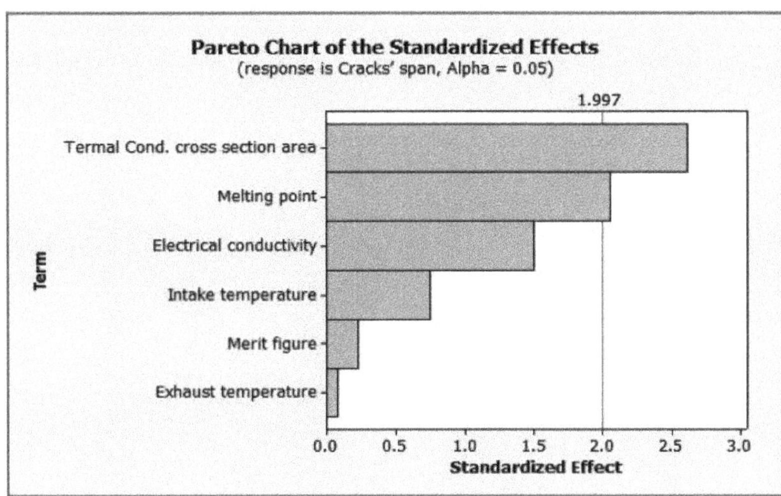

Figure 10. The significant parameters affected by power generation and cracks' span.

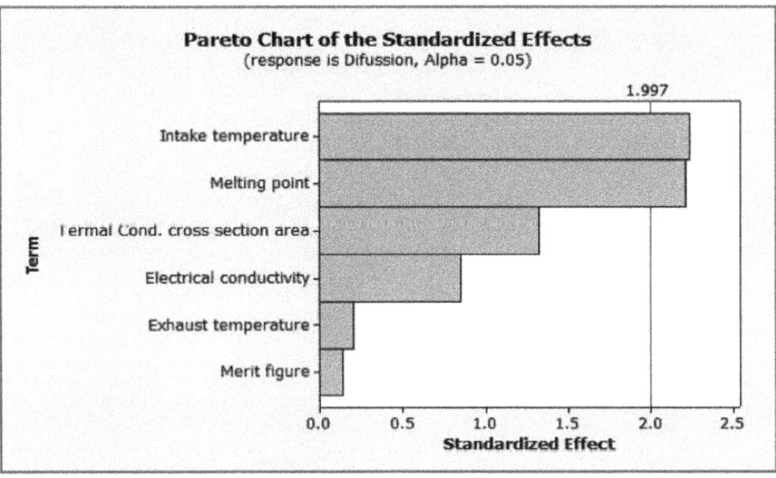

Figure 11. The significant parameters affected by the thermal conductor's failure.

The equivalent radius of failure position expressed by R_{cr} in Equation (5), which must install the TEG legs far away from the perimeter of these circles to guarantee electrical power more than 50 W 10^3hr^{-1}, where R_{cr} is the radius of surrounding circle of all cracks positions form closed shape, and h_r is the radius of hotspot ramifications [44,45]. The E (ξ) represents the parameter rate ξ for the exponential distribution, while $R_0 < 1$ ($b < 50.0$).

$$R_{cr} = R_0 \left(\frac{bh_r}{R_0} \right)^{\frac{1}{b}} \tag{5}$$

Figure 12 illustrates the final decision for setting the gasket and thermal conductors using conditions and cracks' span and must be controlled in a minimum width of less than 12 mm $\times 10^{-2}$ and the cracks' diffusion are less than five cracks' positions per mm $\times 10^{-2}$. Therefore, the optimizer of the digital-twin simulator indicates the difference between hot and cold plates and must exceed 105 °C. The merit figure of Al must be 4.3915 times the other materials in the alloys to reduce the cracks' growth for more than 5184 working hours.

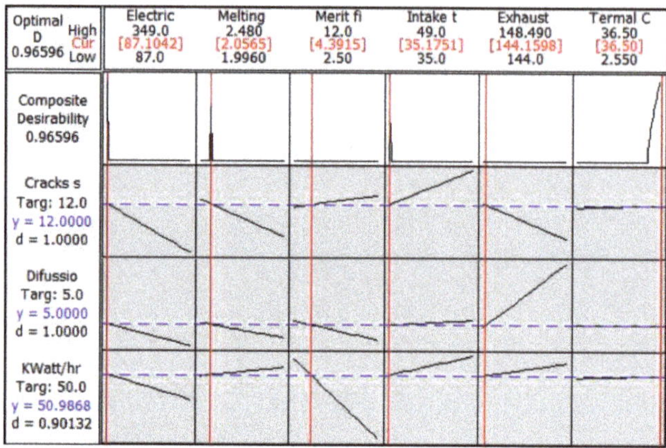

Figure 12. The significant parameters affected by the thermal conductor's quick failure on the gasket and terminals strips (TEG legs).

3.2. Measure the Generated Electricity

When setting the significant parameters as shown in Figure 12, the authors have divided the heat transmission into three phases. The first phase is through the thermal conductors installed in the hot gasket texture (exhaust flow) far away from the cracks' positions. the second phase is through the TEG legs' strips toward the cold gasket gate. The third phase is in the cold gasket texture. The gasket texture and TEG legs are heavily influenced by temperature, and changes in that temperature have a substantial impact on performance (e.g., merit figure) [46–49]. The authors show the thermal relationships based on the Seebeck coefficient α, voltage, and thermal conductivity β of both the hot side position (beginning heat source called $\alpha_{hp(i,j)}$), and cold side position (end heat destination called $\alpha_{cp(i,j)}$) in Equations (6)–(11). The first phase of heat transmission is shown in Equations (6) and (7) and the voltage is expected via relation (7a) for the generated electricity from the heat difference between the hot terminal $\alpha^1_{hp(i,j)}$ and cold terminal $\alpha^1_{cp(i,j)}$ as illustrated in Figure 13, which calculates the number of HVs' battery cells.

$$\alpha_{hp(i,j)} = 5.9214 \times 10^{-13}T^3 - 3.274 \times 10^{-9}T^2 + 2.42 \times 10^{-6}T^1 - 2.744 \times 10^{-4}T^0 \dots \quad (6)$$

$$\alpha_{cp(i,j)} = 1.292 \times 10^{-13}T^3 + 1.074 \times 10^{-9}T^2 - 9.272 \times 10^{-7}T^1 + 8.96 \times 10^{-6}T^0 \dots \quad (7)$$

$$E = LiFePO_4 + 6C \rightarrow LIC_6 + FePO_4 = 3.2 \, volt \dots \quad (7a)$$

The second phase of heat transmission through the thermal conductor strip is shown by Equations (8) and (9), which contribute to the voltage required for electrical conductivity and is illustrated in Figure 14. Therefore, the cross section will be designed between [1.14: 1.31] mm.

$$\beta_{hp(i,j)} = 1/\left(2.25 \times 10^{-14}T^3 - 1.074 \times 10^{-9}T^2 - 9.272 \times 10^{-7}T^1 + 8.96 \times 10^{-6}T^0\right) \dots \quad (8)$$

$$\beta_{cp(i,j)} = 1/\left(-1.25 \times 10^{-14}T^3 - 6.43 \times 10^{-11}T^2 + 9.1 \times 10^{-8}T^1 - 1.06 \times 10^{-5}T^0\right) \dots \quad (9)$$

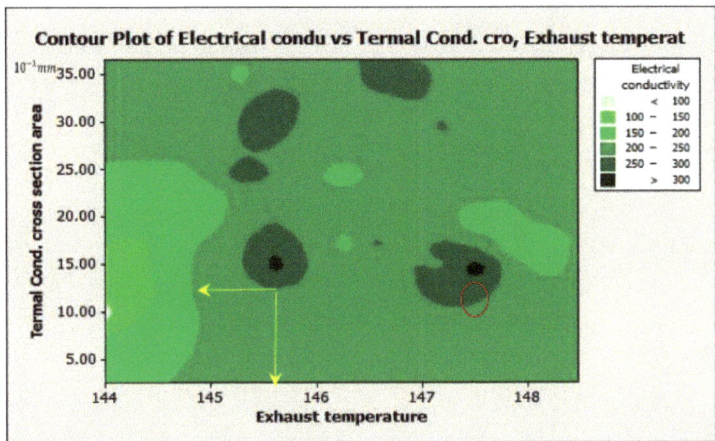

Figure 13. The exhaust gasket thickness affects the electrical conductivity at the highest available level. "Hint: the red circle points to the preferred significant parameters values selection".

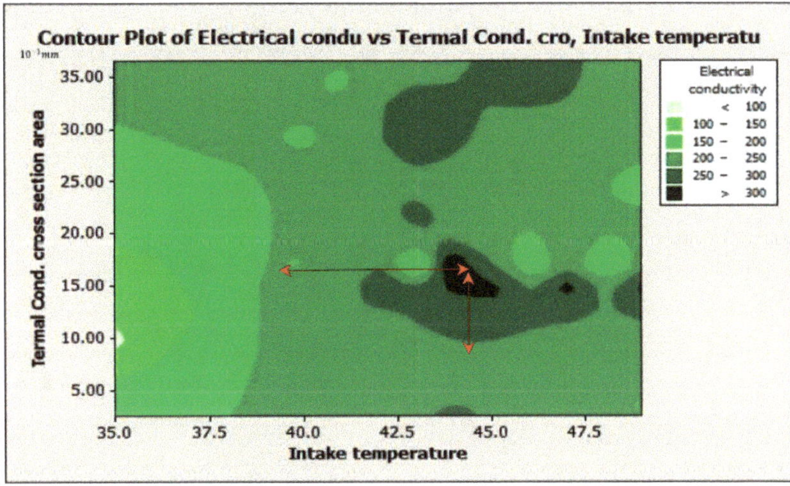

Figure 14. The strip cross section according to the intake affects the electrical conductivity at the highest available level. Hint: → points to the preferred significant values.

The cross sections of the TEG legs have a direct impact on electricity affected by the cracks span and their intensity Q_P as illustrated in Figures 15 and 16, which illustrates the cracks' intensity at specific positions to avoid installing the thermal conductors through these positions to guarantee a long life of the power generated. The regression of cracks formed vs. working hours for the different alloys is expounded in the conclusion section.

By the same Equation (7a), computing the voltage affected by the low heat transmission toward a destination side and is shown in Equations (10) and (11).

$$k_{hp(i,j)} = 1.25 \times 10^{-7} T^3 - 1.27 \times 10^{-4} T^2 + 3.87 \times 10^{-2} T^1 - 2.36 \times T^0 \ldots \tag{10}$$

$$k_{cp(i,j)} = -1.6 \times 10^{-8} T^3 + 2.91 \times 10^{-5} T^2 - 1.58 \times 10^{-2} T^1 + 3.73 \times T^0 \ldots \tag{11}$$

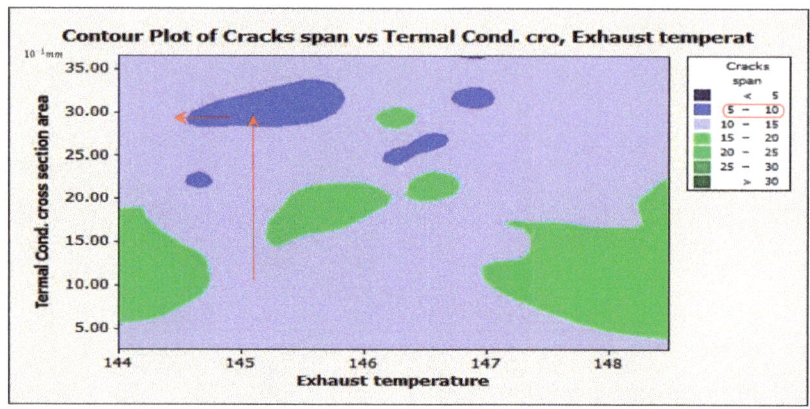

Figure 15. The strip cross section for TEG legs affects the cracks' span. Hint: → points to the preferred significant values.

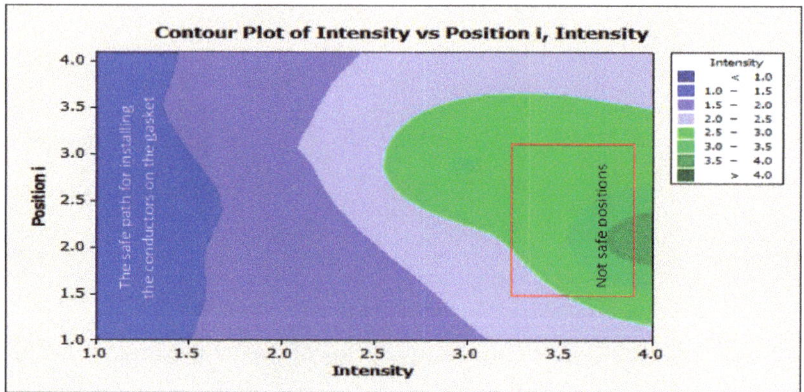

Figure 16. The intensity of the cracks' positions for the thermal conductor terminals.

4. Stage (2): Predicting the Integration Efficiency

The digital twin works on enhancing the neural network model mechanism by using the advanced artificial ant colony (ACO) setting and relies on data extracted from the analysis of the STTF images. The surface of the gasket was divided into small square areas, each of which has a center that we express by $P_{i,j}$. The main objective is to sustain heat transfer from the hot source (e.g., exhaust gasket) after the thermal saturation case, then allow to direct the heat via strips to arrive at the cold plate (intake gasket). The experimental observations reveal that cracks are the worst cause that hinders the success of the electrothermal transformation of the TEG-ICE system and leads to their integration failure. Therefore, tracking the cracks and installing the TEG legs far away from the cracks guarantees the continuity of electrothermal transformation. The work was divided into two stages. The first stage is discussed above and expounds on the tracking failure causes and their positions by using spatiotemporal thermal filming (STTF). In the second stage, the authors push ants to move to seek food (i.e., cracks' positions; $P_{i,j}$) as expressed in Equation (12) and leave concentrated pheromones equal to their intensities (Q_P; numbers of cracks' ramifications; b) as discussed in Equation (13). The evaporation was complete if the line connected between the first trigger point and the cracks' positions illustrated in Figure 4 did not create a path over all damaged positions and is expressed by a line fitting error,

$S_{wt}(i)$. The authors seek to declare the cracks' positions precisely to install the TEGs' legs and achieve continuous electrothermal transformation.

$$P_{ij}^{nl}(t) = \begin{cases} \dfrac{[Q_p(t)]^{\alpha}\,[\eta_{ij}(t)]^{\beta}}{\sum_{j=1}^{d}[Q_p(t)]^{\alpha}\,[\eta_{ij}(t)]^{\beta}}, & if\ j \in N_i^l \leq 20 \\ 0, & otherwise \end{cases} \tag{12}$$

$$\tau_{ijk}(t+1) = (\rho)\tau_{ij}(t) + (\rho)\tau_{jk}(t) + \Delta\tau^l \tag{13}$$

where Q_P is the concentration of pheromone in position (i, j). N_i^{nl}, represents the area not assigned. η_{ij} The visibility provides valuable information about the problem when searching Dorigo and Gambardella (1997) as cited in Hong et al. [50]. The search parameters, which specify the relative significance of the pheromone trail and heuristic data are: $\alpha = 4$, and $\beta = 1$. Additionally, $\rho = 0.8917$ is the pheromone trail evaporation rate, as shown in the first and second terms of Equation (13), which reflects the local and global updates of the pheromone in working hours $t + 1$ based on the solution of working hours t, respectively. The term $\Delta\tau^{nl} \approx 0.0501$ refers to the pheromone increment (the number of connected cracks' positions) and is expressed in Equation (14).

$$\Delta\tau^{nl} = \begin{cases} \rho \times \dfrac{(Z_{Q-max}-Z_{ij})}{(Z_{Q-max}-Z_{Q-min})} + 0.1, & if\ ant'nl'\ discover\ \text{the crack} \\ 0, & otherwise \end{cases} \tag{14}$$

In each working hour t, N ants generate N feasible solutions (monitor all positions that are out of service). Z_l is the solution generated by ant $nl \in \{1, 2, \dots, N\}$ as cited in [50,51]. $Z_{min} = min(Z_l)$ and $Z_{max} = max(Z_l)$, are the best (TEG generates full electrical power) and worst (failed TEG in electrothermal transforming due to crack spreads) total costs of the solutions generated in working hours t, where Z_{min} pheromones and temperature are updated to increase the density of pheromones associated with best solutions and decrease for worse solutions. The transportation cost relies on fuel consumption up to working hours t based on Equation (15), which is used to calculate the visibility (η_{ij}) for assigning a downstream to an upstream (16).

$$\eta_{ij} = \frac{1}{EAC_{ij}} \tag{15}$$

$$\eta_{jk} = \frac{1}{EAC_{jk}} \tag{16}$$

where EAC_{ij} and EAC_{jk} are the electrothermal transformation failure costs which are proportional to fuel consumption cost and the amount of carbon emissions and are expressed by Equation (17).

$$EAC = G_t\ cost \times avilable\ working\ time + (electrical\ conductivity\ losses\ cost \times working\ uptime\,) \tag{17}$$

When comparing the best solution of working hours with the best global solution, both are the same in the first working hours. After completing working hours t, where $\lambda = nl_s(t)$ is the ants that are planned to move to collect food from different places, $P_{i,j}$, and ϑ is a binary parameter (0,1), and illustrates the importance of the period of cycle time searching. Divergence is exacerbated by lower values while being reduced by higher values. Equation (18) demonstrates how to enhance the solutions by fading any ant behavior pattern out of the planned pattern. The Δ is the discrete Laplacian operator, and γ is the running time by seconds for ants' acceleration as shown in Figure 17 and illustrates the catalyst that depends on the placement and timing of the actuator's ants. As a result, this may explain the likelihood that a deviation will occur at the decisive point $(P_{i,j})$ by $\rho(S,\ t) = \vartheta\gamma^2$. $d_s(t)$ is subtracted from both sides of the equation and then divided by δt, [52].

$$\rho_s(t + \delta t) = \ln\left[(1 - \vartheta)\rho_s(t) + \frac{\vartheta^2}{\lambda}\Sigma_{S'\sim S}\Delta\alpha_{s'}(t)\right](1 - \omega\delta t) + \beta n_s(t)p_s(t) \qquad (18)$$

$$\rho_s(t + \delta t) = \ln\left[\rho_s(t) + \frac{\vartheta\gamma^2}{\lambda}\Delta\rho_s(t)\right](1 - \omega\delta t) + \beta n_s(t)p_s(t) \qquad (19)$$

$$\Delta\rho_s(t) = \frac{\left(\Sigma_{s'\sim s}\Delta\rho_{s'}(t) - \lambda\rho_s(t)\right)}{\gamma^2} \qquad (20)$$

Take the limit as δt and ρ^2 both decrease in size, and the ratio $\rho^2/\delta t$ constant remaining with a value specified as $P_{i,j}$ (i.e., the predicted position of food at each phase), and the amount $\beta\delta t$ likewise remaining constant. The suggested equation provides the pheromone-based dynamics of the behavior of finding food sites.

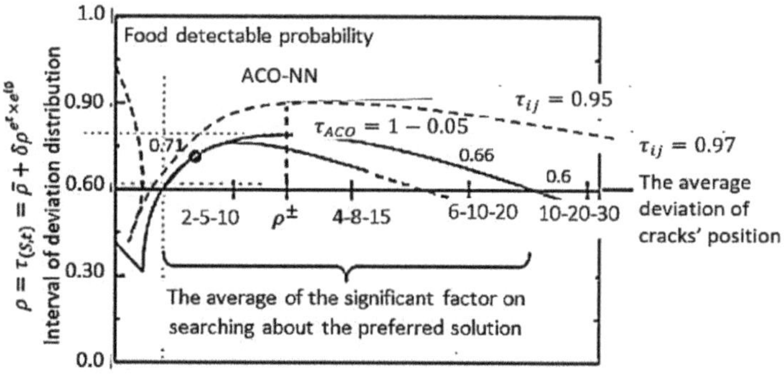

Figure 17. The predictable ρ of pheromone in Equations (19) and (20).

The failure permeates the sites, withers (evaporating) over time, and interacts with the ants to encourage them to seek out other locations. Ants react to failure by consistently performing, which causes them to become drained (i.e., inactive), when $Z_{i+1} - Z_i = 0$. $\tau \equiv 1/\omega$ and establishes the timing for the proposed STTF-NN-ACO. Since the failed propagation is either zero or periodic, the fail rate density is matched with the stationary of Equations (19) and (20) which are integrated across the full phase.

The STTF-NN-ACO methodology strengthens the metaheuristic technique with mathematical equations to quickly search for the optimal values for controllable variables discussed in stage (1) to be fed into the digital simulator twin model. Some of the uncontrollable inputs (uncertain) are hard to measure as cited by [53]. Modern simulation resorted to constructing a suitable digital twin for the studied system to be trusted to get the outputs as discussed by Shen et al. [54] and Gan et al. [55] through using spatiotemporal thermal filming which feeds the model continuously [56]. At the same time, with the help of the optimization algorithms, the digital twins' development changes from merely descriptive to becoming actionable. The estimate in the real system resorts to uncertain parameters with their predicted values and may result in a cautious out-put with unnecessarily high operating costs [57]. The stochastic approach was developed by Kalantari (2020) when highlighting the significance of deterministic meaning based on a fuzzy neural network for reducing the error in the searched optimal value [58]. A mathematical optimization enhances the metaheuristic searching by hybridizing with another artificial technique as discussed by Yang et al. and Qiu et al. [59,60]. Stage (1) in this work indicates that if the working parameters are optimized, excellent results are achieved. Therefore, Han et al. discussed the importance of parameter optimization based on metaheuristic methods [61]. SDS has become a robust and efficient global search and optimization technique that was theoretically defined extensively, especially when combined with other swarm intelligence (SI) algorithms, such as artificial bee colony (ABC) to robust searching techniques jumping over the local optimal solution to global one [62,63]. There is no one optimization technique that is best suited for a wide range of applications or handling diverse sorts of problems. Therefore, the combination and setting of the initial search values for parameters is a very urgent necessity as discussed in stage (1) of this work and is based on Khosravy et al. [64] in analyzing the spatiotemporal thermal filming images for the TEG-ICE component behavior to deduce the preferred working conditions. The challenge is adapting the search mechanism rapidly by predicting specific significant variables as discussed in the cause-and-effect diagram and having the most impact on the results. In the 21st century, most of the current studies adopt an innovative improvement of neural networks by interfering with one of the metaheuristic methods such as the gravitational emulation local search and applied in the electricity-producing optimization in 2010 [65]. In the case of the multi-objective optimization model using the linear decreasing particle swarm optimization algorithm, which is recommended and advised with using the weighted superposition attraction algorithm (WSA) when a solution must be chosen in a minimum time for problems that have multi-pass [66], which was used for the parameter selection and appeared excellent over the native PSO. This thinking approach of the "need-based" must be a guide matched with the understanding of the behavior of the operating parameters through real operating, mainly if applied considering multiple constraints. Therefore, the authors began with laboratory observation by STTF.

4.1. The Virtual Suggested TEG Design

The review revealed unanimous in candidate the hybridization between the whale optimization algorithm (WOA) with the fuzzy neural network (FNN), and ACO with GELS (i.e., gravitational emulation local search) to tackle the multi-objective optimization and rapidly gain the global solution [65]. The proposed TEG-ICE mechanism (i.e., digital simulated twin) as illustrated in Figure 18 combines the ACO with NN model which is supported by mathematical equations that are fed data inputs via spatiotemporal thermal analysis through a neural network model to describe the whole behavior of TEG-ICE integration [STTF-NN-ACO] under deduced conditions from the stage (1). The proposed [STTF-NN-ACO] variables are indicated in Table 2. The objective is to accurately predict the crack positions, $P_{i,j}$, and their intensity, Q_P, to evaluate the risk of damage level by selecting the best locations sitting on the thermal conductors. These goals are divided into two sub-objectives. The first is tracking the cracks' creation positions, $P_{i,j}$, and their intensity, Q_P [66]. The neural network extracts data images ($P_{i,j}$, Q_P, r, ω, h_r, R_{cr}, R_0, d, , L_i, d_{sr}) through many iterations approximate to 1000 to obtain less deviation. Therefore, the initial search values rely on significant parameter values extracted via the output of the experiments illustrated in Figure 12, enabling prediction of the optimal solution with minimal error and time accurately. The mathematical equations focus on significant factors cut by the vertical red line and affect conductor lifetime, as illustrated in Figures 9–11. The next equations describe the designed digital twin to enhance the tracking of the TEG-ICE integration. Equation (21) is used as a reference to check efficiency of generated power as discussed by Shen et al. [64].

$$\delta \leq P_w \leq 2.5\, P_{min},\ KW \tag{21}$$

The downtime generated by cracks appearing in gaskets or legs of the proposed TEG-ICE (i.e., failure area) leads to integration failure and can be expressed by Equation (22):

$$T_u = \sum_{i=1}^{m} t_{tc}^h + \sum_{i=1}^{m} t_s K_{1i} V_i \ \ f_i^{-1} + \sum_{i=1}^{m} t_{tc}^c \tag{22}$$

Phases of heat transition: hot gasket, legs' strips,   cold gasket

The influence of the thermal potential difference induced by the gasket alloy texture (Al-Fe) or (Al-Cu) and the temperature 'Th' and 'Tc' between the TEG terminals are equal to those of the hot source and heat sink. All fields rely exclusively on the coordinates 'r' and 'z' due to the geometry configuration's axial symmetry of the TEG leg and the temperature load. The axisymmetric specific application of steady-state heat flow for the p-type leg is thus [67] suggested in Equations (23)–(27).

$$\frac{\partial^2 T(r,z)}{\partial r^2} + \frac{1}{r}\frac{\partial T(r,z)}{\partial r} + \frac{\partial^2 T(r,z)}{\partial z^2} + \frac{I^2}{\lambda_p A_z^2(\sigma_p)} = 0 \tag{23}$$

$$q_r(r,z) = -\lambda_p \frac{\partial T(r,z)}{\partial r} \tag{24}$$

$$q_z(r,z) = -\lambda_p \frac{\partial T(r,z)}{\partial z} \tag{25}$$

$$T(r,0) = T_h,\ T(r,L) = T_c \tag{26}$$

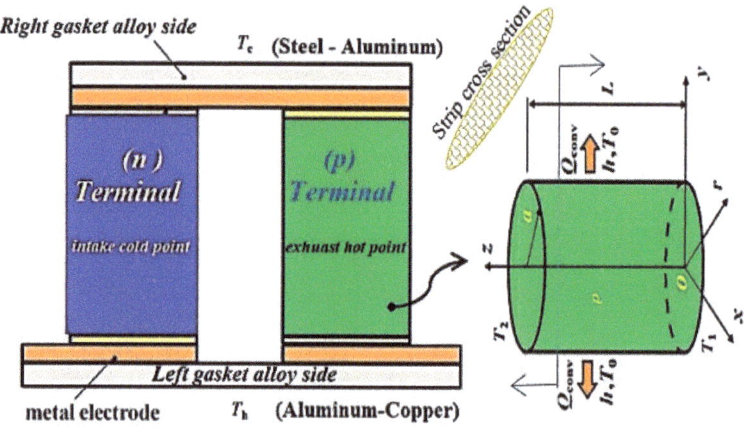

Figure 18. A thermoelectric generator with ellipsoid-cross section terminals.

It is assumed that the ambient temperature will always be constant at the maximum value, T_0, and the heat convection coefficient between the gasket surface of the p-type leg and its surroundings is h. The temperature field's boundary conditions are provided as follows:

$$q_r(a,z) = q_{pconv}(a,z) \tag{27}$$

The solution of Equation (23) for the constraints (26) and (27), using the separation of variables and eigenfunction expansion procedures [68], and is expressed as Equation (28):

$$T(r,z) = -\frac{I^2}{2\sigma_p A_z^2 \lambda_p}z^2 + \left(-\frac{T_h - T_c}{L} + \frac{I^2 L}{2\sigma_p A_z^2 \lambda_p}\right)z + T_h + \sum_{n=1}^{\infty}\frac{h}{\pi}.\Psi I_0(k_b r)\sin(k_b z) \tag{28}$$

where, $\Psi = \dfrac{[\cos(b\pi)-1]I^2\lambda_p/(\sigma_p)k_n^3\ L^3 + (T_c - T_0)\cos(b\pi) - (T_h - T_0)}{\lambda_p(b\pi)L^{-1}I_1(b\pi L^{-1}a) + hI_0(b\pi L^{-1}a)}$ and consider $I_0(b\pi L^{-1}a)$ and $I_1(b\pi L^{-1}a)$ are the zero-order and first-order modifications of the Bessel functions, respectively. Consequently, it is possible to determine the heat transfer rate at the gasket terminal ends as Equations (29) and (30). There is no heat loss between the gasket terminals and the ambient environment.

$$Q_{pz}(0) = \alpha_p I T_h + \int_0^a 2\pi r q_z(r,0)dr = \alpha_p I T_h + K_{pz}(T_h - T_c) - 0.5I^2 R_p - 2\pi a\lambda_p \sum_{n=1}^{\infty} B_b I_1(k_b a) \tag{29}$$

$$Q_{pz}(L) = \alpha_p I T_c + \int_0^a 2\pi r q_z(r,L)dr = \alpha_p I T_c + K_{pz}(T_h - T_c) + 0.5I^2 R_p - 2\pi a\lambda_p \sum_{n=1}^{\infty} B_n I_1(k_b a)\cos(k_b L) \tag{30}$$

The sum of the heat received at the hot source (gasket terminal) $Q_z(0)$, lost by convection at Q_{conv}, and dissipated at $Q_Z(L)$ is calculated as expressed in Equations (31)–(33) [38,68]. Where B_b is a constant of the Bessel function that varies according to alloy material, surface to surrounding heat convection, and contact resistance, which has values between 0 to the difference between the hot and cold terminals to go to zero. Additionally, the Q_{pz} *and* Q_{nz} are the heat transfer rate between positive gasket and negative gasket in the z-axis direction.

$$Q_z(0) = Q_{pz}(0) + Q_{nz}(0) = \alpha I T_h + K_{pz}(T_h - T_c) - 0.5 I^2 R - 2\pi a \lambda_p \sum_{n=1}^{\infty} B_n I_1(k_b a) \tag{31}$$

$$Q_z(L) = Q_{pz}(L) + Q_{nz}(L) = \alpha I T_C + K_z(T_h - T_c) + 0.5 I^2 R - 2\pi a \lambda_p \sum_{n=1}^{\infty} B_n I_1(k_n a) \cos(k_b L) \tag{32}$$

$$Q_{conv} = Q_{pconv} + Q_{nconv} = 4\pi a h \left\{ \frac{I^2 R_p L}{12 K_{pz}} + \frac{L}{2}(T_h + T_c - 2T_0) - \sum_{n=1}^{\infty} \frac{B_n}{k_n} I_0(k_n r)[\cos(k_b L) - 1] \right\} \tag{33}$$

The power, P_W, and the efficiency, η, of energy conversion from the TEG may be assessed as energy conservation efficiency η principle using Equations (34) and (35) as the predicted responses:

$$P = Q_z(0) - Q_z(L) - Q_{conv} = \alpha I(T_h - T_c) - I^2 R \tag{34}$$

$$\eta = \frac{P_{out}}{Q_z(0)}$$
$$= \frac{\alpha I(T_h - T_c) - I^2 R}{\alpha I T_h + k_z(T_h - T_c) - 0.5 I^2 R \left(1 + \frac{8h}{aL^2}\sum_{n=1}^{\infty} \frac{I_1(k_b a)[\cos(k_b L)-1]}{\lambda_p k_b^4 I_1(k_b a) + h k_b^3 I_0(k_b a)}\right) - k_z T_h \frac{4h}{aT_h}\sum_{n=1}^{\infty} \frac{I_1(k_b a)(T_c - T_0)[\cos(k_b L)-(T_h - T_0)]}{k_b \lambda_p k_b^1 I_1(k_b a) + h I_0(k_b a)}} \tag{35}$$

The authors discovered that the heat convection has no impact on the power generated according to Equation (34). The maximum generated power and efficiency is expressed in Equations (36) and (37):

$$I_P = \frac{\alpha(T_h - T_c)}{2R} \tag{36}$$

$$I_\eta = \frac{\alpha(T_h - T_c)}{R(1 + \sqrt{1 + H.Z\overline{T}})} \tag{37}$$

where the influence of heat convection at the change in the level of thermoelectric terminals is measured by the impact factor H, which is calculated by Equation (38), while $\overline{T} = (T_h + T_c)/2$, and treated as dimensionless and approximated to 0.9368 when the heat convection effect is neglected.

$$H = \frac{\frac{(T_h - T_c)}{2\overline{T}}\left[1 - \frac{8h}{aL^2}\left[\sum_{n=1}^{\infty} \frac{I_1(k_b a)[\cos(k_b L) \; 1]}{\lambda_p k_n^4 I_1(k_b a) + h k_b^3 I_0(k_b a)}\right](T_h - T_c)\right]}{1 - \left[\frac{4h}{aT_h}\left[\sum_{n=1}^{\infty} \frac{I_1(k_b a)(T_c - T_0)[\cos(k_b L) - (T_h - T_0)]}{k_b \lambda_p k_b^1 I_1(k_b a) + h I_0(k_b a)}\right]T_h\right]} \tag{38}$$

The greatest power efficiency η_{max} through the conversion and maximum generated power P_{max} are expressed in Equations (39) and (40), respectively:

$$\eta_{max} = \frac{T_h - T_c}{T_h} \cdot \frac{\sqrt{1 + H.Z\overline{T}} - 1}{\left(\sqrt{1 + H.Z\overline{T}} + \frac{T_C}{T_h}\right) - \frac{8h}{aL^2}\left[\sum_{b=1}^{\infty} \frac{I_1(k_b a)[\cos(k_b L)-1]}{\lambda_p k_b^4 I_1(k_b a) + h k_b^3 I_0(k_b a)}\right](T_h - T_c)/T_h} \tag{39}$$

$$P_{max} = \frac{\alpha^2(T_h - T_c)^2}{4R} \tag{40}$$

In this paper, a simulated digital-twin model studied the effectiveness of thermal conductor candidates in transferring thermal energy to electricity through the TEG by forming legs to be ellipsoid in a cross section, and maintain the temperature difference between the exhaust gasket plate and cold intake gasket plate up to 105 °C. The heat transfer equation is nonlinear and untraceable because of temperature-dependent properties such the Seebeck coefficient, and electrical and thermal conductivity [69]. The proposed methodology predicts with the positions and the intensity of cracks to avoid the installation of thermal conductors in these positions to guarantee TEG working $> t_m$: 6421 h for the intake and exhaust gaskets as illustrated in Figure 19.

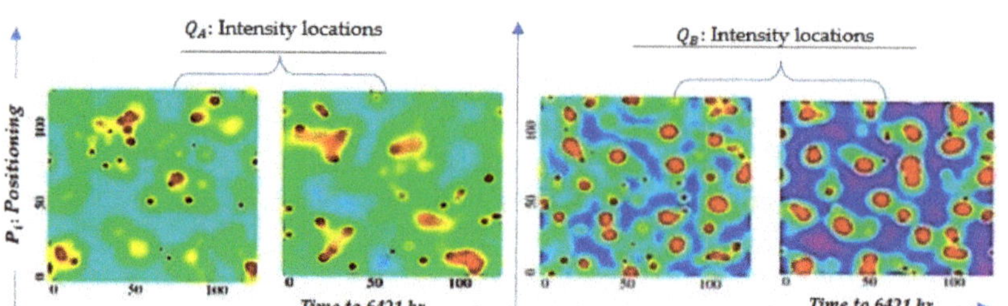

Figure 19. The intake (Al-Fe) and the exhaust (Al-Cu) gaskets' conductors before failure.

The predicted efficiency η in the case of using the candidate alloy strips and fixed in the safe path as identified by STTF and the proposed STTF-NN-ACO methodology is illustrated in Figure 20a. At the same time, the deviation of the actual trace after releasing the gasket and connected conductors to sketch the failure path thermally (stop condition) has a low value. In contrast, the η of electricity generated from the rejected alloy (Cu-Fe) gave the wrong initial values to feed the methodology, resulting in the maximum deviation and a wrong prediction path as illustrated in Figure 20b. The gained power P_{max} by the proposed STTF-NN-ACO when lined by (Al-Fe) and (Al-Cu) sketch a safe path for the conductors far away from the crack positions $P_{i,j}$ and their intensity, Q_P.

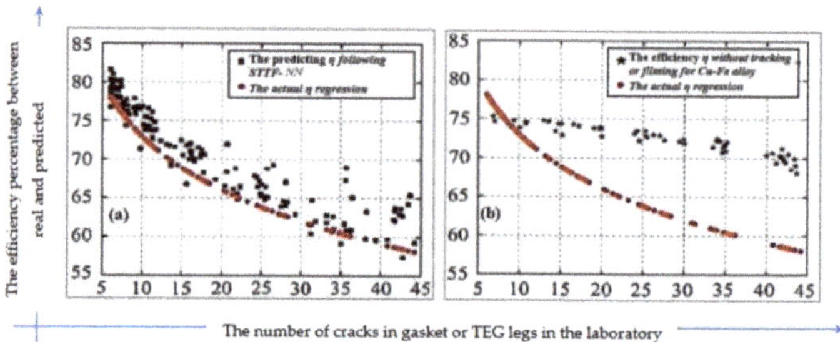

Figure 20. Comparison between the digital twin [STTF-NN] utilization results and the exact measurements for the behavior of cracks effects on generator efficiency η according to their $P_{i,j}$ and Q_P in two cases: (**a**) Al-Fe and/or Al-Cu alloys; (**b**) rejected Cu-Fe alloy.

4.2. The Experimental Measurements' to HVs Batteries

The thermal conductors' strips need to give the required voltage based on the recharging battery specification [69], for instance if the wanted power capacity is 85 KW/hr. Alloyed by Al-Cu with specifications of battery voltage of 350 volt, 3.2 volt for the strip as shown in Equation (7a) and 3.25 Ah. Therefore, the required strip package = 350/3.2 = 110 strips as shown in Equation (7a). The feeding electrical current will be $110 \times 3.2 \times 3.25 =$ 1144 W. However, 85 KW hr./1144 = 75 parallel strips will extract $75 \times 3.25 = 244$ Ah. While using Al-Fe, 108 parallel strips are needed. If an initial efficiency of 25% can save from fossil fuels $0.25 \times 12000 = 3000$ W per kg of energy used. While in the case of the battery, the efficiency is higher, and therefore $150 \times 0.9 = 135$ W per kg of energy can be obtained, which increases the battery usage rate 80 times. The cost of the electrical loss due to cracks and conductivity failure $watt.h^{-1}$ can be expressed in Equations (41) and (42) [67–70]:

$$c_d = \frac{Q_{df} \cdot C_{fd}}{P_w} \tag{41}$$

$$C_\sigma = c_{mat} + (c_\nabla + c_{fd} + c_d + c_t)t_s + \sum_{i=1}^{m}\left(c_\nabla + c_{fd} + c_d + c_t\right)K_{1i}V_i^{-1}f_i^{-1} + \sum_{i=1}^{m} c_{ti}K_{3i}V_i^{(\frac{1}{b})-1}f_i^{[\frac{\omega+gs}{b}]-1} + \sum_{i=1}^{m}\left(c_\nabla + c_{fd} + c_d + c_t\right)t_{tci} \tag{42}$$

The proposed mechanism was evaluated to measure effectiveness by tracking 45 (one exhaust hot gaskets/day) of ICE in the heat lab, studying the fuel consumption, carbon emissions, and generated electric power which is directly reflected on transportation costs as shown in Table 4 and illustrated in Figure 21, and comparing the results with the trial of Hong et al. and Abed et. al. [50,52].

Table 4. TEG-IEC fixed and variable-cost ranges.

A Fixed-Cost $	Avg. Fuel Cost $+Variable_Cost	Avg. Fuel Cost $+Variable_Cost
Whole setup experiment parameters	From gasket 1 to 23	From gasket 24 to 45
(30–50) × Avg. variable-cost	0.808 + (10:30)	0.0808 + (10:50)

The fuel consumption costs are summarized in Table 5 by recording the whole transportation cost as discussed in Hong et al., and Abed, et al. and separating the fuel cost to study the electrothermal transformation efficiency for suitable batteries and calculate the relative percentage enhancing (*RPD*). *RPD* is calculated in Equation (43) and reveals that TEG-ICE integration achieves a 19.63% reduction in fuel consumption.

$$RPD_{ACO-NN} = \frac{Fuel\ Cost_{LINGO} - Fuel\ Cost_{(ACO-NN)}}{Fuel\ Cost_{LINGO}} \times 100 \qquad \forall\, i = 1,2,3 \tag{43}$$

Table 5. Laboratory observations for gasoline consumption using LINGO and STTF-NN-ACO for transportation cost.

Number of Gasket /Day	Total Cost $f(Z)$ LINGO	Total Cost $f(Z)$ Mat-ACO	Total Cost $f(Z)$ STTF-NN-ACO	RPD % Mat-ACO	RPD % STTF-NN-ACO	Number of Gasket /day	Total Cost $f(Z)$ LINGO	Total Cost $f(Z)$ Mat-ACO	Total Cost $f(Z)$ STTF-NN-ACO	RPD % Mat-ACO	RPD % STTF-NN-ACO
1	128,524	128,524	120,812.56	0.00%	6.00%	24	160,355	149,618	101,590.622	−6.70%	36.65%
2	114,997	114,999	104,649.09	0.00%	9.00%	25	137,353	137,292	106,950.468	−0.04%	22.13%
3	150,646	150,647	132,569.36	0.00%	12.00%	26	129,044	140,546	109,485.334	8.91%	15.16%
4	130,997	131,006	123,145.64	0.01%	5.99%	27	135,362	160,355	124,916.545	18.46%	7.72%
5	121,825	123,098	113,250.16	1.04%	7.04%	28	101,255	137,360	111,536.32	35.66%	−10.15%
6	115,001	115,005	108,104.7	0.00%	6.00%	29	144,241	129,051	109,951.452	−10.53%	23.77%
7	158,396	158,359	148,857.46	−0.02%	6.02%	30	142,535	135,369	109,919.628	−5.03%	22.88%
8	157,268	158,439	141,010.71	0.74%	10.34%	31	101,177	101,262	81,110.862	0.08%	19.83%
9	142,211	142,220	129,420.2	0.01%	8.99%	32	143,779	144,051	113,656.239	0.19%	20.95%
10	127,857	127,877	112,915.39	0.02%	11.69%	33	142,535	143,014	99,394.73	0.34%	30.27%
11	122,273	122,124	100,752.3	-0.12%	17.60%	34	132,535	132,842	91,660.98	0.23%	30.84%
12	139,617	139,630	131,252.2	0.01%	5.99%	35	130,535	131,341	94,959.543	0.62%	27.25%
13	138,248	138,264	124,299.33	0.01%	10.09%	36	131,535	132,044	102,730.232	0.39%	21.90%
14	150,085	150,086	136,578.26	0.00%	9.00%	37	177,854	178,113	131,447.394	0.15%	26.09%
15	118,701	118,713	107,316.55	0.01%	9.59%	38	209,060	209,839	128,001.79	0.37%	38.77%
16	134,079	119,207	103,710.09	−11.09%	22.65%	39	167,758	168,004	104,162.48	0.15%	37.91%
17	157,904	118,712	104,466.56	−24.82%	33.84%	40	179,626	179,907	105,965.223	0.16%	41.01%
18	168,949	117,311	106,166.45	−30.56%	37.16%	41	202,973	203,680	106,524.64	0.35%	47.52%
19	155,656	134,093	110,894.91	−13.85%	28.76%	42	166,785	168,884	89,508.52	1.26%	46.33%
20	170,284	157,921	153,499.21	−7.26%	9.86%	43	182,584	185,931	106,166.601	1.83%	41.85%
21	149,618	168,976	151,571.47	12.94%	−1.31%	44	154,217	154,540	92,414.92	0.21%	40.07%
22	137,285	155,663	146,323.22	13.39%	−6.58%	45	179,461	180,016	104,949.328	0.31%	41.52%
23	140,544	170,291	149,685.78	21.17%	−6.50%						

Since it is more typical to record fuels and lubricants in liters, fuel consumption is also calculated in liters per hour for operating vehicles or autonomous engines and can be expressed in Equation (44) [71].

$$c_{fd} = q_e \cdot N_e / 1000 \cdot \rho_t \tag{44}$$

where: q_e—effective fuel consumption, g . $(\mathrm{kW}^{-1} \cdot \mathrm{h}^{-1}) = (3600 \cdot 10^3) / (\eta_e \cdot H_n)$

N_e - effective power, kW;
ρ_t - gasoline density, 0.76 g/cm^3 (kg/Litre);
η_e - ICE efficiency $= \eta_i \cdot \eta_m$;
H_n - gasoline calorific value, kJ/kg;
η_i - ICE efficiency indicator $= [(P_i \cdot L_0 \cdot R \cdot T) / (H_n \cdot \eta_v \cdot P)] \cdot \alpha$;
η_m - mechanical ICE efficiency $= P_e / (P_e + P_m)$;
P_i - regular indicator pressure, kPa $= P_e / \eta_m$;
P_e - regular effective pressure, kPa $= (N_e \cdot 30 \cdot \tau \cdot 10^3) / (V_h \cdot n)$;
P_m - pressure of mechanical losses, kPa $= a_m + b_m \cdot W_n$;
τ - ICE cycle = 4;
V_h - ICE displacement (all cylinders), 1.6;
L_0 - stoichiometric amount of gasoline -air mixture, 0.5119 kmol/kg;
n - ICE rotation speed, rpm ($n_{min} = 800\ rpm$);
R   - universal gas constant, = 8.31 J/(mol · K);
T   - air temperature, 0.346 K;
η_v - ICE cylinder fill ratio $= B_\eta \cdot N_1 + C_\eta$;
B_η - empirical coefficients depending on ICE type approximate to 0.17;
P   - air pressure, kPa;
α - excess air ratio $= A_\alpha \cdot N_1^2 + B_\alpha \cdot N_1 + C_\alpha$;
N_1 - power utilization percentage, % $= (N_e \cdot 100) / N_{e\ max}$;
$A_\alpha, B_\alpha, C_\alpha$ empirical coefficients depending on ICE type, gasoline ICE are -1.1^{-4}, 0.012, 0.85, respectively.
$N_{e\ max}$—maximum effective ICE power, kW;
W_n - average piston speed, m/s $= (30 \cdot S_n) / n$;
a_m, b_m - mechanical loss factors in the engine;
S_n - cylinder height (distance from TDC to BDC), m;

The equation for calculating the hourly fuel consumption in liters per hour is obtained via substitution in Equation (45) [68]:

$$c_{fd} = \frac{0.12 \cdot P \cdot V_n \cdot n}{L_0 \cdot R \cdot T \cdot \tau \cdot \rho_t} \cdot \frac{B_\eta + \frac{10^2 \cdot A_n \cdot N_e}{N_{e\ max}}}{C_\alpha + \frac{10^2 \cdot B_\alpha \cdot N_e}{N_{e\ max}} + \frac{10^4 \cdot A_\alpha \cdot N_e^2}{N_{e\ max}}} = 0.00185 \cdot V_h \cdot \frac{P}{T} \cdot n = 0808\ Lh^{-1}. \tag{45}$$

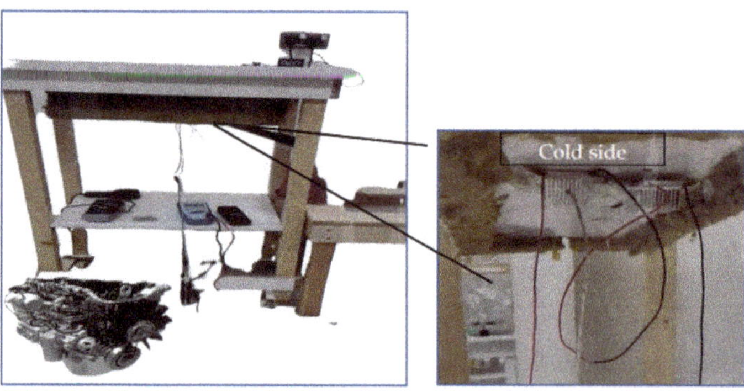

Figure 21. The TEG-ICE integration laboratory and observe 45 gaskets through 23 weeks.

Figure 22 illustrates the fuel consumption reflected on generated electricity power (W) and shown in Figure 23, which if reduced as appeared after 5 working weeks, the experiment stops and unfixes the gasket and records the cracks' position and intensity by filming and installing the thermal conductivity strips in another path far away from the direction of the cracks. Therefore, an increase in power is observed until week #9, which is observed as a steady-state behavior to week #20, then reduced again to week #23. The experiment again stops and analyzes all filming shots to discover the cracks' behavior.

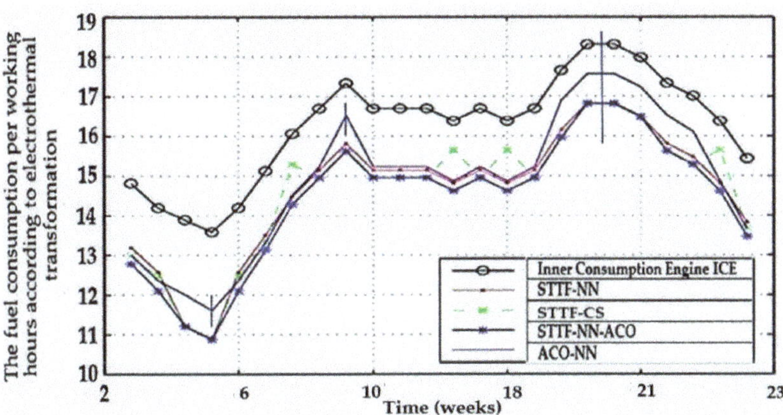

Figure 22. The daily fuel consumption for TEG-ICE is affected by the cracks' prediction and reinstalling the thermal conductors in places far away from cracked positions.

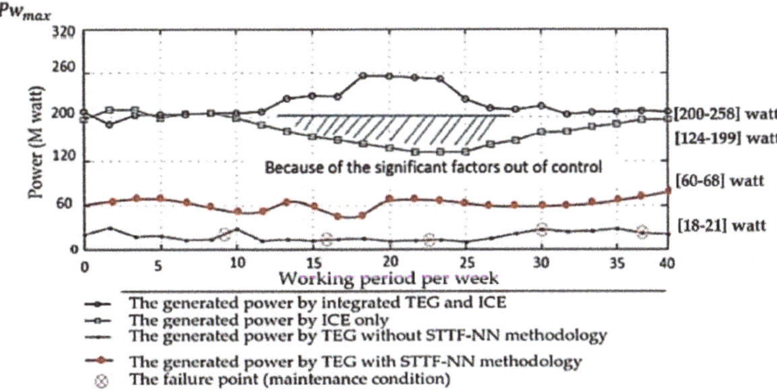

Figure 23. The generated electrical power for the diesel and thermoelectric generators through the digital twin simulator through 6241 working hours.

5. Conclusions

The main objective of the paper is to track the heat transfer, which is due to the difference in temperature between the hot and cold gaskets via the thermal conductors representing the TEG legs, which must be installed in a safe path on the gasket surface without any obstacles preventing heat transfer (e.g., cracks). The main obstacle is the failure of some positions and their precise intensity, as illustrated in Figures 5–8 on the gasket texture, to install the thermal conductors far away from these positions. When the installation of electro thermal conductors began to measure the amount of heat transfer and generated an output of power during a long working time extended to more than 6241 h. All equations are formulated to serve these two objectives through the digital simulator twin which tests the efficiency of hybridization of TEG and another ICE generator which is illustrated in Figure 18. The behavior of heat transferred is illustrated in Figure 20a,b, and the prediction efficiency developed in the laboratory as illustrated in Figure 21 by tracking and testing 45 gaskets over 23 weeks (3360 hr.) and training the NN to extract data from the STTF images to enhance the ACO and cuckoo search techniques in predicting the cracks' breeding locations, which gives the model an advantage for working and extracting power for more than 6241 continuous hours. The cracks' diffusion is proportional to the failure of thermal conductors' terminals fixed on the gaskets' slots and formed from the same alloyed material in strip shape and affects the electrically generated power and the transfer efficiency η as illustrated in Figure 22. Identifying the safe track to install the thermal conductor strips on the gasket surface increases the continuous working hours of generating electricity from 5184 to tm: 6241 h (20.39%). Table 6 shows the average deviation from the actual observation values along tm: 6241 working hours via the proposed predicting method and the three others based on the same data extracted from filming the cracks' growth [70–74]. The efficiency of prediction for cracks' positions and their intensity is positively reflected on fuel consumption cost because of the reliance of the vehicle on electricity generated, where Table 5 shows the superiority of STTF-NN-ACO over STTF-CS and Mat-ACO algorithms by 19.43% and 18.95% respectively. However, the interference of extracting the data using the neural network empowers the prediction training to precisely determine the best path for installing the thermal conductivity strips.

Table 6. The average deviation from the actual observation values along tm: 6241 working hours.

Working Hr.	Optimization Prediction Algorithms	Generated Power Per Hour	number of Terminals' Cracks	Fuel Consumed Per Week	Working Weeks	C_σ: Costs Per Week	Equation (4) Output	Cracks' Position Deviation	Cracks' Intensity mm^2	η
Native TEG		132 KW	216	160 Liter	23 weeks	1.026×10^3	——	——	9	77%
Controlling the significant parameters		148 KW	142	112 Liter	31 weeks	0.826×10^3	Along 23 weeks	——	6	81%
Optimization Improve $\geq$		+12.15%	−1.33%	−1.83%	+2.52%	−2.09%	——	±1.32%	−3.11%	+2.21%
24–3360	STTF-ACO-NN	96.72%	0.42%	1.03%	1.06%	1.04%	1.0	0.16%	0.16%	99.84%
24–3360	Mat-ACO	88.40%	0.90%	2.50%	2.20%	2.30%	1.5	0.96%	2.43%	98.30%
3361–5184	STTF-ACO-NN	97.50%	0.45%	1.06%	1.09%	1.07%	1.6	0.17%	0.17%	99.83%
3361–5184	Mat-ACO	89.25%	1.00%	2.50%	1.20%	2.40%	2.1	0.92%	2.34%	98.37%
5185–6241	STTF-ACO-NN	99.06%	0.48%	1.09%	1.12%	1.10%	2.5	0.18%	0.18%	99.82%
5185–6241	Mat-ACO	91.29%	1.10%	3.50%	1.20%	2.40%	1.6	0.92%	2.34%	98.37%
The Results		165.982	95.14	92.544	35.989	0.808×10^3		±1.32%	4	87.81%

The proposed model inputs fed by STTF with a minimum deviation compared to the power during the actual working conditions are illustrated in Figure 23, which emphasizes that the thermal conductor's efficiency continues for more than 37 weeks, thanks to the support of TEG for ICE generator.

The authors' work in the future is to replace the generator legs with tubes that have engine oil-based nano-fluids absorbed from the engine base, which contains a variety of nanomaterials using a partial model to inspect the thermal aspect of a Brinkman-type nano-fluid composed of molybdenum disulfide (MOS2) and graphene oxide (GO) nanoparticles. These particles are flowing on an oscillating infinite inclined plate and to classify the asymmetrical fluid even when the magnetism, slip boundary conditions, and engine oil sensitive to the Newtonian heating effect were considered. Additionally, future work will test the YbAl3 (Density: $\rho = 5.68$ Mg·m^{-3}) to study the generated power and its ability of service the electrical stations and to charge the EVs.

Author Contributions: "Conceptualization, A.M.A. and S.E.; methodology, A.M.A.; software, A.M.A.; validation, A.M.A., L.F.S. and S.E.; formal analysis, A.M.A.; investigation, A.M.A. and L.F.S.; resources, A.M.A. and S.E.; data curation, A.M.A., L.F.S.; writing—original draft preparation, A.M.A. and S.E.; writing—review and editing, A.M.A. and S.E.; visualization, A.M.A. and L.F.S.; supervision,

A.M.A.; project administration, A.M.A. and S.E.; funding acquisition, A.M.A. All authors have read and agreed to the published version of the manuscript.

Funding: This project was funded by the Deanship of Scientific Research at Prince Sattam bin Abdulaziz University, award number IF-PSAU-2022/01/22745.

Data Availability Statement: http://idealstandarddeviati.wixsite.com/leansixsigma accessed on 10 December 2022.

Acknowledgments: The authors extend their appreciation to the Deputyship for Research and Innovation, Ministry of Education in Saudi Arabia for funding this research through the project number IF-PSAU-2022/01/22745.

Conflicts of Interest: The authors declare no conflict of interest.

References

1. Ritzman, L.P.; Malhorta, M.K. *Operations Management Processes and Value Chains*, 7th ed.; Pearson Prentice Hall: Upper Saddle River, NJ, USA, 2004; pp. 535–581.
2. Scrosati, B.; Garche, J.; Tillmetz, W. (Eds.) *Advances in Battery Technologies for Electric Vehicles*, 1st ed.; Woodhead Publishing: Sawston, UK, 2015; eBook ISBN: 9781782423980; Hardcover ISBN: 9781782423775.
3. Chian, T.Y.; Wei, W.; Ze, E.; Ren, L.; Ping, Y.; Bakar, N.A.; Faizal, M.; Sivakumar, S. A Review on Recent Progress of Batteries for Electric Vehicles. *Int. J. Appl. Eng. Res.* **2019**, *14*, 4441–4461. Available online: http://www.ripublication.com (accessed on 31 December 2019).
4. Dai, Q.; Kelly, J.C.; Gaines, L.; Wang, M. Life Cycle Analysis of Lithium-Ion Batteries for Automotive Applications. *Batteries* **2019**, *5*, 48. [CrossRef]
5. Winslow, K.M.; Laux, S.J.; Townsend, T.G. An economic and environmental assessment on landfill gas to vehicle fuel conversion for waste hauling operations. *Resour. Conserv. Recycl.* **2018**, *142*, 155–166. [CrossRef]
6. Kim, J.; Oh, J.; Lee, H. Review on battery thermal management system for electric vehicles. *J. Appl. Therm. Eng.* **2019**, *149*, 192–212. [CrossRef]
7. BU-214: Summary Table of Lead-based Batteries—Battery University. Available online: https://batteryuniversity.com/learn/article/bu_214_summary_table_of_lead_based_batteries (accessed on 8 November 2019).
8. Indukala, M.P.; Bincy, M.M. A study on electric vehicle battery. *Int. Res. J. Eng. Technol.* **2019**, *6*, 309–314.
9. Miao, Y.; Hynan, P.; von Jouanne, A.; Yokochi, A. Current Li-Ion Battery Technologies in Electric Vehicles and Opportunities for Advancements. *Energies* **2019**, *12*, 1074. [CrossRef]
10. Characteristics of Lead Acid Batteries. Available online: http://pvcdrom.pveducation.org/BATTERY/charlead.htm (accessed on 4 November 2019).
11. How to Prolong and Restore Lead-Acid Batteries—Battery University. Available online: https://batteryuniversity.com/learn/article/how_to_restore_and_prolong_lead_acid_batteries (accessed on 22 November 2019).
12. McNutt, R.L., Jr.; Wimmer-Schweingruber, R.F.; Gruntman, M.; Krimigis, S.M.; Roelof, E.C.; Brandt, P.C.; Vernon, S.R.; Paul, M.V.; Lathrop, B.W.; Mehoke, D.S.; et al. Near-term interstellar probe: First step. *Acta Astronaut.* **2019**, *162*, 284–299. [CrossRef]
13. Eldesoukey, A.; Hassan, H. 3D Study: Model of Thermoelectric Generator (TEG) Case Effect of Flow Regime on the TEG Performance. *Energy Convers. Manag.* **2019**, *180*, 231–239. [CrossRef]
14. Sharma, A.; Saxena, A.; Dinkar, S.K.; Kumar, R.; Al-Sumaiti, A.S. Process Optimization of BioICE Production Using the Laplacian Harris Hawk Optimization (LHHO) Algorithm. *Model. Simul. Eng.* **2022**, *2022*, 6766045.
15. Gubbi, J.; Buyya, R.; Marusic, S.; Palaniswami, M. Internet of Things (IoT): A vision, Architectural Elements, and Future Directions. *Future Gener. Comput. Syst.* **2013**, *29*, 1645–1660. [CrossRef]
16. Alyunov, A.; Vyatkina, O.; Smirnov, I.; Nemirovskiy, A.; Gracheva, E. Assessment of efficiency of diesel generators use in distributed energy industry. *E3S Web Conf.* **2020**, *178*, 01086. [CrossRef]
17. Lao, L.; Li, Z.; Hou, S.; Xiao, B.; Guo, S.; Yang, Y. A Survey of IoT Applications in Blockchain Systems. *ACM Comput. Surv.* **2020**, *53*, 18. [CrossRef]
18. Yang, X.S.; Deb, S. Engineering Optimisation by Cuckoo Search. *Int. J. Math. Model. Numer. Optim.* **2010**, *1*, 330–343. [CrossRef]
19. Saka, M.; Hasançebi, O.; Geem, Z. Metaheuristics in structural optimization and discussions on harmony search algorithm. *Swarm Evol. Comput.* **2016**, *28*, 88–97. [CrossRef]
20. Tsvyashchenko, A.; Fomicheva, L.; Brudanin, V.; Kochetov, O.; Salamatin, A.; Velichkov, A.; Wiertel, M.; Budzynski, M.; Sorokin, A.; Ryasny, G.; et al. The TDPAC study of the hyperfine interactions at 111Cd nuclei in RAl3 compounds synthesized under high pressure. *Solid State Commun.* **2007**, *142*, 664–669. [CrossRef]
21. Hotta, S.K.; Sahoo, N.; Mohanty, K. Comparative assessment of a spark ignition engine fueled with gasoline and raw biogas. *Renew. Energy* **2019**, *134*, 1307–1319. [CrossRef]
22. Massaguer, A.; Massaguer, E.; Comamala, M.; Pujol, T.; González, J.R.; Cardenas, M.D.; Carbonell, D.; Bueno, A.J. A method to assess the fuel economy of automotive thermoelectric generators. *Appl. Energy* **2018**, *222*, 42–58. [CrossRef]

23. Zhao, Y.H.; Zhu, Y.T.; Liao, X.; Horita, Z.; Langdon, T.G. Tailoring stacking fault energy for high ductility and high strength in ultrafine grained Cu and its alloy. *Appl. Phys. Lett.* **2006**, *89*, 121906. [CrossRef]
24. Hoang, D.C.; Yadav, P.; Kumar, R.; Panda, S.K. Real-Time Implementation of a Harmony Search Algorithm-Based Clustering Protocol for Energy-Efficient Wireless Sensor Networks. *IEEE Trans. Ind. Informatics* **2014**, *10*, 774–783. [CrossRef]
25. Wen, H.; Topping, T.D.; Isheim, D.; Seidman, D.N.; Lavernia, E.J. Strengthening mechanisms in a high-strength bulk nanostructured Cu–Zn–Al alloy processed via cryomilling and spark plasma sintering. *Acta Mater.* **2013**, *61*, 2769–2782. [CrossRef]
26. Grieves, M.; Vickers, J. Digital Twin: Mitigating Unpredictable, Undesirable Emergent Behavior in Complex Systems. In *Transdisciplinary Perspectives on Complex Systems*; Springer: Cham, Switzerland, 2017; pp. 85–113.
27. Cao, Q.; Luan, W.; Wang, T. Performance enhancement of heat pipes assisted thermoelectric generator for automobile exhaust heat recovery. *Appl. Therm. Eng.* **2018**, *130*, 1472–1479. [CrossRef]
28. Lan, S.; Yang, Z.; Chen, R.; Stobart, R. A dynamic model for thermoelectric generator applied to vehicle waste heat recovery. *Appl. Energy* **2018**, *210*, 327–338. [CrossRef]
29. Gupta, N.; Gupta, S.; Khosravy, M.; Dey, N.; Joshi, N.; Crespo, R.G.; Patel, N. Economic IoT strategy: The future technology for health monitoring and diagnostic of agriculture vehicles. *J. Intell. Manuf.* **2020**, *32*, 1117–1128. [CrossRef]
30. Parthasarathy, P.; Vivekanandan, S. A Typical IoT Architecture-Based Regular Monitoring of Arthritis Disease Using Time Wrapping Algorithm. *Int. J. Comput. Appl.* **2018**, *42*, 222–232. [CrossRef]
31. Abed, A.M.; Seddek, L.F.; Elattar, S.; Gaafar, T.S. The digital twin model of vehicle containers to provide an ergonomic handling mechanism. *South Fla. J. Dev.* **2022**, *3*, 1971–1992. [CrossRef]
32. Jagannathan, J.; Udaykumar, U. Predictive Modelling for Improving Healthcare Using IoT: Role of Predictive Models in Healthcare Using IoT. In *Incorporating the Internet of Things in Healthcare Applications and Wearable Devices*; Pankajavalli, P., Karthick, G., Eds.; IGI Global: Hershey, PA, USA, 2020; Chapter 15; pp. 243–254.
33. Ateeq, K.; Altaf, S.; Aslam, M. Modeling and Bayesian Analysis of Time between the Breakdown of Electric Feeders. *Model. Simul. Eng.* **2022**, *2022*, 5830945. [CrossRef]
34. Machura, P.; Li, Q. A critical review on wireless charging for electric vehicles. *Renew. Sustain. Energy Rev.* **2019**, *104*, 209–234. [CrossRef]
35. Ding, Y.; Cano, Z.P.; Yu, A.; Lu, J.; Chen, Z. Automotive Li-Ion Batteries: Current Status and Future Perspectives. *Electrochem. Energy Rev.* **2019**, *2*, 1–28. [CrossRef]
36. Alshahrani, S.; Khalid, M.; Almuhaini, M. Electric Vehicles Beyond Energy Storage and Modern Power Networks: Challenges and Applications. *IEEE Access* **2019**, *7*, 99031–99064. [CrossRef]
37. Pang, D.; Zhang, A.; Wen, Z.; Wang, B.; Wang, J. Energy Conversion Efficiency of Thermoelectric Power Generators with Cylindrical Legs. *J. Energy Resour. Technol.* **2022**, *144*, 032104. [CrossRef]
38. Goswami, R.; Das, R. Experimental Analysis of a Novel Solar Pond Driven Thermoelectric Energy System. *ASME J. Energy Resour. Technol.* **2020**, *142*, 121302. [CrossRef]
39. Khalil, H.; Hassan, H. Enhancement of Waste Heat Recovery from Vertical Chimney via Thermoelectric Generators by Heat Spreader. *Process Saf. Environ. Prot.* **2020**, *140*, 314–329. [CrossRef]
40. Admiral, F.; Jsseling, B.; Kolster, J.; Vander, V. Influence of Temperature on Corrosion Product Film Formation on CuNi19Fe in Low Temp. Range. *Br Corr. J.* **1986**, *21*, 33–43. [CrossRef]
41. Richard, W.; Hertzberg, G. *Deformation and Fracture Mechanics of Engineering Materials*, 4th ed.; John Wiley & Sons, Inc.: Hoboken, NJ, USA, 1996.
42. Li, C.; Negnevitsky, M.; Wang, X.; Yue, W.L.; Zou, X. Multi-criteria analysis of policies for implementing clean energy vehicles in China. *Energy Policy* **2019**, *129*, 826–840. [CrossRef]
43. Qiao, Q.; Zhao, F.; Liu, Z.; He, X.; Hao, H. Life cycle greenhouse gas emissions of Electric Vehicles in China: Combining the vehicle cycle and fuel cycle. *Energy* **2019**, *177*, 222–233. [CrossRef]
44. Scrosati, B.; Garche, J.; Sun, Y. Recycling lithium batteries. In *Advances in Battery Technologies for Electric Vehicles*; Woodhead Publishing: Sawston, UK, 2015. [CrossRef]
45. Khalil, H.; Hassan, H. 3D Study of the Impact of Aspect Ratio and Tilt Angle on the Thermoelectric Generator Power for Waste Heat Recovery from a Chimney. *J. Power Sources* **2019**, *418*, 98–111. [CrossRef]
46. Wen, Z.F.; Sun, Y.; Zhang, A.B.; Wang, B.L.; Wang, J.; Du, J.K. Performance Analysis of a Segmented Annular Thermoelectric Generator. *J. Electron. Mater.* **2020**, *49*, 4830–4842. [CrossRef]
47. Khan, U.; Zaib, A.; Ishak, A.; Alotaibi, A.M.; Elattar, S.; Pop, I.; Abed, A.M. Impact of an Induced Magnetic Field on the Stagnation-Point Flow of a Water-Based Graphene Oxide Nanoparticle over a Movable Surface with Homogeneous–Heterogeneous and Chemical Reactions. *Magnetochemistry* **2022**, *8*, 155. [CrossRef]
48. Raza, A.; Khan, U.; Raizah, Z.; Eldin, S.M.; Alotaibi, A.M.; Elattar, S.; Abed, A.M. Numerical and Computational Analysis of Magnetohydrodynamics over an Inclined Plate Induced by Nanofluid with Newtonian Heating via Fractional Approach. *Symmetry* **2022**, *14*, 2412. [CrossRef]
49. Raza, A.; Khan, U.; Eldin, S.M.; Alotaibi, A.M.; Elattar, S.; Prasannakumara, B.C.; Akkurt, N.; Abed, A.M. Significance of Free Convection Flow over an Oscillating Inclined Plate Induced by Nanofluid with Porous Medium: The Case of the Prabhakar Fractional Approach. *Micromachines* **2022**, *13*, 2019. [CrossRef]

50. Hong, J.; Diabat, A.; Panicker, V.V.; Rajagopalan, S. A two-stage supply chain problem with fixed costs: An ant colony optimization approach. *Int. J. Prod. Econ.* **2018**, *204*, 214–226. [CrossRef]
51. Ashour, M.; Elshaer, R.; Nawara, G. Ant Colony Approach for Optimizing a Multi-Stage Closed-Loop Supply Chain with a Fixed Transportation Charge. *J. Adv. Manuf. Syst.* **2021**, *21*, 473–796. [CrossRef]
52. Abed, A.M.; Seddek, L.F.; AlArjani, A. Enhancing Two-Phase Supply Chain Network Distribution via three meta-heuristic Optimization Algorithms subsidized by Mathematical procedures. *J. Adv. Manuf. Syst.* **2022**, *22*, 1–32. [CrossRef]
53. Yin, T.; He, Z. Analytical Model-Based Optimization of the Thermoelectric Cooler with Temperature-Dependent Materials under Different Operating Conditions. *Appl. Energy* **2021**, *299*, 117340. [CrossRef]
54. Shen, Z.-G.; Wu, S.-Y.; Xiao, L.; Yin, G. Theoretical modeling of thermoelectric generator with particular emphasis on the effect of side surface heat transfer. *Energy* **2016**, *95*, 367–379. [CrossRef]
55. Gan, Y.; Liu, C.; He, Z.; Li, H.; Liu, Z. Digital Camouflage Pattern Design Based on the Biased Random Walk. *Model. Simul. Eng.* **2022**, *2022*, 2986346. [CrossRef]
56. Hossain, S.; Haque, M.M.; Kabir, M.H.; Gani, M.O.; Sarwardi, S. Complex Spatiotemporal Dynamics of a Harvested prey–predator Model with Crowley–Martin Response Function. *Results Control. Optim.* **2021**, *5*, 100059. [CrossRef]
57. Rossetti, M.D.; Hill, R.R.; Johansson, B.; Dunkin, A.; Ingalls, R.G. Introduction To Simulation. In Proceedings of the 2009 Winter Simulation Conference, Austin, TX, USA, 13–16 December 2009.
58. Kalantari, K.R.; Ebrahimnejad, A.; Motameni, H. A fuzzy neural network for web service selection aimed at dynamic software rejuvenation. *Turk. J. Electr. Eng. Comput. Sci.* **2020**, *28*, 2718–2734. [CrossRef]
59. Yang, X.S. *Nature-Inspired Metaheuristic Algorithms*, 2nd ed.; Luniver Press, 2010; ISBN 978-1-905986-10-1. Available online: https://www.researchgate.net/publication/235979455_Nature-Inspired_Metaheuristic_Algorithms (accessed on 1 December 2022).
60. Qiu, T.; Li, B.; Zhou, X.; Song, H.; Lee, I.; Lloret, J. A Novel Shortcut Addition Algorithm With Particle Swarm for Multisink Internet of Things. *IEEE Trans. Ind. Inform.* **2020**, *16*, 3566–3577. [CrossRef]
61. Han, F.; Li, L.; Cai, W.; Li, C.; Deng, X.; Sutherland, J.W. Parameters optimization considering the trade-off between cutting power and MRR based on Linear Decreasing Particle Swarm Algorithm in milling. *J. Clean. Prod.* **2020**, *262*, 121388. [CrossRef]
62. Ebrahimnejad, A.; Tavana, M.; Alrezaamiri, H. A novel artificial bee colony algorithm for shortest path problems with fuzzy arc weights. *Measurement* **2016**, *93*, 48–56. [CrossRef]
63. Alrezaamiri, H.; Ebrahimnejad, A.; Motameni, H. Parallel multi-objective artificial bee colony algorithm for software requirement optimization. *Requir. Eng.* **2020**, *25*, 363–380. [CrossRef]
64. Khosravy, M.; Gupta, N.; Patel, N.; Dey, N.; Nitta, N.; Babaguchi, N. Probabilistic Stone's Blind Source Separation with application to channel estimation and multi-node identification in MIMO IoT green communication and multimedia systems. *Comput. Commun.* **2020**, *157*, 423–433. [CrossRef]
65. Hosseinabadi, A.A.R.; Vahidi, J.; Balas, V.E.; Mirkamali, S.S. OVRP_GELS: Solving open vehicle routing problem using the gravitational emulation local search algorithm. *Neural Comput. Appl.* **2018**, *29*, 955–968. [CrossRef]
66. Abed, A.M.; AlArjani, A. The Neural Network Classifier Works Efficiently on Searching in DQN Using the Autonomous Internet of Things Hybridized by the Metaheuristic Techniques to Reduce the EVs' Service Scheduling Time. *Energies* **2022**, *15*, 6992. [CrossRef]
67. Pang, D.; Zhang, A.; Wang, B.; Li, G. Theoretical Analysis of the Thermoelectric Generator Considering Surface to Surrounding Heat Convection and Contact Resistance. *J. Electron. Mater.* **2018**, *48*, 596–602. [CrossRef]
68. Shen, Z.-G.; Tian, L.-L.; Liu, X. Automotive exhaust thermoelectric generators: Current status, challenges and future prospects. *Energy Convers. Manag.* **2019**, *195*, 1138–1173. [CrossRef]
69. Araiz, M.; Casi, Á.; Catalán, L.; Martinez, A.; Astrain, D. Prospects of waste-heat recovery from a real industry using thermoelectric generators: Economic and power output analysis. *Energy Convers. Manag.* **2019**, *205*, 112376. [CrossRef]
70. Mirhosseini, M.; Rezania, A.; Rosendahl, L.A. Power optimization and economic evaluation of thermoelectric waste heat recovery system around a rotary cement kiln. *J. Clean. Prod.* **2019**, *232*, 1321–1334. [CrossRef]
71. Hahn, D.W.; Özişik, M.N. *Heat Conduction*, 3rd ed.; John Wiley & Sons Inc.: Hoboken, NJ, USA, 2012.
72. Krivoshapov, S.I.; Nazarov, A.I.; Mysiura, M.I.; Marmut, I.A.; Zuyev, V.A.; Bezridnyi, V.V.; Pavlenko, V.N. Calculation methods for determining of fuel consumption per hour by transport vehicles. *IOP Conf. Series: Mater. Sci. Eng.* **2020**, *977*, 012004. [CrossRef]
73. Eddine, A.N.; Chalet, D.; Faure, X.; Aixala, L.; Chessé, P. Optimization and characterization of a thermoelectric generator prototype for marine engine application. *Energy* **2018**, *143*, 682–695. [CrossRef]
74. Vieira, F.; Soares, A.; Herbut, P.; Vismara, E.; Godyń, D.; dos Santos, A.; Lambertes, T.; Caetano, W. Spatio-Thermal Variability and Behaviour as Bio-Thermal Indicators of Heat Stress in Dairy Cows in a Compost Barn: A Case Study. *Animals* **2021**, *11*, 1197. [CrossRef] [PubMed]

processes

MDPI

Article

Toward Net-Zero: The Barrier Analysis of Electric Vehicle Adoption and Transition Using ANP and DEMATEL

Tsai-Chi Kuo [1,2], Yung-Shuen Shen [3,*], Napasorn Sriwattana [1] and Ruey-Huei Yeh [1]

[1] Department of Industrial Management, National Taiwan University of Science and Technology, Taipei 10607, Taiwan
[2] Artificial Intelligence for Operations Management Research Center, National Taiwan University of Science and Technology, Taipei 10607, Taiwan
[3] Institute of Geriatric Welfare Technology & Science, Mackay Medical College, New Taipei City 25245, Taiwan
* Correspondence: ysshen@mmc.edu.tw

Abstract: Global greenhouse gas emissions must be reduced to achieve net-zero carbon emissions. One of the solutions for the reduction of greenhouse gas emissions is the adoption and transition from conventional vehicles to electrical vehicles (EVs). Previously, most research on EVs have been from a consumer adoption perspective, few of them are from industry transition and consumer adoption perspectives simultaneously. This also highlights the importance of SDG 12 (responsible for consumption and production). Additionally, the analyses were mostly obtained using one methodology and demonstrated only by weighting without relationships among factors. To consider the problem of adoption and transition, a systematic method should be developed. Therefore, this study intends to identify, prioritize, and display the relationship between EV adoption barriers from an automotive industry perspective using an analytic network process (ANP) and the decision-making trial and evaluation laboratory (DEMATEL) method. The research results show two contributions: First, the identified and prioritized barriers that automakers encounter in EV transition also explored the interrelationships among these barriers. Second, a model comparison of two multicriteria decision-making approaches was conducted to prioritize and identify the interlinkages among EV uptake barriers.

Keywords: electrical vehicle transition; multi-criteria decision making; ANP; DEMATEL; SDG 12

Citation: Kuo, T.-C.; Shen, Y.-S.; Sriwattana, N.; Yeh, R.-H. Toward Net-Zero: The Barrier Analysis of Electric Vehicle Adoption and Transition Using ANP and DEMATEL. *Processes* **2022**, *10*, 2334. https://doi.org/10.3390/pr10112334

Academic Editor: Zhou Li

Received: 15 September 2022
Accepted: 25 October 2022
Published: 9 November 2022

Publisher's Note: MDPI stays neutral with regard to jurisdictional claims in published maps and institutional affiliations.

1. Introduction

Climate change caused by migration, which is widely acknowledged by scientists, is driven by an increased level of greenhouse gases (GHGs). According to the Paris Agreement [1], many governments from different countries have set a target to achieve net-zero by 2050 [2]. Net-zero considers total emissions, permitting the elimination of any inevitable emissions, such as those from transportation or industry.

To achieve net-zero emissions [3], GHG emissions must be reduced rapidly and significantly, and removal must be scaled up [4]. The transportation sector contributes over 16.2% to worldwide GHG emissions, and this percentage is likely to rise in the future [5]. Despite breakthroughs in biofuels and electricity, transportation has generally been based on oil fuels, which accounted for more than 90% of the transport sector's power needs in 2020 [6].

Williams et al. [7] suggested the electrification of transportation required to meet an 80% reduction target. Additionally, an adaptation from internal combustion engine (ICE) vehicles to higher levels of sustainable transportation could be a major contributor towards decreasing global GHG emissions [8]. Electric vehicles (EVs) are emerging as an important substitute solution with conventional vehicles for transportability, contributing to the largest share of transportation emissions [8]. If efficiently and consistently implemented, electric vehicle usage will lead to significant decarbonization, and lead to improved environmental quality as a result, particularly if EVs are combined with a low-carbon electricity generation system [9]. Furthermore, EVs also provide several advantages besides decarbonization, such as no tailpipe emissions, thereby preventing air pollution and exposure to volatile organic compounds, nitrogen oxides, and carbon monoxide in residential areas [10].

Many countries are currently working to lower emissions from the transportation sector. In 2019, governments worldwide expended approximately USD 14 billion to promote the sales of electric vehicles, up 25% from the previous year, potentially resulting in greater incentives in Europe. However, over the last five years, the share of government subsidies in total spending on electric cars has declined, demonstrating that EVs are becoming more appealing to customers [3].

The desire for environmentally friendly vehicles to run on alternate fuels is increasing significantly. As a result, the automobile industry is rapidly transitioning to something that is more environmentally sustainable, and the world's main automobile makers, as well as many countries, are attempting to gain a competitive advantage in electric vehicles and innovative technology [11]. The increased sales trends for EVs in the foreseeable future are possible. In the first quarter of 2021, worldwide EV sales rose by approximately 140 percent compared to the same period in 2020, owing to the sales of around 500,000 cars in China and around 450,000 in Europe. From a lower foundation, sales in the United States more than doubled in the first quarter of 2020 [3].

Despite their potential usefulness, considerable obstacles or barriers to widespread EV diffusion technology remain, and EVs now represent only a small percentage of all vehicles. Various barriers, causes, and difficulties related to diffusion of electric vehicles have been identified and reported in previous studies [12–14]. She et al. [15] classified the barriers into three categories: economic, efficiency, and facilities scoping in China. A recent study [16,17] divided these barriers into five categories using the analytic hierarchy process (AHP).

EV transition is a significant global issue. Several industrial sectors should be carefully studied and supported to achieve EV transition. However, in earlier EV transition-related studies, only the consumer perspective was widely studied. In the case of successful EV transition, apart from focusing on consumer desire, the automotive industry perspective is also pivotal to the study. In previous research that focused on barriers to EV transition, the indicated and prioritized results were computed using only one methodology. Moreover, these outputs only display barrier categories and the weight of each barrier, but do not display a relationship between the barriers. However, the linkage between barriers or their influence on each other is necessary in real-world implementation, which existing research rarely identifies. This highlights the research goal of systematically evaluating the barriers to EV transition, not only from the perspective of consumers, but also EV companies. This paper is organized as follows:

(1) Identification of critical barriers that have the ability to influence electric vehicle transition from an automotive industry perspective.
(2) Analysis and prioritization of the barriers due to their particular effect on EV transition using the analytic network process (ANP) and decision-making trial and evaluation laboratory (DEMATEL) methodology.
(3) Analysis of the linkage between each barrier of EV transition.

The remainder of this paper is organized as follows: A literature review is presented in Section 2, Section 3 describes multi-criteria decision making (MCDM) solutions for the barriers to EV transition, and Section 4 provides the results. Finally, the conclusions and opportunities for future studies are presented in Section 5.

2. Materials and Methods

Despite progress so far, EV transition still has a long way to go before it reaches a large scale, which is dependent on the transformation of energies, technologies, and customer behaviors. According to the literature review, these include technologies regarding electric vehicles, energy transformation of electricity, and electric vehicle barriers.

2.1. Technology regarding Electric Vehicles (EVs)

Internal combustion engines (ICEs) in conventional cars (CVs) burn fossil fuels, operate ineffectively, and produce a large quantity of greenhouse gases. Alternative fuel vehicles (AFVs), including electric cars, fuel cell vehicles, and biofuel vehicles, are designed to run on at least one alternative to petroleum and diesel. Currently, there are three types of electric vehicles: battery electric vehicles (BEVs), plug-in hybrid electric vehicles (PHEVs), and hybrid electric vehicles (HEVs). Table 1 lists the categories of EVs. The driving range, charging duration, refueling costs, and engineering and design of each vehicle type varies significantly. In all countries, BEV models have been presented in most vehicle categories, and PHEVs have leaned towards larger vehicle segments. In all markets, sports utility vehicle (SUV) types contribute to half of the EVs models available [3].

Table 1. Electric vehicle category.

Vehicle Type	Engine Description	Advantages	Disadvantages
Conventional	Internal combustion engine.	Relatively high-power motor and fast acceleration, starts quickly.	Emissions are generally high compared to external combustion engine.
Hybrid electric vehicles (HEVs)	Separated electric motor with internal combustion engine.	Compared to similarly conventional vehicles, has better fuel saving, cheaper operating costs, and lower pollutants.	Higher purchasing cost and complex hybrid technology.
Plug-in hybrid electric vehicles (PHEVs)	Powerful rechargeable battery, smaller internal combustion engine, and bigger electric motor.	Compared to HEVs and conventional vehicles, has better fuel efficiency, cheaper operating costs, and reduced pollutants. Also provides fuel source adaptability.	Not easy to maintain.
Battery electric vehicles (BEVs)	Chargeable battery packs which can be plugged into an electrical socket.	No liquid resources and pollutants. Compared to HEVs and conventional vehicles, it is less expensive to operate.	Battery waste generates environmental problems.

EVs rely on electricity for part- or all of their propulsion, and they come in a variety of forms [12]. On the other hand, in hybrid electric vehicles (HEVs), a battery and an electric engine are added to a vehicle with an internal combustion engine (ICEs) to achieve better fuel efficiency than comparable-sized vehicles. Hybrid electric vehicles (HEVs) such as the Toyota Prius, Honda Insight, and Honda Civic Hybrid, have been commercially successful [18]. Plug-in hybrid electric vehicles (PHEVs) have a more efficient and bigger battery capable of energizing the car for approximately 20 to 60 miles, compared to HEVs. PHEVs reduce the size of the internal combustion engine to be smaller than that of HEVs. Moreover, the batteries of PEVs are rechargeable, and they can be fully charged by plugging into an external power source. PHEVs have superior fuel economy over EVs, but they also have the option of using conventional fuels for prolonged journeys [12]. Another alternative for EVs is battery electric vehicles (BEVs), which have been actively promoted because of their ability to lower local CO_2 and other hazardous air pollutants, while also reducing automobile noise. BEVs can significantly reduce emissions during operation, despite having higher emissions during manufacturing than ICEVs. BEVs are completely powered by a rechargeable electric battery and typically have larger batteries than PHEVs. They can reach up to 100 miles on a single charge [19].

2.2. Energy Transformation for Electricity

Gasoline price is a noticeable factor in EV adoption. Van Bree et al. [20] discovered that rising gas prices influence customers. Gallagher et al. [21] revealed that customers generally decide to purchase EVs because of higher fuel prices and government subsidies. Together with pricing, non-economic considerations, particularly those related to the environment and energy, can affect customers' alternatives to EVs. Moreover, according to Kahn [22], environmentalists are more likely to buy EVs than those who are not interested in environmental protection. Additionally, the study also identified that social preferences for environmental protection and energy policies were significant considerations in EV adoption [21].

During 2007–2008, gasoline price increases were caused by crude oil price rises and refining deficiency. An increased EV market share would immediately reduce fuel consumption, lessen petroleum refining shortages, and likely lower prices. Furthermore, consumers perceive that fuel prices will continue to increase in the years ahead. According to a survey, as fuel prices increase, more people believe that EVs have become a good investment [23].

Issues regarding rising GHG emissions and oil security have driven policymakers worldwide to prioritize low-carbon development and innovation for zero-carbon transportation. The prediction for increasing GHG emissions in the next 20 years is up to 45%, and this concern has compelled many states in the United States to accept regulations requiring a net-zero standard [24]. According to the IEA [3], the automotive sector accounts for 16.2% of global energy consumption and 25% of CO_2 emissions from energy-related sectors. If present trends persist, the worldwide transportation demand for energy and CO_2 emissions from energy is forecasted to double by 2050.

There is an issue that the electricity used to power electric vehicle is not completely clean since it mainly made from fossil fuels like coal and gas. Egbue [12] found that even electricity for EVs is not entirely carbon-free, but compared to conventional vehicles such as diesel or gasoline vehicles, EVs emit less than 50% GHG. Hassan [25] also illustrated that EVs that are fully powered by coal-refined electricity produce less than 25% GHG per mile compared to conventional vehicles.

2.3. EV Adoption Barriers

An obvious barrier is the high purchase price of EVs when considering conventional ICE vehicle prices. Additionally, from a consumer perspective, the total cost of ownership is also included in making purchase decisions, although when considering the total life cycle, EVs are less expensive than conventional vehicles. In addition, technical barriers are also a main factor that indicate the expansion of the EV market in the future. For example, battery performance is one of the most important factors, since electric vehicles are mostly powered by batteries, and consumers are concerned about the exact driving distance of EVs [26]. Charging time is also mentioned as a barrier to EV transition because users desire the shortest charging duration for daily travel [23]. Another major type of barrier are social barriers such as public perception and environmental awareness. The next barrier is infrastructure, which is resultant of several issues, including charging networks and battery recycling systems, both of which require collaboration between the private and government sectors to overcome obstacles. Finally, policy barriers have been mentioned in several previous studies.

Several previous studies have identified barriers. Haddadian et al. [27] categorized the barriers to worldwide EV adoption into four essential categories: EV technology; financial, social, and customer perception; and innovative business models. For the EV technology barrier, there are two sub-barriers: battery technology that involves high battery cost, driving range, and battery safety, and the charging station, which is an important barrier to widening EV adoption. Financial barriers focus on total cost of ownership (TCO) and financial mechanisms. Finally, innovative business models that attempt to identify perceived challenging aspects of EV holders could provide a creative approach to create and capture EV values, facilitating an easier transition to wider EV adoption. She [15] classified and examined China's [17] economic barriers that consist of rising purchase values, higher initial cost of batteries, a lack of understanding of fuel prices, and maintenance expenses. Similarly, efficiency barriers include vehicle trustworthiness, driving distance, battery recharging duration, and EV power. Facility barriers include lack of infrastructure in public locations, places of business, residences, and motorways. In order to accelerate the

EV transition, the optimal planning of electric vehicle charging station is conducted [28–31]. Adhikari et al. [16] studied the EV transition limitation using the analytic hierarchy process (AHP). The identified barriers were separated into five categories: economic, social, policy, infrastructure, and technical. Their survey revealed expert opinions. For the barrier category, the top-ranked barrier was insufficient infrastructure, followed by policy, economic, and technical barriers. The social barrier is the least concerning among the five in the list. Moreover, the study reveals that insufficient goal setting and long-term planning from the government, charging station shortage, and EVs' higher purchase price were identified as the three highest barriers to the adoption of EVs. Tarei et al. [14] mainly studied the relationship between EV barriers with one another. They also focused on the bonds between automakers and the government for strategic planning. Their methodology integrates two MCDM tools, which are the BWM best–worst method (BWM) and interpretive structural modeling (ISM). This study has five barrier categories: technical, infrastructure, financial, behavioral, and external barriers. The results showed that infrastructure barriers were highlighted as the highest based on expert evaluation, followed by financial and behavioral barriers.

3. Research Method

This study proposes two multicriteria decision methodologies to identify and prioritize the barriers to electric vehicle transition, as illustrated in Figure 1. The first step was a literature review that was a comprehensive review of previously published research, articles, and government publications. This process can reveal multiple aspects, difficulties, and inadequacies with regard to EVs. Moreover, national transportation policy and strategy, market trend, geography, economic status, and renewable energy resource readiness might be appropriate variables for EV barrier identification and categorization. The ANP and DEMATEL methodologies were selected because considering the nature of the problem in the real world, each barrier definitely has an influence on each other. The ANP methodology constructs a problem as a network model that is suitable for problems, and the DEMATEL methodology identifies influences through the analysis of elements in cause-and-effect relations. It has the advantages of not depending on a large data sample and simplifying the correlation analysis of factors. However, the classical DEMATEL ignores the vagueness and uncertainties of human judgement which widely exist in the real world. ANP considers the interdependencies and feedback among factors. Both selected methods use pairwise comparison to measure the weights of the components, which is reasonable for result comparisons.

3.1. Criterial Selection

The adoption of electric vehicles is limited by various real and perceived barriers. These barriers were revealed after a comprehensive literature review that included an investigation of relevant online articles, as well as previous research papers. An exhaustive literature search was carried out using keywords such as electric vehicles and barriers. The literature review was conducted using online search tools such as Google Scholar and the Science Direct website. Table 2 lists these barriers and their associated classes.

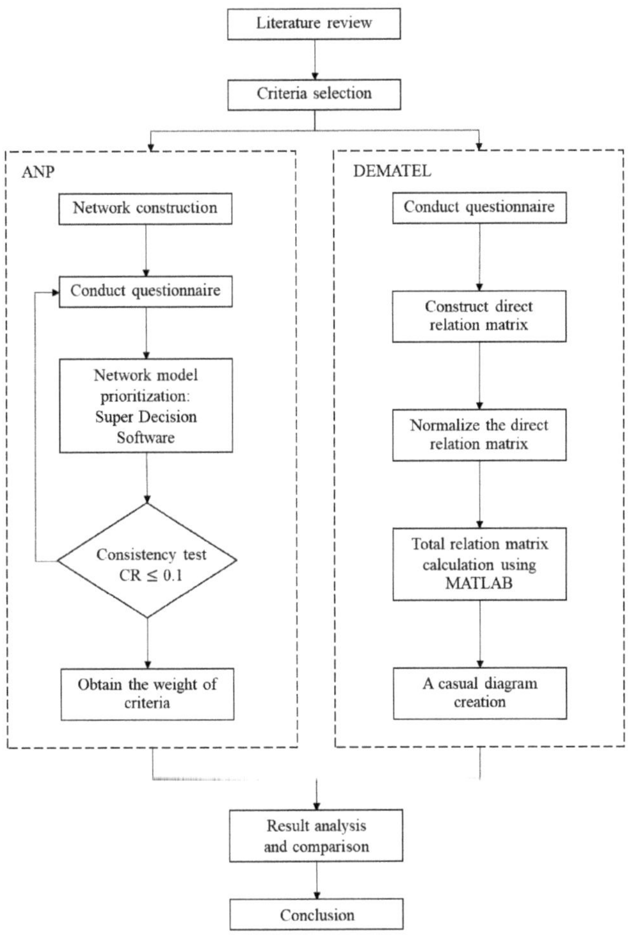

Figure 1. Research flowchart.

Table 2. Barriers of EV transition.

Barriers	Sub Barriers	Index	References
Financial	High manufacturing cost	F_1	[14,26]
	High selling price	F_2	[32,33]
Infrastructure	Shortage of charging station	I_1	[34,35]
	Lack of battery recycling system	I_2	[9,26,36]
	Shortage of Maintenance shop	I_3	[37]
Technology	EV Performance	T_1	[14,38,39]
	Battery capacity and lifespan	T_2	[27,36]
Policy	Government support	P_1	[40,41]
	Impacts of tax and subsidy policies	P_2	[42,43]
	Renewable energy ecosystem	P_3	[40,44]
Customer behavior	Customer awareness	C_1	[16,45]
	Range anxiety	C_2	[46,47]

As a result, 12 barriers to EV adoption were highlighted. Afterwards, the barriers were divided into five categories: financial, infrastructural, technological, customer behavior, and policy.

(1) Financial:

    (a) High cost: High initial costs, especially for batteries, are due to the limited availability of technology and raw material issues that restrict electric vehicle adoption [14,26].

    (b) High prices: Conventional vehicles have a cost advantage that makes consumers more hesitant to buy EVs, which are often more expensive [32]. From the consumer perspective, price plays a major role in car purchase decisions [32,33].

(2) Infrastructure:

    (a) Shortage of charging stations: Slow charging is suitable for home and work, whereas fast charging is best for places such as highways and commercial areas where cars stop for shorter periods of time. A shortage of charging stations has been identified as a barrier to consumers purchasing electric vehicles [34,35]

    (b) Lack of a battery recycling system: An EV's battery material consists of chemical elements such as titanium, nickel, and lithium. Although this increases the cost-effectiveness of the supply chain, it remains an environmental concern [48]. It is essential to develop biologically degradable batteries and recycling systems for them; otherwise, they might hold back the long-term sustainability of electric vehicles [9,26,36].

    (c) Shortage of maintenance shop: A service professional should be able to fix, maintain, and troubleshoot an EV to take appropriate care of it [48]. Current electric vehicle owners are dissatisfied with the lack of support centers or workshops for EV repair and maintenance [37].

(3) Technology:

    (a) EV performance: At present, PHEVs that operate with ICEs provide driving distances of approximately 500 km. Most BEVs have a range of less than 250 km per charge. Nevertheless, some of the most recent models provide a range of up to 400 km. The driving range of a fully charged EV is regarded as a significant disadvantage in terms of EV adoption worldwide [48]. Charging duration is also a main cause for purchase consideration. A conventional vehicle is powered by fuel, which has a short duration for filling the gas tank. However, EVs remain uncertain as to how long they need to recharge the battery in case of a long journey.

    (b) Battery capacity and lifespan: Electric automobiles are typically constructed by substituting the fuel tank of a normal vehicle with batteries and chargers. Because EV batteries are built to last for a long period of time, they will eventually wear out at a specific time. Most suppliers currently provide an 8-year or 100,000-mile warranty [48]. Limited battery life requires frequent replacement, which is a major burden for EV users [27,36].

(4) Customer behavior:

    (a) Customer awareness: Consumer awareness is essential to bringing new customers and maintaining current ones. Despite the expanding range of electric vehicles on the market, the choice of buying an electric car remains limited and is estimated to remain so in the future. As a result, automobile companies should be aware of the power of consumer offerings through marketing, social networks, or any other channels. Potential users' awareness of the benefits of EVs such as financial incentives, infrastructure availability, and potential fuel-related savings are likely to be essential factors affecting their uptake [16,45].

    (b) Range anxiety: Consumers' range anxiety is important in making decisions about charging station placement, and they are influenced by driving range when they have to decide how much they are willing to drive to reach another charging station [46,47]. The driver needs to carefully organize their journey and may not be able to drive long distances.

(5) Policy:

    (a) Government support: Government policy, city planning, and power utilities play a major role in the EV charging infrastructure and consumer awareness of the environment [40].

    (b) Impact of tax and subsidy policies: Vehicle taxes and purchase subsidies are frequently used to provide incentives for the adoption of EVs. Compared to the consumer side, governmental tax and subsidy policies for manufacturers have a better effect on EV diffusion [42].

    (c) Renewable energy ecosystem: However, current renewable energies are insufficient for electricity demand. Government strategy is an essential part of supporting renewable energy ecosystems by focusing on some of the most promising alternatives [40,44].

3.2. ANP

Step 1. Problem structuring and network model construction

According to the ANP procedure, weightings for pairwise comparison completion were given to each factor via a questionnaire issued to experts to collect opinions related to the relative importance of different factors. The weights of the criteria and alternatives can be evaluated from the comparison matrices that were created using a 1–9 scale.

Step 2. Creating pairwise comparison matrices

Pairwise comparison was completed by collecting experts' opinions through a questionnaire. Following the gathering of the experts' assessment results and preferences, we constructed a comparison matrix of clusters and criteria. Afterward, a review of the consistency ratio ($C.R.$) of the matrices was required. The pairwise comparison is considered acceptable when it is less than 0.1 for every comparison matrix. In the case that $C.I. > 0.1$, it means that decision-makers and experts have differing viewpoints. It is necessary to alter their assessments to create a new and consistent comparison matrix [49]. The consistency ratio ($C.R.$) is defined as follows:

$$C.R. = \frac{C.I.}{R.I.}$$

where $C.I. = \frac{\lambda_{max}-n}{n-1}$, represents the consistency index and $R.I.$ represents random index. Table 3 shows the value of $R.I.$

Table 3. The value of $R.I.$

	1	2	3	4	5	6	7	8	9
$R.I.$	0	0	0.58	0.9	1.12	1.24	1.32	1.41	1.45

Step 3. Super-matrix formation and ranking the criteria

Following the completion of the pairwise matrix comparison, a super-matrix was built using prioritized vectors derived from the previous step for interdependency. A normalizing process was constructed to build a weighted super-matrix. Then, for the limited matrix, a raw value of the weighted matrix equal to each column of the super-matrix using the vectors calculated from the previous step to conduct the super-matrix columns was made, where each segment of the partitioned matrix denotes a relationship between two clusters. A considered network is broken down into N clusters, displayed by $C_1, C_2, \ldots, C_N$ and the elements in $Ck, 1 \leq k \leq N$ are e_{k1}, $e_{k2}, \ldots, e_{knk}$ where n_k represents the number of elements in the Ck cluster. These processes were carried out using Super Decisions, an ANP developed software tool. The synthesis process determines the overall priority of each alternative. The results of each subnetwork are merged to assign the final priorities of the alternatives, which are then ranked.

3.3. DEMATEL

Step 1. Conduct questionnaire

A questionnaire with previously identified criteria and collaboration with experts was conducted to complete the direct relation matrix. The respondents were requested to identify the influence of each criterion on others using nxn pairwise comparisons. The questionnaire used a numerical score that ranged from 0 to 4. The score representative sequent is shown as follows: 0, no influence; 1, low influence; 2, medium influence; 3, high influence; and 4, very high influence.

Step 2. Construct direct relation matrix

Assume that, for this study, there were H experts and n factors for consideration. All respondents were required to indicate the level at which they thought a factor, say i, influences another factor j. X_{ij}^{K} represents pairwise comparisons of the opinion provided by kth experts about the ith and jth factors which received a score in the previous step. The score given by the respondent was assigned to the $n \times n$ nonnegative response matrix $X^{k} = [x_{ij}^{k}]$ with $k = 1, 2, \ldots, H$. Therefore, for each H expert, the answer matrices were $X^1, X^2, \ldots, X^H$. Moreover, each component of $X^k = [x_{ij}^k]$ is an integer indicated by X_{ij}^{K}. Each $X^k = [x_{ij}^k]$ response was diagonal, while the other elements were defined as zero. Then, it was possible to calculate $n \times n$ average matrix A for all respondent evaluations by equalizing the given degree by H respondents as follows:

$$[a_{ij}]_{nxn} = \frac{1}{H} \sum_{k=1}^{H} [x_{ij}^{k}]_{nxn}$$

The average direct relation matrix $A = [x_{ij}^k]_{nxn}$ is also called the initial direct relation. Matrix A illustrates the factor's initial direct effects on and obtained from the other factors. Moreover, an influence map was drawn to visualize the causal effect of each aspect of the system. Each node in the map represents a system factor, with arrows corresponding to the interactions between them. As an example, an arrow from C2 to C4 indicates the impact that C2 has over C4, and the effect strength is 1. Using DEMATEL, an intelligible map of the system was created from the structural relations among the factors in the system.

Step 3. Normalize the direct relation matrix

The normalized initial direct-relation matrix $D = [d_{ij}]_{nxn}$ was acquired by applying the following formula to normalize the average matrix A:

$$S = max\left\{ \max_{1 \le i \le n} \sum_{j=1}^{n} aij, \max_{1 \le j \le n} \sum_{i=1}^{n} aij \right\}, D = \frac{A}{S}$$

Consequently, the direct effects of each factor on other factors are represented by the total of each row j of matrix A, and $\max_{1 \le i \le n} \sum_{j=1}^{n} aij$ indicates the factor with the largest direct influence on the other factors. The direct effects received by factor i are represented by the sum of each column i of matrix A, and $\max_{1 \le j \le n} \sum_{i=1}^{n} aij$ indicates the factor that is most influenced by other factors. The larger of the two extreme sums is equal to the positive scalar s. By dividing each element of A by the scalar s, the matrix D was constructed. Each d_{ij} element in matrix D had a value between 0 and 1.

Step 4. Total relation matrix calculation using MATLAB

The power of the normalized initial direct relation matrix D is represented by D^m, which is an m-indirect effect that can be used to illustrate the length effect, or the influence propagated by m-1 intermediates. Convergent solutions to the matrix inversion are guaranteed by a continuous reduction in the indirect effects of problems other than the powers of matrix D, such as an engrossing Markov chain matrix. The total influence or relationship can be obtained by summing up $D^1, D^2, D^3, \ldots, D^\infty$,

$$\lim_{m \to \infty} D^m = [0]_{nxn},\tag{1}$$

where $[0]_{nxn}$ is a nxn null matrix.

The total relation matrix T_{nxn} is accomplished as follows:

$$\sum_{m=1}^{\infty} D_i = D + D^2 + D^3 \ldots D^M$$
$$= D(I + D + D^2 + \ldots + D^{M-1})$$
$$= D(I - D)^{-1}(I - D)(I + D + D^2 + \ldots + D^{M-1})$$
$$= D(I - D)^{-1}(I - D^m) = D(I - D)^{-1}$$

T is total relation matrix $([T]_{nxn})$. I is identity matrix.

The total relation matrix's sum of rows and columns is generated as r and c $nx1$ vectors.

$$r = [r_i]_{nx1} = (\textstyle\sum_{j=1}^{n} t_{ij})_{nx1} \quad, c = [c_j]_{1xn} = (\textstyle\sum_{i=1}^{n} t_{ij})_{1xn}$$

$[r_i]_{nx1}$ is the sum of the ith row of matrix T and identifies the overall effect, including indirect and direct, indicated to other factors by factor I. Additionally, the sum of the jth column of matrix T is $[c_j]_{1xn}$ and both the indirect and direct of the total effects obtained by factor j from other factors can be denoted by $[c_j]_{1xn}$. Furthermore, the total effects, including those received and given by i-th factor, can be represented by the sum $(r_i + c_i)$. Otherwise, the value of $(r_i + c_i)$ indicates the importance of the i-th factor in the system (received and given effects in total). Moreover, the net effect that factor i contributes to the system can be represented by the difference (the relation which is $r_i - c_i$). Factor i is a net causer if $(r_i - c_i)$ is positive. On the other hand, factor i is a net receiver if $(r_i - c_i)$ is negative [50,51].

4. Results and Discussion

4.1. ANP Priority Analysis

The network model was built by considering the relationship between the real world and existing research. After the criteria selection was complete, Super Decision software was used to execute the pairwise comparison process. Table 4 presents the unweighted super-matrix of this study. Then the weighted super-matrix and limited super-matrix show as Tables A1 and A2 in Appendix B, respectively. Table 5 shows the weight of EV barriers.

4.2. DEMATEL Priority Analysis

Through the literature review, 12 criteria for EV barriers were identified. A 12 × 12 pairwise comparison questionnaire was administered to analyze the interrelationships among criteria. After collecting all questionnaires from the experts, the relationships between the criteria were assessed using a 0–4 scale. An average direct relation matrix was constructed, as shown in Tables A3–A5 in Appendix B. For this step, the matrix was computed by using MATLAB software.

Table 4. Unweighted super-matrix.

Barrier	Sub-Barrier	Financial		Infrastructure			Technology		Customer Behavior		Policy		
		F1	F2	I1	I2	I3	T1	T2	C1	C2	P1	P2	P3
Financial	F1	0.000	1.000	0.857	0.000	0.875	0.857	0.857	0.833	0.000	0.857	0.857	0.000
	F2	1.000	0.000	0.143	1.000	0.125	0.143	0.143	0.167	0.000	0.143	0.143	0.000
Infrastructure	I1	0.571	0.000	0.000	1.000	0.000	0.000	1.000	0.750	0.750	1.000	0.000	0.000
	I2	0.143	0.000	0.000	0.000	0.000	0.000	0.000	0.000	0.000	0.000	0.000	0.000
	I3	0.286	0.000	0.000	0.000	0.000	0.000	0.000	0.250	0.250	0.000	0.000	0.000
Technology	T1	0.111	0.000	0.000	0.000	0.000	0.111	1.000	0.111	0.111	0.000	0.111	0.000
	T2	0.889	0.000	1.000	0.000	0.000	0.889	0.000	0.889	0.889	0.000	0.889	0.000
Customer behavior	C1	0.000	1.000	0.889	0.000	0.889	0.000	0.000	0.000	1.000	1.000	0.000	1.000
	C2	0.000	0.000	0.111	0.000	0.111	1.000	0.000	1.000	0.000	0.000	0.000	0.000
Policy	P1	0.667	0.667	0.667	0.000	0.000	0.667	0.667	0.588	0.000	0.000	1.000	0.000
	P2	0.333	0.333	0.333	0.000	0.000	0.333	0.333	0.323	0.000	1.000	0.000	0.000
	P3	0.000	0.000	0.000	0.000	0.000	0.000	0.000	0.089	0.000	0.000	0.000	0.000

Table 5. Weights of EV barriers.

Barrier	Weight	Sub-Barrier	Normalized by Cluster	Limiting	Rank
Financial	0.194	High cost (F1)	0.762	0.148	3
		High price (F2)	0.238	0.046	7
Infrastructure	0.072	Shortage of charging stations (I1)	0.898	0.065	6
		Lack of battery recycling system (I2)	0.030	0.002	11
		Shortage of maintenance shops (I3)	0.072	0.005	10
Technology	0.365	EV performance (T1)	0.356	0.130	5
		Battery capacity and lifespan (T2)	0.644	0.235	1
Customer behavior	0.049	Customer awareness (C1)	0.728	0.036	8
		Range anxiety (C2)	0.272	0.013	9
Policy	0.319	Government support (P1)	0.548	0.175	2
		Impacts of tax and subsidy policies (P2)	0.449	0.143	4
		Renewable energy ecosystem (P3)	0.003	0.001	12

Using the matrix results, the final result of DEMATEL analysis using MATLAB software was calculated. The mean value for all criteria was computed as the threshold value, which was 0.4116. The value of $D + R$ and $D - R$ are calculated and illustrated in Table 6 and Figure 2. The $D + R$ stands for the degree of central role that the factor plays in the system. Similarly, the vertical axis vector called "Relation" shows the net effect that the factor contributes to the system. In addition, the $D + R$ and $D - R$ could be identified as cause factors of perceived benefits, effect factors of perceived benefits, cause factors of perceive risks, and effect factors of perceived risks. In addition, a threshold value is set in many studies to filter out negligible effects. That is, only the elements of the matrix whose influence levels are greater than the value of are selected and converted. If the threshold value is too low, too many factors are included and the IRM is too complex to comprehend. In contrast, some important factors may be excluded if the threshold value is too high. In the literature, the threshold value is usually determined by experts through discussions.

Table 6. Final results of DEMATEL analysis.

No.	Criteria	D	R	D−R	D+R
1	High cost (F1)	5.438	5.185	0.252	10.623
2	High price (F2)	5.049	5.433	−0.384	10.482
3	Shortage of charging stations (I1)	5.131	4.891	0.240	10.021
4	Lack of battery recycling system (I2)	3.597	4.333	−0.736	7.930
5	Shortage of maintenance shops (I3)	4.128	4.337	−0.209	8.466
6	EV performance (T1)	5.236	4.827	0.409	10.063
7	Battery capacity and lifespan (T2)	5.938	5.007	0.930	10.945
8	Customer awareness (C1)	4.880	5.713	−0.833	10.593
9	Range anxiety (C2)	5.064	4.377	0.687	9.440
10	Government support (P1)	5.685	5.241	0.445	10.926
11	Impacts of tax and subsidy policies (P2)	5.500	5.419	0.081	10.919
12	Renewable energy ecosystem (P3)	3.624	4.506	−0.882	8.130

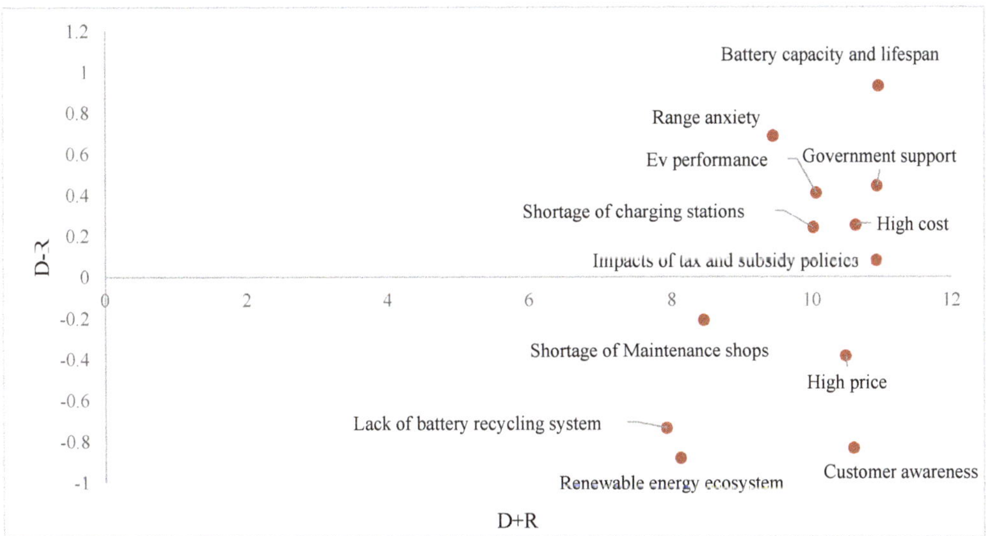

Figure 2. The casual diagram.

4.3. Result Analysis

From the 12 selected barriers to EV transition, this study compared the application of Ctwo different multicriteria decision methodologies for the analysis. Table 7 displays the summary results of the weighting of EV transition barriers using the ANP and DEMATEL methods.

The results illustrate that "battery capacity and lifespan" from the EV technology category is the most important barrier to EV transition from both methodologies, which implies that the largest obstacle to EV adoption for the automotive sector is battery technology development. Next, the second most important barrier from ANP and DEMATEL methodology is "government support" from the policy category which includes long-term policy, city planning, power utilities etc.

Table 7. Result comparison.

Rank	ANP	DEMATEL
1	Battery capacity and lifespan	Battery capacity and lifespan
2	Government support	Government support
3	High cost	Impacts of tax and subsidy policies
4	Impacts of tax and subsidy policies	High cost
5	EV performance	Customer awareness
6	Shortage of charging stations	High price
7	High price	EV performance
8	Customer awareness	Shortage of charging stations
9	Range anxiety	Range anxiety
10	Shortage of maintenance shops	Shortage of maintenance shops
11	Lack of battery recycling system	Renewable energy ecosystem
12	Renewable energy ecosystem	Lack of battery recycling system

When the sequence between the fifth place and the eighth place of the summary result is compared, it can be seen that the importance sequence of each barrier slightly fluctuates between "EV performance", "shortage of charging stations", "high price", and "customer awareness". Table 7 shows that "lack of battery recycling system", "renewable energy ecosystem", and "shortage of maintenance shops" are the final important barriers to EV transition. However, the priority of the last two barriers, "lack of battery recycling system" and "renewable energy ecosystem" are swapped because of the different stages of the utilized methods.

4.4. Summary of the Results

The result comparison of the two multicriteria decision-making (MCDM) methods were applied in this study. The summarized outputs are presented. Among the 5 barrier categories and 12 sub-barriers analyzed, the EV technology barrier category was prioritized, after first referring to the opinions and analyses of participating experts. Within the major category, battery capacity and lifespan play a pivotal role among the sub-barriers, and not only has the highest weighting but also the highest influence on other barriers according to the DEMATEL result.

Notably, the policy barrier category was the second-highest-ranked barrier in this study. Within the policy category, the government support barrier had the highest weighting and second most important rank among all sub-barriers. This output strengthens the research findings, which highlight that government reinforcement is a critical determinant of consumers' adoption of EVs. Additionally, a study in China also emphasizes the government's role in various aspects that support the achievement of EVs, such as R&D investment, EV demonstration programs, and electric vehicle promotion strategies [52]. Moreover, another sub-barrier listed in the policy category and ranked in the top four of the overall result is the impact of tax and subsidy policy barriers. This sub-barrier has been mentioned and investigated in several previous studies. For example, research in Europe [42] concluded that tax and subsidy impacts at various points include increasing EV sales by reducing total ownership costs and increasing environmental benefits by implementing a green tax. Another study [53] also highlighted the importance of subsidy policy on EV adoption; the subsidy policy from the public sector increases EV cost compatibility with ICE vehicles.

From a manufacturer's perspective, a shortage of charging stations is not very important. It is completely different from the consumer perspective in that the shortage of charging stations is the main concern of potential EV purchasers. In addition, this study surveyed most data in Thailand. When compared with related topics in Taiwan [54], the direction of EV policy is considerably influenced by the leading manufacturer, which differs from the results of this study, indicating the importance of policy over the enterprise. Moreover, the results from the two different methods were similar owing to the consistent estimation of experts. It is worth mentioning, although in this study, only the emission values of the vehicles at the time of use were considered. However, the effect of these vehicles on GHG during the production phase should also be investigated.

5. Conclusions

The near-term future of EV sales is high. In the first quarter of 2021, global electric car sales rose by approximately 140% compared to the same period in 2020 [3]. To summarize, as mentioned in the previous section, previous studies were considered and analyzed to identify barriers to EV transition. After identifying barriers related to EV diffusion, the ANP and DEMATEL methodologies were applied.

The first research output was a list of 12 barriers to EV transition under five categories identified from a comprehensive review, including financial, infrastructure, technology, customer behavior, and policy. The second output research was the prioritized result from the analytic network process (ANP) and the decision-making trial and evaluation laboratory (DEMATEL) technique performed concurrently. The highest weighted barrier was battery capacity and lifespan, followed by government support, the impacts of tax and subsidy policies, and high costs. Moreover, the outcome from the DEMATEL methodology showed that high weighing barriers such as the battery capacity, lifespan, and government support barriers have influence on other barriers such as high price barriers, customer awareness barriers, and lack of battery recycling systems.

Managerial implications for the EV industry: This study's findings emphasize the urgent barrier to electric vehicle transition for the automotive sector and improve the understanding of private and public sector corporations. Battery capacity and lifespan require major concentration and enhancement from manufacturers, which greatly influence electric vehicle performance, range anxiety, and awareness of EV customers. Moreover, government support plays a key role. The scarcity of sales offices and charging time are both also barriers. Policy makers need to pay attention to EVs supporting regulations such as driving up the number of public charging stations, decreasing electricity fees, and encouraging renewable energy enterprises. Eventually, the successful adoption of electric vehicles will result in lower consumption of fossil fuels, natural resource usage optimization, and sustainability.

The main contributions of this study are as follows: First, the research identified and prioritized the barriers that manufacturers confront in making the transition to electric vehicles, as well as the interrelationships between these barriers. Second, this study proposed a comparison model of two multi criteria decision methodologies for prioritizing and identifying the interrelation of the EV adoption barrier, which potentially provides more robust results than applying only one multi criteria decision methodology.

A limitation of this study is that the methodology input greatly depends on the judgement and estimation of experts. However, the number of respondents was limited because of the high level of experience required by the respondents.

In future research, other barriers against EV transition that may spring up in the years ahead could be analyzed. To define such challenges consistently, regular and continuous literature reviews and interrelations with consumers, automakers, specialists, and policymakers are necessary. This study can be applied to EV barrier analysis. Therefore, model applicability should be reviewed before it can be used in a specific context. Different MCDM tool combinations can be used to examine the robustness of the study findings.

Author Contributions: Conceptualization, T.-C.K. and N.S.; methodology and writing, T.-C.K. and N.S.; validation, Y.-S.S.; formal analysis, N.S.; investigation, T.-C.K., N.S. and R.-H.Y. All authors have read and agreed to the published version of the manuscript.

Funding: This research was funded by the National Science and Technology Council of the Republic of China, Taiwan, grant number 110-2621-M-011-002-MY3 and NTUST- Kind-9399.

Conflicts of Interest: The authors declare no conflict of interest.

Appendix A. Survey

Instruction: Please rate each question in the horizontal row based on your experience using the following scale

Importance Scale	Definition of Importance Scale
1	Equal importance preferred
2	Equal to moderate importance preferred
3	Moderate importance preferred
4	Moderate to strong importance preferred
5	Strong importance preferred
6	Strong to very strong importance preferred
7	Very strong importance preferred
8	Very strong to extreme importance preferred
9	Extreme importance preferred

0. With respect to infrastructure, which is more important to electric vehicle transition barriers between:

No	Barrier	More Importance on the Left								More Importance on the Right								Barrier
		9	8	7	6	5	4	3	2	1	2	3	4	5	6	7	8	9	
0	Customer behavior													√					Financial

With respect to Infrastructure, if you think "Financial" is strongly importance to electric vehicle transition barrier more than "Customer behavior" then check 5 on the right

1. With respect to finance, which is more important to electric vehicle transition barrier between

No	Barrier	More Importance on the Left								More Importance on the Right								Barrier
		9	8	7	6	5	4	3	2	1	2	3	4	5	6	7	8	9	
1	Customer behavior																		Financial
2	Customer behavior																		Infrastructure
3	Customer behavior																		Policy
4	Customer behavior																		Technology
5	Financial																		Infrastructure
6	Financial																		Policy
7	Financial																		Technology
8	Infrastructure																		Policy
9	Infrastructure																		Technology
10	Policy																		Technology

2. With respect to infrastructure, which is more important to electric vehicle transition barriers between:

No	Barrier	More Importance on the Left									More Importance on the Right									Barrier
		9	8	7	6	5	4	3	2	1	2	3	4	5	6	7	8	9	
1	Customer behavior																		Financial
2	Customer behavior																		Infrastructure
3	Customer behavior																		Policy
4	Customer behavior																		Technology
5	Financial																		Infrastructure
6	Financial																		Policy
7	Financial																		Technology
8	Infrastructure																		Policy
9	Infrastructure																		Technology
10	Policy																		Technology

3. With respect to technology, which is more important to electric vehicle transition barriers between:

No	Barrier	More Importance on the Left									More Importance on the Right									Barrier
		9	8	7	6	5	4	3	2	1	2	3	4	5	6	7	8	9	
1	Customer behavior																		Financial
2	Customer behavior																		Infrastructure
3	Customer behavior																		Policy
4	Customer behavior																		Technology
5	Financial																		Infrastructure
6	Financial																		Policy
7	Financial																		Technology
8	Infrastructure																		Policy
9	Infrastructure																		Technology
10	Policy																		Technology

4. With respect to customer behavior, which is more important to electric vehicle transition barriers between:

No	Barrier	More Importance on the Left									More Importance on the Right									Barrier
		9	8	7	6	5	4	3	2	1	2	3	4	5	6	7	8	9	
1	Customer behavior																		Financial
2	Customer behavior																		Infrastructure
3	Customer behavior																		Policy
4	Customer behavior																		Technology
5	Financial																		Infrastructure
6	Financial																		Policy
7	Financial																		Technology
8	Infrastructure																		Policy
9	Infrastructure																		Technology
10	Policy																		Technology

5. With respect to policy, which is more important to electric vehicle transition barriers between:

No	Barrier	More Importance on the Left									More Importance on the Right								Barrier
		9	8	7	6	5	4	3	2	1	2	3	4	5	6	7	8	9	
1	Customer behavior																		Financial
2	Customer behavior																		Infrastructure
3	Customer behavior																		Policy
4	Customer behavior																		Technology
5	Financial																		Infrastructure
6	Financial																		Policy
7	Financial																		Technology
8	Infrastructure																		Policy
9	Infrastructure																		Technology
10	Policy																		Technology

6. With respect to high cost, which is more important to electric vehicle transition barriers between:

No	Barrier	More Importance on the Left									More Importance on the Right								Barrier
		9	8	7	6	5	4	3	2	1	2	3	4	5	6	7	8	9	
1	Shortage of charging stations																		Lack of battery recycling system
2	Shortage of charging stations																		Shortage of maintenance shops
3	Lack of battery recycling system																		Shortage of maintenance shops
4	EV performance																		Battery capacity and life span
5	Government support																		Impact of tax and subsidy policies

7. With respect to high price, which is more important to electric vehicle transition barriers between:

No	Barrier	More Importance on the Left									More Importance on the Right								Barrier
		9	8	7	6	5	4	3	2	1	2	3	4	5	6	7	8	9	
1	Government support																		Impact of tax and subsidy policies

8. With respect to shortage of charging stations, which is more important to electric vehicle transition barriers between:

No	Barrier	More Importance on the Left									More Importance on the Right								Barrier
		9	8	7	6	5	4	3	2	1	2	3	4	5	6	7	8	9	
1	High cost																		High price
2	Customer awareness																		Range anxiety
3	Government support																		Impact of tax and subsidy policies

9. With respect to shortage of maintenance shops, which is of more importance to electric vehicle transition barriers between:

No	Barrier	More Importance on the Left									More Importance on the Right								Barrier
		9	8	7	6	5	4	3	2	1	2	3	4	5	6	7	8	9	
1	High cost																		High price
2	Customer awareness																		Range anxiety

10. With respect to EV performance, which is of more importance to electric vehicle transition barriers between:

No	Barrier	More Importance on the Left									More Importance on the Right								Barrier
		9	8	7	6	5	4	3	2	1	2	3	4	5	6	7	8	9	
1	High cost																		High price
2	EV performance																		Battery capacity and life span
3	Government support																		Impact of tax and subsidy policies

11. With respect to battery capacity and life span, which is of more importance to electric vehicle transition barriers between:

No	Barrier	More Importance on the Left									More Importance on the Right								Barrier
		9	8	7	6	5	4	3	2	1	2	3	4	5	6	7	8	9	
1	High cost																		High price
2	Government support																		Impact of tax and subsidy policies

12. With respect to customer awareness, which is of more importance to electric vehicle transition barriers between:

No	Barrier	More Importance on the Left									More Importance on the Right								Barrier
		9	8	7	6	5	4	3	2	1	2	3	4	5	6	7	8	9	
1	High cost																		High price
2	EV performance																		Battery capacity and life span
3	Shortage of charging stations																		Shortage of maintenance shops
4	Government support																		Impact of tax and subsidy policies

13. With respect to range anxiety, which is of more importance to electric vehicle transition barriers between:

No	Barrier	More Importance on the Left									More Importance on the Right								Barrier
		9	8	7	6	5	4	3	2	1	2	3	4	5	6	7	8	9	
2	EV performance																		Battery capacity and life span
3	Shortage of charging stations																		Shortage of maintenance shops

14. With respect to government support, which is of more importance to electric vehicle transition barriers between:

No	Barrier	More Importance on the Left								More Importance on the Right								Barrier
		9	8	7	6	5	4	3	2	1	2	3	4	5	6	7	8	9	
1	High cost																		High price

15. With respect to impacts of tax and subsidy policies, which is of more importance to electric vehicle transition barriers between:

No	Barrier	More Importance on the Left								More Importance on the Right								Barrier
		9	8	7	6	5	4	3	2	1	2	3	4	5	6	7	8	9	
1	High cost																		High price
2	EV performance																		Battery capacity and life span

Appendix B

Table A1. Weighted super-matrix.

Barrier	Sub-Barrier	Financial		Infrastructure			Technology		Customer Behavior		Policy		
		F1	F2	I1	I2	I3	T1	T2	C1	C2	P1	P2	P3
Financial	F1	0.000	0.302	0.135	0.000	0.570	0.164	0.159	0.110	0.000	0.219	0.149	0.000
	F2	0.149	0.000	0.023	0.850	0.081	0.027	0.027	0.022	0.000	0.037	0.025	0.000
Infrastructure	I1	0.058	0.000	0.000	0.150	0.000	0.000	0.095	0.045	0.075	0.176	0.000	0.000
	I2	0.015	0.000	0.000	0.000	0.000	0.000	0.000	0.000	0.000	0.000	0.000	0.000
	I3	0.029	0.000	0.000	0.000	0.000	0.000	0.000	0.015	0.025	0.000	0.000	0.000
Technology	T1	0.050	0.000	0.000	0.000	0.000	0.051	0.445	0.050	0.083	0.000	0.059	0.000
	T2	0.400	0.000	0.499	0.000	0.000	0.406	0.000	0.402	0.666	0.000	0.469	0.000
Customer behavior	C1	0.000	0.094	0.075	0.000	0.310	0.000	0.000	0.000	0.151	0.128	0.000	1.000
	C2	0.000	0.000	0.009	0.000	0.039	0.072	0.000	0.091	0.000	0.000	0.000	0.000
Policy	P1	0.199	0.402	0.172	0.000	0.000	0.187	0.182	0.155	0.000	0.000	0.299	0.000
	P2	0.099	0.201	0.086	0.000	0.000	0.093	0.091	0.085	0.000	0.441	0.000	0.000
	P3	0.000	0.000	0.000	0.000	0.000	0.000	0.000	0.023	0.000	0.000	0.000	0.000

Table A2. Limited super matrix.

Barrier	Sub-Barrier	Financial		Infrastructure			Technology		Customer Behavior		Policy		
		F1	F2	I1	I2	I3	T1	T2	C1	C2	P1	P2	P3
Financial	F1	0.148	0.148	0.148	0.148	0.148	0.148	0.148	0.148	0.148	0.148	0.148	0.148
	F2	0.046	0.046	0.046	0.046	0.046	0.046	0.046	0.046	0.046	0.046	0.046	0.046
Infrastructure	I1	0.065	0.065	0.065	0.065	0.065	0.065	0.065	0.065	0.065	0.065	0.065	0.065
	I2	0.002	0.002	0.002	0.002	0.002	0.002	0.002	0.002	0.002	0.002	0.002	0.002
	I3	0.005	0.005	0.005	0.005	0.005	0.005	0.005	0.005	0.005	0.005	0.005	0.005
Technology	T1	0.130	0.130	0.130	0.130	0.130	0.130	0.130	0.130	0.130	0.130	0.130	0.130
	T2	0.235	0.235	0.235	0.235	0.235	0.235	0.235	0.235	0.235	0.235	0.235	0.235

Table A2. *Cont.*

Barrier	Sub-Barrier	Financial		Infrastructure			Technology		Customer Behavior		Policy		
		F1	F2	I1	I2	I3	T1	T2	C1	C2	P1	P2	P3
Customer behavior	C1	0.036	0.036	0.036	0.036	0.036	0.036	0.036	0.036	0.036	0.036	0.036	0.036
	C2	0.013	0.013	0.013	0.013	0.013	0.013	0.013	0.013	0.013	0.013	0.013	0.013
Policy	P1	0.175	0.175	0.175	0.175	0.175	0.175	0.175	0.175	0.175	0.175	0.175	0.175
	P2	0.143	0.143	0.143	0.143	0.143	0.143	0.143	0.143	0.143	0.143	0.143	0.143
	P3	0.001	0.001	0.001	0.001	0.001	0.001	0.001	0.001	0.001	0.001	0.001	0.001

Table A3. Average direct relation matrix.

Criteria	F1	F2	I1	I2	I3	T1	T2	C1	C2	P1	P2	P3
F1	0.000	3.556	2.000	1.444	1.889	2.111	2.667	2.333	1.778	2.111	2.889	1.444
F2	2.000	0.000	1.889	1.222	2.111	2.000	2.333	2.889	1.778	2.556	2.889	1.444
I1	2.222	1.889	0.000	1.000	1.889	1.778	2.778	2.333	2.778	2.111	2.111	1.667
I2	1.667	1.000	1.000	0.000	1.111	1.111	1.222	1.333	0.778	1.444	1.667	1.667
I3	1.778	1.556	1.333	1.333	0.000	1.889	1.667	2.111	1.000	1.444	1.222	1.222
T1	2.444	2.778	2.111	1.444	2.000	0.000	2.667	2.222	2.333	1.556	1.667	1.667
T2	2.778	2.778	2.556	1.889	2.111	3.111	0.000	2.444	2.889	2.111	2.333	1.778
C1	1.556	2.000	2.333	1.111	2.000	2.000	1.778	0.000	2.222	2.556	2.444	1.889
C2	2.222	1.778	2.222	1.222	1.889	2.667	2.333	2.778	0.000	1.556	1.444	1.222
P1	2.444	3.222	2.667	2.444	2.000	1.444	1.444	2.778	1.889	0.000	3.222	3.000
P2	2.222	3.222	2.111	2.444	1.889	1.889	1.556	2.667	1.778	2.889	0.000	2.333
P3	1.667	1.444	0.778	1.667	1.000	1.444	2.000	1.889	0.889	2.000	1.889	0.000

Table A4. Normalized direct relation matrix.

Criteria	F1	F2	I1	I2	I3	T1	T2	C1	C2	P1	P2	P3
F1	0.000	0.133	0.075	0.054	0.071	0.079	0.100	0.087	0.066	0.079	0.108	0.054
F2	0.075	0.000	0.071	0.046	0.079	0.075	0.087	0.108	0.066	0.095	0.108	0.054
I1	0.083	0.071	0.000	0.037	0.071	0.066	0.104	0.087	0.104	0.079	0.079	0.062
I2	0.062	0.037	0.037	0.000	0.041	0.041	0.046	0.050	0.029	0.054	0.062	0.062
I3	0.066	0.058	0.050	0.050	0.000	0.071	0.062	0.079	0.037	0.054	0.046	0.046
T1	0.091	0.104	0.079	0.054	0.075	0.000	0.100	0.083	0.087	0.058	0.062	0.062
T2	0.104	0.104	0.095	0.071	0.079	0.116	0.000	0.091	0.108	0.079	0.087	0.066
C1	0.058	0.075	0.087	0.041	0.075	0.075	0.066	0.000	0.083	0.095	0.091	0.071
C2	0.083	0.066	0.083	0.046	0.071	0.100	0.087	0.104	0.000	0.058	0.054	0.046
P1	0.091	0.120	0.100	0.091	0.075	0.054	0.054	0.104	0.071	0.000	0.120	0.112
P2	0.083	0.120	0.079	0.091	0.071	0.071	0.058	0.100	0.066	0.108	0.000	0.087
P3	0.062	0.054	0.029	0.062	0.037	0.054	0.075	0.071	0.033	0.075	0.071	0.000

Table A5. Total relation matrix.

Criteria	F1	F2	I1	I2	I3	T1	T2	C1	C2	P1	P2	P3
F1	0.378	0.539	0.423	0.339	0.400	0.430	0.460	0.505	0.402	0.447	0.494	0.373
F2	0.430	0.401	0.404	0.319	0.392	0.410	0.432	0.503	0.386	0.444	0.476	0.360
I1	0.430	0.457	0.331	0.305	0.378	0.397	0.440	0.477	0.412	0.421	0.442	0.359
I2	0.278	0.281	0.241	0.169	0.234	0.247	0.259	0.294	0.225	0.269	0.289	0.247
I3	0.322	0.343	0.290	0.245	0.230	0.311	0.313	0.364	0.270	0.308	0.316	0.265
T1	0.438	0.487	0.404	0.319	0.383	0.336	0.439	0.474	0.399	0.405	0.429	0.359
T2	0.504	0.548	0.470	0.376	0.435	0.491	0.402	0.543	0.466	0.476	0.507	0.410
C1	0.396	0.446	0.398	0.300	0.370	0.390	0.395	0.382	0.382	0.423	0.439	0.357
C2	0.410	0.431	0.389	0.295	0.361	0.406	0.408	0.468	0.301	0.383	0.399	0.327
P1	0.481	0.549	0.461	0.388	0.421	0.425	0.442	0.542	0.421	0.396	0.527	0.443
P2	0.455	0.528	0.426	0.373	0.400	0.421	0.425	0.516	0.400	0.473	0.399	0.405
P3	0.322	0.344	0.275	0.261	0.268	0.299	0.325	0.360	0.268	0.330	0.342	0.226

References

1. UN the Paris Agreement. Available online: https://unfccc.int/process-and-meetings/the-paris-agreement/the-paris-agreement (accessed on 24 March 2021).
2. IEA. Net Zero by 2050: A Roadmap for the Global Energy Sector—A special report by the International Energy Agency, 2021. p. 224. Available online: https://unfccc.int/documents/278467?gclid=Cj0KCQjw--2aBhD5ARIsALiRlwBxB9Z-zhGeCtvzTNZiF1 agJSMDofOsA3ktVeguD7fwIR8PbJoJWugaAt_rEALw_wcB (accessed on 24 March 2021).
3. IEA. Global EV Outlook 2021. International Energy Agency. 2021. Available online: https://iea.blob.core.windows.net/assets/ ed5f4484-f556-4110-8c5c-4ede8bcba637/GlobalEVOutlook2021.pdf (accessed on 24 March 2021).
4. Zhongming, Z.; Linong, L.; Wangqiang, Z.; Wei, L. The Emissions Gap Report 2020. 2020. Available online: https: //www.unep.org/resources/emissions-gap-report-2022?gclid=Cj0KCQjwk5ibBhDqARIsACzmgLRi0T3kw2XdIWBmrRlec4 nI5Lp86Q4oNiDrk87LlZDqJOoWIbvnMysaAmI9EALw_wcB, (accessed on 24 March 2021).
5. Ritchie, H.; Roser, M. CO$_2$ and Greenhouse Gas Emissions. Available online: https://ourworldindata.org/co2-and-other-greenhouse-gas-emissions (accessed on 24 March 2021).
6. Bouckaert, S.; Pales, A.F.; McGlade, C.; Remme, U.; Wanner, B.; Varro, L.; D'Ambrosio, D.; Spencer, T. Net Zero by 2050: A Roadmap for the Global Energy Sector. 2021. Available online: https://www.ourenergypolicy.org/resources/net-zero-by-2050-a-roadmap-for-the-global-energy-sector/, (accessed on 24 March 2021).
7. Williams, J.H.; DeBenedictis, A.; Ghanadan, R.; Mahone, A.; Moore, J.; Morrow, W.R.; Price, S.; Torn, M.S. The Technology Path to Deep Greenhouse Gas Emissions Cuts by 2050: The Pivotal Role of Electricity. *Science* **2012**, *335*, 53–59. [CrossRef] [PubMed]
8. Metais, M.O.; Jouini, O.; Perez, Y.; Berrada, J.; Suomalainen, E. Too much or not enough? Planning electric vehicle charging infrastructure: A review of modeling options. *Renew. Sustain. Energy Rev.* **2022**, *153*, 111719. [CrossRef]
9. Wang, L.; Wang, X.; Yang, W. Optimal design of electric vehicle battery recycling network–From the perspective of electric vehicle manufacturers. *Appl. Energy* **2020**, *275*, 115328. [CrossRef]
10. Hawkins, T.R.; Singh, B.; Majeau-Bettez, G.; Strømman, A.H. Comparative environmental life cycle assessment of conventional and electric vehicles. *J. Ind. Ecol.* **2013**, *17*, 53–64. [CrossRef]
11. Ju, N.; Lee, K.-H.; Kim, S.H. Factors Affecting Consumer Awareness and the Purchase of Eco-Friendly Vehicles: Textual Analysis of Korean Market. *Sustainability* **2021**, *13*, 5566. [CrossRef]
12. Egbue, O.; Long, S. Barriers to widespread adoption of electric vehicles: An analysis of consumer attitudes and perceptions. *Energy Policy* **2012**, *48*, 717–729. [CrossRef]
13. Goel, P.; Sharma, N.; Mathiyazhagan, K.; Vimal, K.E.K. Government is trying but consumers are not buying: A barrier analysis for electric vehicle sales in India. *Sustain. Prod. Consum.* **2021**, *28*, 71–90. [CrossRef]
14. Tarei, P.K.; Chand, P.; Gupta, H. Barriers to the adoption of electric vehicles: Evidence from India. *J. Clean. Prod.* **2021**, *291*, 125847. [CrossRef]
15. She, Z.-Y.; Sun, Q.; Ma, J.-J.; Xie, B.-C. What are the barriers to widespread adoption of battery electric vehicles? A survey of public perception in Tianjin, China. *Transp. Policy* **2017**, *56*, 29–40. [CrossRef]
16. Adhikari, M.; Ghimire, L.P.; Kim, Y.; Aryal, P.; Khadka, S.B. Identification and Analysis of Barriers against Electric Vehicle Use. *Sustainability* **2020**, *12*, 4850. [CrossRef]

17. Ghimire, L.P. Two Essays on Electric Vehicle Use in Nepal. Seoul National University Graduate School. 2019. Available online: https://www.semanticscholar.org/paper/Two-Essays-on-Electric-Vehicle-Use-in-Nepal-Ghimire/f3a942ab91152f601 3ff1f9c255d02df602048ff (accessed on 24 March 2021).
18. Sovacool, B.K.; Hirsh, R.F. Beyond batteries: An examination of the benefits and barriers to plug-in hybrid electric vehicles (PHEVs) and a vehicle-to-grid (V2G) transition. *Energy Policy* **2009**, *37*, 1095–1103. [CrossRef]
19. Zheng, G.; Peng, Z. Life Cycle Assessment (LCA) of BEV's environmental benefits for meeting the challenge of ICExit (Internal Combustion Engine Exit). *Energy Rep.* **2021**, *7*, 1203–1216. [CrossRef]
20. Van Bree, B.; Verbong, G.P.; Kramer, G.J. A multi-level perspective on the introduction of hydrogen and battery-electric vehicles. *Technol. Forecast. Soc. Chang.* **2010**, *77*, 529–540. [CrossRef]
21. Gallagher, K.S.; Muehlegger, E. Giving green to get green? Incentives and consumer adoption of hybrid vehicle technology. *J. Environ. Econ. Manag.* **2011**, *61*, 1–15. [CrossRef]
22. Kahn, M.E. Do greens drive Hummers or hybrids? Environmental ideology as a determinant of consumer choice. *J. Environ. Econ. Manag.* **2007**, *54*, 129–145. [CrossRef]
23. Yildiz, B. Assessment of Policy Alternatives for Mitigation of Barriers to EV Adoption. Ph.D. Thesis, Portland State University, Portland, OR, USA, 2018.
24. Nakicenovic, N. World Energy Outlook 2007: China and India Insights. IEA/OECD. 2007. Available online: https://www.oecd-ilibrary.org/energy/world-energy-outlook-2007_weo-2007-en (accessed on 24 March 2021).
25. Hassan, M.N.A.; Jaramillo, P.; Griffin, W.M. Life cycle GHG emissions from Malaysian oil palm bioenergy development: The impact on transportation sector's energy security. *Energy Policy* **2011**, *39*, 2615–2625. [CrossRef]
26. König, A.; Nicoletti, L.; Schröder, D.; Wolff, S.; Waclaw, A.; Lienkamp, M. An Overview of Parameter and Cost for Battery Electric Vehicles. *World Electr. Veh. J.* **2021**, *12*, 21. [CrossRef]
27. Haddadian, G.; Khodayar, M.; Shahidehpour, M. Accelerating the Global Adoption of Electric Vehicles: Barriers and Drivers. *Electr. J.* **2015**, *28*, 53–68. [CrossRef]
28. Kaya, Ö.; Tortum, A.; Alemdar, K.D.; Çodur, M.Y. Site selection for EVCS in Istanbul by GIS and multi-criteria decision-making. *Transp. Res. Part D Transp. Environ.* **2020**, *80*, 102271. [CrossRef]
29. Bouguerra, S.; Layeb, S.B. Optimal Locations Determination for an Electric Vehicle Charging Infrastructure in the City of Tunis, Tunisia. In *Recent Advances in Environmental Science from the Euro-Mediterranean and Surrounding Regions*; Springer: Berlin/Heidelberg, Germany, 2018; pp. 979–981.
30. Cui, F.-B.; You, X.-Y.; Shi, H.; Liu, H.-C. Optimal Siting of Electric Vehicle Charging Stations Using Pythagorean Fuzzy VIKOR Approach. *Math. Probl. Eng.* **2018**, *2018*, 9262067. [CrossRef]
31. Kaya, Ö.; Alemdar, K.D.; Çodur, M.Y. A novel two stage approach for electric taxis charging station site selection. *Sustain. Cities Soc.* **2020**, *62*, 102396. [CrossRef]
32. Cherchi, E. A stated choice experiment to measure the effect of informational and normative conformity in the preference for electric vehicles. *Transp. Res. Part A Policy Pract.* **2017**, *100*, 88–104. [CrossRef]
33. Krishna, G. Understanding and identifying barriers to electric vehicle adoption through thematic analysis. *Transp. Res. Interdiscip. Perspect.* **2021**, *10*, 100364. [CrossRef]
34. Sierzchula, W.; Bakker, S.; Maat, K.; Van Wee, B. The influence of financial incentives and other socio-economic factors on electric vehicle adoption. *Energy Policy* **2014**, *68*, 183–194. [CrossRef]
35. Tanaka, M.; Ida, T.; Murakami, K.; Friedman, L. Consumers' willingness to pay for alternative fuel vehicles: A comparative discrete choice analysis between the US and Japan. *Transp. Res. Part A Policy Pract.* **2014**, *70*, 194–209. [CrossRef]
36. Manzetti, S.; Mariasiu, F. Electric vehicle battery technologies: From present state to future systems. *Renew. Sustain. Energy Rev.* **2015**, *51*, 1004–1012. [CrossRef]
37. Quak, H.; Nesterova, N.; van Rooijen, T. Possibilities and Barriers for Using Electric-powered Vehicles in City Logistics Practice. *Transp. Res. Procedia* **2016**, *12*, 157–169. [CrossRef]
38. Mruzek, M.; Gajdáč, I.; Kučera, Ľ.; Gajdošík, T. The Possibilities of Increasing the Electric Vehicle Range. *Procedia Eng.* **2017**, *192*, 621–625. [CrossRef]
39. Hardman, S. Investigating the decision to travel more in a partially automated electric vehicle. *Transp. Res. Part D Transp. Environ.* **2021**, *96*, 102884. [CrossRef]
40. Colmenar-Santos, A.; Muñoz-Gómez, A.-M.; Rosales-Asensio, E.; López-Rey, Á. Electric vehicle charging strategy to support renewable energy sources in Europe 2050 low-carbon scenario. *Energy* **2019**, *183*, 61–74. [CrossRef]
41. Li, Y.; Zhan, C.; de Jong, M.; Lukszo, Z. Business innovation and government regulation for the promotion of electric vehicle use: Lessons from Shenzhen, China. *J. Clean. Prod.* **2016**, *134*, 371–383. [CrossRef]
42. Yan, S. The economic and environmental impacts of tax incentives for battery electric vehicles in Europe. *Energy Policy* **2018**, *123*, 53–63. [CrossRef]
43. Li, J.; Jiao, J.; Tang, Y. An evolutionary analysis on the effect of government policies on electric vehicle diffusion in complex network. *Energy Policy* **2019**, *129*, 1–12. [CrossRef]
44. Åhman, M. Government policy and the development of electric vehicles in Japan. *Energy Policy* **2006**, *34*, 433–443. [CrossRef]

45. Jin, L.; Slowik, P. Literature Review of Electric Vehicle Consumer Awareness and Outreach Activities. International Council on Clean Transportation. Available online: https://www.theicct.org/sites/default/files/publications/Consumer-EV-Awareness_ICCT_Working-Paper_23032017_vF.pdf (accessed on 24 March 2017).
46. Pevec, D.; Babic, J.; Carvalho, A.; Ghiassi-Farrokhfal, Y.; Ketter, W.; Podobnik, V. Electric Vehicle Range Anxiety: An Obstacle for the Personal Transportation (R)evolution? In Proceedings of the 2019 4th International Conference on Smart and Sustainable Technologies (SpliTech), Split, Croatia, 18–21 June 2019; pp. 1–8.
47. Neubauer, J.; Wood, E. The impact of range anxiety and home, workplace, and public charging infrastructure on simulated battery electric vehicle lifetime utility. *J. Power Sources* **2014**, *257*, 12–20. [CrossRef]
48. Goel, S.; Sharma, R.; Rathore, A.K. A review on barrier and challenges of electric vehicle in India and vehicle to grid optimisation. *Transp. Eng.* **2021**, *4*, 100057. [CrossRef]
49. Wind, Y.; Saaty, T.L. Marketing applications of the analytic hierarchy process. *Manag. Sci.* **1980**, *26*, 641–658. [CrossRef]
50. Falatoonitoosi, E.; Ahmed, S.; Sorooshian, S. Expanded DEMATEL for Determining Cause and Effect Group in Bidirectional Relations. *Sci. World J.* **2014**, *2014*, 103846. [CrossRef]
51. Lee, H.-S.; Tzeng, G.-H.; Yeih, W.; Wang, Y.-J.; Yang, S.-C. Revised DEMATEL: Resolving the infeasibility of DEMATEL. *Appl. Math. Model.* **2013**, *37*, 6746–6757. [CrossRef]
52. Du, J.; Ouyang, M. Review of Electric Vehicle Technologies Progress and Development Prospect in China. In Proceedings of the 2013 World Electric Vehicle Symposium and Exhibition (EVS27), Barcelona, Spain, 17–20 November 2013; pp. 1086–1093.
53. Lévay, P.Z.; Drossinos, Y.; Thiel, C. The effect of fiscal incentives on market penetration of electric vehicles: A pairwise comparison of total cost of ownership. *Energy Policy* **2017**, *105*, 524–533. [CrossRef]
54. Cheng, L.-M. Electric Vehicle Promotion Policy in Taiwan. *Energy Manag. Sustain. Dev.* **2018**, *69*.

Article

Zero-Waste Watermelon Production through Nontraditional Rind Flour: Multiobjective Optimization of the Fabrication Process

Juan Pablo Capossio [1], María Paula Fabani [2], María Celia Román [3], Xin Zhang [4], Jan Baeyens [5,6,*], Rosa Rodriguez [3] and Germán Mazza [1,*]

[1] Instituto de Investigación y Desarrollo en Ingeniería de Procesos, Biotecnología y Energías Alternativas, PROBIEN (CONICET-UNCo), Neuquén 8300, Argentina
[2] Instituto de Biotecnología, Facultad de Ingeniería, Universidad Nacional de San Juan, San Juan J5400ARL, Argentina
[3] Instituto de Ingeniería Química, Facultad de Ingeniería, Universidad Nacional de San Juan—Grupo Vinculado al PROBIEN (CONICET-UNCo), San Juan J5400ARL, Argentina
[4] Beijing Academy of Food Sciences, Fengtai District, Beijing 100068, China
[5] Process and Environmental Technology Lab, Department of Chemical Engineering, KU Leuven, 2860 Sint-Katelijne-Waver, Belgium
[6] Beijing Advanced Innovation Centre for Soft Matter Science and Engineering, Beijing University of Chemical Technology, Beijing 100029, China
* Correspondence: jan.baeyens@kuleuven.be (J.B.); german.mazza@probien.gob.ar (G.M.)

Abstract: Watermelon is a fruit produced around the world. Unfortunately, about half of it—the rind—is usually discarded as waste. To transform such waste into a useful product like flour, a thermal treatment is needed. The drying temperature for the rind that produces flour with the best characteristics is most important. A multiobjective optimization (MOO) procedure was applied to define the optimum drying temperature for the rind flour fabrication to be used in bakery products. A neural network model of the fabrication process was developed with the drying temperature as input and five process indicators as outputs. The group of process indicators comprised acidity, pH, water-holding capacity (WHC), oil-holding capacity (OHC), and batch time. Those indicators represent conflicting objectives that are to be balanced by the MOO procedure using the weighted distance method. The MOO process showed that the temperature interval from 67.3 °C to 73.1 °C holds the compromise solutions for the conflicting indicators based on the stakeholder's preferences. Optimum indicator were 0.12–0.19 g malic acid/100 g dwb (acidity), 5.7–5.8 (pH), 8.93–9.08 g H_2O/g dwb (WHC), 1.46–1.56 g oil/g dwb (OHC), and 128–139 min (drying time).

Keywords: watermelon rind; by-product; waste valorization; multiobjective optimization; neural network modeling

Citation: Capossio, J.P.; Fabani, M.P.; Román, M.C.; Zhang, X.; Baeyens, J.; Rodriguez, R.; Mazza, G. Zero-Waste Watermelon Production through Nontraditional Rind Flour: Multiobjective Optimization of the Fabrication Process. *Processes* **2022**, *10*, 1984. https://doi.org/10.3390/pr10101984

Academic Editor: Adam Smoliński

Received: 8 September 2022
Accepted: 27 September 2022
Published: 1 October 2022

1. Introduction

Watermelon (*Citrullus lanatus*, family Cucurbitaceae) is a fruit with a high amount of water (91%) and sugar (6%). Half of a watermelon fruit—the pulp—is consumed, whereas the other half, consisting of rind and seeds (about 40–45%), is generally discarded, being considered solid waste. Nevertheless, it can be transformed into a more useful form and used in pharmaceutical, wastewater treatment, and food applications. The main regions producing watermelons in the world are Asia (79.5%), Africa (7.5%), and the Americas (6.9%). China has the largest volume of watermelon production (about 55% of total volume), reaching 60.9 M tons/year [1], being also a major consumer of watermelon. On the other hand, Argentina produces 127,000 tons/year of watermelon mainly cultivated in San Juan Province (Cuyo region).

The watermelon rind contains natural antioxidants, proteins, minerals, and fibers [2], helping the digestive system and producing a feeling of satiety. The raw watermelon rind is consumed in some countries as a vegetable by stir-frying, stewing, pickling [3], and cooking with sugar to make jam [4]. Concerning the use of dry watermelon rind as an ingredient in the formulation of baking products, several authors reported its utilization in bread [5,6], cakes [7], and noodles [8]. Waghmare et al. [9] developed a new bulk-forming laxative from watermelon rind flour and Ho et al. [10] used it as a natural food additive. Liu et al. [11] reported that watermelon rind can be considered an eco-friendly and economic biosorbent for removing Pb from water and wastewater, while Petkowicz et al. [12] and Bhattacharjee et al. [13] cited its potential for use as an absorbent for the removal of heavy metals.

The main problem in processing watermelon rind into powder is its high water content (90–95%). Drying is the most common method used to extend the shelf life of raw fruits and vegetables with high water content, which in itself reduces the need for packaging and storage [14]. However, in the food industry, excessive heating should be avoided because it is detrimental to food quality, provoking potentially irreversible changes in the nutritional values of the products, and underutilizes plant capacity [15]. The watermelon rind must have a final humidity of 15% to be used as flour in several food applications.

According to the authors' bibliographic research, there are some weaknesses in existing works about watermelon rind drying. Oberoi and Sogi [16] studied the drying of watermelon rind in a fluidized-bed dryer (FBD) and cabinet drying chambers and their effect on the lycopene content at temperatures of 40 °C and 60 °C only. Similar investigations were reported by Ho and Dahri [8], while Ho et al. [10] cited the effects on physicochemical and functional properties, only at these two temperatures. In this context, it can be inferred that the drying kinetics at several temperatures (60 °C to 95 °C) of the watermelon rind has not previously been studied. The multiobjective optimization of the drying temperature and main operating variables (like pH, acidity, water-holding capacity, and oil-holding capacity, among others) has not been reported in the literature either. Such understanding of the process would allow obtaining flour with optimal functional characteristics to be employed in the food industry as an ingredient in bakery products.

The use of nontraditional flour in baking is based on the fact that it represents an interesting alternative in the field of healthy eating, since bread is part of the first level of the nutritional pyramid and one of the most consumed foods in many countries [17]. The US Department of Agriculture (USDA) introduced in the set of recommended dietary allowances (RDAs) [18] the "basic seven" to help people manage their food rationing. This list included bread and flour, among others.

Fabani et al. [18] recently reported the development of an artificial neural network (ANN) model used to simulate the convective and solar convective drying process of watermelon rind pomace applied to the fabrication of nontraditional flour. This information could be used to optimize the drying process, which is involved in the quality parameters of flour, due to technofunctional properties depending on it, and these are essential to assess the baking process to obtain a final product with the highest quality. According to Bennion [19], the pH of the flour influences the capacity of gluten to form the spongy network, and a pH lower than 3.4 will cause an alteration due to acetic and butyric microorganisms. To obtain a greater fermentative development and maximum production of CO_2 in the formed bread, it is necessary to have pH values between 5 and 6, the best being between 5.4 and 5.8. Concerning the acidity of the flour, it must not be higher than 0.25% because greater values can modify the physical, chemical, and rheological properties of the dough. Otherwise, the water-holding capacity (WHC) and the oil-holding capacity (OHC) should be as high as possible within the previously established values for the pH and acidity indicators.

In the literature, there are not many studies reporting MOO of drying processes and even fewer referring to watermelon rind. Chauhan et al. [20] applied multicriteria decision-making to the optimization of pineapple drying. This involved the optimization of three indicators: energy consumption, nutrient retention, and drying time. Shivapour et al. [5] reported the optimization of the formulation and processing parameters involved in bread baking by response surface methodology (RSM). Sendín et al. [21] described and applied an efficient and robust multicriteria optimization method for the thermal sterilization of canned food. The authors emphasize that multicriteria optimization was much more realistic than the more common single-objective approach, although the complexity of the problem was increased. An example of this complexity was shown by Goñi and Salvadori [22], who initially developed a first-principle model for beef roasting in a convective oven and then solved it using the finite element method. The optimization process included conflicting objectives, such as temperature, time, and weight loss. This last factor was directly linked to the proteins, lipids, texture, and juiciness of the final product. The mathematical model was developed, simulated, and then validated with experiments.

Another important aspect of the optimization problem is the search algorithm to be applied. Abakarov et al. [23] proposed a modification of the random search algorithm that showed advantages over the classic formulation of the search strategy. The algorithm was also tested with the optimization of the thermal processing of canned food problem. Such problems include both maximization of quality and minimization of lethality.

Several authors have recently applied the multiattribute decision-making strategy based upon the AHP/TOPSYS principles for building-scale heating systems [24], physical adsorption of CO_2 in carbon-capture processes [25,26], and other processes [27,28]. The difference between this partly fuzzy and the present approach is related to the different procedures in applying weighting factors to the selected process/product parameters in a higher/better or lower/better assessment.

Objectives of This Work

The present work's objectives were as follows: (1) to develop a data-based model of the watermelon rind fabrication process based on experimental data (search space sampling), (2) to determine the Pareto-optimal front, which encompasses all the optimal solutions to the multiobjective optimization problem, and (3) to determine, based on literature criteria, the most appropriate temperature operating conditions within the Pareto-optimal front for obtaining the best possible watermelon rind flour.

The results obtained from previous work (drying kinetics at 70, 75, 80, 85, 90, and 95 °C) [17] complemented with new data, and the technofunctional properties of the watermelon rind flour were used to train the ANN model. Figure 1 shows a logic diagram of the present work.

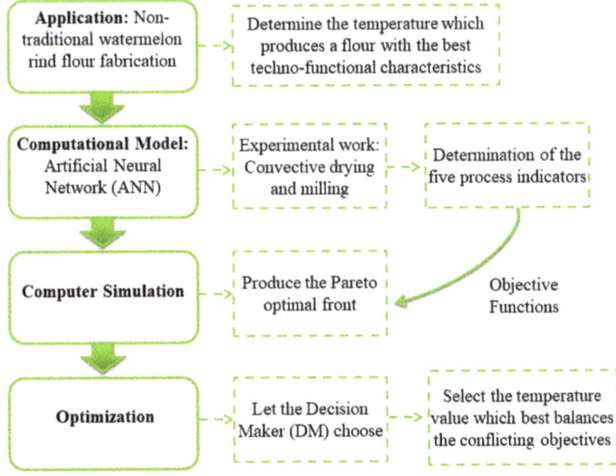

Figure 1. Logic diagram.

2. Materials and Methods

2.1. Feedstock Samples

Fresh watermelon samples (*Citrullus lanatus* cv. "crimson sweet") were collected in January–February 2020 from farms located in the Department of Iglesia (province of San Juan, Argentina) and stored at 20–25 °C in the dark until analysis within 2–4 days after sampling. The watermelons were washed, manually peeled and rinds were cut into cubes (1 cm^3, preserving epicarp and mesocarp).

2.2. Drying of the Samples

The watermelon rind flour samples were obtained by drying fresh cubes of rind in an oven described previously by Baldan et al. [14] at temperatures between 60 and 100 °C, with 5 °C increments for lower temperatures and 2.5 °C increments for higher temperatures. The drying equipment consisted of a stainless-steel cylinder of 50 mm ID and 1000 mm in height, which was heated by an 850 Watt electrical resistance. It is coupled to an analytical balance, a temperature controller, and a computer (Figure 2) in order to register the time, temperature, and mass data. The airflow was set to 100 mL/min. The drying interval of 60–95 °C was chosen since below this temperature range, the drying time is too long. On the other hand, when considering temperatures higher than 95 °C, changes in the color and flavor of the rind were observed. Samples were dried until moisture content was lower than 15%, according to the Codex standard for wheat flour [29] so as to prevent any degradation of the flour.

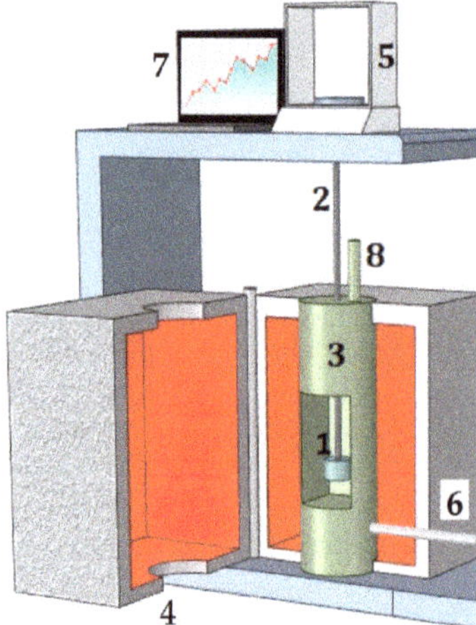

Figure 2. Drying apparatus. 1—crucible; 2—metal wire; 3—internal tube; 4—heater; 5—electronic balance; 6—air inlet; 7—data acquisition system; 8—air outlet. Reprinted from [15], Copyright 2020, Elsevier.

2.3. Preparation and Analysis of Watermelon Rind Flour

Each sample of watermelon rind was milled in a stainless-steel knife mill (TecnoDalvo, model TDMC, Argentina) until reaching a grain size of between 0.10 and 0.20 mm. The flours were stored in plastic bags at 20 °C until further use.

The moisture content of the feedstock watermelon rind and the obtained flours was determined using an infrared moisture analyzer (Radwag PMR50) with a halogen energy source at a temperature of 105 °C [30]. The flour sample (2.0 g) was weighed into a 100 mL beaker and 40 mL of distilled water was added and stirred for 30 min. Then, the mixture was filtered and the pH was measured in the filtrate. The total titratable acidity of the flour sample was determined in the filtrate by titration with sodium hydroxide (NaOH, 0.1 N), using 5 drops of phenolphthalein indicator until the mixture turned pink. The total titratable acidity in the watermelon flour sample is reported as grams malic acid/100 g sample (acid factor used: 0.067) [30].

The technofunctional properties of the watermelon, water-holding capacity (WHC), and oil-holding capacity (OHC) were determined as described by Garau et al. [31] with slight modifications. For the WHC determination, ground samples of watermelon flour ($\pm$0.5 g) were hydrated in excess for 24 h in 15 mL Falcon tubes and then were centrifuged at 2000$\times$ g for 30 min. The supernatant was eliminated and water retention was expressed as grams of water held per gram of watermelon rind flour sample on a dry weight basis (dwb; g water/g sample dwb).

The OHC analyses were performed by mixing 0.5 g of sample with sunflower oil (10 mL) in 15 mL Falcon tubes. After 24 h, tubes were centrifuged at 2000$\times$ g for 30 min, the supernatant was eliminated and oil retention was reported as grams of oil held per gram of watermelon rind flour sample on a dry weight basis (g oil/g sample dwb).

2.4. Neural Network Modeling

The goal of the drying process is to output a raw material that will later produce flour with the required quality characteristics and at the lowest possible cost. To do so, first, a data-based model of the fabrication process was developed based on experimental determinations made on the flours obtained after milling the dried rinds. One of the advantages of having a model is that the optimization process can be carried out offline, i.e., without having to try every possible temperature value in the laboratory. The most conventional form of the data-based model to implement is linear or polynomial regression. However, nowadays access to more complex data-based models is available researchers. Such is the case of neural networks (NNs), which are computational structures capable of modeling high-order polynomials with lower errors and overfit. The neural model does not require large computational resources to run, as it is very fast, even on a normal personal computer. Another advantage of the NN data-based model is that no formal (explicit) set of equations is needed to solve the optimization problem because the model is only a software structure. No validated theoretical, semitheoretical, or empirical model is needed. The training of the NN is performed with the backpropagation algorithm and all the training sets are obtained from the laboratory experiments. Examples of similar models can be found in the literature, most only applied to modeling the drying process, e.g., Topuz [32] and Chokphoemphun and Chokphoemphun [33], and some like Kalathingal et al. [34] incorporating quality characteristics, such as total phenolic content (TPC) and total color difference (TCD).

ANNs are computational structures capable of adapting to fit any type of process without having explicit knowledge of the principles that apply. That is why they are regarded as universal approximators and the multilayer perceptron in particular (Hornik et al., 1989 [35], Pinkus et al., 1999 [36]), that is, if given at least one hidden layer with sufficient neurons in it and a nonlinear activation function, they are capable of mapping any input–output relationship, regardless of the linear or nonlinear nature of it. As mentioned earlier, the ANN model has to have a nonlinear activation function in the hidden layer or layers. The main options available are ReLu, hyperbolic tangent, sigmoid, and others. There is no rule for choosing one. For this application in particular, the sigmoid gave the best results in terms of fitting performance, as in many other works (for instance, see [32]). Another general guideline applied in this work was to keep the ANN model as small as possible in terms of neurons and layers. After a short trial-and-error process, adding more neurons to the hidden layer only led to overfitting the data. There was no further need to use more complex optimization.

The software utilized to design, train weights and biases, test, and implement the ANN was MATLAB® Neural Fitting Tool. More details of the ANN are presented in Table 1. The total number of training sets used was 13, one for each temperature experiment, and one ANN was trained for every process indicator (five models in total). One of the resulting ANN models is presented in Figure 3.

Table 1. Neural model summary.

Parameter	Value
Network type	Multilayer feed-forward (MLFF)
Neuron type	Perceptron
Inputs	1 (temperature)
Outputs	1 per network (acidity, pH, WHC, OHC, time)
Normalization type	Min–Max (-1 to $+1$)
Activation function	Sigmoid (hidden) and linear (output)
Training algorithm	Bayesian regularization backpropagation
Training sets	13
Number of hidden layers	1
Number of neurons per layer	2

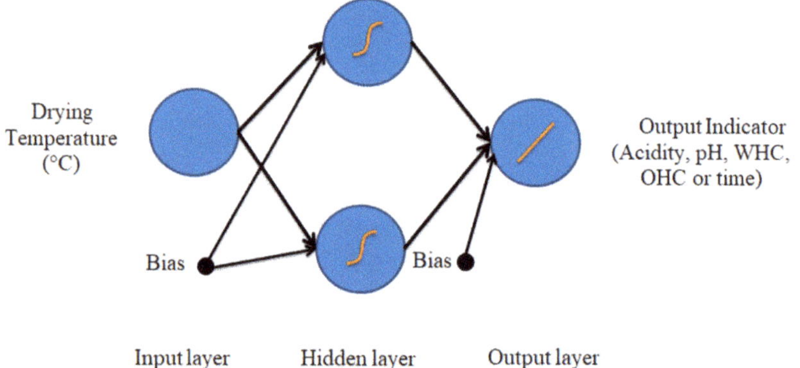

Figure 3. Artificial neural network data-based model.

The evaluation of the ANNs' performance was based on two statistical indicators: the root mean squared error, RMSE (Equation (1)) and the coefficient of determination, R^2 (Equation (2)):

$$\text{RMSE} = \sqrt{\frac{1}{N} \times \sum_{i=1}^{N} \left(y_{pred,i} - y_{exp,i} \right)^2} \tag{1}$$

$$R^2 = 1 - \left[\frac{\sum_{i=1}^{N} \left(y_{pred,i} - y_{exp,i} \right)^2}{\sum_{i=1}^{N} \left(\overline{y}_{pred} - y_{exp,i} \right)^2} \right] \tag{2}$$

R^2 values closer to unity are more desirable, show better fitting performance, while the RMSE should be as low as possible.

2.5. Multiobjective Optimization

Multiobjective optimization (MOO) [37], also known as multicriteria decision-making or multicriteria optimization, is a decision-making process in which a person, called a decision-maker (DM), or group of some responsibility has to make one or several choices to achieve a certain level of process efficiency and final product quality. In modern industry, most likely situations are those of the so-called Pareto efficiency—a concept that comes from the field of economics—where, for instance, a certain indicator "A" can only be improved to the detriment of another indicator "B" of the same product and vice versa. In other words, neither indicator can be optimized (minimized or maximized) concurrently or at the same time, so they are said to be conflicting objectives. This can also be extended to any number of objectives. In these cases, a necessary compromise solution or middle point has to be determined utilizing a methodological approach together with some criteria and available information. Such is the concept of an acceptable solution, where there is no single set of independent variables (solution) that can optimize all objectives concurrently, but there is a group of solutions that can achieve various levels of optimality (compromise). Those solutions are called Pareto-optimal since they are the only acceptable ones for a MOO problem and any other solution is improvable. Additionally, those points are plotted together to form what is called the Pareto-optimal front.

Sometimes, just the minimization or maximization of a certain objective function may not be the DM's desired goal. That is because there are problems where the decision has been made, based on information, to achieve a specific level of precision that does not necessarily require a minimization or maximization of an objective.

Another useful concept is the definition of a utopian vector. Such a vector contains all the individual global optima for each objective function and it represents the place that each one could reach if there were no conflicting objectives. The problem is that it is seldom known where those points lay precisely. The location of those points will probe valuable in the present work. Lastly, the feasible set, S, is defined as the decision variable interval that yields a solution that is physically attainable and satisfies the problem's constraints. So, as explained in Section 2.2, a first approximation of the feasible set is restricted to the 60–95 °C interval.

Now, let us formulate the multiobjective optimization problem mathematically in its general form with Equations (3) and (4), where $f_{i(X)}$ is an individual objective function or performance indicator and X is a real n-dimensional vector of controlled or decision variables. The performance indicators and the controlled variables constitute the system's output and input variables respectively. Figure 4 shows some examples of input and output variables belonging to several different food processing systems. As seen, the system can have multiple inputs and outputs; also, it can have a single input and multiple outputs or any combination. Ideally, finding a decision variable X that concurrently minimizes each objective function and, consequently, minimizes the global optimization function $\Phi_{(X)}$, would be preferable.

$$\Phi_{(X)} = \left\langle f_{1(X)},\ f_{2(X)},\ f_{3(X)}, \ldots, f_{l(X)} \right\rangle \underset{X \in\, S\, \in\, \mathbb{R}^n}{\rightarrow \min} \tag{3}$$

$$X = (x_1,\ x_2,\ x_3, \ldots, x_n) \tag{4}$$

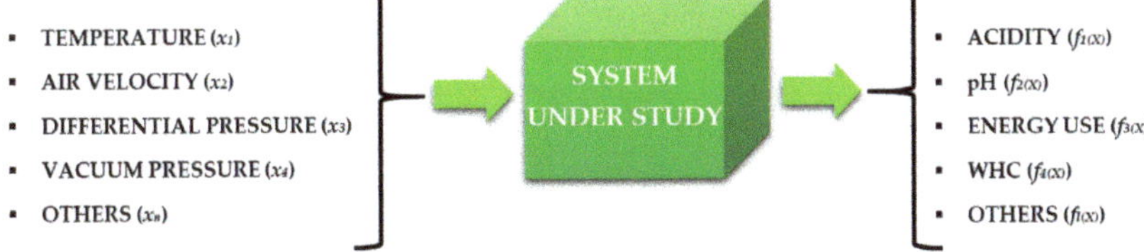

Figure 4. Input and output variables correspond with several different food-processing systems.

However, in overwhelmingly most realistic situations, it is impossible to optimize all individual objective functions at the same time. Here, the concept of an acceptable solution is applied. Consequently, to calculate the Pareto-optimal front, $P_{(f)}$, an aggregating function, which transforms the multiobjective optimization problem into a single global optimization problem, is generally applied. For example, the linear weighted sum aggregating function [38], as seen in Equations (5) and (6), where w_i is the weight corresponding to the i-th particular objective function, can be solved using a search algorithm. Here, it is important to point out that those weights are ultimately determined by the DM.

$$\Phi_{(X)} = \sum_{i=1}^{l} w_i \cdot f_{i(X)} \underset{X \epsilon\, S\, \epsilon\, \mathbb{R}^n}{\rightarrow \min} \tag{5}$$

$$\sum_{i=1}^{l} w_i = 1 \ \wedge \ w_i \geq 0 \tag{6}$$

If among the objectives there is a mixture of minimizations and maximizations, then a homogenization is required according to $max\ \{f_{i(X)}\} = -\min\ \{-f_{i(X)}\}$. The main drawback of the weighted sum aggregating function is that it is impossible to obtain points on nonconvex (concave) segments of the front. It is possible to obtain an incomplete front if the solution space is not entirely convex. Another drawback of the method is that changing the weight values might not yield an even distribution of the Pareto-optimal points, making it difficult for the DM to have an accurate representation of the optimal set [39]. The consequence could be that the weight values might end up being meaningless to the DM, who will not comprehend why a certain change in the weights produces unexpected effects in the outcome. A better approach would be to use the weighted distance method [40], also called a method of the weighted metrics (L_p metrics) in which the idea is to minimize the distance to each objective's utopia point, as seen in Equations (7) and (8), provided those points are known. To obtain the utopia point, Z_i^*, optimization of each objective function is necessary by applying the expression in Equation (9) and/or Equation (10), depending on if the aim is to minimize or maximize.

$$\Phi_{(X)} = \left(\sum_{i=1}^{l} w_i \cdot \left| f_{i(X)} - Z_i^* \right|^{p}\right)^{1/p} \underset{X \epsilon\, S\, \epsilon\, \mathbb{R}^n}{\rightarrow \min} \tag{7}$$

$$\sum_{i=1}^{l} w_i = 1 \ \wedge \ w_i \geq 0 \tag{8}$$

$$Z_i^* = \underset{X \epsilon\, S\, \epsilon\, \mathbb{R}^n}{\min}\, f_{i(X)} = f_{i(X^*)} \tag{9}$$

$$Z_i^* = \underset{X \epsilon\, S\, \epsilon\, \mathbb{R}^n}{max}\, f_{i(X)} = f_{i(X^*)} \tag{10}$$

The exponent p can adopt different values ($1 \leq p < \infty$). For instance, if $p = 1$ the sum of the weighted distances to the utopia points is minimized, but as the p exponent is increased, the minimization of the largest errors becomes ever more important. Its relevance lies in the fact that a small value of the exponent p may not yield all the Pareto-optimal solutions if the front is not convex and, also, there may not be a p-value that provides all optimal solutions without convexity. In other words, there may be no weight value combination that will produce some of the Pareto-optimal points. One approach for selecting the p exponent is to use larger values when there are few objective functions and lower values when there are many objective functions.

In the present work, the only controlled variable of interest is the temperature due to the characteristics of the drying equipment and due to its most significant impact on the final product. At the other end, the objective functions can be divided into two groups: the flour's quality characteristics and the process's efficiency indicators. The first group is comprised of four indicators: acidity, pH, WHC, and OHC. Finally, the second group is formed by batch time. Five objectives in total, as shown in Equation (11), along with five weights.

$$\Phi_{(T)} = Acidity_{(T)},\ pH_{(T)},\ WHC_{(T)}, OHC_{(T)}, Time_{(T)} \underset{T \epsilon S}{\rightarrow \min} \tag{11}$$

Search Method for Obtaining the Pareto-Optimal Set

To find the Pareto-optimal set, different combinations of the weights are tried to identify all the temperatures belonging to the front given a certain aggregating function. The most direct and error-free method for finding all temperatures belonging to the Pareto-optimal front would be to try every possible weight value. In numerical terms that is called uninformed search, brute-force search, or manual search, where "no stone is left unturned" [39]. However, that is, of course, very costly from a computational standpoint because the search space is vast. A more efficient way of doing this is utilizing a search heuristic like a random search in its adaptive form or not; also, a grid search could be used.

Figure 5 shows the algorithm implemented to obtain the Pareto-optimal set. It is based on the pure random search (PRS) algorithm, which is actually a family of numerical optimization algorithms that are independent of the gradient of the problem to be maximized or minimized. Hence, these search methods can be applied to undifferentiable and discontinuous problems. It is based on generating the weights (hence, it is also called the "generative method") from a given probability distribution, for example, a normal distribution, but, in this case, a uniform distribution was used. In this way, the weight's space is randomly sampled and the Pareto-optimal set is obtained if sufficient samples are taken. Undersampling the search space could be detected by small gaps in the Pareto set. Lastly, to make sure the Pareto-optimal front is complete, the algorithm has to be run a few times and, if all the resulting Pareto sets are equal, most likely there are no new Pareto-optimal points to be found and the front is complete. In every new run, different weights are used but the resulting front has to be the same.

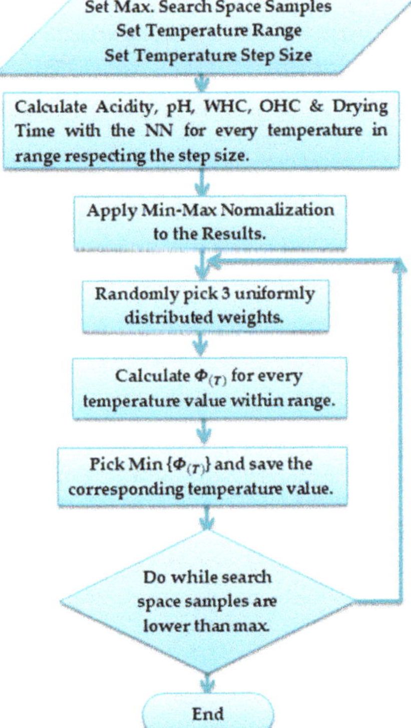

Figure 5. Flowchart describing the search algorithm used to determine the Pareto-optimal set. All code inside the while loop may be executed in different threads since every iteration is independent.

In every iteration, Min $\{\Phi_{(T)}\}$ is calculated with a uniformed search, that is, in other words, every temperature value within the feasible set is tried and the global minimum is obtained for each set of randomly picked weights. This less efficient search method was used because the temperature search space is much smaller than the weight search space. For more complex problems (more inputs, larger domains, between others), a genetic algorithm or an adaptive random search would be more suitable methods [41].

It is also important to notice that all the above calculations were performed on normalized values (min–max normalization) because the five process indicators have different units and very dissimilar orders of magnitude that can affect the overall MOO results.

Finally, since in the PRS algorithm every candidate solution (i.e., set of weights) is independent, the algorithm can be executed in parallel with a multicore/thread computer, thus greatly speeding up the search process. In the present work's case, a mid-range laptop computer with four cores was utilized to perform the calculations. The computer has 4 physical cores, each split in 2 if required that share each core's resources. They are commonly referred to as computational threads.

2.6. Problem-Solving Strategy

Firstly, it can be observed that there are mainly two groups of indicators that lean towards opposite sides of the temperature spectrum. On one hand, the flour's quality indicators (acidity, pH, WHC, and OHC) are optimized with lower temperatures. On the other hand, the process batch time is minimized with higher temperatures. Lastly, a subgroup of the former is optimized with temperature values that slightly tilt toward the center of the spectrum. Consequently, there is no doubt of the difficulty to deal with or manage the number of indicators to optimize. In this context, being sure to find the complete Pareto-optimal set is a very difficult task to perform with a search algorithm because of the number of different weight combinations to try and graphically too, due to the high dimensionality. Accordingly, in the next sections, we will apply the weighted distance method to find the Pareto-optimal front, where the main idea will be to reduce the weight and temperature search space and, at the same time, reduce the multidimensional objective function space identifying redundant objectives.

2.6.1. Redundant Objectives

As mentioned above, three most important points can be identified in the temperature range: (1) at the low end the acidity is minimized; (2) at the high end the batch time is minimized, and (3) at the middle-low part WHC and OHC are maximized. A redundant objective function [42] is defined in terms of conflict between objectives, then an objective function is redundant if it is not in conflict with any other objective function. It is important to identify redundant objective functions to reduce the number of weights the DM has to pick. In other words, this means a reduction of the search space.

An objective function is redundant if it is not in conflict with any other objective function, so it can be eliminated ($w_i = 0$). On the other hand, an objective function is relatively redundant if it is only in conflict with some of the remaining objective functions. For instance, as described in Gal and Leberling [42], if two objective functions, f_x and f_y, are nonconflicting, one of them can be removed without changing the Pareto-optimal front, $P_{(f)}$, but not both at the same time (Equations (12) and (13)).

$$P_{(f)} = P_{(f-f_x)} \vee P_{(f)} = P_{(f-f_y)} \tag{12}$$

$$P_{(f)} \neq P_{(f-f_x-f_y)} \tag{13}$$

Consequently, a decision was made to eliminate some objective functions that are relatively and absolutely redundant and leave only three conflicting objective functions, $f_{1(T)}, f_{2(T)}$, and $f_{3(T)}$. Because visualization of more than two or three objective functions is not a trivial matter, this also helps with the visualization of the Pareto front, since it lowers the order of the problem and allows plotting a tridimensional trajectory.

2.6.2. Final Decision Determination

There are, mainly, two methods or approaches to determining the final decision: a priori and a posteriori [43]. In the a priori method, the decision-maker (DM) specifies hopes, preferences, and/or opinions beforehand, although he or she does not know how realistic those expectations are. On the other hand, a posteriori methods try to find the Pareto-optimal front, present it to the DM, and let him or her decide among the optimal points. In the case presented here, there is some a priori information that can reduce the search space of the variables pH and acidity (Section 1). According to the neural network's predictions, those two conditions reduce the search space to approximately 60–75.5 °C. Temperature cannot be increased any further because that would mean having an acidity value greater than 0.25 g malic acid/100 g. The pH values within that same temperature interval are among the desired best ones of 5.4 to 5.8. This makes the pH objective function absolutely redundant and, in consequence, it can be removed. Alternatively, those intervals can be disregarded and the entire Pareto-optimal front can be determined within the whole width of the feasible set (60–95 °C) in which case the DM will have more options. In Equation (14), the redundancy-free aggregating function is presented.

$$\Phi_{(T)} = \left(w_1 . \left| Acidity_{(T)} - Acidity^* \right|^p + w_2 . \left| WHC_{(T)} - WHC^* \right|^p + w_3 . \left| Time_{(T)} - Time^* \right|^p \right)^{\frac{1}{p}} \tag{14}$$

In Table 2, a summary of the selection process applied to keep or remove a certain objective function is presented. It is worth reiterating that this elimination process is key for solving the MOO problem in a reasonable time and computing effort.

Table 2. Summary of the selection process.

Objective Function $f_{i(T)}$	Redundancy Status	Possible Action	Final Action
Acidity (g malic acid/100 g dwb)	Not redundant	Keep	Keep
pH	Absolute redundant. Optimum values coincide with the 60–75.5 °C range	Remove	Remove
WHC (g water/g dwb)	Relative redundant	Keep only one of the group	Keep
OHC (g oil/g dwb)	Relative redundant	Keep only one of the group	Remove
Drying time (min)	Not redundant	Keep	Keep

3. Results and Discussion

3.1. Watermelon Rind Flour

The Codex standard for wheat flour [29] defines moisture content as a quality factor and its value has to be lower than 15.5%. The flours for baking mixes must have a moisture content of less than 10%, as this value improves the quality and shelf life of the mix. Consequently, the watermelon rinds were dried to a moisture content of less than 10%. Al-Sayed et al. [6] reported a moisture content of 10.61% for watermelon rind flour.

Figure 6 shows the acidity and pH values of the raw watermelon rinds and the flour obtained by drying them at the analyzed temperatures. The acidity of the rind increased after drying, showing the highest value at 85 °C. Then, a slight decrease in acidity was observed, which then increased again with the temperature rise. The pH value decreases as the drying temperature of the watermelon rind increases, showing behavior that correlates with acidity values.

The technofunctional properties of flours obtained at different temperatures evaluated are presented in Figure 7. The results showed that the WHC and the OHC were higher for the flour obtained at 70 and 75 °C. The values of WHC and OAC were higher than those reported by Al-Sayed et al. [6] for watermelon rind dried at 50 °C, being 7.13 (g water/g) and 1.65 (g oil/g), respectively. These results indicate that between 70 and 75 °C watermelon rind flour holds more water and oil compared to rinds dried at 50 °C.

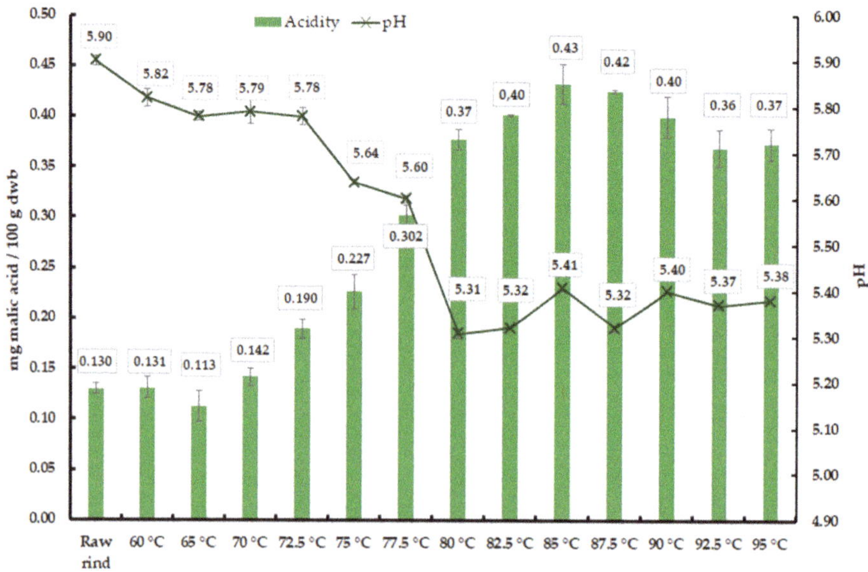

Figure 6. Acidity (mg malic acid/100 g dwb) and pH values of the raw watermelon rind and flours obtained by convective drying at different temperatures.

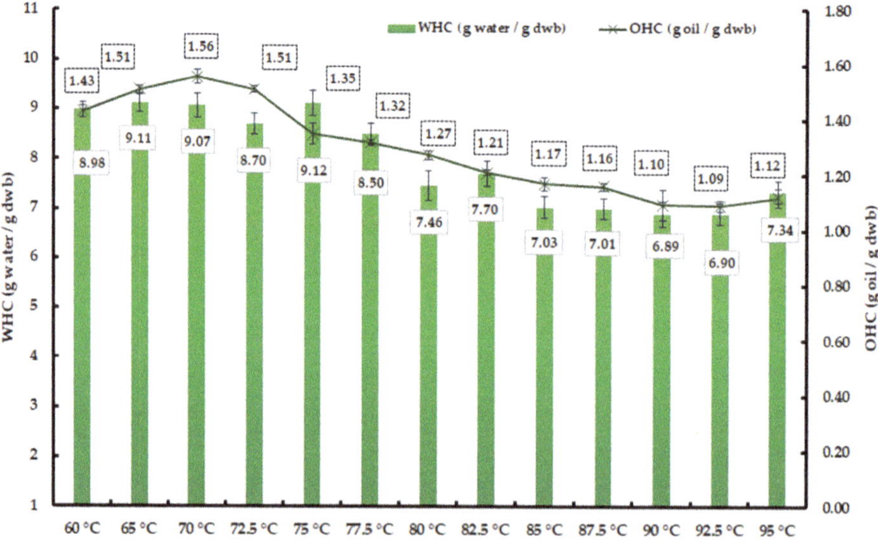

Figure 7. WHC and OHC of flours obtained by convective drying at different temperatures.

3.2. Neural Network Data-Based Models Training and Regression Results

The neural network's training results are presented in Table 3. The average R^2 was 0.988, with the lowest score of 0.973 for WHC and the highest of 0.998 for acidity. The fitting performance of the ANN was better than the performance of a comparable fourth-order polynomial (0.988 to 0.966). The average RMSE was 0.772 with a lowest score of 0.009 (best) for acidity and the highest of 3.592 (worst) for batch time.

Table 3. Individual and average regression coefficients for the training and testing lots.

Indicator	R^2	RMSE
Acidity	0.998	0.009
pH	0.991	0.025
WHC	0.973	0.210
OHC	0.989	0.024
Time	0.988	3.592
Average	0.988	0.772

In Figure 8, the ANN predictions are plotted together with each of the experimental results. The ANN models allowed us to perform simple minimizations and maximizations to find the individual optima of each of the objective functions and to use them to obtain the Pareto-optimal front with the weighted distance method. In Table 4, the resulting utopia vector is shown. The necessary temperature values for each indicator to reach its utopian objective are also presented.

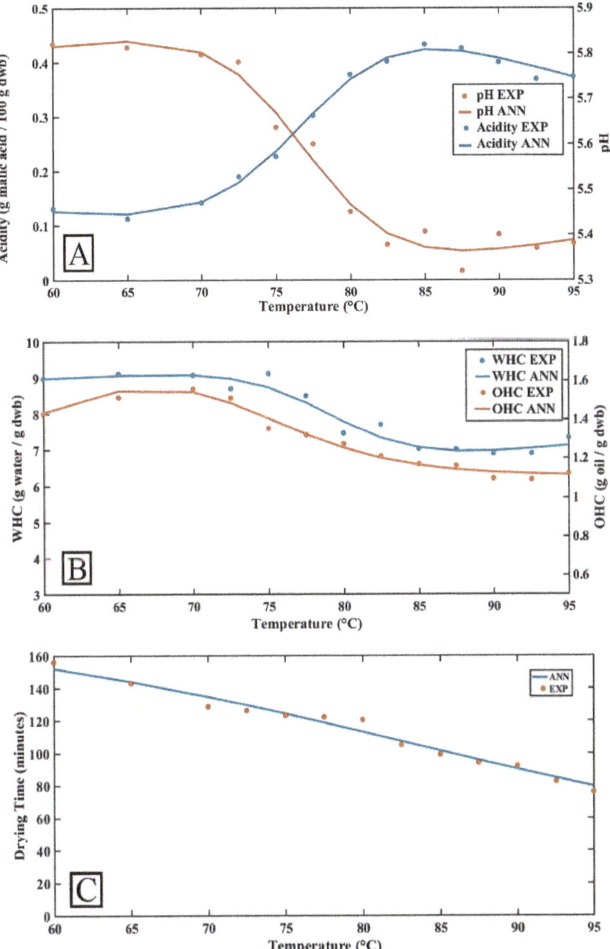

Figure 8. Predictions (ANN) and experimental results (EXP). (**A**) Acidity and pH, (**B**) WHC and OHC, (**C**) Batch time.

Table 4. Utopia vector.

Indicator $f_{i(T)}$	Goal	f_i^*	T (°C)
Acidity (g malic acid/100 g dwb)	Min	0.12	64.5
pH	Margin	5.4–5.8	60–75.5
WHC (g water/g dwb)	Max	9.09	68.20
OHC (g oil/g dwb)	Max	1.56	67.40
Drying time (min)	Min	123.04	75.5

Analyzing these results illustrates the conflicting goals because lowering the oven temperature will improve WHC, OHC—factors that improve dough formation—also acidity will improve but, at the same time, the pH value climbs and batch time could be too long for economic feasibility. On the other hand, the temperature increase will lower the pH value and shorten the batch time. Therefore, it is of the utmost importance to find the necessary trade-offs and, in turn, this is accomplished only by providing subjective input into the process.

3.3. Optimization Results

In this section, first, the Pareto-optimal front corresponding to the selected three objective functions will be presented and, second, the data-driven model plus the random search algorithm and the defined redundancy-free aggregating function will be put together to obtain a complete and well-distributed Pareto-optimal front. In this manner, the DM will only have to choose the relevant weights according to his or her own desired outcomes. Also, a couple of problems have to be avoided for the weights to be truly meaningful: first, very similar weights cannot generate diametrically different Pareto-optimal points and, alternatively, utterly different weights cannot produce very similar Pareto-optimal points. These two goals (front completeness and weight coherence) will be tested by comparing the weight distribution versus optimal temperature of the weighted distance method and the weighted sum method.

Figure 9 shows in a four-dimensional plot the Pareto-optimal front in the 64.5 to 75.5 °C interval. The three objective functions are plotted spatially and the temperature corresponding to each point is represented by a color scale. As it can be observed the front is convex. Then, a priori, the weighted sum aggregating function could be a good choice because it produces all points belonging to the front. The 60 to 64.5 °C interval is suboptimal because all objective functions will worsen if the temperature is lowered from 64.5 °C.

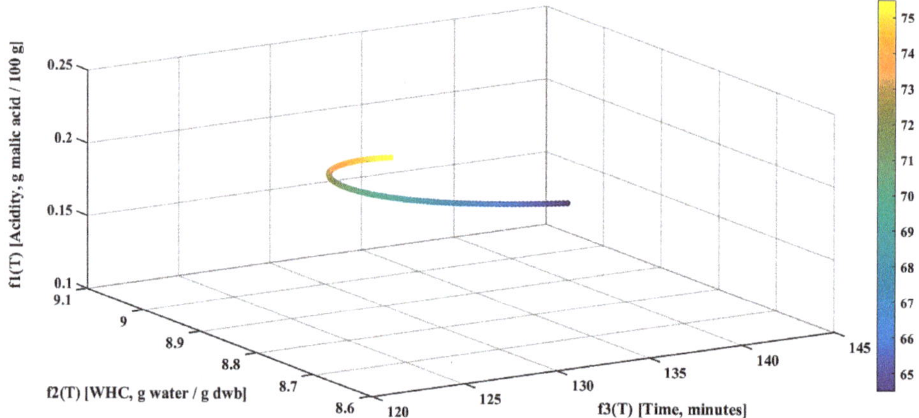

Figure 9. A four-dimensional plot of the Pareto-optimal front in the 64.5 to 75.5 °C interval.

Figure 10 shows the X–Z plane for acidity and time, respectively, and temperature in the color scale. There, the evolution of every one of the indicators can be observed. All trade-offs are located along with the 64.5 to 75.5 °C interval. Increasing temperature will be to the detriment of acidity and OHC, but at the same time, batch time will improve and WHC will also improve, but only until the 68.2 °C value is reached. Beyond that point, only batch time will improve. Then, lowering the temperature will produce alternative effects.

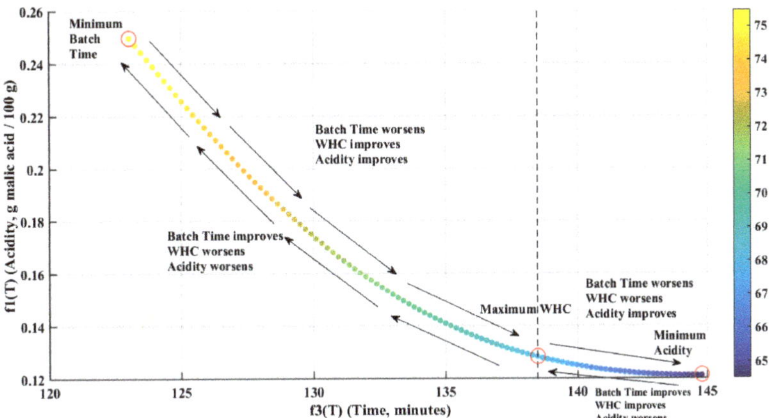

Figure 10. X–Z plane for time and acidity, respectively, of the Pareto-optimal front.

Figure 11 shows the X–Y plane for time and WHC, respectively, and temperature in the color scale. There, the three utopia points can be seen encircled in red and the point of maximum WHC can be better appreciated.

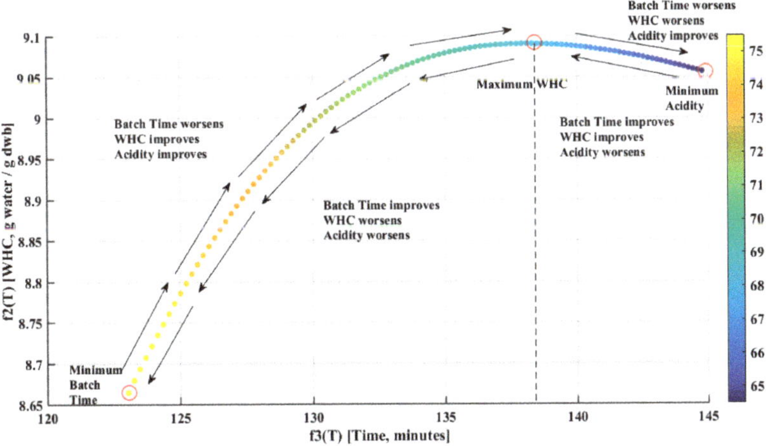

Figure 11. X–Y plane for time and WHC, respectively, of the Pareto-optimal front.

It is worth mentioning that even though the pure adaptive search algorithm has, under certain conditions, a linear expected number of iterations, while the PRS algorithm has an exponential expected number of iterations [44], the applied search strategy converged repeatedly with 10,000 iterations in about 25 min per run. This is mainly because the PRS algorithm is highly parallel (i.e., its iterations are independent) while the adaptive form is not and it is also due to the straightforward approach of the ANN model. Of course, one cannot always make use of these advantages and the simplest solution may not be the best one. Additionally, the before-mentioned running time also includes the uninform search method used to obtain Min $\{\Phi_{(T)}\}$ in every iteration. In more complex problems, this task can be given to a genetic algorithm (evolutionary multiobjective optimization, EMO). However, in this case, such a thing is unfeasible because we deal with a single input model, a condition that limits the crossover operator.

The numerical results of the random search process using the weighted distance aggregating function yielded a complete front for $p = 1$. As expected, the same outcome is obtained with the weighted sum aggregating function, but these results only mean that there exists at least one weight vector that produces every temperature within the optimal front. Now, it remains to be seen if the weights can make sense to the DM. Table 5 shows a simulated sweep along the weight's space where the relative importance of each objective function is varied from 0.8 to 0.1. The resulting optimal temperatures for the weighted distance aggregating function range from 65.9 °C ($w_1 = 0.8$) to 75.5 °C ($w_3 = 0.8$) passing through 68.4 °C ($w_2 = 0.8$) in a very linear fashion; while with the weighted sum aggregating function the optimal temperatures are mostly 75.5 °C plus suboptimal temperatures also appear. No other temperatures can be obtained with these weight combinations. It is clear now that the DM cannot make a good decision based on the weighted sum aggregating function.

Table 5. Simulated sweep along the weight's space.

w_1	w_2	w_3	Optimal Temperature (°C)	
			Weighted Distance Method	**Weighted Sum Method**
0.8	0.1	0.1	65.9	65
0.7	0.2	0.1	66.4	64.1
0.6	0.3	0.1	66.8	61.7
0.5	0.4	0.1	67.1	60
0.4	0.5	0.1	67.5	75.5
0.3	0.6	0.1	67.8	75.5
0.2	0.7	0.1	68.1	75.5
0.1	**0.8**	0.1	68.4	75.5
0.1	0.7	0.2	68.8	75.5
0.1	0.6	0.3	69.3	75.5
0.1	0.5	0.4	69.9	75.5
0.1	0.4	0.5	70.6	75.5
0.1	0.3	0.6	71.6	75.5
0.1	0.2	0.7	73	75.5
0.1	0.1	**0.8**	75.5	75.5

This can be explained using a graph of the plane $w_1 + w_2 + w_3 = 1$. Figure 12a shows the weight plane for the weighted distance aggregating function with 10,000 different weight combinations and a color scale for each point that represents the optimal temperature for that particular weight combination. As it can be observed the optimal temperatures are well distributed along the plane and there is no marked preponderance of any optimal temperature(s). Contrarily, Figure 12b shows the weight plane for the weighted sum aggregating function with the same number of points. It is remarkable to find in it a big preponderance of the 75.5 °C optimal temperature along the outstanding majority of the weight combinations. The rest of the optimal temperatures are distributed along a small portion of the weight plane. So, in large portions of the weight space very similar weight combinations can yield completely different optimal temperatures. A similar conclusion can be drawn when observing the histograms of the optimal temperature distribution of both methods (Figure 12c,d). In the weighted sum method, weight vectors other than those that produce the 75.5 °C optimal temperature are very infrequent. On the other hand, the weighted distance method distributes the weight vectors along with a wider range of optimal temperatures, but weights that produce optimal temperatures from 66 to 72 °C (approximately) are more frequent. Nonetheless, it is enough to present meaningful scenarios to the DM.

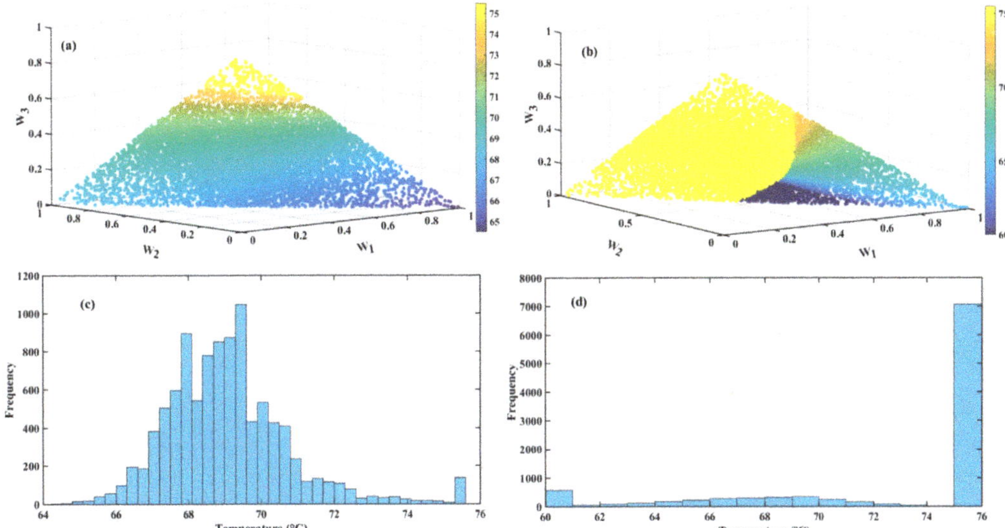

Figure 12. (**a**) Weight plane for the weighted distance aggregating function with 10,000 points. (**b**) Weight plane for the weighted sum aggregating function with the same number of points (**c**) Histogram of the optimal temperature distribution for the weighted distance aggregating function (**d**) Histogram of the optimal temperature distribution for the weighted sum aggregating function.

3.4. Possible Scenarios for Making a Final Decision

Although the present work is not intended as a group decision analysis, in this section some examples or scenarios are presented where two stakeholders [45] involved in the decision-making process apply different preference information (weights) to rate the importance of the same criteria (objective functions). In any industry or market, several stakeholders can be identified, but the paradigmatic example is the producer/consumer relationship, where the producer will surely favor lowering the batch time to maximize yield and, on the other hand, the consumer (internal or external) will surely favor maximizing the product quality factors. A final decision could be a middle ground between those two interested parties.

The weight's relative value directly reflects the relative importance of each particular objective function within the optimization problem. Accordingly, in the first three scenarios, the consumer's standpoint will be represented by setting the batch time indicator to have the lowest relative weight, specifically: $w_3 = 0.1$. Then, for the first scenario, the DM will favor the indicators optimized at 64.5 °C ($w_1 = 0.7$, $w_2 = 0.2$, and $w_3 = 0.1$); secondly, the DM will favor the indicators optimized around 68.2 °C ($w_1 = 0.2$, $w_2 = 0.7$, and $w_3 = 0.1$); thirdly, the DM will consider that the indicators optimized at 64.5 °C and around 68.2 °C are equally important ($w_1 = 0.45$, $w_2 = 0.45$ and $w_3 = 0.1$).

Lastly, in the fourth scenario, the producer could consider that given the fact that both acidity and pH indicators are within acceptable values in the 64.5 to 75.5 °C interval, the most important thing is to minimize batch time without setting a marked preference on the rest of the indicators ($w_1 = 0.2$, $w_2 = 0.1$, and $w_3 = 0.7$).

Table 6 presents all results for every weight combination scenario set by the DMs, including the decision variable T and the ANN predictions for each indicator corresponding to that temperature value. Those outcomes show that the most preferred alternative for the producer is to apply a 73.1 °C drying temperature. On the other hand, the consumer prefers lower drying temperatures, which range from 66.4 to 68.1 °C, depending on the relative weight.

Table 6. Possible scenario outcome summary.

Scenario: DM's Weight Choices ($p = 1$)				Pareto-Optimal Point Decision Variable T (°C)	Neural Network's Predictions				
n°	w_1	w_2	w_3		Acidity$_{(T)}$ (g Malic Acid/ 100 g dwb)	pH$_{(T)}$ (Dimensionless)	WHC$_{(T)}$ (g of H_2O/ g dwb)	OHC$_{(T)}$ (g of oil/g dwb)	Drying Time$_{(T)}$ (min)
1	0.7	0.2	0.1	66.4	0.123	5.825	9.080	1.561	141.449
2	0.2	0.7	0.1	68.1	0.129	5.819	9.090	1.562	138.294
3	0.45	0.45	0.1	67.3	0.125	5.823	9.087	1.564	139.797
4	0.2	0.1	0.7	73.1	0.190	5.735	8.935	1.466	128.226

Finally, if we assume that the third scenario best represents the consumer's standpoint, then any temperature between 67.3 and 73.1 °C would be an acceptable compromise between the two parties since any temperature within that range is an efficient solution.

4. Conclusions

In the present work, the MOO of the watermelon rind nontraditional flour fabrication is presented. First, a data-based model of the watermelon rind drying and milling processes based on experimental data was developed. Such a model presented very good performance indicators with an average regression coefficient, R^2, of 0.988.

Then, the data-based model (ANN) and a pure random search algorithm were used to find the Pareto-optimal front. Before that, the elimination of redundant objectives was necessary to reduce the complexity of the problem. The remaining objective functions were only three: acidity, WHC, and drying time.

Finally, four scenarios for making a final decision were analyzed based on the producer–consumer relationship. In conclusion, it was found that any temperature between 67.3 and 73.1 °C would be an acceptable compromise solution between the two decision-makers because any temperature within that range is an efficient one. Optimum indicators' values ranged from 0.12–0.19 g malic acid/100 g dwb for acidity; 5.7–5.8 for pH; 8.93–9.08 g of H_2O/g dwb for WHC; 1.46–1.56 g of oil/g dwb for OHC; and 128–139 min for drying time.

Future work includes the optimization of processes involving other wastes suitable for nontraditional flour fabrication, such as local microbrewery's spent grains and pomegranate shells. A comparative study is also considered for a follow-up paper.

Author Contributions: Conceptualization, J.P.C., M.P.F., M.C.R., X.Z. and R.R.; methodology, M.P.F. and X.Z.; software, J.P.C.; formal analysis, J.P.C., M.P.F., M.C.R., R.R., G.M. and J.B.; investigation, J.P.C., M.P.F., M.C.R., R.R. and X.Z.; resources, R.R. and G.M.; writing—original draft preparation, J.P.C. and M.P.F.; writing—review and editing, G.M., R.R. and J.B.; visualization, G.M.; supervision, G.M., R.R. and J.B.; project administration, G.M., R.R. and J.B.; funding acquisition, G.M., R.R. and J.B. All authors have read and agreed to the published version of the manuscript.

Funding: This research was funded by the National Scientific and Technical Research Council, CONICET, Argentina [grant number PUE PROBIEN 22920150100067 and PIP 2021–2023—No. 11220200100950CO]; Universidad Nacional de San Juan, Argentina [grant number PDTS Res. 1054/18]; Universidad Nacional del Comahue, Neuquén, Argentina [grant number PIN 2022-04/I260]; FONCYT- ANPCyT (National Agency for Scientific and Technological Promotion, Argentina) [grant numbers PICT 2017-1390 and PICT 2019-01810]; San Juan Province, Argentina [grant number IDEA -Res. N° 0272-SECITI-2019].

Institutional Review Board Statement: Not applicable.

Informed Consent Statement: Not applicable.

Data Availability Statement: Not applicable.

Acknowledgments: The authors would like to express their gratitude to FECOAGRO SRL and Cooperativa VALLES IGLESIANOS for providing the watermelon samples. F. Ronchetti is acknowledged for his scientific advise.

Conflicts of Interest: The authors declare no conflict of interest.

References

1. FAO. Production Quantities of Watermelons by Country. Last Update 22 December 2020. Food and Agriculture Organization of the United Nations. FAOSTAT (Food and Agriculture). Available online: http://www.fao.org/faostat/en/#data/QC/visualize (accessed on 22 January 2021).
2. Fila, W.; Itam, E.; Johnson, J.; Odey, M.; Effiong, E.; Dasofunjo, K.; Ambo, E. Comparative proximate compositions of watermelon Citrullus lanatus, Squash Curcubita pepo'l and Rambutan Nephelium lappaceum. *Int. J. Sci. Tech.* **2013**, *2*, 81–88.
3. Souad, A.M.; Jamal, P.; Olorunnisola, K.S. Effective jam preparations from watermelon waste. *Int. Food Res. J.* **2012**, *19*, 1545–1549.
4. Badr, S. Quality and antioxidant properties of pan bread enriched with watermelon rind powder. *Curr. Sci. Int.* **2015**, *4*, 117–126.
5. Shivapour, M.; Yousefi, S.; Ardabili, S.M.; Weisany, W. Optimization and quality attributes of novel toast breads developed based on the antistaling watermelon rind powder. *J. Agric. Food Res.* **2020**, *2*, 100073. [CrossRef]
6. Al-Sayed, H.; Ahmed, A. Utilization of watermelon rinds and sharlyn melon peels as a natural source of dietary fiber and antioxidants in cake. *Ann. Agric. Sci.* **2013**, *58*, 83–95. [CrossRef]
7. Hoque, M.; Iqbal, A. Drying of watermelon rind and development of cakes from rind powder. *Int. J. Novel Res. Life Sci.* **2015**, *2*, 14–21.
8. Ho, L.; Dahri, N. Effect of watermelon rind powder on physicochemical, textural, and sensory properties of wet yellow noodles. *CyTA—J. Food* **2016**, *14*, 465–472. [CrossRef]
9. Waghmare, A.G.; Patil, S.U.; Arya, S.S. Novel bulk forming laxative from watermelon rind. *Int. J. Pharm. Sci. Res.* **2015**, *6*, 3877. [CrossRef]
10. Ho, L.; Suhaimi, M.; Ismail, I.; Mustafa, K. Effect of different drying conditions on proximate compositions of red- and yellow-fleshed watermelon rind powders. *J. Agric.* **2016**, *7*, 1–12.
11. Liu, C.; Ngo, H.H.; Guo, W. Watermelon Rind: Agro-waste or Superior Biosorbent? *Appl. Biochem. Biotechnol.* **2012**, *167*, 1699–1715. [CrossRef]
12. Petkowicz, C.L.O.; Vriesmann, L.C.; Williams, P.A. Pectins from food waste: Extraction, characterization and properties of watermelon rind pectin. *Food Hydrocoll.* **2017**, *65*, 57–67. [CrossRef]
13. Bhattacharjee, C.; Dutta, S.; Saxena, V. A review on biosorptive removal of dyes and heavy metals from wastewater using watermelon rind as biosorbent. *Environ. Adv.* **2020**, *2*, 100007. [CrossRef]
14. Baldán, Y.; Fernandez, A.; Reyes Urrutia, A.; Fabani, M.P.; Rodriguez, R.; Mazza, G. Non-isothermal drying of bio-wastes: Kinetic analysis and determination of effective moisture diffusivity. *J. Environ. Manag.* **2020**, *262*, 110348. [CrossRef]
15. Simpson, R.; Almonacid, S.; Teixeira, A. Optimization criteria for batch retort battery design and operation in food canning-plants. *J. Food Process Eng.* **2003**, *25*, 515–538. [CrossRef]
16. Oberoi, S.; Sogi, D. Drying kinetics, moisture diffusivity and lycopene retention of watermelon pomace in different dryers. *J. Food Sci. Technol.* **2015**, *52*, 7377–7384. [CrossRef]
17. Fabani, M.P.; Capossio, J.P.; Román, M.C.; Zhu, W.; Rodriguez, R.; Mazza, G. Producing non-traditional flour from watermelon rind pomace: Artificial neural network (ANN) modeling of the drying process. *J. Environ. Manag.* **2021**, *281*, 111915. [CrossRef]
18. NRC. *Recommended Dietary Allowances*; Reprint and Circular Series; National Research Council: Washington, DC, USA, 1948; Volume 129.
19. Bennion, E.B. *Bread Fabrication*, 1st ed.; Acribia: Zaragoza, Spain, 1969.
20. Chauhan, A.; Singh, S.; Dhar, A.; Powar, S. Optimization of pineapple drying based on energy consumption, nutrient retention, and drying time through Multi-Criteria Decision-Making. *J. Clean. Prod.* **2021**, *292*, 125913. [CrossRef]
21. Sendín, J.; Alonso, A.; Banga, J. Efficient and robust multi-objective optimization of food processing: A novel approach with application to thermal sterilization. *J. Food Eng.* **2010**, *98*, 317–324. [CrossRef]
22. Goñi, S.M.; Salvadori, V.O. Model-based multi-objective optimization of beef roasting. *J. Food Eng.* **2012**, *111*, 92–101. [CrossRef]
23. Abakarov, A.; Sushkov, Y.; Almonacid, S.; Simpson, R. Thermal processing optimization through a modified adaptive random search. *J. Food Eng.* **2009**, *93*, 200–209. [CrossRef]
24. Deng, Y.; Dewil, R.; Baeyens, J.; Ansart, R.; Zhang, H. The "Screening Index" to select building-scale heating systems. *IOP Conf. Ser. Earth Environ. Sci.* **2020**, *586*, 012004. [CrossRef]
25. Seville, J.P.K.; Deng, Y.; Dawn Bell, S.; Dewil, R.; Appels, L.; Ansart, R.; Leadbeater, T.; Parker, D.; Zhang, H.; Ingram, A.; et al. CO2 positron emission imaging reveals the in-situ gas concentration profile as function of time and position in opaque gas-solid contacting systems. *Chem. Eng. J.* **2021**, *404*, 126507. [CrossRef]
26. Liu, J.; Baeyens, J.; Deng, Y.; Tan, T.; Zhang, H. The chemical CO2 capture by carbonation-decarbonation cycles. *J. Environ. Manag.* **2020**, *260*, 110054. [CrossRef] [PubMed]
27. Deng, Y.; Ansart, R.; Baeyens, J.; Zhang, H. Flue gas desulphurization in circulating fluidized beds. *Energies* **2019**, *12*, 3908. [CrossRef]
28. Liu, J.; Baeyens, J.; Deng, Y.; Wang, X.; Zhang, H. High temperature Mn_2O_3/Mn_3O_4 and Co_3O_4/CoO systems for thermo-chemical energy storage. *J. Environ. Manag.* **2020**, *260*, 110582. [CrossRef]
29. Alimentarius. Codex Standard for Wheat Flour. CXS 152-1985. International Food Standards, 1985. Revised in 1995. Amended in 2016. 2019. Available online: http://www.fao.org/fao-who-codexalimentarius (accessed on 3 January 2021).
30. AOAC. *Official Methods of Analysis*, 8th ed.; The AOAC International: Rockville, MD, USA, 2010.

31. Garau, M.C.; Simal, S.; Rossello, C.; Femenia, A. Effect of air-drying temperature on physico-chemical properties of dietary fibre and antioxidant capacity of orange (*Citrus aurantium* v. Canoneta) by-products. *Food Chem.* **2007**, *104*, 1014–1024. [CrossRef]
32. Topuz, A. Predicting moisture content of agricultural products using artificial neural networks. *Adv. Eng. Softw.* **2010**, *41*, 464–470. [CrossRef]
33. Chokphoemphun, S.; Chokphoemphun, S. Moisture content prediction of paddy drying in a fluidized-bed drier with a vortex flow generator using an artificial neural network. *Appl. Therm. Eng.* **2018**, *145*, 630–636. [CrossRef]
34. Kalathingal, M.S.H.; Basak, S.; Mitra, J. Artificial neural network modeling and genetic algorithm optimization of process parameters in fluidized bed drying of green tea leaves. *J. Food Process Eng.* **2020**, *43*, e13128. [CrossRef]
35. Hornik, K.; Stinchcombe, M.; White, H. Multilayer feedforward networks are universal approximators. *Neural Netw.* **1989**, *2*, 359–366. [CrossRef]
36. Pinkus, A. Approximation theory of the MLP model in neural networks. *Acta Numer.* **1999**, *8*, 143–195. [CrossRef]
37. Miettinen, K. Chapter 2: Concepts. Nonlinear Multiobjective Optimization. In *International Series in Operations Research & Management Science*; Springer: Boston, MA, USA, 1998; Volume 12, pp. 5–36. [CrossRef]
38. Chang, K. Chapter 5, Multiobjective Optimization and Advanced Topics. In *Design Theory and Methods using CAD/CAE*; Academic Press: Cambridge, MA, USA, 2015; pp. 325–406. [CrossRef]
39. Russell, S.; Norvig, P. *Artificial Intelligence: A Modern Approach*, 3rd ed.; Prentice Hall: Hoboken, NJ, USA, 2010.
40. Zeleny, M. A concept of compromise solutions and the method of the displaced ideal. *Comput. Oper. Res.* **1974**, *1*, 479–496. [CrossRef]
41. Abakarov, A.; Nuñez, M. Thermal food processing optimization: Algorithms and software. *J. Food Eng.* **2013**, *115*, 428–442. [CrossRef]
42. Gal, T.; Leberling, H. Redundant objective functions in linear vector maximum problems and their determination. *Eur. J. Oper. Res.* **1977**, *1*, 176–184. [CrossRef]
43. Miettinen, K. Chapter 3: A Posteriori Methods. Nonlinear Multiobjective Optimization. In *International Series in Operations Research and Management Science*; Springer: Boston, MA, USA, 1998; Volume 12, pp. 77–113. [CrossRef]
44. Zabinsky, Z.B. Chapter 2: Pure Random Search and Pure Adaptive Search. In *Stochastic Adaptive Search for Global Optimization. Nonconvex Optimization and Its Applications*; Springer: Boston, MA, USA, 2003; Volume 72, pp. 25–54. [CrossRef]
45. Su, Y.; Jin, S.; Zhang, X.; Shen, W.; Eden, M.R.; Ren, J. Stakeholder-oriented multi-objective process optimization based on an improved genetic algorithm. *Comput. Chem. Eng.* **2020**, *132*, 106618. [CrossRef]

 *processes*

Article

Project Management Maturity and Business Excellence in the Context of Industry 4.0

Angela Fajsi *, Slobodan Morača, Marko Milosavljević and Nenad Medić

Faculty of Technical Sciences, University of Novi Sad, 21000 Novi Sad, Serbia; moraca@uns.ac.rs (S.M.); milosavljevic.di12.2018@uns.ac.rs (M.M.); medic.nenad@uns.ac.rs (N.M.)
* Correspondence: angela.fajsi@uns.ac.rs; Tel.: +381-60-446-47-13

Abstract: Even though Industry 4.0 is primarily focused on the implementation of advanced digital technologies, this is not the only aspect that should be considered. One of the aspects that calls for attention is the ability to create a sustainable and agile industrial environment. In this sense, the role of project management is crucial for achieving business excellence in a new industrial paradigm. The main goal of this paper was to determine the impact of different levels of project management maturity on business excellence in the context of Industry 4.0. The research in the paper was made using a sample of 124 organizations, differing in industry type and size, and recognized through the business excellence awards or recognitions given by European Foundation for Quality Management (EFQM). Using the Project Management Maturity Model (ProMMM), a significant connection was found between project management maturity and business excellence. Considering technology advances, these relationships were further examined in the context of Industry 4.0. Empirically based conclusions were drawn, which contribute to the literature on project management and business excellence in the context of Industry 4.0. Practitioners can implement them for more effective project management with the intention of bringing excellence into the organization's operations and results. Additionally, they can be useful to help organizations better cope with changing technology trends.

Keywords: project management maturity; business excellence; Industry 4.0; EFQM

Citation: Fajsi, A.; Morača, S.; Milosavljević, M.; Medić, N. Project Management Maturity and Business Excellence in the Context of Industry 4.0. *Processes* **2022**, *10*, 1155. https://doi.org/10.3390/pr10061155

Academic Editor: Jorge Cunha

Received: 13 April 2022
Accepted: 6 May 2022
Published: 9 June 2022

Publisher's Note: MDPI stays neutral with regard to jurisdictional claims in published maps and institutional affiliations.

1. Introduction

Strict rules and requirements regarding the knowledge economy and the modern industrial paradigm make organizations strive towards higher business excellence levels. The 'quality management' paradigm is moving towards 'managing quality', which is the basis of the business excellence concept that organizations strive for. Porter and Tanner [1] stated that 'the concept of business or organizational excellence provides support for the absolute integration of improvement initiatives within the organization'. It is based on the philosophy of continuous improvement, directing all organization's activities to enhance business performance, stakeholder satisfaction, corporate social responsibility, and environmental protection [2,3]. Toma and Marinescu [4] stated that there is a growing interest among companies in implementing business excellence strategies, which lead to increased quality of their business philosophy and improved business performances [4,5]. Effective formulation and implementation of these strategies have motivated organizations to change their way of doing business, and in this respect, to adopt various tools, methods, and techniques, such as enterprise resource planning (ERP) [4,6], balanced scorecard [7,8], lean or six sigma practices [9–11], and project management approaches [12–16], etc.

According to Kerzner [12], one of the main characteristics of organizations that were awarded the prestigious Malcolm Baldrige Business Excellence Award, was the existence of a project management system, which indicates a strong relationship between project management and business excellence. Effective project management at the organisation level does not just involve the application of software or the use of a specific tool [17].

To effectively implement this practice, which is thought to deliver sustainable project results, it is necessary to have acceptance and a positive attitude towards the project approach at all levels in the organisation, followed by the establishment of stable and long-term processes and competencies that will support its implementation and ensure excellence in their operations and results.

A key component of today's economy is a greater reliance on intelligence and intellectual abilities, rather than physical or natural resources [18], which contributes to the accelerated pace of scientific and technological progress related to Industry 4.0. This concept, also known as the fourth industrial revolution, helps "in implementing innovative technologies to improve productivity and working system" [19]. Jally et al. [19] also stated that the approaches to managing a project will be significantly altered due to the creation of these changes. The central aspect of the implementation of Industry 4.0 is the initiation of "smart business" and the acceleration of innovations through continual advancements where projects have a crucial role. Bag et al. [20] highlighted the role of project management in the process of Industry 4.0 integration and in achieving sustainable business. Considering everything aforementioned, the following research question arises: how does project management maturity affect organizational business excellence in the context of Industry 4.0?

The main goal of the study was to examine the impact of different levels of project management maturity on organizational business excellence in the context of Industry 4.0. For this purpose, an online questionnaire was distributed to organizations awarded or recognized for business excellence by the EFQM. The purpose of the questionnaire was to examine the levels of project management maturity in these organizations and their relationships with business excellence, considering their Industry 4.0 readiness. The ProMMM methodology [21] was used to assess the maturity level of project management, and the maturity level of Industry 4.0 was examined using the attributes defined by the authors Schumacher et al. [22].

In the next section, the theoretical background for project management and business excellence in the context of Industry 4.0 will be covered. Section 3 includes quantitative research, which encompasses a sample of 124 respondents and assesses the impact of project management maturity on business excellence and considers the readiness level for Industry 4.0 as a mediation variable. This section is concluded with a discussion to summarize and evaluate the research results. In the final section, managerial implications are presented, with potential limitations of the study and recommendations for future research.

2. Theoretical Background

2.1. Evaluating the Relationship between Project Management Maturity and Business Excellence

According to the EFQM [2], excellent organizations 'achieve and sustain outstanding levels of performance'. The study introduced by Talwar [23] acknowledges a positive relationship between business excellence implementation and organizational performance. Nowadays, several models are used to measure business excellence within an organization. The most cited models in the literature are the Malcolm Baldrige Criteria for Performance Excellence (CPE) and the European Foundation for Quality Management (EFQM). These models are based on TQM principles, and they cover topics such as customer focus, leadership, people involvement, continuous improvement, etc. The purpose of any business excellence model is to help organizations to sustain flexibility and embrace changes that could have a positive impact on their competency in the digital business environment. Achieving excellence in business activities implies adopting Deming's continuous improvement approach: plan, do, check, and act [24]. Many management practices support this approach. For example, project management [12–16] can be seen as a complementing part to the organisation's practice while reaching business excellence [1], even in the uncertain conditions that characterize technological changes. Planning, implementing, and controlling changes effectively are crucial in the process of implementing continuous improvements within organizations that aim to achieve business excellence. Vora [25] stated that only 30%

of organizational change programs are considered successful. One of the main reasons why most change management efforts fail is ineffective project management [25].

Project management is essential in today's business world—it is an approach that promotes continuous improvement through different types of projects that lead to improved organizational performance [26]. As already noted, Kerzner [13] emphasized the strong relationship between project management and business excellence, indicating that all organizations involved in the research which won the Malcolm Baldrige Award for Excellence also had a high level of project management implemented. In addition, Craddock [14] has proved that project success and sustainability are directly related to business excellence. As the business excellence models are based on the TQM principles, it is important to make links between TQM and project management approaches. The TQM is a fundamental concept of continuous improvement, within which organizations constantly review and enhance their business processes. Bryde and Robinson [27] emphasized that the TQM principles are important for maintaining effective project management, especially in customer service, failure prevention, professional development of employees, and strong leadership.

To measure an organization's project management effectiveness, different project management maturity models can be deployed. According to Kerzner [13], maturity in project management can be defined as 'the development of systems and processes that are repetitive in their nature to provide a high probability that each project will be successful'. The most used maturity models in the literature are the PMMM model [28], PM2-Project Management Process Maturity Model [29], and the Kerzner's project management maturity model [30]. On the other side, some models have moved from a strict relationship between CMMI and PMBOK group processes. Pennypacker and Grant [31] stated that one of these models is the ProMMM model [21], which is also based on the CMMI model, but instead of PMBOK elements, relationships from the EFQM Model are taken. Most existing models test the maturity of the project management processes, while using this model, organizations assess other attributes and provide a true picture of their project management capability. Therefore, the ProMMM model has a wide application in practice and empirical studies [32–35].

Achieving a satisfactory level of maturity is a continuous and long-term process. However, due to built-in constraints and environmental factors, many organizations are not able to reach the highest levels of maturity during their existence [21]. Andersen and Jessen [35] stated that fully matured organizations do not exist in the real world, so considering different levels of maturity is a reasonable task for any organization. Research presented by Backlund et al. [36] revealed that higher levels of project management maturity led to success in project implementation, which further leads to improved organisation's processes in their road to bring excellence [37]. Therefore, the main hypothesis is proposed:

H1. *A high level of project management maturity has a positive impact on business excellence.*

In modern business conditions, the area of project management faces a much more complex and dynamic environment as a characteristic of the new industrial revolution more generally known as Industry 4.0 [38].

2.2. Project Management Maturity and Business Excellence in the Context of Industry 4.0

Determining a relationship between project management maturity and business excellence is a complex issue affected by many factors related to Industry 4.0 and digitalization that comes with it. Raj et al. [39] opined that there is a growing need for implementation of standards and government regulations to accelerate the process of adoption of Industry 4.0 digital technologies. They also asserted that the "lack of a digital strategy alongside resource scarcity" followed by a "lack of standards, regulations, and forms of certification", constrains companies from strengthening their capabilities in the process of fully leveraging Industry 4.0 digital technologies. This concept is especially applicable to the manufacturing and IT industry, while Al Amri et al. [40] stated that its applicability to measure was still uncertain for other areas. On the contrary, there are studies that confirm the importance of Industry 4.0 for service organizations [41,42].

The modern business excellence paradigm is strongly oriented 'to the necessity to transform the current organization for the future' [3]. Gunasekaran et al. [43] stated that it is important to define 'what might be the future of excellence'. The term 'future' relates in this context to digital transformation, Industry 4.0, and organizational agility with special emphasis on technology and human capacity development. Fonseca [44] made the comparison between the EFQM 2013 and EFQM 2020 models and stated that the new model has "a focus on the futuristic requirements of the organizations rather than merely a business excellence model and/or just a quality award enablement model".

According to the EFQM 2020 model, both concepts of business excellence and Industry 4.0 share a common goal to improve organizational operations and results. The Singapore Smart Industry Readiness EDB report [45] indicated that business excellence is directly related to human resource ability to adopt a range of different approaches, methods and tools promoted within Industry 4.0.

Industry 4.0 promotes the adoption of new organizational models but also the adaptation of existing ones to achieve excellence in the conditions set by the new industrial paradigm. Accordingly, project managers are looking for different ways to understand technological change and its impact on project management processes. Moreover, the role of project management in the development of Industry 4.0 is essential for its success and vice versa [46]. Therefore, the authors state that traditional project management systems should be analysed and updated according to the requirements of the new industrial revolution, which will help reduce the complexity of projects [21] and increase the likelihood of projects succeeding.

The above discussions lead to the following hypothesis:

H2. *Industry 4.0 readiness level is a mediator between project management maturity and business excellence.*

Due to a lack of research on the relationship between project management maturity and business excellence in the context of Industry 4.0, this issue requires further empirical analysis.

3. Research Methodology

3.1. Data Collection and Sample

Data collection began in January 2021 and continued through March 2021. The questionnaire was distributed in electronic format, via the Google Forms platform to organizations that have received awards and recognitions for business excellence [47]. Besides, an invitation letter and a survey were sent to the National Representatives for EFQM, so that they would be aware of the research to influence their members to participate in it.

The EFQM 2020 model was launched in November 2019 and when data collection began there was only a small number of organizations that followed the new model framework. For this reason, the sample included organizations that have achieved awards and recognitions for business excellence according to the EFQM 2013 model. It was emphasized that the survey should be filled in by a person who deals with project management or development processes. Their task was to assess the project management maturity and Industry 4.0 readiness levels within their organization.

The total number of participants who took part in the research was 130. Of those, six had missing data, so the final number of participating organizations was 124. The total number of relevant research organizations as of January 2021 was 1293, thus the response rate was 10.05%. Rogelberg and Stanton [48] stated that a response rate of 10% should not be ignored; rather it should be examined as to whether it has a substantial impact on the conclusions, considering that a lower response rate is important to understand topics that are insufficiently researched in the literature.

The greatest portion of respondents were employed in top management positions (37.90%), followed by middle-management (27.42%) and project management (11.29%). The participating organizations varied in size (Figure 1) and originated from 27 countries (Table 1). The following table shows the respondents by the type of activity they performed (Figure 2).

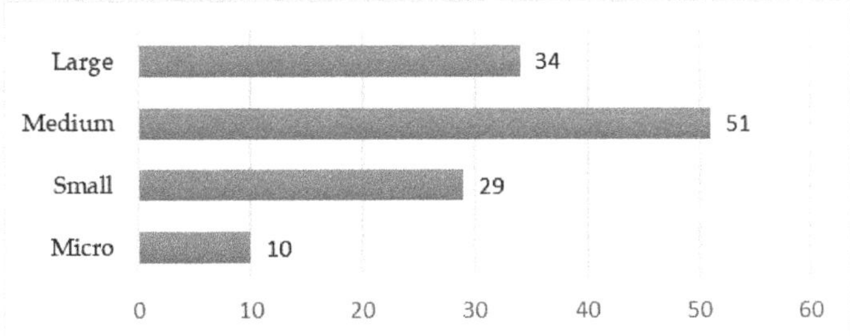

Figure 1. Profile of organizations by size.

Table 1. Profile of organizations by country.

Country	N	%	Country	N	%
Spain	28	22.58	Finland	2	1.61
Switzerland	13	10.48	Greece	2	1.61
United Kingdom	11	8.87	Netherlands	2	1.61
Turkey	9	7.26	Jordan	2	1.61
Austria	6	4.84	Hungary	2	1.61
Portugal	6	4.84	Saudi Arabia	2	1.61
Ireland	5	4.03	Sweden	2	1.61
Germany	5	4.03	United Arab Emirates	1	0.81
Belgium	4	3.23	Italy	1	0.81
Ecuador	4	3.23	Peru	1	0.81
Colombia	3	2.42	Poland	1	0.81
Czech Republic	3	2.42	Russia	1	0.81
France	3	2.42	Slovenia	1	0.81
Lithuania	3	2.42	Missing data	1	0.81

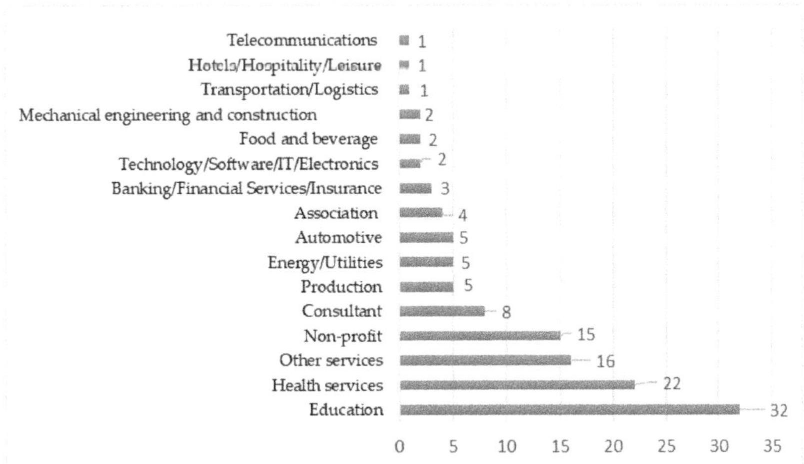

Figure 2. Profile of organizations by industry type.

Looking at the profile of participating organizations, most of them were medium-sized (41.13%) and had their headquarters in Europe because the EFQM model is most represented there. Most respondents were from Spain (22.58%), which had the most organizations with EFQM awards, followed by Switzerland (10.48%) and the United Kingdom (8.87%).

When it comes to the type of industry, most responses were from organizations engaged in education (25.81%), followed by health services (17.74%) and other service activities (12.90%). These numbers can be explained by the fact that these sectors/areas had the largest number of organizations with recognition for business excellence.

3.2. Research Instruments

3.2.1. Project Management Maturity

The aforementioned literature review showed different models for measuring project management maturity. The ProMMM model is based on the CMMI model [49] and the EFQM model was used to measure project management maturity within the organization. It describes four levels of project management maturity: Naïve, Novice, Normalized and Natural. They were further defined in terms of four attributes: culture, process, experience, and application. Each of those dimensions contained a set of five items that were measured using a 5-Point Likert scale. There was no significant difference between the average mark of the dimensions of project management maturity. Research has shown that organizations that are excellent in business have a slightly more developed project management culture compared to other attributes (average mark: culture 2.91, processes 2.79, experience 2.53, and applications 2.67).

3.2.2. Business Excellence

Business excellence levels were defined according to the prescribed criteria established by the EFQM, which were contained within the EFQM Excellence Model, defined as a 'framework for measuring the strengths and areas for improvement of an organization across all of its activities' [2]. The EFQM process recognition is a complex assessment. It is carried out by independent EFQM assessors, and organizations that have won EFQM recognition were taken as a sample. Different levels of recognition were presented in the form of the 7-Point Likert Scale. The majority of respondents were in the category Recognized for Excellence with 4 stars (24.19%), followed by Committed to Excellence (19.35%), Recognised for Excellence 5 stars (16.94), Committed to Excellence 2 stars (15.32%), Recognised for Excellence 3 stars (9.68%), EFQM Award Finalist (8.06%), EFQM Excellence Award/Prize Winner (6.45%).

3.2.3. Industry 4.0 Readiness

Industry 4.0 readiness was assessed based on two sets of questions:

(1) Stages of technological development were measured by a 4-Point Likert scale. A total of 32.26% of organizations stated that they used only existing, well-established, and mature technologies, and the same percentage of organizations stated that they used many new and recently developed technologies. Limited new technology, or 'a new feature', were used by 30.64% of respondents, while new, unproven technological concepts were used by only 4.84% of respondents.
(2) Dimensions and items of the Industry 4.0 Readiness model were measured by a 4-Point Likert scale (Table 2) [22].

3.3. Research Results and Discussion

3.3.1. Preliminary Analysis

The objectives of the preliminary analysis were to check the reliability of measures and to obtain insights into the dataset. Internal structure validity, reliability analysis and descriptive statistics were done for the purpose of this research.

To examine the internal structure of the test and the reliability of individual dimensions of project management maturity, analyses were performed on culture, processes, experience, application, exploratory factor analysis, and reliability.

The reliability of those dimensions was adequate ($\alpha > 0.70$) [50]. When it came to the Process dimension, it was noticeable that the item examining the degree of formality of the project management process had a very low loading. Removing this item would lead to an increase in dimension reliability. As expected, both an exploratory factor analysis and a reliability analysis showed increased values of relevant coefficients after excluding this dimension. The results of the exploratory factor analysis and the reliability analysis for the Industry 4.0 Readiness dimension were assessed as adequate.

Table 2. Industry 4.0 Readiness model.

Areas of Industry 4.0	Does Not Exist or It Is at a Very Low Level (%)	Low-Level (%)	Medium-Level (%)	High-Level (%)
Industry 4.0 strategy	31.45	23.39	33.06	12.1
Leadership	16.13	26.61	37.1	20.16
Customers	8.87	31.45	40.32	19.35
Products and services	8.06	30.64	41.13	20.16
Operations	8.87	37.1	45.16	8.87
Culture	8.06	34.68	38.71	18.55
People	8.06	31.45	45.97	14.52
Governance	12.90	31.45	44.35	11.3
Technology	8.06	33.87	41.13	16.93

3.3.2. Descriptive Statistics

The parameters distribution shape, skewness, and flatness, showed that the distribution had the typical bell curve pattern of normal distributions (Table 3). The normal distribution, according to the conventional criterion, has the value of the stated parameters in the range ± 1.5 [51].

Table 3. Descriptive statistics.

Variable	Min	Max	AS	SD	Sk	Ku
Culture	1	20	14.63	3.70	−0.89	1.27
Processes	4	20	13.98	3.17	−0.46	0.29
Experience	0	20	12.71	3.91	−0.61	0.50
Application	0	20	13.42	3.78	−0.83	0.67
Business excellence	1	7	3.56	1.82	0.08	−1.00
Industry 4.0 readiness	10	38	25.54	6.28	−0.27	−0.42

Legenda. Min—minimum value. Max—maximum value. AS—arithmetic mean. SD—standard deviation. Sk—skewness. Ku—kurtosis.

3.3.3. Mediation Analysis—Effects of Project Management Maturity on Business Excellence in the Context of Industry 4.0

Hayes' macro 'process' v4.0 software [52] was used to test the mediation effect of Industry 4.0 readiness on the relationship between project management maturity and business excellence. A conceptual diagram of mediation analysis was shown in Figure 3.

The analysis was conducted using 5000 bootstrap samples and with 95 confidence intervals, in line with Hayes's [52] suggestion. Overall, the mediation model was significant ($F_{(2, 121)} = 5.36$, $p < 0.01$, $R^2 = 0.081$). Individual relations between variables are presented in the Table 4.

Significant effects were found for the a-path (direct effect from Project management maturity on Industry 4.0 readiness) and the c'-path (direct effect from Project management maturity on Business excellence). On the other hand, the c-path (indirect effect of Project management maturity on Business excellence through Industry 4.0 readiness) and the b-path (direct effect from Industry 4.0 readiness to Business excellence) were not significant. As the results suggested:

H1. *A high level of project management maturity has a positive impact on business excellence—was accepted.*

H2. *Industry 4.0 readiness level is a mediator between the project management maturity and business excellence—was rejected.*

In addition to providing better explanations for these relationships, the authors examined which Industry 4.0 technologies have been used within respondent organizations. The discussion chapter will include the importance and relevance of these findings.

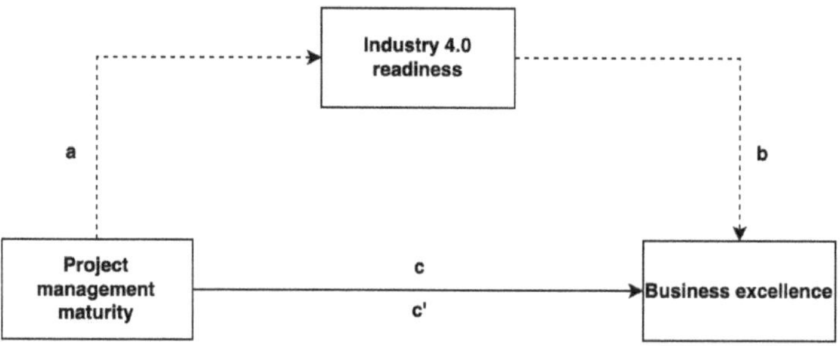

Figure 3. Conceptual diagram of mediation analysis.

Table 4. Description of individual relations between variables.

Path	Description	Parameters
a	Direct effect from Project management maturity on Industry 4.0 readiness	$\beta = 0.28$; 95 CI = 0.106–0.452
b	Direct effect from Industry 4.0 readiness to Business excellence	$\beta = 0.04$; 95 CI = -0.003–0.111
c	Indirect effect of Project management maturity on Business excellence through Industry 4.0 readiness	$\beta = 0.04$; 95 CI = -0.003–0.111
c'	Direct effect from Project management maturity on Business excellence	$\beta = 0.19$; 95 CI = 0.056–0.373

4. Discussion

This study presents empirical evidence linking project management maturity and business excellence in the context of Industry 4.0. The hypothesis, which claims that a higher level of project management maturity in organizations has a positive effect on business excellence, was confirmed by using Hayes' macro "process" v4.0 software. Business excellence is achieved through continuous improvement, innovation, and learning [2], and, importantly, the project management approach is in line with those principles [53–55].

There were no statistically significant differences between the individual dimensions of project management maturity and their impact on business excellence. All dimensions had almost the same effect in synergy, and therefore, organizations should understand and develop project management culture, establish processes, educate people, and effectively apply project management methods and tools. Nevertheless, culture has a slightly higher level of maturity compared to other attributes (2.91), which indicates that excellent organizations clearly define and support the "corporate culture" for project management [56].

In this study, there was no evidence of a mediating effect of Industry 4.0 readiness on the relationship between project management maturity and business excellence. Strong statistical significance was found for the effect of project management maturity on Industry 4.0 readiness, which is in line with previous studies [19,57–61]. The Industry 4.0 variable had no statistically significant impact on business excellence, which contradicts previous studies [43,44,46]. The authors explain that this was due to the specificity of the sample, because mainly the organizations that were engaged in the service industry participated in the research. Although the literature proves the importance and necessity of applying the concept of Industry 4.0, industry-related institutes define its application, and the progress is very slow [34], especially when it comes to the service industry. Most previous studies covered the topic of Industry 4.0 in manufacturing companies [39,62], while Bodrow [41] and Rennung et al. [42] showed that the service industry becomes an important element of the Industry 4.0 concept, mainly for the reason that most products and services are connected into the integrative offering.

In addition, the advanced technologies used in organizations were examined. The use of more advanced programs in the IT system such as ERP, CRM, and the use of mobile technologies are mostly represented; they were used in almost 70% of organizations. Jally et al. [19] defined Industry 4.0 technologies that were significant for project management, such as additive manufacturing, IoT, Autonomous systems, Big data, which are presented in more than 20% of respondent organizations. These data indicate that more than 20% organizations have technologies which can be successfully integrated with the project management approach. Jun et al. [63] identified these technologies as important for useful and accurate quality management. Other technologies, such as artificial intelligence and blockchain, are less represented because they are primarily related to the manufacturing industry. The development of smart services is expected in the near future, and it is evident that these will not only influence but also facilitate project management within the organization, with the aim of achieving and/or maintaining business excellence.

4.1. Theoretical Implications

This study aimed to test the theory of project management that links the level of project management maturity with business excellence in the context of Industry 4.0. It has been proven that project management has a strong positive impact on business excellence, and it provides empirical evidence in theory that previous studies support. Furthermore, the model developed for this research raises the possibility for other researchers in the field to incorporate specificities into their studies of the new industry trends imposed by Industry 4.0.

4.2. Practical Implications

In a practical manner, the findings can help organizations to define strategies for more effective implementation of project management approach to achieve and/or maintain business excellence within the new industrial paradigm.

The finding suggests that a balanced development of project management culture, processes, people, methods, and tools for application leads to excellence in business operations and results. Adopting an organization's project management culture helps organizations to understand and adapt their core activities to different norms, regulations, and behaviours. Furthermore, it influences employee's expertise and commitment, project management processes and its application by using a variety of methods and tools such as requirement analysis, timeline frameworks, agile methods, specific software to support project management, etc. [64].

The literature review found that technologies, such as additive manufacturing, IoT, Autonomous systems, and Big data, have a positive impact on both project management and business quality management, which indicates to practitioners the importance of their more intensive use.

As a final practical implication, the authors suggest that organisations operating in emerging economies, where the conditions for achieving awards for excellence have not yet been met, should consider applying the modified ProMMM model. With several limitations, primarily caused by unfavourable environmental factors, the proposed model can provide a clear direction regarding the organization of project activities to improve their business results, stakeholder satisfaction, socially responsible business, and environmental protection.

5. Conclusions, Limitations and Future Research

The effective deployment of project management approach helps organizations to deal with the issues of achieving business excellence. A novel modified ProMMM model was proposed to access project management maturity within organizations. The empirical evidence found in this study shows that higher levels of project management maturity led to more recognition and awards for business excellence.

Examining the mediation role of Industry 4.0, no significant statistical differences were observed. This contradicts previous studies, which have emphasized the importance of Industry 4.0 in the context of project management and business excellence, but there are no studies that have examined Industry 4.0 as a mediation effect in the aforementioned relationship.

One of the major limitations of this study was the limited number of participants in the research. Given that the sample consisted of a specific population of respondents that included organizations that exclusively have some form of EFQM recognition, it is considered that a sample of 100 or more respondents is acceptable for valid results [65,66]. Additionally, the invitation letter invited respondents engaged in project management or development processes, which further narrowed the sampled population. It is necessary to involve a larger number of people from organizations where the maturity of the process will be viewed from different aspects. The creator of the original questionnaire suggested that information needs to be obtained from a wide range of staff to avoid responses from specific individuals [21]. This limitation can be overcome by implementing qualitative methods that can perform a more detailed analysis and verify the results obtained in quantitative research. Hillson [21] suggested methods such as interviews and case studies.

The authors emphasized that there is a need for further development in this area through empirical studies on this topic, especially including manufacturing companies that are closely related to Industry 4.0 practices. Further researches can investigate relationships between project management maturity and business excellence model dimensions (such as leadership, strategy, customers, etc.) to determine the level of project management impact on individual dimensions. Furthermore, Malcolm Baldrige Criteria for Performance Excellence (CPE) can be used as criteria for some future research to verify results obtained in this research where the EFQM model was used.

Author Contributions: Conceptualization, A.F., S.M. and N.M.; methodology, A.F., S.M. and N.M.; validation, S.M.; formal analysis, A.F.; investigation, A.F. and S.M.; resources, S.M.; data curation, M.M.; writing—original draft preparation, A.F.; writing—review and editing, M.M., S.M. and N.M.; visualization, A.F. and M.M.; supervision, S.M.; project administration, A.F. All authors have read and agreed to the published version of the manuscript.

Funding: This research received no external funding.

Acknowledgments: The results presented in this paper are part of the research within the project "Improvement of teaching processes at DIEM through the implementation of the results of scientific research in the field of Industrial engineering and management", Department of Industrial Engineering and Management, Faculty of Technical Sciences in Novi Sad, University of Novi Sad, Republic of Serbia.

Conflicts of Interest: The authors declare no conflict of interest.

References

1. Porter, L.; Tanner, S. *Assessing Business Excellence*; Routledge: London, UK, 2012.
2. EFQM. *EFQM Excellence Model 2013*; EFQM Publications: Bruxelles, Belgium, 2013.
3. EFQM. *EFQM Excellence Model 2020*; EFQM Publications: Bruxelles, Belgium, 2019.
4. Toma, S.G.; Marinescu, P. Business Excellence Models: A Comparison. In Proceedings of the International Conference on Business Excellence, Bucharest, Romania, 22–23 March 2018; Volume 12, pp. 966–974.

5. Dale, B.G.; Bamford, D.; van der Wiele, T. *Managing Quality: An Essential Guide and Resource Gateway*, 6th ed.; Wiley: Chichester, UK, 2016.

6. Jha, V.S.; Joshi, H. Relevance of Total Quality Management (TQM) or Business Excellence Strategy Implementation for Enterprise Resource Planning (ERP)–A Conceptual Study. In Proceedings of the 12th International Conference on Information Quality, Cambridge, MA, USA, 9–11 November 2007; pp. 1–16.

7. Odríguez-González, C.G.; Sarobe-González, C.; Durán-García, M.E.; Mur-Mur, A.; Sánchez-Fresneda, M.N.; Pañero-Taberna, M.D.L.M.; Pla-Mestre, R.; Herranz-Alonso, A.; Sanjurjo-Sáez, M. Use of the EFQM excellence model to improve hospital pharmacy performance. *Res. Soc. Adm. Pharm.* **2019**, *16*, 710–716. [CrossRef]

8. Vokurka, R.J. Operationalising the balanced scorecard using the Malcolm Baldrige Criteria for Performance Excellence (MBCPE). *Int. J. Manag. Enterp. Dev.* **2004**, *1*, 208–217. [CrossRef]

9. Campatelli, G.; Citti, P.; Meneghin, A. Development of a simplified approach based on the EFQM model and Six Sigma for the implementation of TQM principles in a university administration. *Total Qual. Manag. Bus. Excel.* **2011**, *22*, 691–704. [CrossRef]

10. Shahin, A.; Pourbahman, R. Integration of EFQM and Ultimate Six Sigma: A Proposed Model. *Int. Bus. Res.* **2010**, *4*, 176. [CrossRef]

11. Tabari, M.; Gholipour-Kanani, Y.; Tavakkoli-Moghaddam, R. Application of the six sigma methodology in adopting the business excellence model for a service company—A case study. *World Appl. Sci. J.* **2012**, *17*, 1066–1073.

12. Kerzner, H. *Advanced Project Management: Best Practices on Implementation*; John Wiley & Sons: Hoboken, NJ, USA, 2003.

13. Craddock, W. How Business Excellence Models Contribute to Project Sustainability and Project Success. In *Sustainability Integration for Effective Project Management*; IGI Global: Hershey, PA, USA, 2013; pp. 1–19.

14. Bryde, D.; Leighton, D. Improving HEI Productivity and Performance through Project Management. *Educ. Manag. Adm. Leadersh.* **2009**, *37*, 705–721. [CrossRef]

15. Calvo-Mora, A.; Navarro-García, A.; Periañez-Cristobal, R. Project to improve knowledge management and key business results through the EFQM excellence model. *Int. J. Proj. Manag.* **2015**, *33*, 1638–1651. [CrossRef]

16. Digehsara, A.A.; Rezazadeh, H.; Soleimani, M. Performance evaluation of project management system based on combination of EFQM and QFD. *J. Proj. Manag.* **2018**, *3*, 171–182. [CrossRef]

17. Carstens, D.S.; Richardson, G.L. *Project Management Tools and Techniques: A Practical Guide*; CRC Press: Boca Raton, FL, USA, 2019.

18. Powell, W.W.; Snellman, K. The knowledge economy. *Annu. Rev. Sociol.* **2004**, *30*, 199–220. [CrossRef]

19. Jally, V.; Kulkarni, V.N.; Gaitonde, V.N.; Satis, G.J.; Kotturshettar, B.B. A Review on Project Management Transformation Using Industry 4.0. In *AIP Conference Proceedings*; AIP Publishing LLC.: Melville, NY, USA, 2021; Volume 2358, p. 100014.

20. Bag, S.; Yadav, G.; Dhamija, P.; Kataria, K.K. Key resources for industry 4.0 adoption and its effect on sustainable production and circular economy: An empirical study. *J. Clean. Prod.* **2020**, *281*, 125233. [CrossRef]

21. Hillson, D. Assessing organizational project management capability. *J. Facil. Manag.* **2003**, *2*, 298–311. [CrossRef]

22. Schumacher, A.; Erol, S.; Sihn, W. A maturity approach for assessing Industry 4.0 readiness and maturity of manufacturing enterprises. *Procedia Cirp* **2016**, *52*, 161–166. [CrossRef]

23. Talwar, B. Business excellence models and the path ahead. *TQM J.* **2011**, *23*, 21–35. [CrossRef]

24. Deming, W. *Elementary Principles of the Statistical Control of Quality: A Series of Lectures*; Nippon Kagaku Gijutsu Remmei: Tokyo, Japan, 1952.

25. Vora, M.K. Business excellence through sustainable change management. *TQM J.* **2013**, *25*, 625–640. [CrossRef]

26. Radujković, M.; Sjekavica, M. Project Management Success Factors. *Procedia Eng.* **2017**, *196*, 607–615. [CrossRef]

27. Bryde, D.J.; Robinson, L. The relationship between total quality management and the focus of project management practices. *TQM Mag.* **2007**, *19*, 50–61. [CrossRef]

28. Project Management Institute. *PMBOK Guide*, 5th ed.; Project Management Institute: Newtown Square, PA, USA, 2013.

29. Ibbs, C.W.; Kwak, Y.H. Assessing Project Management Maturity. *Proj. Manag. J.* **2000**, *31*, 32–43. [CrossRef]

30. Kerzner, H. *Strategic Planning for Project Management Using a Project Management Maturity Model*; John Wiley & Sons: Hoboken, NJ, USA, 2000.

31. Pennypacker, J.S.; Grant, K.P. Project Management Maturity: An Industry Benchmark. *Proj. Manag. J.* **2003**, *34*, 4–11. [CrossRef]

32. Karlsen, J.T. Supportive culture for efficient project uncertainty management. *Int. J. Manag. Proj. Bus.* **2011**, *4*, 240–256. [CrossRef]

33. Simangunsong, E.; Da Silva, E.N. Analyzing Project Management Maturity Level in Indonesia. *South East Asian J. Manag.* **2013**, *7*, 72. [CrossRef]

34. Rezaeean, A.; Falaki, P. Agile Project Management. *Int. Res. J. Appl. Basic Sci.* **2012**, *34*, 698–707.

35. Andersen, E.; Jessen, S. Project maturity in organisations. *Int. J. Proj. Manag.* **2003**, *21*, 457–461. [CrossRef]

36. Backlund, F.; Chronéer, D.; Sundqvist, E. Project management maturity models–A critical review: A case study within Swedish engineering and construction organizations. *Procedia-Soc. Behav. Sci.* **2014**, *119*, 837–846. [CrossRef]

37. Alshawi, M.; Ingirige, B. Web-enabled project management: An emerging paradigm in construction. *Autom. Constr.* **2003**, *12*, 349–364. [CrossRef]

38. Ribeiro, A.; Amaral, A.; Barros, T. Project Manager Competencies in the context of the Industry 4.0. *Procedia Comput. Sci.* **2021**, *181*, 803–810. [CrossRef]

39. Raj, A.; Dwivedi, G.; Sharma, A.; de Sousa Jabbour, A.B.L.; Rajak, S. Barriers to the adoption of industry 4.0 technologies in the manufacturing sector: An inter-country comparative perspective. *Int. J. Prod. Econ.* **2020**, *224*, 107546. [CrossRef]

40. Al Amri, T.; Khetani, K.P.; Marey-Perez, M. Towards Sustainable I4.0: Key Skill Areas for Project Managers in GCC Construction Industry. *Sustainability* **2021**, *13*, 8121. [CrossRef]
41. Bodrow, W. Impact of Industry 4.0 in service oriented firm. *Adv. Manuf.* **2017**, *5*, 394–400. [CrossRef]
42. Rennung, F.; Luminosu, C.T.; Draghici, A. Service Provision in the Framework of Industry 4.0. *Procedia—Soc. Behav. Sci.* **2016**, *221*, 372–377. [CrossRef]
43. Gunasekaran, A.; Subramanian, N.; Ngai, W.T.E. Quality management in the 21st century enterprises: Research pathway towards Industry 4.0. *Int. J. Prod. Econ.* **2018**, *207*, 125–129. [CrossRef]
44. Fonseca, L.; Amaral, A.; Oliveira, J. Quality 4.0: The EFQM 2020 Model and Industry 4.0 Relationships and Implications. *Sustainability* **2021**, *13*, 3107. [CrossRef]
45. EDB Singapore. The Singapore smart industry readiness index. In *Catalysing the Transformation of Manufacturing*; EDB Singapore: Singapore, 2018.
46. Nenadál, J. The New EFQM Model: What is Really New and Could Be Considered as a Suitable Tool with Respect to Quality 4.0 Concept? *Qual. Innov. Prosper.* **2020**, *24*, 17–28. [CrossRef]
47. EFQM Recognition Database. Available online: https://shop.efqm.org/recognition-database/ (accessed on 15 January 2021).
48. Rogelberg, S.; Stanton, J. Introduction: Understanding and dealing with organizational survey nonresponse. *Organ. Res. Methods* **2007**, *10*, 195–209. [CrossRef]
49. Paulk, M.; Curtis, B.; Chrissis, M.; Weber, C. Capability maturity model, version 1.1. *IEEE Softw.* **1993**, *10*, 18–27. [CrossRef]
50. DeVellis, R. *Scale Development: Theory and Applications*; Sage Publications: Thousand Oaks, CA, USA, 2016.
51. Tabachnick, B.G.; Fidell, L.S. *Using Multivariate Statistics*, 6th ed.; Pearson: Boston, MA, USA, 2013.
52. Hayes Processes: A Versatile Computational Tool for Observed Variable Mediation, Moderation, and Conditional Process Modeling [White Paper]. 2012. Available online: http://www.afhayes.com/public/process2012.pdf (accessed on 15 February 2022).
53. Shenhar, A.; Dvir, D. *Reinventing Project Management: The Diamond Approach to Successful Growth and Innovation*; Harvard Business Review Press: Cambridge, MA, USA, 2007.
54. Mahmoud-Jouini, S.; Christophe, M.; Silberzahn, P. Contributions of design thinking to project management in an innovation context. *Proj. Manag. J.* **2016**, *47*, 144–156. [CrossRef]
55. Cavaleri, S.A.; Fearon, D.S. Integrating organizational learning and business praxis: A case for intelligent project management. *Learn. Organ.* **2000**, *7*, 251–258. [CrossRef]
56. Kerzner, H. *Applied Project Management: Best Practices on Implementation*; Wiley: New York, NY, USA, 2000.
57. Helfer, D.G.; Kipper, L.M.; Lemões, S.E. Scientific Mapping of Project Management and Educational Perspectives for Industry 4.0. In Proceedings of the International Conference on Industrial Engineering and Operations Management, Singapore, 7–11 March 2021.
58. Simion, C.P.; Popa, S.C.; Albu, C. Project Management 4.0–Project Management in the Digital Era. In Proceedings of the 12th International Management Conference (str. 93–100), Bucharest, Romania, 1–2 November 2018.
59. Vrchota, J.; Řehoř, P.; Maříková, M.; Pech, M. Critical Success Factors of the Project Management in Relation to Industry 4.0 for Sustainability of Projects. *Sustainability* **2020**, *13*, 281. [CrossRef]
60. López-Robles, J.R.; Otegi-Olaso, J.R.; Cobo, M.; Bertolin-Furstenau, L.; Kremer-Sott, M.; López-Robles, L.; Gamboa-Rosales, N.K. The Relationship between Project Management and Industry 4.0: Bibliometric Analysis of Main Research Areas through Scopus. In Proceedings of the 3rd International Conference on Research and Education in Project Management—REPM 2020, Bilbao, Spain, 20–21 February 2020.
61. Marnewick, C.; Marnewick, A.L. The demands of industry 4.0 on project teams. *IEEE Trans. Eng. Manag.* **2019**, *3*, 941–949. [CrossRef]
62. Frank, A.G.; Dalenogare, L.S.; Ayala, N.F. Industry 4.0 technologies: Implementation patterns in manufacturing companies. *Int. J. Prod. Econ.* **2019**, *210*, 15–26. [CrossRef]
63. Jun, J.-H.; Chang, T.-W.; Jun, S. Quality Prediction and Yield Improvement in Process Manufacturing Based on Data Analytics. *Processes* **2020**, *8*, 1068. [CrossRef]
64. Turner, R.; Ledwith, A.; Kelly, J. Project management in small to medium-sized enterprises: Tailoring the practices to the size of company. *Manag. Decis.* **2012**, *50*, 942–957. [CrossRef]
65. Mundfrom, D.J.; Shaw, D.G.; Ke, T.L. Minimum Sample Size Recommendations for Conducting Factor Analyses. *Int. J. Test.* **2005**, *5*, 159–168. [CrossRef]
66. Kock, N.; Hadaya, P. Minimum sample size estimation in PLS-SEM: The inverse square root and gamma-exponential methods. *Inf. Syst. J.* **2016**, *28*, 227–261. [CrossRef]

Article

Application Research of Soft Computing Based on Machine Learning Production Scheduling

Melinda Timea Fülöp [1,*], Miklós Gubán [2], Ákos Gubán [2] and Mihály Avornicului [2]

[1] Faculty of Economics and Business Administration, Babeş-Bolyai University, 400591 Cluj-Napoca, Romania
[2] Faculty of Finance and Accountancy, Budapest Business School, 1149 Budapest, Hungary; guban.miklos@uni-bge.hu (M.G.); guban.akos@uni-bge.hu (Á.G.); avornicului.mihalyszilard@uni-bge.hu (M.A.)
* Correspondence: melinda.fulop@econ.ubbcluj.ro

Abstract: An efficient and flexible production system can contribute to production solutions. These advantages of flexibility and efficiency are a benefit for small series productions or for individual articles. The aim of this research was to produce a genetic production system schedule similar to the sustainable production scheduling problem of a discrete product assembly plant, with more heterogeneous production lines, and controlled by one-time orders. First, we present a detailed mathematical model of the system under investigation. Then, we present the IT for a solution based on a soft calculation method. In connection with this model, a computer application was created that analyzed various versions of the model with several practical problems. The applicability of the method was analyzed with software specifically developed for this algorithm and was demonstrated on a practical example. The model handles the different products within an order, as well as their different versions. These were also considered in the solution. The solution of this model is applicable in practice, and offers solutions to better optimize production and reduce the costs of production and logistics. The developed software can not only be used for flexible production lines, but also for other problems in the supply chain that can be employed more widely (such as the problem of delivery scheduling) to which the elements of this model can be applied.

Keywords: soft computing; genetic algorithms; product scheduling; heuristic methods

Citation: Fülöp, M.T.; Gubán, M.; Gubán, Á.; Avornicului, M. Application Research of Soft Computing Based on Machine Learning Production Scheduling. *Processes* **2022**, *10*, 520. https://doi.org/10.3390/pr10030520

Academic Editor: Anna Trubetskaya

Received: 19 February 2022
Accepted: 3 March 2022
Published: 5 March 2022

Publisher's Note: MDPI stays neutral with regard to jurisdictional claims in published maps and institutional affiliations.

1. Introduction

In view of demographic change and high-cost pressure, ever greater efforts are being made to automate processes. On the one hand, there is an enormous combinatorial variety in process design. This includes dividing the tasks between available robots, determining the processing sequence, and adjusting the process settings. On the other hand, due to narrow and dynamic robot workspaces, motion planning proves to be extremely computationally intensive. The aim of this work was therefore to use heuristic algorithms to make processes more efficient to save costs and time. In this work, a new implementation of the genetic algorithm is used for process optimization.

Industrial production has been going through a major change for several years, also known as the fourth industrial revolution [1–3]. An essential part of this change is the complete penetration of processes with the help of virtual methods. At the same time, due to increasing complexity and decreasing batch sizes in production, there is an ever-increasing need for software for planning, automation, and optimization. To meet these increasing requirements, new methods and technologies have to be developed that enable optimized planning and control under different target values [4–7].

The sustainability-oriented production scheduling problem has been researched by many authors in the field, and there is ample literature on the topic [8–14]. However, there are circumstances which have led to changes in production processes, such as changes in customers' habits in the past decade, especially in post-crisis times. In this article, we deal with the sustainable scheduling of generalizable systems with a structure similar to the production scheduling problem of a discrete product assembly plant with more heterogeneous production lines, and controlled by one-time orders. Some examples of such systems are digital server channels and parallel transport routes, multichannel IT, and so on. Herein, for simplicity, we will be dealing with sustainable production scheduling, and rely on the concepts used therein when modeling our system.

Several solutions have been applied to the scheduling problem. In addition to linearizing the task, linear programming methods and heuristic methods for the problem have been developed at the University of Miskolc. In practice, other queuing solutions have been used as well.

Several studies [8–12,14] have already shown that one of the best results of most heuristic methods of soft computing is the GA (genetic algorithm) for logistics systems. Following these studies, we chose this method. The advantages of this solution are that it is fast and accurate, especially in comparison to heuristic solutions. By setting the parameters of the objective function to the current production target, production can be quickly redesigned according to the current targets. The algorithm is very simple, and the runtime is better than for LP (linear programming) solutions. In addition, many aspects can be easily considered by the developed solution method. The model is more adaptable. In the daily, weekly, and monthly breakdown of the production organization, the schedule must be available during production. Thus, there is no need for real-time results. Experience has shown that, in the event of a production line failure, the production schedule can be easily rescheduled in a short time.

International competition is intensifying, and many companies are feeling the pressure to shorten their innovation cycles while at the same time further individualizing their customer approach. In order to continue to operate successfully on the market, innovative and intelligent solutions for comprehensive process optimization are required. Intelligent automation, also known as hyperautomation, offers a very good solution here [15–17].

Production planning and control are often characterized by laborious problem solving. A plan often has to be changed at short notice because of a lack of material or employees, or because a machine has broken down. As a rule, there is a lack of transparency about the impact of these adjustments on relevant key figures such as delivery reliability and productivity. Heuristic methods are used for the approximate solution of complex decision and optimization problems or associated optimization models. Opening methods construct a (first) feasible solution, while improvement methods lead to improved (locally optimal) solutions through successive solution transformation. Metastrategies drive improvement processes with a view to investigating promising solution areas and overcoming local optimality. The main contribution of this study is the development of a mathematical model for production scheduling, for which a new, modified GA solution is given. Software has been developed and tested in a practical environment. This can be integrated into a production management system to make the scheduling production processes on flexible production lines more efficient. During the solution, in addition to the appropriate optimal composition of the services, the shorter lead time also ensures fewer shifts, so the use of harmful substances used during the changeover is also less. In addition, less energy consumption helps sustainability, as the optimal solution leaves a smaller footprint.

In this article, we present a detailed mathematical model of the system under investigation. Then, we compile the IT for a solution based on a soft calculation method. We have created a computer application for this model, which we have run multiple times. Runtime results are also presented in this study.

2. Review of the Literature

In the relevant literature, there are many studies on how to improve the efficiency of production schedules. An optimum or close-to-optimum solution results from the use of the genetic algorithm. The genetic algorithm produces new individuals by crossing individual pairs, whose positive properties may improve [18]. During selection, the individuals with the better properties remain. Thus, the solution gradually improves and approaches the optimum [12]. The efficiency of the genetic algorithm is greatly influenced by the applied crossover operator. Today, several crossovers have been developed. Crossbreeding simulates the analogy of genetic crossbreeding in nature. This is a procedure that generates a child entity from two selected parent entities [9]. They differ in various ways, but each is similar in that they either change just the order or just the position [19]. Mutation can do a lot to help eliminate the stagnation of the procedure, making mutations just as important as the crossover operator.

Other heuristic solutions to this problem are suggested in the literature, and the proposed method is also unique in that it can be effectively applied to rescheduling—even in the case of dynamic production scheduling—because it provides a very fast run. No such approach was found with GA [10,20].

Due to rapid technological development, automation has improved significantly in many areas of production. This leads to a complicated situation where decisions need to be made within a short timeframe and from a number of possible cases. Wadhwa and his colleagues suggest a flexible system, where n independent workplace products need to be processed on m machines, and each workplace has the same processing order on the machines [20]. It is important that the system finds jobs on the machines that minimize the make-span. The objective is achieved by the evolutionary heuristics of the genetic algorithm on the sustainable scheduling problem of the flow line. The advantage of the genetic algorithm is that it manages constraints and goals easily, thus facilitating the adaptation of the GA scheduler to a wide range of possible scheduling problems. The results of the study show that the implementation of the genetic algorithm is very effective compared to standard sequencing rules, such as a shorter processing time, the total processing time, and so on.

He and Hui [10] presented a heuristic genetic algorithm for the parallel one-step multipurpose scheduling (SMSP) of large units. First, they proposed a random search based on heuristic rules. By crawling through a set of random solutions, they obtained more feasible solutions. To improve the quality of the solutions, a genetic algorithm under heuristic rule was proposed. Because the run time of the genetic algorithm drastically increased due to certain limitations, a penal procedure was introduced. Thus, the proposed algorithm became effective, and can be used to deal with highly limited, large scheduling problems.

The authors studied the reactive sustainable scheduling method in a previous study [21]. In this previous study, a genetic algorithm-based reactive scheduling method was proposed. When dealing with aggregate production schedules, it is difficult to change the initial schedule due to unknown factors in the manufacturing system. In their research, they modified only a portion of the initial scheduling that sets the appropriate scheduling range.

The studies [10–12] on the field of the genetic algorithm also describe the network scheduling of a typical multipurpose batch plant. Multipurpose process scheduling is more difficult to handle than one-step or multistep process scheduling [14,19,20]. Most authors apply mathematical programming (MP) methods to solve this problem. However, these adjustments result in a very long calculation time. The genetic algorithm proposed by the authors selected a small part of the binary variables to encode into binary chromosomes, which is key to identifying the tasks. The genetic algorithm was first developed with a separate crossover to minimize the makespan and maximize the production.

The researchers of [8] developed a module based on genetic algorithm scheduling based on the priority rules (PRGA-Sched) module that provides shorter completion times in the production scheduling. The module was integrated into the Faborg-Sim simulation tool. Using production data from the Faborg-Sim PRGA Module, a heating boiler production system was analyzed and simulated using six products and orders from customers. Their results showed that a better finishing time and starting position can be achieved through the PRGA-Schedule module.

Dao and Marian [22] presented a genetic algorithm for the integrated optimization of precedent-limited production sequences and scheduling in several production line environments. This group of problems is an NP-hard combinatorial problem, requiring triple optimization: allocation of resources to each production line, production line layout, and production line scheduling. Due to the nature of the constraints, the length of the problem varies. To overcome this variability and the global optimum search, new resource allocation strategies, chromosome, crossover, and mutation were encoded.

Zhang [14] proposed a solution to reduce the pollutants generated through fabric production. First, a three-way model was drawn that included both the traditional delays and the environmental aspects. Then, he presented an innovative solution for the sustainable scheduling problem, namely a multi-objective genetic algorithm with a taboo-enhanced, iterated greedy local search strategy.

We noticed that the issue related to production programs has been approached from several perspectives, as is a complex optimization problem that has been analyzed from different perspectives by different researchers. For example, models such as backpack modelling [23–25], the PSO-GA hybrid algorithm [26–28], the tabu search algorithm [29–32], the improved cuckoo search (ICS) [33–35], Lagrangian heuristic algorithm and other heuristic algorithms [36–40], the mixed model [36–43], and so on, are other types of models used in the literature.

Based on the characteristics of these genetic algorithms, the scope is expanding, and research on the optimization and improvement of genetic algorithms is becoming more sophisticated. In recent years, production planning based on genetic algorithms has been analyzed in the literature using various software [44–54]. Our article contributes to the existing literature in that we did not use existing software. Instead, we developed software based on the genetic algorithm to solve this problem. We consider that this model is easy to adopt for different production problems, and can be customized by the software that we developed in comparison to the models using standard software.

3. Research Methodology

When investigating the problem, we concluded that, after the elaboration of the detailed mathematical model, the evolutionary algorithm was the most efficient solution to solving the problem associated with the model.

An exact solution may be given for the problem (as for all mathematical problems with a finite number of elements), but the number of steps is of a magnitude (appearing as an NP-problem) that precludes practical application, especially for the solution of real-time problems [55,56]. This is already a deterrent in terms of exact solutions. Our practical tests have already shown significant runtimes for small tasks, and the use of the trunk function during the solution presented a further problem (which we simplified in our sample tasks). To solve the problems emerging in practice, an efficient algorithm must be developed.

During our initial attempts, the exact solutions were discarded, and the heuristic methods were unreliable. The evolutionary algorithms were successfully applied to other problems. As such, for the next step in finding a solution, the choice of method seemed to be obvious. Implementing evolutionary algorithms is simple, and we can easily create applications for unique problems.

Evolutionary algorithms have already been used efficiently in practice for solving several problems [57]. Genetic and evolutionary algorithms gained significant popularity in the 1990s, and several studies have analyzed these methods [58–60]. In addition to its simplicity and ease of implementation, adequate modifications could provide even more efficient solutions.

4. Modelling

Investigating the problem reveals an optimization problem. In such cases, the mathematical modelling of the problem is the most convenient way of creating a computational model of it. When preparing the model, one should take care that the level of detail of the model is optimally suited to the problem. All notations of the model are presented in Appendix A. The model does not have to be too detailed, because then it would be unmanageable, while a too broad model would produce inaccurate results [61].

4.1. The Tested System

To demonstrate the operation of the system we have developed, we first define what is meant by the problem of sustainable production scheduling of the order-driven discrete product assembly system with heterogeneous production lines. The discrete product system described above means that the product can be measured in pieces, and the production time of the same product n is n times the product, and its production cost is also n times of the product, and $n \in N^+$. The system is controlled by orders; that is, prior to the operation of the system, orders are known for a production (assembly) period of the plant. We know what kind and how many parts the orders must be produced. To maintain compatibility with other systems (e.g., packet data transfer), all identical products from the (virtual) orders will be sequentially produced on the same production line. Obviously, this should not be a requirement for a conventional schedule. Inhomogeneity of the production lines means that both the lead time and the production cost of a product depend on the product and the production line.

After clarifying our basic concepts, we formulated the task set. The plant had an $n \in N^+$ flexible production line and, for further reference, C_j ($1 \leq j \leq n$) designates a production line. The production line is characterized by its ability to manufacture every product. If this is not the case, then the product on this production line has a high lead time and cost, which will never be allocated. The plant can produce (assemble) different product types ($m \in N^+$) on these production lines. We did not deal with deadlines because we assumed that all acceptable solutions met the ordering deadlines.

Formally, it can be described as the following: $l \in N^+$ orders waiting for manufacturing be in the system.

$$S_{ij} : 1 \leq j \leq k_i \tag{1}$$

$$s = \sum_{i=1}^{l} \sum_{j=1}^{k_i} |S_{ij}|. \tag{2}$$

For example, $S_{ij} := \left\{ \begin{matrix} e_k e_k e_k \dots e_k \\ 1\,2\,3 \dots |S_{ij}| \end{matrix} \right\}$. This kind of series cannot be divided into further parts. $\sigma(S_{ij}) = p$ is the number of series S_{ij}. It will be used the function $\tau(p) := j$ for further reference. The system model uses three essential matrices, all of which create a link between the product and production lines. One is the lead time of one type of product on one production line, and the other is the production cost of the product associated with the production line or the switch time between the products on the production line. These formally mean:

$\mathbf{P}_{n \times m} = \left[p_{ij}\right]$ is manufacturing cost of the product j on line i.

$\mathbf{T}_{n \times m} = \left[t_{ij}\right]$ is turnaround time of product j on line i.

$\mathbf{G}_{n \times s \times s} = \left[g_{i\tau(j)\tau(l)}\right]$, $\mathbf{D}_{n \times s \times s} = \left[d_{i\tau(j)\tau(l)}\right]$ is resetting time and cost from series j to series l on line i. $(1 \leq i \leq n), (1 \leq j \leq s), (1 \leq l \leq s)$.

Our task was to find an optimal or close-to-optimal layout that complied with the set of conditions outlined below. As can be seen from the objective, a dual and counterpart parameter appear in the optimization targeting function of optimization. That is, if we are trying to minimize the cost, the best solution is to assign each series to its optimal production line. If this is achieved through a model in which all series designate a single production line as optimal, the lead time will be the maximum. In the same way, if it is possible to assign each sequence to a line with the optimum lead time, each series may be ordered at the highest cost line. Thus, it appears that the solution is somewhere between these two models.

For these cases, it is important to know the total leading time (T_j) and the manufacturing cost (K_j) of the production line j for an assignment X. For easier handling, suppose that $t_{i\varnothing j} := 0$, meaning before the first series is not reset.

An optimization task formulates the above system which will manufacture all series on a production line, then at least one following objectives are met:

1. Minimal lead time;
2. Minimal manufacturing cost;
3. The compromise objective function between time and cost.

In this paper, the third case will be examined, as the following objective function handles cases 1 and 2:

$$\lambda \cdot \max_{j} T_j + \mu \cdot \sum_{j} K_j \to \min \tag{3}$$

In the first case, $\mu = 0$, in the second case $\lambda = 0$.

4.2. The Mathematical Model

4.2.1. Constraints of the Model

This is the assignment hypermatrix (see Figure 1):

$$\mathbf{X}_{s \times n \times s} = \left[x_{ijk} \right] \tag{4}$$

where this denotes:

$$x_{ijk} = 1 \tag{5}$$

that series i is assigned to manufacturing element k of production line j.

$$x_{ijk} \in \{0;1\} \ (1 \le i \le s; \ 1 \le j \le n; \ 1 \le k \le s) \tag{6}$$

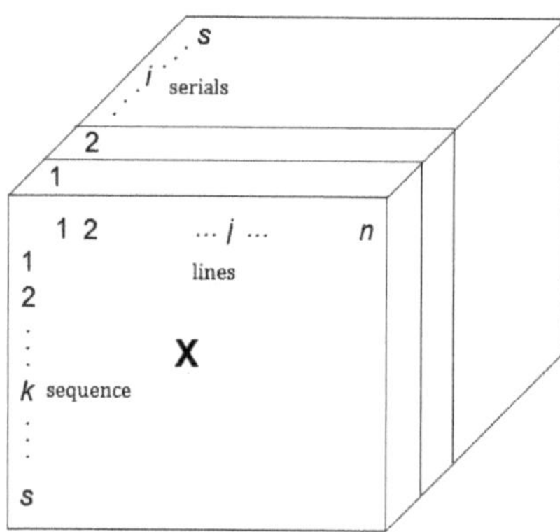

Figure 1. Hypermatrix of the assignment.

Each series is assigned to only one production line and only one manufacturing number:

$$\sum_{j=1}^{n} \sum_{k=1}^{s} x_{ijk} = 1 \ (1 \le i \le s) \tag{7}$$

For each production line, and within a production series, a number is assigned up to one series:

$$\sum_{i=1}^{s} x_{ijk} \le 1 \ (1 \le j \le n, 1 \le k \le s) \tag{8}$$

The series is assigned to the first numbers of the production line. The following condition shows this:

$$\sum_{i=1}^{s} x_{ijk} - \sum_{i=1}^{s} x_{ij(k+1)} \ge 0 \ (1 \le j \le n, 1 \le k < s). \tag{9}$$

4.2.2. The Objective Function

Taking a step forward:

$$trunc(x) = \begin{cases} 1 \text{ if } x \geq 2 \\ 0 \text{ otherswise} \end{cases} \tag{10}$$

The manufacturing cost of the production line j is:

$$K_{p_j}(\mathbf{X}) = \sum_{i=1}^{s} \sum_{k=1}^{s} p_{j\tau(i)} x_{ijk} \tag{11}$$

The resetting cost of the production line j is:

$$K_{t_j}(\mathbf{X}) = \sum_{i=1}^{s} \sum_{k=1}^{s-1} \sum_{r=1}^{s} d_{\tau(i)\tau(r)} \cdot trunc\left(x_{ijk} + x_{rj(k+1)}\right). \tag{12}$$

The total cost of the production line j is:

$$K_j(\mathbf{X}) = K_{p_j}(\mathbf{X}) + K_{t_j}(\mathbf{X})$$
$$= \sum_{i=1}^{s} \sum_{k=1}^{s} p_{j\tau(i)} x_{ijk} + \sum_{i=1}^{s} \sum_{k=1}^{s-1} \sum_{r=1}^{s} d_{\tau(i)\tau(r)} \cdot trunc\left(x_{ijk} + x_{rj(k+1)}\right). \tag{13}$$

The manufacturing time of the production line j is:

$$T_{p_j}(\mathbf{X}) = \sum_{i=1}^{s} \sum_{k=1}^{s} t_{j\tau(i)} x_{ijk}. \tag{14}$$

The reset time of the production line j is:

$$T_{t_j}(\mathbf{X}) = \sum_{i=1}^{s} \sum_{k=1}^{s-1} \sum_{r=1}^{s} g_{\tau(i)\tau(r)} \cdot trunc\left(x_{ijk} + x_{rj(k+1)}\right). \tag{15}$$

The total manufacturing time of the production line j is:

$$T_j(\mathbf{X}) = T_{p_j}(\mathbf{X}) + T_{t_j}(\mathbf{X}) = \sum_{i=1}^{s} \sum_{k=1}^{s} t_{j\tau(i)} x_{ijk} +$$
$$+ \sum_{i=1}^{s} \sum_{k=1}^{s-1} \sum_{r=1}^{s} g_{\tau(i)\tau(r)} \cdot trunc\left(x_{ijk} + x_{rj(k+1)}\right) \tag{16}$$

The total objective function is:

$$f(\mathbf{X}) - \lambda \cdot \max_{j \in \{1,\dots,n\}} T_j(\mathbf{X}) + \mu \cdot \sum_{j=1}^{n} K_j(\mathbf{X})$$
$$= \lambda \cdot \max_{j \in \{1,\dots,n\}} \left(\sum_{i=1}^{s} \sum_{k=1}^{s} t_{j\tau(i)} x_{ijk} + \sum_{i=1}^{s} \sum_{k=1}^{s-1} \sum_{r=1}^{s} g_{\tau(i)\tau(r)} \cdot trunc\left(x_{ijk} + x_{rj(k+1)}\right) \right) \tag{17}$$
$$+ \mu \cdot \sum_{j=1}^{n} \left(\sum_{i=1}^{s} \sum_{k=1}^{s} p_{j\tau(i)} x_{ijk} + \sum_{i=1}^{s} \sum_{k=1}^{s-1} \sum_{r=1}^{s} d_{\tau(i)\tau(r)} \cdot trunc\left(x_{ijk} + x_{rj(k+1)}\right) \right) \to \min.$$

The Complete Model:

$$x_{ijk} \in \{0;1\} \ (1 \leq i \leq s; \ 1 \leq j \leq n; \ 1 \leq k \leq s) \tag{18}$$

$$\sum_{j=1}^{n} \sum_{k=1}^{s} x_{ijk} = 1 \ (1 \leq i \leq s)$$

$$\sum_{i=1}^{s} x_{ijk} \leq 1 \ (1 \leq j \leq n, 1 \leq k \leq s)$$

$$\sum_{i=1}^{s} x_{ijk} - \sum_{i=1}^{s} x_{ij(k+1)} \geq 0 \ (1 \leq j \leq n, 1 \leq k < s)$$

$$f(\mathbf{X}) = \lambda \cdot \max_{j \in \{1,\ldots,n\}} T_j(\mathbf{X}) + \mu \cdot \sum_{j=1}^{n} K_j(\mathbf{X})$$

$$= \lambda \cdot \max_{j \in \{1,\ldots,n\}} \left(\sum_{i=1}^{s} \sum_{k=1}^{s} t_{j\tau(i)} x_{ijk} + \sum_{i=1}^{s} \sum_{k=1}^{s-1} \sum_{r=1}^{s} g_{\tau(i)\tau(r)} \cdot trunc\left(x_{ijk} + x_{rj(k+1)} \right) \right)$$

$$+ \mu \cdot \sum_{j=1}^{n} \left(\sum_{i=1}^{s} \sum_{k=1}^{s} p_{j\tau(i)} x_{ijk} + \sum_{i=1}^{s} \sum_{k=1}^{s-1} \sum_{r=1}^{s} d_{\tau(i)\tau(r)} \cdot trunc\left(x_{ijk} + x_{rj(k+1)} \right) \right) \to \min$$

5. Description of the Heuristic Approach

5.1. Our Previous Solutions and Examinations

The basic problem of optimization is the size of the solution space. If n is the number of production lines and l is the order number, then $D_{li} = \sum_{i=1}^{l}(l - i - 1)! \frac{n!}{(n-i)!}$ is the solution space number. For example, 4 production lines and 50 orders will have a solution space of $1.3 \cdot 10^{65}$ elements. The optimal solution for such a large task is very difficult to ensure, so we have been looking for heuristic solutions for the last few years.

Our first reasoning was built on the greedy algorithm. The solution was based on the fact that the same products were merged and, depending on how the weight values in the target function were configured, the assignment was assigned to the production line with the shortest lead time or with the lowest cost line. Then, with manual tuning, we eliminated the outstanding assignments by smoothing the mergers. This kind of solution provided a usable result. However, the efficiency test found that we did not reach the right target for each order system. Our attention was directed to Ant Colony Optimization (ACO). The results of the completed software were subjected to statistical analysis, and we found that the deterministic sustainable production scheduling tasks realized with the ant algorithm resulted in very large variations. Thus, it was difficult to determine how close this would be to the optimal. However, the advantage of this solution was that it simultaneously handled two parameters—lead time and cost—so the one-on-one contact could be easily controlled.

As expected, with a higher run number and more agents, the average run result yielded a better target than a lower run number or fewer agents. However, the lowest aggregate target value was obtained from the lower run increments. On the contrary, the average target value was close to 1.5% lower for the higher runner step [4]. It was clear that the process did not converge, as can be seen in the examples. Only in the case of a high number of agents and a long run, or an evaporation factor proportional to the maximum path length, could a relatively low target value be found. A detailed examination was unnecessary since, in the examined samples (about 1000 runs), we found that the solution obtained could be very different at the same parameter value. Thus, it was found that usage of the ant algorithm primarily provided good ground for pre-processing other processes. Therefore, we will use the more efficient and verifiable convergent GA solution to play a role in pre-processing [6]. Accordingly, the solutions provided by the ant algorithm form the initial population of the GA. This is positive from the point of view of good-quality initial chromosomes, and, from experience, it is also beneficial that each individual is relatively distant from the other.

5.2. The Structure of the Chromosome

The method is based on a well-designed chromosome. If it is possible to determine the chromosome together with its locus of limbs, such that we can efficiently apply the GA operators to the genes on it, then we can develop a good and usable method. First, we determine the chromosome whose loci number is:

$$g = 2 \cdot s$$

We assign an integer number to every series of the original problem with function (see Table 1) $\sigma(S_{ij}) = p$.

Table 1. $\sigma(i)$ Function.

$\sigma\left(S_{ij}\right)$	1	2	...	k_1	k_1+1	...	k_1+k_2	...	s
S_{ij}	S_{11}	S_{12}	...	S_{1k_1}	S_{22}	...	S_{2k_2}	...	S_{lk_l}

The chromosome will then be structured as follows (see Table 2).

Table 2. Structure of a chromosome.

1		2		...		s	
L_{1j_1}	N_{1j_1}	L_{2j_2}	N_{2j_2}	...		L_{lj_l}	N_{lj_l}

Denote

L_{ij_i}: production line j_i is assigned to series i.

N_{ij_i}: production sequence number of series i on line j_i (production sequence). It is not important that the strict sequence (1,2, ... ,ki) is followed; only the order of the values determines the sequence of series on the line. If two series have the same values, then the series with a lower sequence number will be manufactured first. This consideration will accelerate the algorithm (every result will be a possible result). For example Table 3.

Table 3. An example of a chromosome.

1		2		3		4		5	
0	10	1	4	1	3	0	2	1	4

In Table 3, two lines (0 and 1) are used with 5 locusts of a chromosome. The first gray block shows that series 1 is assigned to line 0.

The series 1 has sequential number 10 and the series 4 has sequential number 2 on line 0. Series 2 has got sequential number 4, series 3 has got sequential number 3, and series 5 has got sequential number 4 on line 1. The allocation will be as follows Table 4.

Table 4. Example of production lines.

Line 0	Series 4	Series 1	
Line 1	Series 3	Series 2	Series 5

The order number and the order series number can be decoded from Table 1. The type of product and the number of an order can be decoded from these numbers with function $\tau\left(S_{ij}\right)$. The number of an order is assigned to series S_{ij}. Resetting cost and leading time can be easily determined from these numbers.

(a) Crossover

P denotes the number of elements in the population.

The first operator of GA is the crossover that creates two new chromosomes by means of any two chromosomes from the population. That is, two new possible solutions are possible in our assignment from two possible assignments. During the crossing, the genes on a given loci interval of one of the parent chromosomes, in which rows are strung to the loci, are exchanged with the genes of this locus domain of the other parent genome, and vice versa. The crossing is based on the starting point and the length of the chromosomes. The procedure is outlined by the following algorithm.

KER be a random integer between 1 and $\frac{p}{2}$ (half of the population's number). The KER represents the number of crosses applied to a particular population.

Select randomly two different individuals from the current population.

Select at random a POZ crossing point (between 1 and 2) and a HOSSZ value (between 1 and 2—of POZ). The POZ shows from which loci to begin the crossing, and the HOSSZ shows the number of those loci on which genes are to be exchanged.

Create two new chromosomes by replacing the genes of the two parents from POZ length to HOSSZ length. See the above example.

Perform the second and third steps KER times.

Consider the following two chromosomes (Table 5).

Table 5. Chromosome of parent entities.

1		2		3		4		5	
0	10	1	4	1	3	0	2	1	4
1		2		3		4		5	
1	7	1	3	0	1	0	4	0	2

Take the starting point 2 and the length 6 (Table 6).

Table 6. Chromosome of successive entities.

1		2		3		4		5	
0	10	1	3	0	1	0	4	1	4
1		2		3		4		5	
1	7	1	4	1	3	0	2	0	2

Note that our crossing solution is flexible. As it is not necessary to replace blocks, it is also possible to replace subblocks. This means that the length can not only be an odd value, but also an odd value, such as 5. In the case of the fourth series, the value determining the production order does not change, only the number of the production line. Based on the above, this still provides a possible solution for all crossings.

(b)   *Mutation*

The role of mutation is to include in the population chromosomes possible solutions that would never enter the optimization through initial design and crossing. Mutations always change the gene of a locus. This, in our case, can be a production line gene, but it can also be a production sequence gene.

MUT is a random integer between 1 and $\frac{p}{2}$ (half of the population's number). This value determines the number of chromosomes that will be allowed mutation in our current population.

Select a random MUT number of individuals from the population.

Take the first selected individual.

Select a locus from this individual randomly. If this is an even-value gene, then generate a value between 1 and s. Change the value of the selected gene to the value obtained.

If it is odd, generate a value between 0 and $n - 1$ (production number 1) and substitute it for the gene's existing value.

After this, select the next individual and perform the steps from step 3 until the mutation is performed for all selected individuals.

(c)   *Selection*

The role of selection is, in addition to the new individuals that have been crossed and mutated, for the somewhat valuable individuals of the present population to be included in the new population (generation) and eventually aid in optimization (the list is in descending order). In our method, we combine the following two selections:

Proportional fitness selection;

Elite list selection.

$p + \left\lceil \dfrac{p}{2} \right\rceil$ be the number of genetically generated entities. The number of all entities is $q = 2p + \left\lceil \dfrac{p}{2} \right\rceil$. Find the best fit for fitness (f). If there is more than one, choose one at random and add this to the new population. $p - 1$ is selected from the $q - 1$ chromosome. Determine the fitness of each individual. This value is weighted using the lead time and total production cost assigned to the solution (evaluation).

5.3. Convergence of the Process

The above procedure converges. Convergence is ensured by the fact that fitness values are limited from the bottom (the value cannot be negative) and because of elite list selection, where at least one of the best individuals of the previous population will be included in the population. Thus, the fitness value for a new population with the smallest fitness value cannot be higher than the best fit in the previous population. Therefore, the smallest fitness value of the population is a monotonous downward series. Thus, it will be a convergent series. The question is whether this series converges to the optimum value. The convergence of the genetic algorithm to optimum is provided in the study [3].

6. Discussions and Results

We have developed a simple application for our investigation, which functions according to the above. During the visualization, we strived for simplicity, since our objective consisted of the adequacy of the model and the applicability of the method.

The examined sample task is presented below. The structure of the sample task corresponds to the outlines of the model. This sample task is from a real company, but it was simplified for reasons of transparency. In some cases, we significantly deviated from the actual numbers to demonstrate the functionality of our developed method.

The task is presented below using the elements of the software setup window.

The Structure of the Sample Task:

The company had four production lines, and our goal was to schedule four production lines (channel).

Our aim was to produce six different products on these production lines.

The number of orders was 40.

The table of the specific orders shows how many units of a specific product had to be produced. The figure shows the products related to the first 16 orders (order) and their number of units.

C#i shows, for the production line i (channel), the turnaround time necessary to produce the product j after the product i (Ti).

The turn-around timetable presents the lead time of specific products (Ti) on the specified production line (Cj).

The cost table shows the production costs of the specific products (Ti) on the specified production line (Cj).

Time is measured in minutes and the cost is measured in euros. In this case, we can dispense the currency and the unit of time, as these do not influence the system (Figure 2).

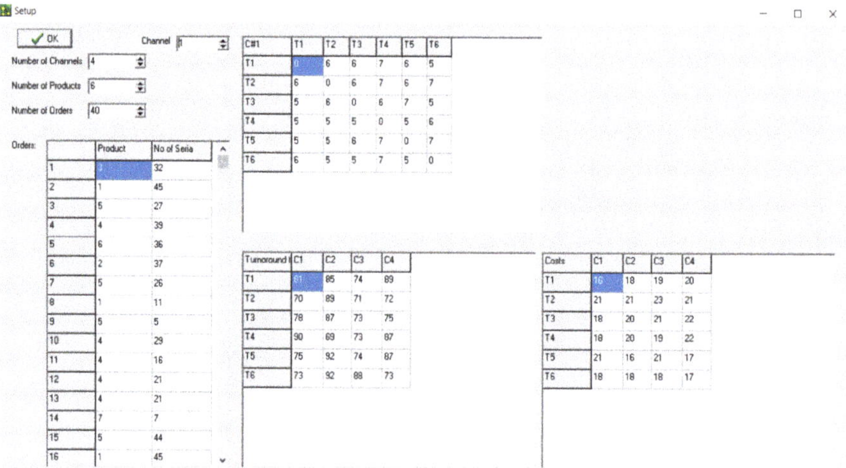

Figure 2. The data of the production scheduling problem (C#i in the computer application means the selected C4 production line).

Our aim was to place the elements in the orders table on the four production lines in such a manner that it was optimal from a perspective of (the list is in descending order):

Lead time;

Minimal costs.

This can be controlled through the alpha and beta parameters shown in Figure 3. If alpha = 1 and beta = 0, then we optimized for lead time. If alpha = 0 and beta = 1, we optimized for costs. The two criteria were considered in weighted form for all subsequent results.

Under the set conditions, along with the two aforementioned cases, we also presented a case in which both criteria were taken into account with different weights.

The run included 200 steps for each investigation. The run results from the following figures had the chromosomes obtained during the selection, with the individual lines representing the chromosomes. The fields for setting the two parameters can be seen in the upper part of Figure 3.

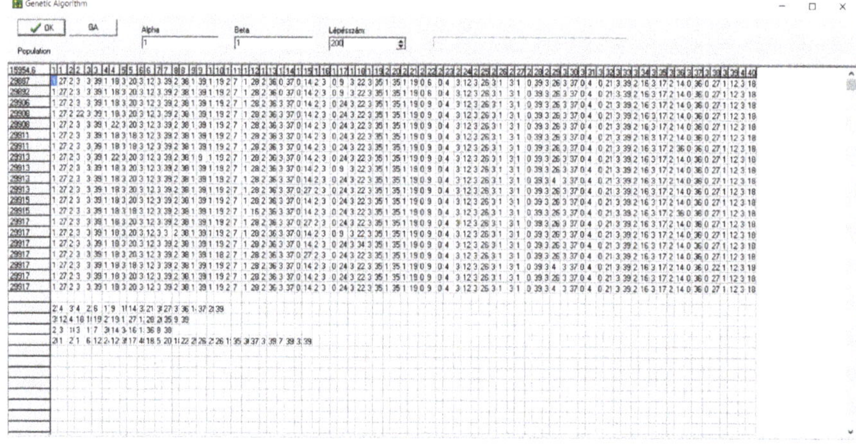

Figure 3. The optimization panel with the parameters.

Depending on the change in parameters, we investigated the result provided by the method. In the case of examined samples, when turnaround time was optimized, the turnaround time could not be less than 16,163 time units, and the cost could not be less than 18,494 cost units.

Figure 4 shows a run result in a case where the lead time parameter alpha = 1, the figure shows that the system seeks equal load. The figure can be interpreted as follows.

Each order is associated with a color. The quantities to be produced for each order are shown by the white numbers. The gray gaps represent inactive times. The breadth of the individual orders symbolize the necessary production time (the broader the order, the longer the production time), and the color yellow represents the production lines (of which there are currently four).

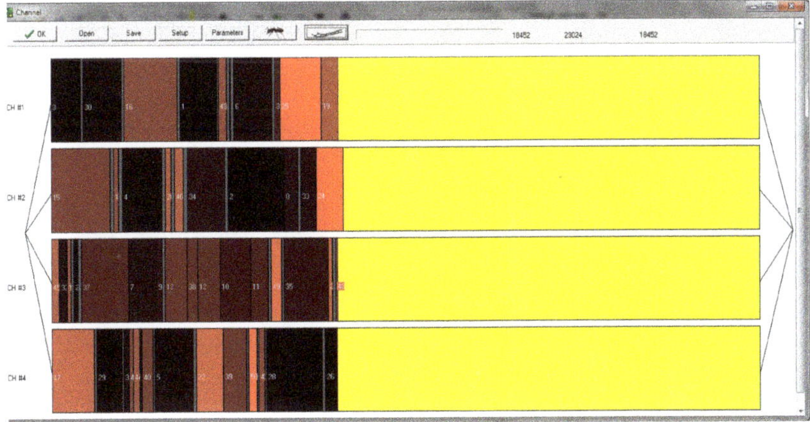

Figure 4. Uniform allocation in case of parameter $\alpha = 1$.

If the cost is considered for the purpose of optimization, the 18,189.75 time units cannot amount to less than 16,273 units of expenditure. The example is examined for the expenditure case in which the parameter of the expenditure (beta) is 1 and the lead time (alpha) is 0 (Figure 5).

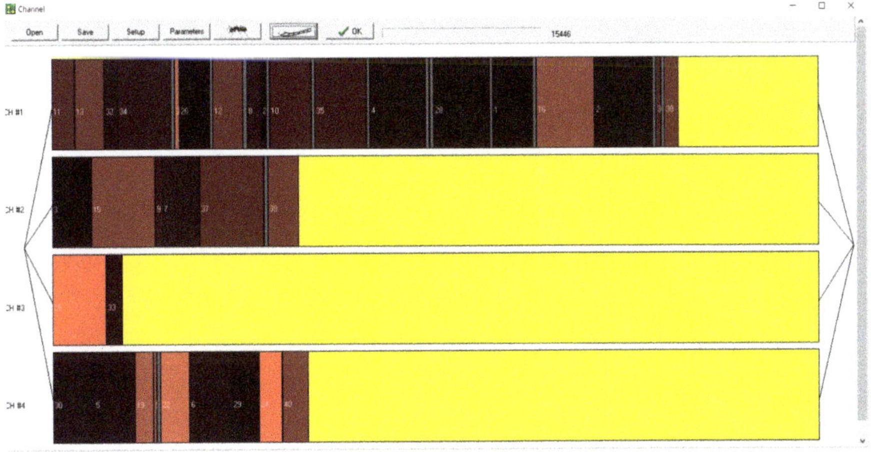

Figure 5. Allocation for high-cost production lines (3 lines).

Based on the description of the sample task, the third production line was among one of the most expensive lines, from the perspective of the products to be manufactured. The solution shows that the assembly line that was expensive relative to the ordered products was only minimally used by the system. This can be seen on the third production line.

In practice, the joint investigation of the two criteria can be conducted by considering the experience. In our examined example, the experiential result was as follows (Figure 6).

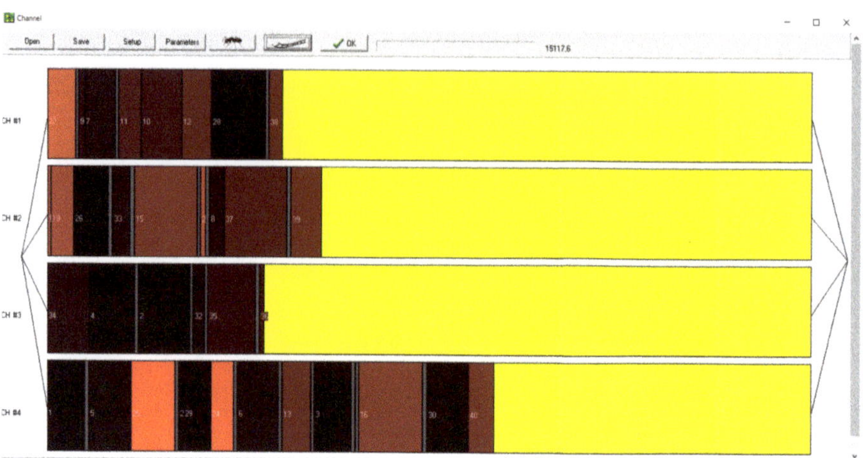

Figure 6. The result of the allocation for the parameters alpha = 0.6 and beta = 0.4.

This was in line with the practical experience that most orders were allocated to the fourth line.

7. Efficiency of the Algorithm

In the examined sample tasks, the program found the near-optimal solution in a maximum of 200 steps.

The run times of the individual cases did not exceed 30 s.

In general, based on experience to date, the program performs thousands of cycles in the optimal search for the most complex task. The running time for complex tasks was no more than 5 min. The program does not require special hardware.

8. Conclusions

8.1. Theoretical and Practical Implication

In this research, we presented the results of the objectives set out in the introduction. The first goal was to create an exact unique mathematical model for the task, so that the GA's IT of the task could be fitted onto this model.

What caused the unreliability of previous heuristic solutions? Studies have shown that, due to the robustness of such tasks, only in certain cases is the solution's optimum closeness of the solution ensured. In some cases, the solutions differed very much from the actual—that is, from the optimum obtained by us through the analysis of a well-analyzed test system—since the methods were largely based on the conditions of an average system. Unfortunately, the ant colony optimization was different in behavior. This solution had the great advantage of avoiding the alpha and beta parameters of the current method. The method itself tried to adjust to optimize costs and lead time together, and that is what caused its failure. As demonstrated by the combined effect of the two objectives, we did not obtain targets changing in the same way as the two parameters analyzed with the methods we employed, so the system could not produce good solutions. Sadly, through its use, we obtained possible solutions but not an optimal solution.

The third objective was to provide a GA model that would fit the task. Basic models can be adapted with little modification, since, in our case, the structure of the chromosomes and the genes to be replaced by the locus used were special in their structure. Genes contain two different elements, so genetic operators must adapt to this. The unique chromosomes and operators created fitted the mathematical model.

In this research, algorithms for generating time-optimized solutions were implemented in a software program. For this purpose, all necessary steps were presented, starting with a probabilistic movement planning algorithm with fast and accurate collision detection, a heuristic solver for process flow optimization and load balancing, up to a practical validation of the methodology. Due to the high complexity, a robust process automation or optimization with common planning and technical methods can only be implemented to a limited extent. Mathematical analyzes and computer-aided methods can make additional contributions to greater robustness and increased efficiency. Time-optimized programs for process optimization are supplied as a practical application. For this, it is necessary to integrate the methods into the production network. The evaluation of the results is based on precise simulations with the developed software. In this paper, we demonstrated that order-driven production can be mathematically described, and be assigned to a well-managed model. The model condition criteria were simple, but the complexity of the target function required that the task related to the model was not solved by using the exact method. Compared to previous studies, we have now solved the problem with a variation of the genetic algorithm. This solution method provided the best results compared to our previous experiments.

In the past, we performed numerous runs and evaluated the results using statistical methods. These findings also supported our hypothesis that soft calculation methods are well suited for production tasks as well.

8.2. Limitations and Future Research

For future research, it will be important to investigate the scalability of the solver so that the entire production line can be optimized, instead of a single process. This is particularly interesting for load balancing because the number of tasks is much larger. A promising approach is to first build a population of good and diverse solutions using heuristics, and then apply artificial intelligence methods to identify a common pattern among them according to the pattern that frequently occurs.

Integer values were four-byte integers, so these were not a limitation in the practical problem. Variables were dynamic, so the hardware set the limits. The results of the tests so far, which were used in a medium-sized production plant, were satisfactory. Very large serial numbers have not been published, but based on the experience so far, the previous day's production plan is prepared quickly—well within deadlines.

Furthermore, in the future, applications are to be examined in which robot programs are created solely from the specifications of product planning. Reinforcement learning can be used for this in order to train a model from the processes. In this way, all relevant process technology requirements and tolerances can be recorded.

Author Contributions: Conceptualization: Á.G., M.G. and M.T.F.; methodology Á.G. and M.G.; writing—original draft preparation, Á.G., M.G., M.T.F. and M.A.; writing—review and editing: M.G. and M.T.F.; visualization M.G. and M.A.; supervision M.T.F.; project administration M.T.F. and M.A.; funding acquisition M.A. All authors have read and agreed to the published version of the manuscript.

Funding: This research was funded by Budapest Business School Research Fund.

Institutional Review Board Statement: Not applicable.

Informed Consent Statement: Not applicable.

Data Availability Statement: Not applicable.

Acknowledgments: This research was carried out with the framework of the Centre of Excellence for Future Value Chains of Budapest Business School.

Conflicts of Interest: The authors declare no conflict of interest.

Appendix A

Notation	Range	Explanation
$n \in N^+$		the number of flexible production lines
$m \in N^+$		the number of different product types
$l \in N^+$		the number of orders waiting for manufacturing
i		the series number of the order
j		the series number of the production line
k		the manufacturing element
k_i		the number of the series of order i
C_j	$(1 \leq j \leq n)$	ID of production line j
S_{ij}	$S_{ij} : 1 \leq j \leq k_i$	the series j of the order i.
s		the total number of series of all orders
$\sigma\left(S_{ij}\right) = p$		the number of the series S_{ij}
$\tau(p) := j$		
$\mathbf{P}_{n \times m} = \left[p_{ij}\right]$		the manufacturing cost of the product j on line i.
$\mathbf{T}_{n \times m} = \left[t_{ij}\right]$		the turnaround time of product j on line i.
$\mathbf{G}_{n \times s \times s} = \left[g_{i\tau(j)\tau(l)}\right],$ $\mathbf{D}_{n \times s \times s} = \left[d_{i\tau(j)\tau(l)}\right]$	$(1 \leq i \leq n), (1 \leq j \leq s),$ $(1 \leq l \leq s)$	the resetting time and cost from series j to serial l on line i.
$\mathbf{X}_{s \times n \times s} = \left[x_{ijk}\right]$	$x_{ijk} \in \{0;1\}$ $\left(\begin{array}{c}1 \leq i \leq s; 1 \leq j \leq n; \\ 1 \leq k \leq s\end{array}\right)$	$x_{ijk} = 1$ that series i is assigned to manufacturing element k of production line j.
$K_{p_j}(\mathbf{X})$		The manufacturing cost of the production line j
$K_{t_j}(\mathbf{X})$		The resetting cost of the production line j
$K_j(\mathbf{X})$		The total cost of the production line j
$T_{p_j}(\mathbf{X})$		The manufacturing time of the production line j
$T_{t_j}(\mathbf{X})$		The reset time of the production line j
$T_j(\mathbf{X})$		The total manufacturing time of the production line j
$f(\mathbf{X})$		The total objective function
$\lambda=$ alpha (α) in the software	$\left[\frac{1}{min}\right]$	Normalizing and at the same time weight factor Lead time parameter
$\mu =$ alpha (β) in the software	$\left[\frac{1}{Euro}\right]$	Normalizing and at the same time weight factor Cost parameter

References

1. Zambon, I.; Cecchini, M.; Egidi, G.; Saporito, M.G.; Colantoni, A. Revolution 4.0: Industry vs. Agriculture in a Future Development for SMEs. *Processes* **2019**, *7*, 36. [CrossRef]
2. Salah, B.; Khan, S.; Ramadan, M.; Gjeldum, N. Integrating the Concept of Industry 4.0 by Teaching Methodology in Industrial Engineering Curriculum. *Processes* **2020**, *8*, 1007. [CrossRef]
3. Sevinç, A.; Gür, Ş.; Eren, T. Analysis of the Difficulties of SMEs in Industry 4.0 Applications by Analytical Hierarchy Process and Analytical Network Process. *Processes* **2018**, *6*, 264. [CrossRef]
4. Liu, Y.; Dong, H.; Wang, S.; Lan, M.; Zeng, M.; Zhang, S.; Yang, M.; Yin, S. An Optimization Approach Considering User Utility for the PV-Storage Charging Station Planning Process. *Processes* **2020**, *8*, 83. [CrossRef]
5. Szentesi, S.; Illés, B.; Cservenák, Á.; Skapinyecz, R.; Tamás, P. Multi-Level Optimization Process for Rationalizing the Distribution Logistics Process of Companies Selling Dietary Supplements. *Processes* **2021**, *9*, 1480. [CrossRef]
6. Liu, W.; Luo, F.; Liu, Y.; Ding, W. Optimal Siting and Sizing of Distributed Generation Based on Improved Nondominated Sorting Genetic Algorithm II. *Processes* **2019**, *7*, 955. [CrossRef]
7. Ehyaei, M.A.; Ahmadi, A.; Rosen, M.A.; Davarpanah, A. Thermodynamic Optimization of a Geothermal Power Plant with a Genetic Algorithm in Two Stages. *Processes* **2020**, *8*, 1277. [CrossRef]
8. Aydemir, E.; Koruca, H.I. A New Production Scheduling Module Using Priority-Rule Based Genetic Algorithm. *Int. J. Simul. Model.* **2015**, *14*, 450–462. [CrossRef]
9. Akram, U.; Fülöp, M.T.; Tiron-Tudor, A.; Topor, D.I.; Căpusneanu, S. Impact of Digitalization on Customers' Well-Being in the Pandemic Period:Challenges and Opportunities for the Retail Industry. *Int. J. Environ. Res. Public Health* **2021**, *18*, 7533. [CrossRef]
10. He, Y.; Hui, C.-W. A binary coding genetic algorithm for multi-purpose process scheduling: A case study. *Chem. Eng. Sci.* **2010**, *65*, 4816–4828. [CrossRef]
11. Fülöp, M.T.; Szora Tamas, A.; Ivan, O.R.; Solovăstur, A.N. Regressive model regarding the necessary profit margin forecast for a new project in the construction field. *Econ. Comput. Econ. Cybern. Stud. Res.* **2020**, *54*, 181–198. [CrossRef]
12. Salido, M.A.; Escamilla, J.; Giret, A.; Barber, F. A genetic algorithm for energy-efficiency in job-shop scheduling. *Int. J. Adv. Manuf. Technol.* **2015**, *85*, 1303–1314. [CrossRef]
13. Gu, Z.; Chen, M.; Wang, C.; Zhuang, W. Static and Dynamic Analysis of a 6300 KN Cold Orbital Forging Machine. *Processes* **2020**, *9*, 7. [CrossRef]
14. Zhang, R. Sustainable Scheduling of Cloth Production Processes by Multi-Objective Genetic Algorithm with Tabu-Enhanced Local Search. *Sustainability* **2017**, *9*, 1754. [CrossRef]
15. Müller, J.M.; Däschle, S. Business Model Innovation of Industry 4.0 Solution Providers Towards Customer Process Innovation. *Processes* **2018**, *6*, 260. [CrossRef]
16. Borowski, P. Innovative Processes in Managing an Enterprise from the Energy and Food Sector in the Era of Industry 4. *Processes* **2021**, *9*, 381. [CrossRef]
17. Anser, M.; Khan, M.; Awan, U.; Batool, R.; Zaman, K.; Imran, M.; Sasmoko; Indrianti, Y.; Khan, A.; Bakar, Z. The Role of Technological Innovation in a Dynamic Model of the Environmental Supply Chain Curve: Evidence from a Panel of 102 Countries. *Processes* **2020**, *8*, 1033. [CrossRef]
18. Shim, S.-O.; Park, K. Technology for Production Scheduling of Jobs for Open Innovation and Sustainability with Fixed Processing Property on Parallel Machines. *Sustainability* **2016**, *8*, 904. [CrossRef]
19. Zhang, R.; Chiong, R. Solving the energy-efficient job shop scheduling problem: A multi-objective genetic algorithm with enhanced local search for minimizing the total weighted tardiness and total energy consumption. *J. Clean. Prod.* **2016**, *112*, 3361–3375. [CrossRef]
20. Wadhwa, S.; Madaan, J.; Raina, R. A Genetic Algorithm Based Scheduling for a Flexible System. *Glob. J. Flex. Syst. Manag.* **2007**, *8*, 15–24. [CrossRef]
21. Sakaguchi, T.; Kamimura, T.; Shirase, K.; Tanimizu, Y. GA Based Reactive Scheduling for Aggregate Production Scheduling. *Manuf. Syst. Technol. New Front.* **2008**, *7*, 275–278. [CrossRef]
22. Dao, S.D.; Marian, R.M. Genetic Algorithms for Integrated Optimisation of Precedence-Constrained Production Sequencing and Scheduling. In *Lecture Notes in Electrical Engineering*; Springer Science and Business Media LLC: Berlin/Heidelberg, Germany, 2012; Volume 130, pp. 65–80.
23. Ning, S.-S.; Wang, W.; Liu, Q.-L. An optimal scheduling algorithm for reheating furnace in steel production. *Control Decis.* **2006**, *21*, 1138–1142.
24. Gong, Z.; Li, J.; Luo, Z.; Wen, C.; Wang, C.; Zelek, J. Mapping and Semantic Modeling of Underground Parking Lots Using a Backpack LiDAR System. *IEEE Trans. Intell. Transp. Syst.* **2021**, *22*, 734–746. [CrossRef]
25. Ren, L.; Howard, D.; Jones, R.K. Mathematical Modelling of Biomechanical Interactions between Backpack and Bearer during Load Carriage. *J. Appl. Math.* **2013**, *2013*, 1–12. [CrossRef]
26. Kim, J.H.; Ma, S.B.; Kim, S.; Choi, Y.S.; Kim, K.Y. Design and verification of a single-channel pump model based on a hybrid optimization technique. *Processes* **2019**, *7*, 747. [CrossRef]
27. Li, C.; Zhai, R.; Liu, H.; Yang, Y.; Wu, H. Optimization of a heliostat field layout using hybrid PSO-GA algorithm. *Appl. Therm. Eng.* **2018**, *128*, 33–41. [CrossRef]

28. Choudhary, A.; Kumar, M.; Gupta, M.K.; Unune, D.K.; Mia, M. Mathematical modeling and intelligent optimization of submerged arc welding process parameters using hybrid PSO-GA evolutionary algorithms. *Neural Comput. Appl.* **2019**, *32*, 5761–5774. [CrossRef]
29. Xu, A.; Lu, Y.; Da, D.; Ti, N.; He, D. Hybrid direct hot charge rolling production for specific reheating furnace mode. *J. Univ. Sci. Technol. Beijing* **2012**, *34*, 1091–1096.
30. Mao, K.; Pan, Q.; Tasgetiren, M.F. Lagrangian heuristic for scheduling a steelmaking-continuous casting process. In Proceedings of the 2013 IEEE Symposium on Computational Intelligence in Scheduling (CISched), Singapore, 16–19 April 2013; pp. 68–74.
31. Hou, N.; He, F.; Zhou, Y.; Chen, Y. An efficient GPU-based parallel tabu search algorithm for hardware/software co-design. *Front. Comput. Sci.* **2020**, *14*, 1–18. [CrossRef]
32. Burduk, A.; Musiał, K.; Kochańska, J.; Górnicka, D.; Stetsenko, A. Tabu search and genetic algorithm for production process scheduling problem. *LogForum* **2019**, *15*, 181–189. [CrossRef]
33. Sun, L.; Luan, F.; Ying, Y.; Mao, K. Rescheduling optimization of steelmaking-continuous casting process based on the Lagrangian heuristic algorithm. *J. Ind. Manag. Optim.* **2017**, *13*, 1431–1448. [CrossRef]
34. Hu, H.X.; Lei, W.X.; Gao, X.; Zhang, Y. Job-Shop Scheduling Problem Based on Improved Cuckoo Search Algorithm. *Int. J. Simul. Model.* **2018**, *17*, 337–346. [CrossRef]
35. Zhang, L.; Yu, Y.; Luo, Y.; Zhang, S. Improved cuckoo search algorithm and its application to permutation flow shop scheduling problem. *J. Algorithms Comput. Technol.* **2020**, *14*, 1748302620962403. [CrossRef]
36. Ghosh, T.K.; Das, S.; Barman, S.; Goswami, R. Job scheduling in computational grid based on an improved cuckoo search method. *Int. J. Comput. Appl. Technol.* **2017**, *55*, 138–146. [CrossRef]
37. Wang, X.; Tang, L. Integration of batching and scheduling for hot rolling production in the steel industry. *Int. J. Adv. Manuf. Technol.* **2008**, *36*, 431–441. [CrossRef]
38. Lambiase, F. Optimization of shape rolling sequences by integrated artificial intelligent techniques. *Int. J. Adv. Manuf. Technol.* **2013**, *68*, 443–452. [CrossRef]
39. Liu, H.; Liu, H.; Sun, F.; Fang, B. Kernel regularized nonlinear dictionary learning for sparse coding. *IEEE Trans. Syst. Man Cybern. Syst.* **2017**, *99*, 1–10. [CrossRef]
40. Liu, H.; Wu, Y.; Sun, F.; Fang, B.; Guo, D. Weakly Paired Multimodal Fusion for Object Recognition. *IEEE Trans. Autom. Sci. Eng.* **2017**, *15*, 784–795. [CrossRef]
41. Leu, S.-S.; Hwang, S.-T. GA-based resource-constrained flow-shop scheduling model for mixed precast production. *Autom. Constr.* **2002**, *11*, 439–452. [CrossRef]
42. Yan, S.; Lai, W. An optimal scheduling model for ready mixed concrete supply with overtime considerations. *Autom. Constr.* **2007**, *16*, 734–744. [CrossRef]
43. Caridi, M.; Sianesi, A. Multi-agent systems in production planning and control: An application to the scheduling of mixed-model assembly lines. *Int. J. Prod. Econ.* **2000**, *68*, 29–42. [CrossRef]
44. de Prada, C.; Grossmann, I.; Sarabia, D.; Cristea, S. A strategy for predictive control of a mixed continuous batch process. *J. Process Control* **2009**, *19*, 123–137. [CrossRef]
45. Silva, A.F.D.; Marins, F.A.S.; Montevechi, J.A.B. Application of mixed binary goal programming in an enterprise in the sugar and energy sector. *Gestão Produção* **2013**, *20*, 321–336. [CrossRef]
46. Lim, M.; Zhang, D. An integrated agent-based approach for responsive control of manufacturing resources. *Comput. Ind. Eng.* **2004**, *46*, 221–232. [CrossRef]
47. Scarlat, E.; Boloş, M.; Popovici, I. Agent-based modeling in decision-making for project financing. *J. Econ. Comput. Econ. Cybern. Stud. Res.* **2011**, 5–10, WOS:000292347200001.
48. Rody, R.; Mahmudy, W.F.; Tama, I.P. Using Guided Initial Chromosome of Genetic Algorithm for Scheduling Production-Distribution System. *J. Inf. Technol. Comput. Sci.* **2019**, *4*, 26–32. [CrossRef]
49. Que, Y.; Zhong, W.; Chen, H.; Chen, X.; Ji, X. Improved adaptive immune genetic algorithm for optimal QoS-aware service composition selection in cloud manufacturing. *Int. J. Adv. Manuf. Technol.* **2018**, *96*, 4455–4465. [CrossRef]
50. Sharma, S.; Chadha, M.; Kaur, H. Multi-step crossover genetic algorithm for bi-criteria parallel machine scheduling problems. *Int. J. Math. Oper. Res.* **2021**, *18*, 71. [CrossRef]
51. Li, S.; Yu, T.; Cao, X.; Pei, Z.; Yi, W.; Chen, Y.; Lv, R. Machine learning-based scheduling: A bibliometric perspective. *IET Collab. Intell. Manuf.* **2021**, *3*, 131–146. [CrossRef]
52. Li, M.W.; Hong, W.C.; Geng, J.; Wang, J. Berth and quay crane coordinated scheduling using multiobjective chaos cloud particle swarm optimization algorithm. *Neural Comput. Appl.* **2017**, *28*, 3163–3182. [CrossRef]
53. Lamghari, A.; Dimitrakopoulos, R.; A Ferland, J. A variable neighbourhood descent algorithm for the open-pit mine production scheduling problem with metal uncertainty. *J. Oper. Res. Soc.* **2014**, *65*, 1305–1314. [CrossRef]
54. Koo, J.; Kim, B.-I. Some comments on "Optimization of production scheduling with time-dependent and machine-dependent electricity cost for industrial energy efficiency". *Int. J. Adv. Manuf. Technol.* **2016**, *86*, 2803–2806. [CrossRef]
55. Aghelinejad, M.; Ouazene, Y.; Yalaoui, A. Production scheduling optimisation with machine state and time-dependent energy costs. *Int. J. Prod. Res.* **2017**, *56*, 5558–5575. [CrossRef]
56. Yao, M.-J.; Huang, J.-X. Solving the economic lot scheduling problem with deteriorating items using genetic algorithms. *J. Food Eng.* **2005**, *70*, 309–322. [CrossRef]

57. Salga, P.; Szilágyi, R.; Herdon, M. Genetikus algoritmus alkalmazása a mezőgazdasági termelés optimalizálásában (The use of genetic algorithms in the optimization of agricultural production). In *E-agrárium & E-vidék: Agrárinformatikai Nyári Egyetem és Agrárinformatikai Fórum*; Magyar Agrárinformatikai Szövetség: Debrecen, Hungary, 2013; pp. 1–7.
58. Capusneanu, S.; Topor, D.I.; Hint, M.S.; Ionescu, C.A.; Coman, M.D.; Paschia, L.; Nicolau, N.L.G.; Ivan, O.R. Mathematical model for identifying and quantifying the overall environmental cost. *J. Bus. Econ. Manag.* **2020**, *21*, 1307–1328. [CrossRef]
59. Coita, I.F.; Cioban, S.; Mare, C. Is Trust a Valid Indicator of Tax Compliance Behaviour? A Study on Taxpayers' Public Perception Using Sentiment Analysis Tools. In Proceedings of the 4th International Conference on Resilience and Economic Intelligence through Digitalization and Big Data Analytics, Bucharest, Romania, 10–11 June 2021; ICESS Springer in Business and Economics: Bucharest, Romania, 2021.
60. Ionescu, C.A.; Fülöp, M.T.; Topor, D.I.; Căpușneanu, S.; Breaz, T.O.; Stănescu, S.G.; Coman, M.D. The New Era of Business Digitization through the Implementation of 5G Technology in Romania. *Sustainability* **2021**, *13*, 13401. [CrossRef]
61. Gubán, M. *Matematikai modellezés Az önfenntartó falugazdaság modellje a hálózati gazdaságban (Mathematical Modelling: The Model of the Self-Sustaining Village Economy in Network Economies)*; Budapesti Gazdasági Főiskola: Salgótarján, Hungary, 2005.

Article

Sustainable Solar Drying of Brewer's Spent Grains: A Comparison with Conventional Electric Convective Drying

Juan Pablo Capossio [1], María Paula Fabani [2], Andrés Reyes-Urrutia [1], Rodrigo Torres-Sciancalepore [1], Yimin Deng [3], Jan Baeyens [3,4,*], Rosa Rodriguez [5] and Germán Mazza [1,*]

[1] Instituto de Investigación y Desarrollo en Ingeniería de Procesos, Biotecnología y Energías Alternativas, PROBIEN (CONICET-UNCo), Neuquén 8300, Argentina; juan.capossio@probien.gob.ar (J.P.C.); andres.reyes@probien.gob.ar (A.R.-U.); rodrigo.torres@probien.gob.ar (R.T.-S.)

[2] Instituto de Biotecnología, Facultad de Ingeniería, Universidad Nacional de San Juan, San Juan 5400, Argentina; paufabani@unsj.edu.ar

[3] Process and Environmental Technology Lab, Department of Chemical Engineering, Katholieke Universiteit Leuven, 2860 Sint-Katelijne-Waver, Belgium; yimin.deng@kuleuven.be

[4] Beijing Advanced Innovation Centre for Soft Matter Science and Engineering, Beijing University of Chemical Technology, Beijing 100029, China

[5] Grupo Vinculado al PROBIEN (CONICET-UNCo), Instituto de Ingeniería Química, Facultad de Ingeniería, Universidad Nacional de San Juan, San Juan 5400, Argentina; rrodri@unsj.edu.ar

*   Correspondence: jan.baeyens@kuleuven.be (J.B.); german.mazza@probien.gob.ar (G.M.)

Citation: Capossio, J.P.; Fabani, M.P.; Reyes-Urrutia, A.; Torres-Sciancalepore, R.; Deng, Y.; Baeyens, J.; Rodriguez, R.; Mazza, G. Sustainable Solar Drying of Brewer's Spent Grains: A Comparison with Conventional Electric Convective Drying. *Processes* **2022**, *10*, 339. https://doi.org/10.3390/pr10020339

Academic Editor: Peter Glavič

Received: 14 December 2021
Accepted: 7 February 2022
Published: 10 February 2022

Publisher's Note: MDPI stays neutral with regard to jurisdictional claims in published maps and institutional affiliations.

Abstract: Spent grains from microbreweries are mostly formed by malting barley (or malt) and are suitable for a further valorization process. Transforming spent grains from waste to raw materials, for instance, in the production of nontraditional flour, requires a previous drying process. A natural convection solar dryer (NCSD) was evaluated as an alternative to a conventional electric convective dryer (CECD) for the dehydration process of local microbrewers' spent grains. Two types of brewer's spent grains (BSG; Golden ale and Red ale) were dried with both systems, and sustainability indices, specific energy consumption (e_C), and CO_2 emissions were calculated and used to assess the environmental advantages and disadvantages of the NCSD. Then, suitable models (empirical, neural networks, and computational fluid dynamics) were used to simulate both types of drying processes under different conditions. The drying times were 30–85 min (depending on the drying temperature, 363.15 K and 333.15 K) and 345–430 min (depending on the starting daytime hour at which the drying process began) for the CECD and the NCSD, respectively. However, e_C and CO_2 emissions for the CECD were 1.68–1.88 · 10^{-3} (kW h)/kg and 294.80–410.73 kg/(kW h) for the different drying temperatures. Using the NCSD, both indicators were null, considering this aspect as an environmental benefit.

Keywords: natural convection solar dryer; electric convective dryer; brewer's spent grains; waste valorization; artificial neural networks; computational fluid dynamics

1. Introduction

The beer production market has been a highly concentrated one worldwide for a long time, with a substantial market share held by a few brewing companies which produce the most consumed industrial lager type of beer. Starting a decade ago, however, local microbreweries have been increasingly capturing the consumer's interest with a wide range of different types of beer. The Argentinian beer industry does not escape this reality, and the craft beer market was growing 30% annually (pre-pandemic) [1]. Beer in all its variants is one of the most popular alcoholic beverages in the world. In Argentina in particular, its consumption began to grow steadily, reaching $2 \cdot 10^{10}$ L/a [2].

Beer is made from sugars acquired from cereals and other grains (mainly barley and wheat), flavored and aromatized mostly with hops, but also with other herbs and additives, which are fermented in water with yeasts of the *Saccharomyces* type. In this process, large amounts of a solid fraction residue are produced, called brewer's spent grains (BSG) [3], which form the most abundant byproduct of the brewing process, representing 85 % of the total residue and, on average, 31 % of the original weight of the malt spent in the process [4,5].

As a consequence of the appearance of microbreweries scattered around the country, large amounts of spent grains have been made available locally. The disposal of this residue is a serious environmental problem for communities and breweries because the volume of BSG generated is around 0.60 kg/L of beer brewed (as reported by the establishment which provided the samples and [6]). BSG are mainly used for animal feed; also, in a low fraction, for biogas generation; or are eliminated in landfills [7]. However, those byproducts potentially have several applications because of their year-round availability, low cost, and valuable chemical composition. The chemical composition is influenced by diverse factors like the grain used (barley, wheat, rice, corn, among others), processing method, harvesting time, and brewing conditions (i.e., the recipe used) [8]. Spent grains mainly consist of 15–26% protein and 70% fiber. The fiber consist of 16–25% cellulose, 28–35% hemicellulose, and 7–28% lignin [5].

Bolwig et al. [7] and Kavalopoulos et al. [9] studied the valorization of BSG as animal feed and biofuel, respectively, in the framework of biorefining and circular bioeconomy. Additionally, Fabani et al. [10] proposed a byproduct development process for watermelon rind in which rinds are transformed into nontraditional flour. Similarly, BSG are materials suitable for further valorization through such a process. For instance, BSG are added as an ingredient to enhance the fiber contents in recipes for pasta [11], bread, and snacks [3,12,13]. Recently, the potential use of BSG as fuel and food in Africa and its global warming capacity were studied by Maqhuzu et al. [14]. In addition, BSG can be used to directly recover fibers and antioxidant compounds [15].

Because of its high initial moisture (60–80%), transforming BSG from waste to raw materials requires a prior drying process. High moisture makes the product vulnerable to fast deterioration and spoilage within a few hours after brewing is over [16], and BSG at a disposal site like a landfill show an emission (expressed in CO_2 equivalents ($CO_{2, eqv}$)) of 513 kg/t [9]. Currently, several devices are used for food dehydration, like hot air oven dryers and natural convection solar dryers (NCSD) [17,18]. The latter are economic to fabricate, reliable, environmentally friendly, and have lower operational costs than the former while being more efficient than direct sun dryers. In addition, solar drying reduces the $CO_{2, eqv}$, for instance, by reducing the use of conventional energy by as much as 80 % if forced convection is used [19]. Its working principle is based on sunrays directly heating the inlet air, thus lowering its humidity. Such an effect makes the air flow toward the dryer's exhaust naturally. The passing air removes the moisture from the product on the trays and releases it into the outer atmosphere. An example can be found in [20] where a mixed-type solar dryer was successfully used for the dehydration of pear slices (*Pyrus communis* L.).

There is little information about BSG drying in the literature. Mallen and Najdanovic-Visak [21] studied the BSG drying kinetics at four temperatures (333.15 K, 343.15 K, 353.15 K, and 363.15 K) to extract biodiesel, while Kavalopoulos et al. [9] dehydrated BSG and milled them at the same time with a rotary drum food waste dryer at 378.15 K to produce biodiesel, bioethanol, and/or biogas.

The conventional modeling procedure for predicting the drying characteristic of a foodstuff is to use an existing empirical correlation based on a large number of experiments, like the Midilli model [10]. However, these correlations are frequently not sufficiently general to cover all particularities of the specific drying system [22]. One of the main problems that these models share is that their parameters have no physical meaning, and the only independent variable is the drying time. Consequently, there is no explicit relationship between the other operational variables (inputs) and the output. On the other hand, more complex models require the use of computational fluid dynamics (CFD). For instance, Sanghi et al. [23] developed a CFD model to simulate the corn drying process in an NCSD. Such a model successfully predicted the internal temperature, humidity, and air velocity profiles of the dryer. It was validated by using experimental information. Although the authors did not detail the simulation times needed to generate results, long computational times are a very important limitation because simulating even a few seconds of the drying process requires high computing power and time [24]. Instead, artificial neural networks (ANN) present an interesting alternative option. The training process of the ANN requires a relatively low number of datasets obtained from experiments, making it very efficient. The prediction of a new drying curve with a trained network is completed in seconds, thus avoiding time-consuming new experiments. Furthermore, expanding the neural model with new inputs and/or outputs is a very straightforward procedure. The most significant drawback of ANNs is that they can present poor results when attempting to perform extrapolations if limited data are available. Of course, these modeling approaches are not mutually exclusive and can be used together in order to achieve a synergetic effect between the local and global phenomena modeling and simulation.

One particular application of process models is the development of soft sensors, which are instruments capable of measuring difficult process variables indirectly [25], like moisture ratio (M_R). Ryniecki et al. [26] used a soft sensor based on a correlation for the determination of malting barley moisture ratio inside a fixed bed dryer. The purpose of the sensor was to allow the automation of the process through the prediction of the drying endpoint.

Objectives of This Work

This work aimed mainly to evaluate the NCSD as an alternative to electric convective dryers for the dehydration of local microbrewers' spent grains. Two kinds of BSGGolden ale (GA) and Red ale (RA) were dried at different daytime hours in an NCSD. Furthermore, BSG were also dried in a conventional electric convectivedryer (CECD) at 333.15 K, 338.15 K, 343.15 K, 348.15 K, 353.15 K, 358.15 K, 363.15 K, and 368.15 K. Kinetic studies were completed for both drying methods. The measured dimensionless M_R values were modeled with eleven semitheoretical and empirical models for the CECD and, in turn, using CFD and ANNs for the NCSD. Finally, sustainability indices, specific energy consumption (e_C), and $CO_{2,\ eqv}$ were calculated and used to consider the environmental impact of the NCSD with both its benefits and drawbacks.

2. Materials and Methods

2.1. Samples

The samples were provided by Cervecería Cumbre, San Juan Province, Argentina. They were collected after being removed from the maceration process of two types of beer: Golden ale (GA) and Red ale (RA) (Figure 1). Then, the samples were stored at 277.15 K in the dark until needed (no more than 1 or 2 days after sampling).

(a)   (b)

Figure 1. Brewer's spent grains (BSG) of both (**a**) Golden ale and (**b**) Red ale varieties used.

2.2. Analysis of the Samples

The moisture content, pH, and titratable acidity (citric acid content) were analyzed in the collected BSG (both raw and dried, i.e., initial and final values after drying) following the methodology of the AOAC (Official Methods of Analysis, 2010). All the analyses in fresh and dried BSG were repeated three times ($N = 3$) and the data gathered from them were expressed as the means $\pm$ standard deviation.

2.3. Drying Equipment and Experimental Procedure

In this subsection, first, both types of dryers are presented; second, the experimental procedures used with each one of them are described. In addition, the most important instrumentation equipment is described.

2.3.1. Conventional Electric Convective Dryer (CECD)

The convective BSG drying experiments were performed without pretreatment at eight different temperatures: 333.15 K, 338.15 K, 343.15 K, 348.15 K, 353.15 K, 358.15 K, 363.15 K, and 368.15 K, in an electric oven with a stainless-steel tube (0.12 m internal diameter and 1 m height), heated by the electrical resistance of 850 W. A complete description of this device was reported by Baldan et al. [27]. The air velocity was constant at 1 m/s during all the drying trials which were carried out until the average moisture content was lower than 10% according to the Codex Standard for wheat flour used for bakery and confectionery products (Codex Alimentarius, 1985 [28]) so as to prevent any degradation of the flour.

2.3.2. Natural Convection Solar Dryer (NCSD)

A solar dehydrator with natural airflow (Figure 2) was used for the BSG drying experiments which lasted until the moisture content was lower than 10% [28]. The product was placed on trays inside the solar dryer in layers of 0.5 cm thickness (Figure 2a). The drying experiments were carried out on two consecutive summertime days, each one in triplicate, with three full trays per trial. The weight of BSG used in each experiment was 0.265 kg per tray. Furthermore, each day, the experiments started at two different hours (8:30 and 11:00 in the morning) to analyze the influence of irradiance and ambient temperature on the drying process of the samples. The morning time was selected considering that at 8:30 a.m. the irradiance value is minimal while at 11:00 a.m. a steady increase in irradiance (slope) begins and is sustained over time. Weight loss data was registered at different time intervals during the drying process until a constant weight was obtained. The measurements were performed by taking the trays from the chamber in regular intervals to be weighted with a balance Compass CR621 (OHAUS, Parsippany, United States). The drying chamber was 1.32 m in height, 0.56 m in width, and 0.33 m in depth. The collector was 1.24 m in length, 0.56 m in width, and 0.15 m in depth (Figure 2b). The cover of the collector (upper part) was transparent and made of a material that allowed the passage of solar energy, minimal loss of sunlight waves, and was resistant to inclement weather. In this case, 150 μm polyethylene was used. The drying system was classified as a natural convection indirect type, and the angle of inclination of the collector was set to 10° (Figure 2C). Although the Sun's position could have allowed an optimum inclination of between 30° and 50° [29], a very conservative value of 10° was selected to not overheat BSG according to the values of irradiance obtained in San Juan Province where the experiments were performed. The maximum and minimum temperatures achievable by the solar dryer were, approximately, between 318.55 K and 291.85 K, respectively.

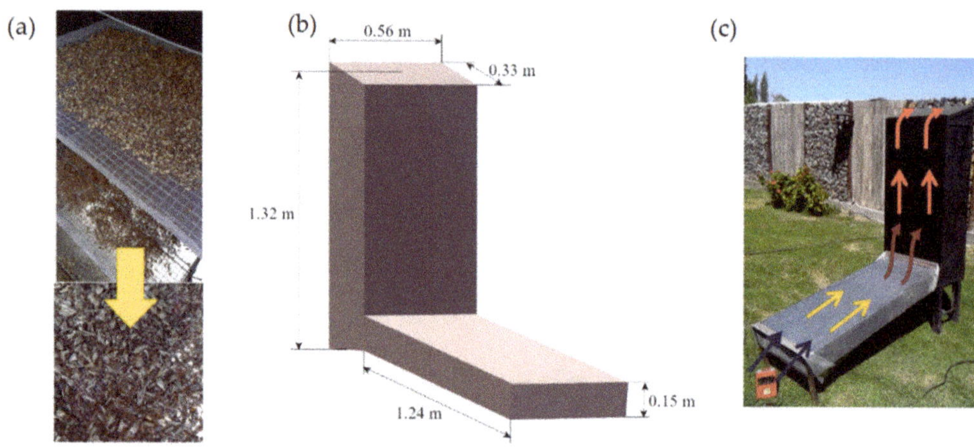

Figure 2. (**a**) Placement of the biomass on trays inside the solar drying chamber. (**b**) Natural convection solar dryer (NCSD) dimensions. (**c**) Heating and air circulation inside the NCSD.

The final moisture content of samples was determined using a PMR50 analyzer (Radwag, Radom, Poland) at 378.15 K [30]. The dried samples were stored at a temperature of around 293.15 K in a dark room until needed for analysis within 14 d after drying. All the analyses were performed in triplicates, and the data gathered from them were expressed as the means $\pm$ standard deviation. The irradiance values were measured with a solar power meter model TM-206 (TENMARS, Taipei, Taiwan).

2.4. Mathematical Modeling of Drying Curves

M_R was calculated with the experimental results obtained from the drying trials [31]. The drying curves of BSG ($M_{R(t)}$) were modeled using 11 different empirical models taken from the literature [32]. To assess the goodness of fit of the applied models with the experimental data, the statistical indicators chi-squared (χ^2), sum of squared errors (S_{SE}), and root-mean-square error (μ_{RSE}) were used. The models with lower values of χ^2, S_{SE}, and μ_{RSE} have a better fitting performance, while a value of R^2 closer to unity is more desirable.

2.5. Specific Energy Consumption and CO_2 Emissions

The high energy consumption of dryers also influences greenhouse gas emissions and has environmental consequences, which is very important [33]. Nowadays, scientists are looking for solutions that reduce energy consumption and CO_2 emissions using different pretreatment methods or adjusting the methods and parameters of drying to obtain the best results [34]. For that, e_C and $CO_{2,\,eqv}$ were used to compare the CECD and the NCSD. Firstly, for e_C, with Equation (1), the energy needed to dry 1 kg of fresh BSG was calculated [35]. Secondly, since in Argentina the electric energy supply for the CECD is provided by a mixture of nonrenewable and renewable sources (natural gas, thermal power, hydropower, and wind, among others) the electricity $CO_{2,\,eqv}$ is 0.358 3 kg/(kW h) according to Climate Transparency 2019 [36]. Thus, to calculate the $CO_{2,\,eqv}$ (expressed in kg/(kW h)) needed to dry 1 kg of BSG, the calculated e_C values have to be multiplied by this constant (Equation (2)).

$$e_C = \left(c_{p,a} + c_{p,v}h_a\right)\frac{qt(T_{in} - T_{am})}{m_v V_h} \tag{1}$$

$$CO_{2,\,eqv} = 0.358\,3 \cdot e_C \tag{2}$$

2.6. Modeling and Simulation of the Solar Drying Process

The main reason for developing computational models of the solar drying process is to be able to predict the behavior of M_R evolution based on the main independent variables of the process like irradiance, initial moisture content (w_{IM}), and others under different conditions without the need for so much experimental work. Needless to say, the conventional electric drying process is less subject to upsets in independent variables because it is carried out in an indoor controlled environment/system, whereas the solar drying process is carried out in an outdoor environment, without control of the main independent variables. The conventional electric drying process modeling is covered in Section 2.4.

2.6.1. CFD–ANN Complementary Modeling Approach

Two complementary techniques were applied to model and simulate the behavior of the solar dryer under different conditions: computational fluid dynamics (CFD) and artificial neural networks (ANN). This is because neither of them can singlehandedly model and simulate both the local and global phenomena of solar dryers efficiently. CFD can accurately and efficiently predict the temperature gradient, air velocity, and other variables inside the solar dryer, whereas it is very costly computationally to model and simulate the global drying phenomenon (i.e., M_R evolution). The latter is easily modeled and simulated by an ANN based on a handful of experimental data, whereas the former would need plenty of experiments to generate sufficient data to train a good ANN model.

Further complementation could also be achieved because the CFD model can generate the irradiance values of each day of the year for any particular latitude and longitude. These data could be applied as training data and simulation input for the ANN model without the need for hourly and daily actual measurements. This kind of complementation would be useful in a future expansion of the ANN model to cover daily and seasonal changes in solar drying curves. Such a project would need more drying experiments distributed around the year.

2.6.2. CFD Modeling

The analysis was performed by solving basic fluid governing equations (continuity, momentum, and energy) by using the finite volume method (FVM). The simulations were carried out with ANSYS Fluent®18.1 (ANSYS, Canonsburg, PA, USA) using a ray-tracing analysis. With this technique, we simulated the interaction between solar rays and the solar collector to quantify the amount of energy that impinges on the receiver at a particular daytime. The governing equations are summarized, for instance, in the work of Sanghi et al. [23]. The solar dryer's geometrical setup and boundary conditions adopted are shown in Figure 3.

A polyhedral mesh was applied with 77,090 cells. Mesh validation with experimental results showed that no significant improvement was obtained by adding more cells to the mesh. A k-epsilon turbulence model was used. All the simulations were performed in the steady state and the reverse flow in the outlet was disabled.

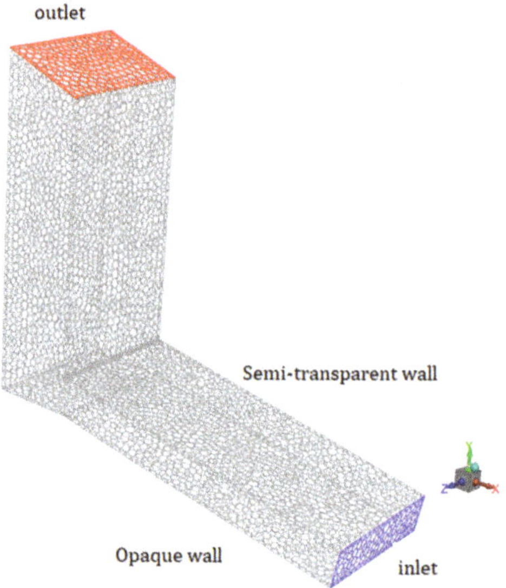

Figure 3. Boundary conditions and mesh employed in the solar dryer simulations.

2.6.3. ANN Modeling

Since San Juan Province has a very dry climate with low humidity, the main drivers of the drying process are the irradiance values and w_{IM} of the lot. A solar drying chamber does not have a fairly constant controlled internal temperature as a conventional electric convective dryer. Instead, it presents a temperature gradient that is subject to the Sun's irradiance variations (as shown in the CFD simulations). These particularities set the ANN's input/output design as shown in Figure 4.

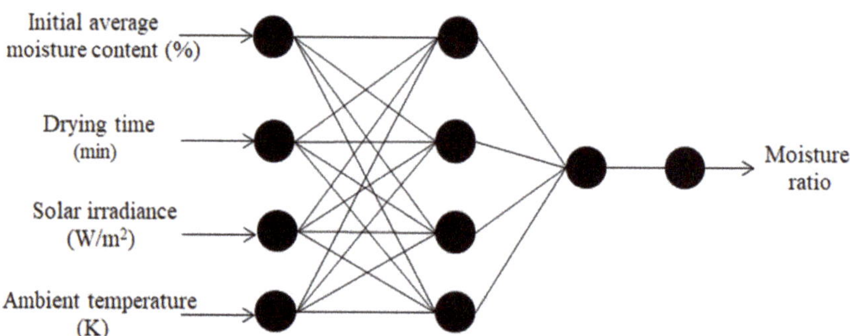

Figure 4. Artificial neural networks (ANN) model with four inputs and one output.

Once the number of inputs and outputs is established based on the experimental data available, the design of the ANN model involves the selection of the network type, the activation functions, the number of hidden layers, and the number of neurons in each one. Regarding the choice of network and neuron types, the universal approximation theorem (Pinkus [37]) establishes that multilayer perceptron networks with at least one hidden layer and a nonlinear activation function are universal approximators that can fit any given polynomial. In addition, for some models, deeper networks can work better than single-hidden-layer networks (no more than two hidden layers, most likely).

The process of selecting the number of hidden layers and the number of neurons in each one was carried out by trial-and-error. First, single-hidden-layer networks with as few as three neurons showed good performance ($R^2 > 0.999$). Increasing the number of hidden neurons to amounts of more than seven proved to show signs of overfitting when prediction tests of drying curves were performed. These prediction tests involved drying curves with different values of w_{IM}, batch starting hour, and batch times. Particularly, adding a second hidden layer helped in fitting the batch times better. The final model selection was performed according to the algorithm presented by Fabani et al. [10]. The complete set of ANN parameters is shown in Table 1. Finally, the MATLAB® (MathWorks®, Natick, MA, USA) release 2016a, Neural Fitting Tool was used to design, train weights and biases, test and implement the ANN model.

Table 1. Complete set of Artificial neural networks (ANN) parameters.

Parameter	Value
Network type	Multilayer feedforward (MLFF)
Neuron type	Perceptron
Inputs	4
Output	1 (moisture ratio)
Normalization type	Min–Max
Activation function	Sigmoid (hidden) and linear (output)
Training algorithm	Bayesian regularization backpropagation *
Training sets	264
Number of hidden layers	2
Number of neurons per layer	4 and 1
Train ratio	97%
Validation ratio	N/A **
Test ratio	3%

* This particular training algorithm's stop condition is the maximum number of epochs. ** N/A: not applicable.

2.6.4. ANN Simulations

One of the main independent variables which directly affect the batch time is the average w_{IM} of the sample. Additionally, the starting daytime hour (SDTH) of the drying process directly affects the drying rate (v_D) because the amount of power the Sun provides is not constant. These two factors can shorten or extend the drying batch duration so it is paramount to determine how they can affect the total yield. The consequences of variations in these two variables were determined using ANN simulations which, in turn, were used to calculate the indicators in Equations (3) and (4): the average first batch time duration (t_{B1}) and the average second batch duration (t_{B2}), both for any given starting daytime hour.

$$\overline{t_{B1}}_{SDTH} = 10^{-1} \cdot \sum_{w_{IM}=64}^{74} t_{B1\,w_{IM}} \tag{3}$$

$$\overline{t_{B1}}_{SDTH} = 10^{-1} \cdot \sum_{w_{IM}=64}^{74} t_{B1\,w_{IM}} \tag{4}$$

The simulations were performed considering that the first batch can start as early as 08:30 in the morning and that the second batch starts a half-hour after the first one finishes. The batch stop condition is M_R lower than 0.1. Finally, due to the experimental results, only two full batches are expected. It is important to mention that since the ANN model is data-based, the validity of its results is limited to the day the drying experiments were carried out. On the other hand, the CFD model can produce simulation results of any day of the year.

2.7. Statistical Performance Indicators

Firstly, the goodness of fit between the empirical models and the experimental data was assessed based on the chi-squared (χ^2) statistical indicator, the sum of squared errors (S_{SE}), and the mean squared error (μ_{SE}):

$$\chi^2 = \sum_{i=1}^{N} \frac{\left(M_{R\ pre,i} - M_{R\ exp,i}\right)^2}{M_{R\ exp,i}} \tag{5}$$

$$S_{SE} = \sum_{i=1}^{N} \left(M_{R\ pre,i} - M_{R\ exp,i}\right)^2 \tag{6}$$

$$\mu_{SE} = \frac{1}{N} \cdot \sum_{i=1}^{N} \left(M_{R\ pre,i} - M_{R\ exp,i}\right)^2 \tag{7}$$

Secondly, the evaluation of the ANNs performance was based on two statistical indicators: μ_{SE} (Equation (7)) and the coefficient of determination R^2:

$$R^2 = 1 - \left[\frac{\sum_{i=1}^{N} \left(M_{R\ pre,i} - M_{R\ exp,i}\right)^2}{\sum_{i=1}^{N} \left(\overline{M_{R}}_{\ pre,i} - M_{R\ exp,i}\right)^2} \right] \tag{8}$$

Unlike χ^2 and μ_{SE}, R^2 values closer to unity are more desirable, show better fitting performance. Additionally, the R^2 indicator is more commonly associated with the software applications used to train ANNs.

Finally, in this work, units and quantities were mentioned according to the International System of units (SI) and compatible units as reported by Glavič [38].

3. Results

3.1. Raw and Dried Moisture, pH and Titratable Acidity of the Samples

The fresh samples of BSG presented high w_{IM} values, which were 64.3 ± 0.2% for BSG-GA and 74.2 ± 0.5% for BSG-RA, all within the previously reported range of 60–90% mass [18]. Secondly, the fresh samples presented pH values of 5.63 ± 0.08 for the GA variety and 6.06 ± 0.05 for the RA variety, while the titratable acidity values (expressed in % citric acid) were 0.40 ± 0.02% and 0.37 ± 0.02%, respectively.

The samples dried in the CECD presented the final moisture content values of less than 7 % and final pH and acidity values which varied according to the temperature value used, ranging from 5.4 to 5.8 and from 0.65% to 1.10%, respectively. On the other hand, the final moisture content of the sun-dried BSG was less than 6–7%, pH was 5.49 ± 0.06 for the GA variety and 5.62 ± 0.02 for the RA variety, and, finally, the titratable acidity values (expressed in % citric acid) were 0.56 ± 0.05% and 0.52 ± 0.01% for GA and RA, respectively.

In Table 2, a summary of the analytical results is shown. Both driers presented similar values for moisture content and pH of the dried samples, while the NCSD presented improved values of titratable acidity compared to the CECD. The titratable acidity values of the samples dried in the NCSD were lower, closer to those of the fresh samples. This particular indicator is important because it is related to the flavor of a foodstuff. The NCSD samples had lower values (improved organoleptic quality indices) than the CECD ones at all the temperatures assessed. Dzelagha et al. [39] obtained similar results when studying different drying technologies and the effect on bean quality parameters. These could be related to the significant increment of free fatty acid and acetic acid levels with the rising temperature which, in turn, was proportional to the corresponding decrease in pH, suggesting the optimal oven temperature of 318.15 K. Moreover, certain vitamins are broken down at high temperatures, such as vitamin C, thus diminishing their content [32].

Table 2. Summary of the analytical results.

Variety	Fresh Samples			Dried Samples: CECD			Dried Samples: NCSD		
	w_M (%)	pH	Titratable Acidity (% Citric Acid)	w_M (%)	pH	Titratable Acidity (% Citric Acid)	w_M (%)	pH	Titratable Acidity (% Citric Acid)
Golden ale	64.3 ± 0.2	5.63 ± 0.08	0.40 ± 0.02	6.52 ± 0.16	5.42 ± 0.08	0.86 ± 0.18	6.42 ± 0.21	5.49 ± 0.06	0.56 ± 0.05
Red ale	74.2 ± 0.5	6.06 ± 0.05	0.37 ± 0.02	6.78 ± 0.13	5.71 ± 0.24	0.55 ± 0.05	6.30 ± 0.27	5.62 ± 0.02	0.52 ± 0.01

3.2. Drying Characteristics of BSG

The drying curves of BSG, GA and RA varieties, in the CECD and the NCSD were modeled with empirical and ANN models, respectively. The CECD experimental drying curves showed an exponential decay with the first phase where the moisture is rapidly lost (high drying rates) and the second phase that presented low drying rates until the samples reached low moisture contents (Figure 5a). In addition, the drying rates increased with temperature due to the increased heat transfer between the circulating air and the BSG, favoring water evaporation. The drying models which fitted the experimental data best were, first, the Midilli model ($S_{SE} = 4.1 \cdot 10^{-5}$; $\mu_{RSE} = 6.2 \cdot 10^{-3}$, and $\chi^2 = 4.3 \cdot 10^{-5}$), and, second, the two-term model ($S_{SE} = 6.6 \cdot 10^{-5}$; $\mu_{RSE} = 7.9 \cdot 10^{-3}$, and $\chi^2 = 6.9 \cdot 10^{-5}$). Therefore, based on the statistical indicators presented, the Midilli model is more accurate to predict the M_R evolution for the CECD.

The NCSD showed greater efficiency when the drying process was started at 11:00 in the morning, achieving a decrease in moisture content in a shorter time interval for the two varieties studied (Figure 5b). The experiments that started at 11:00 achieved a reduced drying time (M_R lower than 0.1) from 420 min to 345 min for the GA variety and from 430 min to 390 min for the RA variety when compared with those that started at 08:00. The ANN model's fitting performance had $R^2 > 0.999$ and $\mu_{RSE} = 4.25 \cdot 10^{-3}$. In the solar bagasse drying, a greater variation in the drying rate (more peaks) was observed compared to the convection bagasse drying.

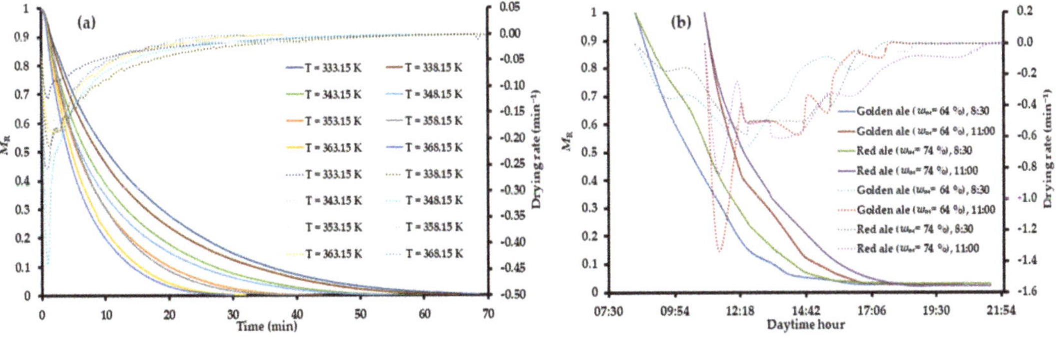

Figure 5. Experimental drying curves. (**a**) Conventional electric convective dryer CECD RA variety; (**b**) NCSD (continuous lines, $M_{R(t)}$; dotted lines, drying rates).

3.3. Comparative Analysis of the CECD and the NCSD

In Tables 3 and 4, the results of the CECD process of the raw product are presented. Both varieties of the BSG were dried at eight different temperatures, from 333.15 K to 368.15 K. As expected, from the environmental impact indicators' point of view, there were no significant differences. Both ales presented similar values for e_C and $CO_{2,\ eqv}$, averaging 346.92 (kW h)/kg and 124.30 kg/(kW h) for e_C and $CO_{2,\ eqv}$, respectively. Similar values of the operational indicators are also shown for both varieties, averaging 51.23 min of batch time and 0.86 USD of electricity per kilogram of raw product. In conclusion, from that single perspective only, higher drying temperatures are more efficient.

Table 3. Results of the conventional convective drying process of the raw matter: Golden ale.

Drying Temperature (K)	Batch Time (min)	w_M (%)	Total Energy Consumption (kW h)	Removed Water (g)	e_C ((kW h)/kg)	$CO_{2,\,eqv}$ (kg/(kW h))	Electricity Cost (USD/kg)
333.15 ± 0.20	84.33 ± 5.48	6.79 ± 0.21	0.29 ± 0.02	1.68 ± 0.03	407.58 ± 26.76	158.14 ± 10.38	1.02 ± 0.07
338.15 ± 0.20	64.50 ± 1.32	6.70 ± 0.11	0.24 ± 0.01	1.60 ± 0.03	376.13 ± 15.34	145.94 ± 5.95	0.94 ± 0.04
343.15 ± 0.20	54.33 ± 1.26	6.60 ± 0.09	0.22 ± 0.01	1.61 ± 0.06	352.48 ± 7.46	136.76 ± 2.89	0.88 ± 0.02
348.15 ± 0.20	50.81 ± 3.01	6.52 ± 0.12	0.22 ± 0.01	1.66 ± 0.02	356.47 ± 16.33	138.31 ± 6.33	0.89 ± 0.04
353.15 ± 0.20	46.17 ± 3.55	6.45 ± 0.08	0.21 ± 0.02	1.73 ± 0.08	341.80 ± 10.81	132.62 ± 4.19	0.85 ± 0.03
358.15 ± 0.20	40.17 ± 2.08	6.40 ± 0.05	0.19 ± 0.01	1.64 ± 0.04	341.72 ± 12.27	132.59 ± 4.76	0.85 ± 0.03
363.15 ± 0.20	35.5 ± 1.04	6.38 ± 0.12	0.18 ± 0.01	1.68 ± 0.01	320.86 ± 7.13	124.49 ± 2.77	0.80 ± 0.05
368.15 ± 0.20	34.17 ± 2.02	6.34 ± 0.09	0.18 ± 0.01	1.88 ± 0.06	296.44 ± 18.89	115.02 ± 7.33	0.74 ± 0.05

Table 4. Results of the conventional convective drying process of the raw matter: Red ale.

Drying Temperature (K)	Batch Time (min)	w_M (%)	Total Energy Consumption (kW h)	Removed Water (g)	e_C ((kW h)/kg)	$CO_{2,\,eqv}$ (kg/(kW h))	Electricity Cost (USD/kg)
333.15 ± 0.20	71.50 ± 3.91	6.94 ± 0.14	0.24 ± 0.01	1.49 ± 0.09	391.95 ± 39.39	152.08 ± 15.28	0.98 ± 0.10
338.15 ± 0.20	69.83 ± 4.25	6.91 ± 0.16	0.26 ± 0.02	1.80 ± 0.01	360.22 ± 23.97	139.76 ± 9.30	0.90 ± 0.06
343.15 ± 0.20	59.50 ± 0.90	6.87 ± 0.13	0.24 ± 0.01	1.89 ± 0.03	331.83 ± 5.12	128.75 ± 1.99	0.83 ± 0.01
348.15 ± 0.20	58.17 ± 0.58	6.82 ± 0.14	0.25 ± 0.01	1.94 ± 0.08	350.23 ± 17.69	135.89 ± 6.86	0.88 ± 0.04
353.15 ± 0.20	45.17 ± 1.89	6.77 ± 0.20	0.20 ± 0.01	1.65 ± 0.20	353.12 ± 29.74	137.01 ± 11.54	0.88 ± 0.07
358.15 ± 0.20	43.00 ± 0.50	6.71 ± 0.12	0.21 ± 0.01	1.69 ± 0.04	354.41 ± 10.63	137.51 ± 4.12	0.89 ± 0.03
363.15 ± 0.20	36.00 ± 3.28	6.67 ± 0.09	0.18 ± 0.02	1.64 ± 0.01	330.27 ± 27.89	128.14 ± 10.82	0.83 ± 0.07
368.15 ± 0.20	33.17 ± 2.75	6.54 ± 0.10	0.17 ± 0.01	1.57 ± 0.03	328.80 ± 17.77	127.57 ± 6.89	0.82 ± 0.04

In Table 5, the solar dryer's results of the drying process of the raw product are shown. The experiments were conducted during the austral summer at two different daytime hours, 08:00 and 11:00, for each type of the BSG. The results showed an average batch time of 396 min, with the RA type exhibiting slightly longer drying times due to its higher w_{IM} compared to the GA variety. Additionally, the NCSD has null operating costs because solar energy is freely available and there is no forced airflow. Consequently, both environmental impact indicators have null values, too. In Table 6, a summary of the most important environmental and process indicators previously explained is presented.

Table 5. Results of the solar drying process of the raw matter for Golden ale and Red ale.

BSG	Daytime Hour	Batch Time (min)	e_C ((kW h)/kg)	$CO_{2,\,eqv}$ (kg/(kW h))	Electricity Cost (USD/kg)
Golden ale	08:30	420	0.00	0.00	0.00
Golden ale	11:00	345	0.00	0.00	0.00
Red ale	08:30	430	0.00	0.00	0.00
Red ale	11:00	390	0.00	0.00	0.00

Table 6. Summary of indicators.

	Conventional Electric Convective Dryer (CECD)	Natural Convection Solar Dryer (NCSD)
Average drying temperature (K)	350.65	319.65
Average batch time (min)	51.23	396.00
Average e_C ((kW h)/kg)	346.92	0.00
Average $CO_{2, eqv}$ (kg/(kW h))	124.30	0.00
Average operating cost (USD/kg)	0.86	0.00

3.4. CFD Simulation Results

Aiming to validate the CFD computational model, temperature measurements were performed inside the empty drying chamber under different irradiance conditions during the springtime, before the experiments. Table 7 shows the temperature measurements performed in the lower section of the chamber (point 1), in the middle section of the chamber (point 2), and the upper section of the chamber near the exit (point 3) at three different daytime hours. Additionally, the corresponding simulation results are shown, indicating a good agreement between the simulated values and the measured ones (Table 7). The minimal differences can be attributed to deviations in the model's irradiance values, and it is also important to consider that the lower wall of the dryer exchanges heat with the surrounding environment (heat transfer coefficient adopted for simulations = 10 W/(m^2 K)), rendering the temperature in the drying chamber lower than the temperature in the solar heating section.

Table 7. CFD simulations' results and experimental validations.

					Temperature (K)					
Daytime Hour	Ambient Temperature (K)		Solar Irradiance (W/m^2)		Point 1		Point 2		Point 3	
	EXP	CFD	EXP	CFD	EXP	CFD	EXP	CFD	EXP	CFD
12:30	299.65	299.15	1005	944.63	305.75	309.28	307.95	309.07	309.65	308.87
14:00	302.75	299.15	1040	945.46	312.65	312.48	313.15	312.15	314.45	311.84
15:30	303.75	299.15	860	928.91	312.75	316.12	312.85	315.82	312.85	315.54

Once the CFD model was validated, simulations were performed to assess how both seasonal changes and daytime hour changes affect the temperature gradient inside the drying chamber. In Figure 6, four longitudinal cutaways of the drying chamber are presented, also under four different irradiance conditions. The calculated mean temperatures were as follows: 309.55 K for a summer morning (11:00); 312.58 K for a summer afternoon (15:00); 306.81 K for a winter morning (11:00); and 309.23 K for a winter afternoon (15:00).

3.5. ANN Simulation Results 1: The Effect of Irradiance and w_{IM} in the Drying Rates

The drying rate of the sample inside the drying chamber is controlled by the rate at which heat is applied to it (i.e., power applied), the rate at which the sample's internal moisture is released from its surface, and the rate at which moist air is removed from the drying chamber. Regarding the power applied, in a conventional electric dryer, once the chamber's internal temperature is constant, the power consumption is constant, too, whereas in the solar dryer, the available power per square meter (irradiance) varies throughout the day and is only controlled by the Earth's rotation. Therefore, the drying rates during the several known stages of the process vary if the same sample starts drying early in the morning or early in the afternoon, for instance.

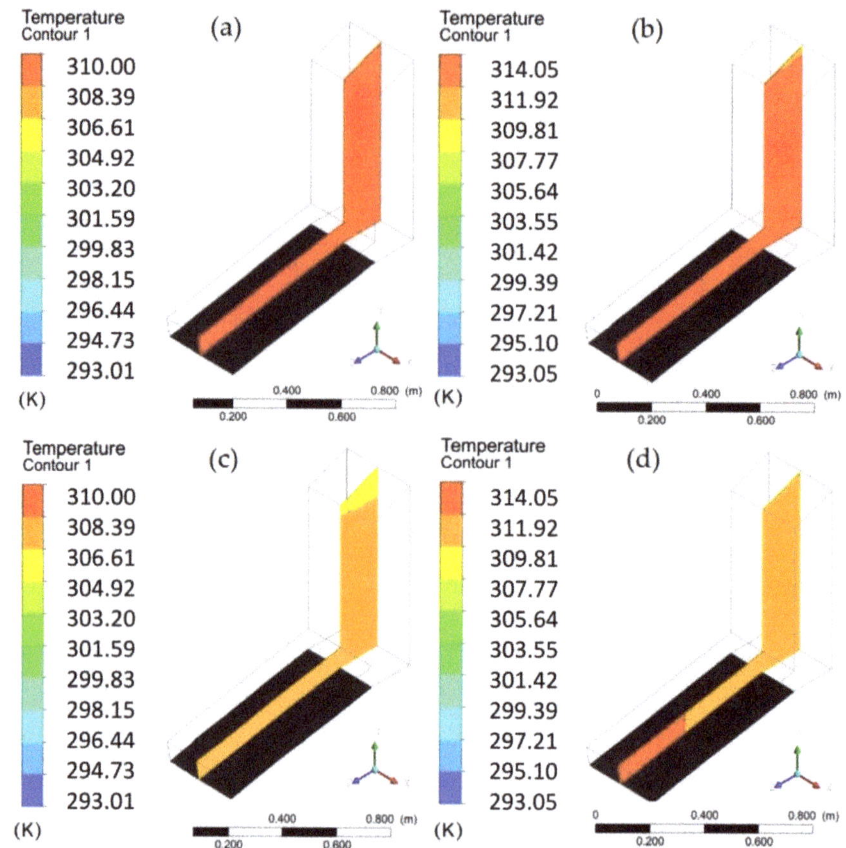

Figure 6. Four cutaways of the drying chamber showing the temperature gradients inside it under different irradiance conditions: (**a**) summer morning, 11:00; (**b**) summer afternoon, 15:00; (**c**) winter morning, 11:00; and (**d**) winter afternoon, 15:00.

It can be seen from the experimental curves (Figure 5b) that the slope of those samples that started drying at 08:30 was less steep than the slope of those that started drying at 11:00 during the constant drying rate stage. Additionally, the ANN predictions in Figure 7 show that the samples which start drying later in the afternoon show an even steeper slope during the constant drying rate stage as the irradiance values continue to rise. This is because the bulk of the water is easier to remove and because the irradiance values are very high. Then, the drying rate slows down because both the remaining water is harder to remove and the irradiance values rapidly start to fall. In comparison, the drying rate of the samples which started at 08:30 is pretty constant throughout because when the irradiance values start to climb, the remaining water is harder to remove, and vice versa.

Finally, the ANN predictions also show the effect of w_{IM} of the samples. In Figure 7, it can be seen that a sample with a w_{IM} of 70% (a value which is in between the GA and the RA) had a batch time that was shorter than the RA but longer than the GA for both the curves which started early in the morning and later in the afternoon. Other curves with w_{IM} values within that same range showed a similar behavior.

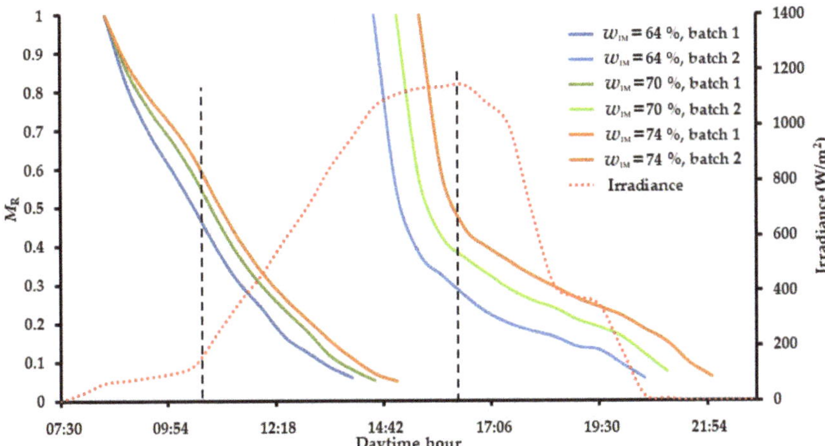

Figure 7. The drying curves of three lots with different w_{IM}. The black dotted vertical lines indicate major changes in the Sun's irradiance slope associated with changes in the drying rates.

3.6. ANN Simulation Results 2: Average Batch Durations

As the experimental results suggest, only two full batches can be completed per day with the measured irradiance values and w_{IM}. Therefore, simulations were performed to determine how w_{IM} and the starting daytime hour affect batch times. This is another indicator of how the irradiance controls the drying process and its potential total yield.

Applying all the simulation results to Equations (3) and (4), the plots of Figure 8 were obtained. The results show that both average batch duration curves—which depend on the starting daytime hour—have the opposite slope of the irradiance curve, meaning that as long as the irradiance rises, the batch durations fall, and vice versa. The total batch time curve (batch 1 plus batch 2) also presents a similar shape. The results also show that the first batch can start as late as 11:30 in the morning to complete two full batches in the same day. Finally, simulations were used to calculate the mean batch time for every w_{IM} in the 64–74% range and for every SDTH. The result was a mean batch time of 366 min as shown in Table 4, 30 min slower than the obtained experimental value. This difference could be because many more curves were used from the simulation to calculate these indicators.

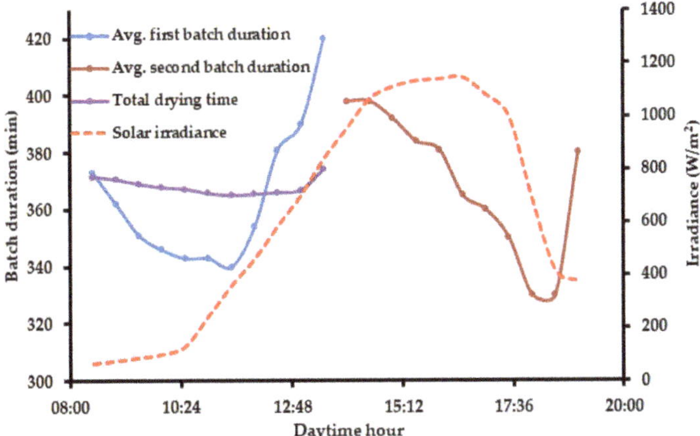

Figure 8. Batch curves calculated based on the simulation results. All of them show a similar shape, opposite to the shape of the irradiance curve.

4. Discussion and Conclusions

A natural convection solar dryer (NCSD) was evaluated as an alternative to a conventional electric convective drying system (CECD) for the dehydration process of local microbrewers' spent grains. For the different drying temperatures (333.15–363.15 K), e_C and $CO_{2,\,eqv}$ for the CECD varied between $1.68 \cdot 10^{-3}$ (kW h)/kg and $1.88 \cdot 10^{-3}$ (kW h)/kg and between 294.80 kg/(kW h) and 410.73 kg/(kW h); however, using the NCSD, both indicators were null. From the environmental management perspective, the NCSD does not need any grid energy to function and eliminates 100% of the $CO_{2,\,eqv}$. Thus, if labor costs are not considered, the overall operating costs are greatly reduced. Secondly, the NCSD produces a dried product with a comparatively better quality, as shown by the lower titratable acidity values. Lastly, another important aspect is the cost of the equipment itself (initial investment). The NCSD is about 40–45% cheaper than the CECD.

On the other hand, the most important disadvantage of the NCSD is the fact that batch times (345–430 min, depending on the starting daytime hour at which the drying process begins) are much longer, consequently reducing the total yield. In light of this, it is of paramount importance to find tools that help operators find the optimal conditions in which to work to make the NCSD more appealing. Through simulations, the importance of the Sun's irradiance was measured under conditions different from those of the experiments. It was established with the ANN model how the irradiance values directly govern the drying rates and the average batch times. Moreover, the developed CFD model can produce irradiance values via simulations for any location on the planet where the same solar dryer could be placed regardless of the day and time of the year. The internal temperature gradients were obtained under several conditions. This information can be used to determine the optimal operating conditions for the solar dryer to get the highest possible yield of dried BSG. These results further show the contrast between both drying methods. In an electric convective dryer, the operating variables adapt themselves to the operator's needs and, on the other hand, in the NCSD, the operator has to adapt to the operating conditions. Consequently, these simulation tools can further assist the solar drying process by using them to develop soft sensors, which are instruments that can, for instance, determine the drying endpoint.

Future work should include, as the first step, the optimization of the drying process based on the technofunctional properties of the flour obtained by milling dried BSG using the response surface methodology.

Author Contributions: Conceptualization, J.P.C., M.P.F., A.R.-U. and R.R.; methodology, Y.D. and R.T.-S.; software, J.P.C.; formal analysis, J.P.C., M.P.F., A.R.-U., R.R., Y.D., R.T.-S., J.B. and G.M.; investigation, J.P.C., M.P.F., A.R.-U., R.R., Y.D. and R.T.-S.; resources, R.R.; writing—original draft preparation, J.P.C. and M.P.F.; writing—review and editing, R.R., J.B. and G.M.; visualization, G.M.; supervision, J.B., G.M. and R.R.; project administration, R.R. and G.M.; funding acquisition, G.M. and R.R. All authors have read and agreed to the published version of the manuscript.

Funding: This research was funded by the National Scientific and Technical Research Council (CONICET), Argentina (grant No. PUE PROBIEN 22920150100067 and PIP 11220200100950CO); Universidad Nacional de San Juan, Argentina (grant No. PDTS Res. 1054/18); Universidad Nacional del Comahue, Neuquén, Argentina (grant No. PIN 2022-04/I260); FONCYT-ANPCyT (National Agency for Scientific and Technological Promotion, Argentina) (grant Nos. PICT 2017-1390 and PICT 2019-01810); San Juan Province, Argentina (grant No. IDEA—Res. N° 0272-SECITI-2019). Maria Paula Fabani, Juan Pablo Capossio, Rosa Rodriguez, and Germán Mazza are Research Members of CONICET.

Institutional Review Board Statement: Not applicable.

Informed Consent Statement: Not applicable.

Data Availability Statement: Not applicable.

Acknowledgments: The authors would like to express their gratitude to brewing company CERVECERÍA CUMBRE for providing the spent grains samples.

Conflicts of Interest: The authors declare no conflict of interest.

Nomenclature

$CO_{2:\ eqv}$	CO_2 emission expressed as equivalence, kg/(kW h) or kg/t when necessary
$c_{p,a}$	Specific heat capacity of air, (kW h)/(kg K)
$c_{p,v}$	Specific heat capacity of vapor, (kW h)/(kg K)
e_C	Specific energy consumption, (kW h)/kg
h_a	Absolute air humidity, 1
M_R	Moisture ratio, 1
m_v	Mass of removed water, kg
Q	Flow rate of inlet air, m^3/min
R^2	Coefficient of determination, 1
S_{SE}	Sum of squared errors, 1
t	Time, s
t_{B1}	First batch duration, min
t_{B2}	Second batch duration, min
T_{am}	Ambient air temperature, K
T_{in}	Inlet air temperature, K
v_D	Drying rate, min^{-1}
V_h	Specific air volume, m^3/kg
w_M	Moisture content expressed as mass fraction of water (%)
w_{IM}	Initial moisture content expressed as mass fraction of water (%)
μ_{SE}	Mean squared error, 1
μ_{RSE}	Root-mean-square error, 1
χ^2	Chi-squared, 1

Abbreviations

ANN	Artificial neural network
BSG	Brewer's spent grains
CECD	Conventional electric convective dryer
CFD	Computational fluid dynamics
EXP	Experimental values (Table 7)
FVM	Finite volume method
GA	Golden ale
NCSD	Natural convection solar dryer
RA	Red ale
SDTH	Starting daytime hour

References

1. Díaz, M.; Cabello, X. (Eds.) *South American Brewery Yearbook*, 1st ed.; Ediciones Inter Andina: San Carlos de Bariloche, Argentina, 2016.
2. FAO. Production Quantities of Beer by Country. Food and Agriculture Organization of the United Nations. FAOSTAT, Food and Agriculture Data. 2021. Available online: https://www.fao.org/faostat/en/#data/FBS (accessed on 12 December 2021).
3. Lynch, K.M.; Steffen, E.J.; Arendt, E.K. Brewers' spent grain: A review with an emphasis on food and health. *J. Inst. Brew.* **2016**, *122*, 553–568. [CrossRef]
4. Nigam, P.S. An overview: Recycling of solid barley waste generated as a by-product in distillery and brewery. *Waste Manag.* **2017**, *62*, 255–261. [CrossRef] [PubMed]
5. Steiner, J.; Procopio, S.; Becker, T. Brewer's spent grain: Source of value-added polysaccharides for the food industry in reference to the health claims. *Eur. Food Res. Technol.* **2015**, *241*, 303–315. [CrossRef]
6. Ministry of Agriculture, Argentine Republic. Available online: http://www.alimentosargentinos.gob.ar/HomeAlimentos/Nutricion/documentos/TendenciaBagazo.pdf (accessed on 10 December 2021).
7. Bolwig, S.; Mark, M.S.; Happel, M.K.; Brekke, A. Beyond animal feed?: The valorisation of brewers' spent grain. In *From Waste to Value. Valorisation Pathways for Organic Waste Streams in Circular Bioeconomies*, 1st ed.; Taylor & Francis: Abingdon, UK, 2019; pp. 107–126. [CrossRef]
8. Mussatto, S.I. Brewer's spent grain: A valuable feedstock for industrial applications. *J. Sci. Food Agric.* **2014**, *94*, 1264–1275. [CrossRef] [PubMed]

9. Kavalopoulos, M.; Stoumpou, V.; Christofi, A.; Mai, S.; Barampouti, E.M.; Moustakas, K. Sustainable valorisation pathways mitigating environmental pollution from brewers' spent grains. *Environ. Pollut.* **2021**, *270*, 116069. [CrossRef] [PubMed]

10. Fabani, M.P.; Capossio, J.P.; Román, M.C.; Zhu, W.; Rodriguez, R.; Mazza, G. Producing non-traditional flour from watermelon rind pomace: Artificial neural network (ANN) modeling of the drying process. *J. Environ. Manag.* **2021**, *281*, 111915. [CrossRef] [PubMed]

11. Nocente, F.; Taddei, F.; Galassi, E.; Gazza, L. Upcycling of brewers' spent grain by production of dry pasta with higher nutritional potential. *LWT* **2019**, *114*, 108421. [CrossRef]

12. Aliyu, S.; Bala, M. Brewer's spent grain: A review of its potentials and applications. *Afr. J. Biotechnol.* **2011**, *10*, 324–331. [CrossRef]

13. Ktenioudaki, A.; Crofton, E.; Scannell, A.G.M.; Hannon, J.A.; Kilcawley, K.N.; Gallagher, E. Sensory properties and aromatic composition of baked snacks containing brewer's spent grain. *J. Cereal Sci.* **2013**, *57*, 384–390. [CrossRef]

14. Maqhuzu, A.B.; Yoshikawa, K.; Takahashi, F. Prospective utilization of brewers' spent grains (BSG) for energy and food in Africa and its global warming potential. *Sustain. Prod. Consum.* **2021**, *26*, 146–159. [CrossRef]

15. Vadivel, V.; Moncalvo, A.; Dordoni, R.; Spigno, G. Effects of an acid/alkaline treatment on the release of antioxidants and cellulose from different agro-food wastes. *Waste Manag.* **2017**, *64*, 305–314. [CrossRef]

16. Fărcaş, A.C.; Socaci, S.A.; Mudura, E.; Dulf, F.V.; Vodnar, D.C.; Tofană, M. Exploitation of Brewing Industry Wastes to Produce Functional Ingredients. *Brew Technol.* **2017**, *1*, 137–156. [CrossRef]

17. Fabani, M.P.; Román, M.C.; Rodriguez, R.; Mazza, G. Minimization of the adverse environmental effects of discarded onions by avoiding disposal through dehydration and food-use. *J. Environ. Manag.* **2020**, *271*, 110947. [CrossRef] [PubMed]

18. Chavan, A.; Vitankar, V.; Mujumdar, A.; Thorat, B. Natural convection and direct type (NCDT) solar dryers: A review. *Dry. Technol.* **2021**, *39*, 1969–1990. [CrossRef]

19. Udomkun, P.; Romuli, S.; Schock, S.; Mahayothee, B.; Sartas, M.; Wossen, T. Review of solar dryers for agricultural products in Asia and Africa: An innovation landscape approach. *J. Environ. Manag.* **2020**, *268*, 110730. [CrossRef]

20. Lopez-Vidaña, E.C.; Cesar-Munguía, A.L.; García-Valladares, O.; Salgado Sandoval, O.; Dominguez Niño, A. Energy and exergy analyses of a mixed-mode solar dryer of pear slices (*Pyrus communis* L.). *Energy* **2021**, *220*, 119740. [CrossRef]

21. Mallen, E.; Najdanovic-Visak, V. Brewers' spent grains: Drying kinetics and biodiesel production. *Bioresour. Technol. Rep.* **2018**, *1*, 16–23. [CrossRef]

22. Topuz, A. Predicting moisture content of agricultural products using artificial neural networks. *Adv. Eng. Softw.* **2010**, *41*, 464–470. [CrossRef]

23. Sanghi, A.; Ambrose, R.P.K.; Maier, D. CFD simulation of corn drying in a natural convection solar dryer. *Dry. Technol.* **2018**, *36*, 859–870. [CrossRef]

24. Malekjani, N.; Jafari, S.M. Simulation of food drying processes by Computational Fluid Dynamics (CFD); recent advances and approaches. *Trends Food. Sci. Technol.* **2018**, *78*, 206–223. [CrossRef]

25. Kadlec, P.; Gabrys, B.; Strandt, S. Data-driven Soft Sensors in the process industry. *Comput. Chem. Eng.* **2009**, *33*, 795–814. [CrossRef]

26. Ryniecki, A.; Gawrysiak-Witulska, M.; Wawrzyniak, J. Correlation for the automatic identification of drying endpoint in near-ambient dryers: Application to malting barley. *Biosyst. Eng.* **2007**, *98*, 437–445. [CrossRef]

27. Baldán, Y.; Fernandez, A.; Reyes Urrutia, A.; Fabani, M.P.; Rodriguez, R.; Mazza, G. Non-isothermal drying of bio-wastes: Kinetic analysis and determination of effective moisture diffusivity. *J. Environ. Manag.* **2020**, *262*, 110348. [CrossRef] [PubMed]

28. Codex Alimentarius. Codex Standard for Wheat Flour. CXS 152-1985.: International Food Standards. Revised in 1995. Amended in 2016. 2019. Available online: http://www.fao.org/fao-who675codexalimentarius (accessed on 13 December 2021).

29. Zhang, H.L.; Baeyens, J.; Degreve, J.; Caceres, G. Concentrated solar power plants: Review and design methodology. *Renew. Sust. Energ Rev.* **2018**, *22*, 466–481. [CrossRef]

30. AOAC. *Official Methods of Analysis of AOAC International*, 18th ed.; AOAC International: Gaithersburg, MD, USA, 2010; ISBN 0-935584-80-3. Available online: http://www.eoma.aoac.org/ (accessed on 3 November 2021).

31. Praveen Kumar, D.G.; Hebbar, H.U.; Ramesh, M.N. Suitability of thin layer models for infrared-hot air-drying of onion slices. *LWT Food Sci. Technol.* **2006**, *39*, 700–705. [CrossRef]

32. Roman, M.C.; Fabani, M.P.; Luna, L.C.; Feresin, G.E.; Mazza, G.; Rodriguez, R. Convective drying of yellow discarded onion (Angaco INTA): Modelling of moisture loss kinetics and effect on phenolic compounds. *Inf. Process Agric.* **2020**, *7*, 333–341. [CrossRef]

33. Nabavi-Pelesaraei, A.; Rafiee, S.; Mohtasebi, S.S.; Hosseinzadeh-Bandbafha, H.; Chau, K. Integration of artificial intelligence methods and life cycle assessment to predict energy output and environmental impacts of paddy production. *Sci. Total Environ.* **2018**, *631–632*, 1279–1294. [CrossRef] [PubMed]

34. Witrowa-Rajchert, D.; Wiktor, A.; Sledz, M.; Nowacka, M. Selected emerging technologies to enhance the drying process: A review. *Dry Technol.* **2014**, *32*, 1386–1396. [CrossRef]

35. Kaveh, M.; Amiri Chayjan, R.; Taghinezhad, E.; Rasooli Sharabiani, V.; Motevali, A. Evaluation of specific energy consumption and GHG emissions for different drying methods (Case study: Pistacia Atlantica). *J. Clean. Prod.* **2020**, *259*, 120963. [CrossRef]

36. Transparency Climate. Brown to Green. Report. 2019. Available online: https://www.climate-transparency.org/g20-climate-performance/g20report2019 (accessed on 10 December 2021).

37. Pinkus, A. Approximation theory of the MLP model in neural networks. *Acta Numer.* **1999**, *8*, 143–195. [CrossRef]
38. Glavič, P. Review of the International Systems of Quantities and Units Usage. *Standards* **2021**, *1*, 2–16. [CrossRef]
39. Dzelagha, B.F.; Ngwa, N.M.; Nde Bup, D. A Review of Cocoa Drying Technologies and the Effect on Bean Quality Parameters. *Int. J. Food Sci.* **2020**, *2020*, 8830127. [CrossRef] [PubMed]

Article

Numerical Study of Electrostatic Desalting Process Based on Droplet Collision Time

Marco A. Ramirez-Argaez [1], Diego Abreú-López [1], Jesús Gracia-Fadrique [1] and Abhishek Dutta [2,*]

[1] School of Chemistry, National Autonomous University of Mexico (UNAM), Av. Universidad #3000, Cd. Universitaria, Mexico City 04510, Mexico; marco.ramirez@unam.mx (M.A.R.-A.); diego.abreu05@gmail.com (D.A.-L.); jgraciaf@unam.mx (J.G.-F.)

[2] Department of Chemical Engineering, Izmir Institute of Technology, Gülbahçe Campus, Urla, Izmir 35430, Turkey

* Correspondence: abhishekdutta@iyte.edu.tr; Tel.: +90-(232)-7506617

Abstract: The desalting process of an electrostatic desalting unit was studied using the collision time of two droplets in a water-in-oil (W/O) emulsion based on force balance. Initially, the model was solved numerically to perform a process analysis and to indicate the effect of the main process parameters, such as electric field strength, water content, temperature (through oil viscosity) and droplet size on the collision time or frequency of collision between a pair of droplets. In decreasing order of importance on the reduction of collision time and consequently on the efficiency of desalting separation, the following variables can be classified such as moisture content, electrostatic field strength, oil viscosity and droplet size. After this analysis, a computational fluid dynamics (CFD) model of a biphasic water–oil flow was developed in steady state using a Eulerian multiphase framework, in which collision frequency and probability of coalescence of droplets were assumed. This study provides some insights into the heterogeneity of a desalination plant which highlights aspects of design performance. This study further emphasizes the importance of two variables as moisture content and intensity of electrostatic field for dehydrated desalination by comparing the simulation with the electrostatic field against the same simulation without its presence. The overall objective of this study is therefore to show the necessity of including complex phenomena such as the frequency of collisions and coalescence in a CFD model for better understanding and optimization of the desalting process from both process safety and improvement.

Keywords: electrostatic desalting; droplet collision; mathematical model; emulsion breakage

Citation: Ramirez-Argaez, M.A.; Abreú-López, D.; Gracia-Fadrique, J.; Dutta, A. Numerical Study of Electrostatic Desalting Process Based on Droplet Collision Time. *Processes* **2021**, *9*, 1226. https://doi.org/10.3390/pr9071226

Academic Editor: Peter Glavič

Received: 10 June 2021
Accepted: 13 July 2021
Published: 15 July 2021

Publisher's Note: MDPI stays neutral with regard to jurisdictional claims in published maps and institutional affiliations.

1. Introduction

Crude oil is a non-renewable and a naturally occurring unrefined petroleum product that consists mainly of hydrocarbons and other inorganic substances [1]. These hydrocarbons have to be separated before they can be used as a fuel. Crude oil is extracted mostly from underground oil fields and also from under the sea bed. It is extracted from wells containing gases, water and a small amount of inorganic salts. The presence of saline water in oil causes numerous problems in an oil refining plant such as corrosion, fouling and formation of scales on the walls of equipment and reduction in activity of refinery catalyst [2–4] and thus needs to be removed. The process of removal of saline water from raw crude oil is called desalting. There are several methods for desalting, the most efficient and common being the electrostatic desalting [2]. The main objective of such a unit is to minimize the salt and water content in the oil stream and also to reduce the requirement of energy for pumping and transportation of oil [5,6]. The unit is a complex system involving many phenomena during the operation of the desalting process, which includes fluid flow with droplet–droplet and liquid–droplet interactions, heat transfer and electromagnetic phenomena. Additionally, the geometry of such electrostatic desalting unit is extremely complex. Desalting of crude oil involves two steps, the first is to mix the oil stream with a stream of freshwater intensively, which results in reducing the stability of the water-in-oil emulsion and the second is to separate the saline water present in the emulsion [7,8]. Depending on the amount of gas, salt and water present in the extracted crude oil, two different separation techniques are used. If salt water content is much less, then two-phase separators are used to separate gas from the raw oil stream. If a significant amount of salt water is produced in the process of extraction of raw crude oil, then three-phase separators are used to separate gas from the liquid phase and also water from the oil stream. The oil that leaves the separator contains finely divided water droplets of average diameter ranging from 10 to 20 μm [9]. These finely divided water droplets get emulsified and form water-in-oil (W/O) emulsions. These emulsions consist of water droplets dispersed in oil and the third component being the emulsifying agent to stabilize the system. The volume of water that is present in the W/O emulsion is usually 7% of the extracted hot crude oil [7]. Therefore, electrostatic desalting is one of the most popular methods to remove saline water from crude oil by applying an external electrostatic field. This increases the rate of collision of water droplets thereby improving the coalescence.

Treatment of saline water present in emulsified form (desalting) is a sophisticated process as it depends on various factors such as the concentration of demulsifying agents, wash water, concentration of salt in water, rate of mixing with wash water [2], temperature and rate of dilution [10]. Several researchers [5,7,10,11] have studied the effect of these factors on desalting efficiency. Aryafard et al. [12] simulated the process using a single stage desalting process in which a demulsifier is added to the raw crude oil stream at the beginning. They used population balance equations to account for the interactions between the dispersed phase (water droplets) and the continuous phase (oil) and also for breaking and coalescence of water droplets. Results from their simulation showed that an increase in the inlet flow rate of fresh water from 3% to 6% increases the water separation efficiency from 96.5% to 98.5% which they validated with industrial data. Aryafard et al. [13] further developed a mathematical model in order to determine the efficiency of removal of salt and water from raw crude oil stream in an industrial two-staged desalting process. Their study provided a trend, based on the effect of inlet flow rate of fresh water, pressure drop and magnitude of applied electric field on water and salt removal efficiencies which matched their previous study [12]. A similar model was developed by Kakhki et al. [14] consisting of a mixing valve and an electrostatic drum using the same operating conditions as in [12]. They noted that an increase in the rate of collision between water droplets promotes coalescence. The effect on the inlet flow rate of fresh water showed similar results to that of Aryafard et al. [13]. Bresciani et al. [7] proposed a mechanism for droplet–droplet collision and coalescence in the presence of an external electric field. A model was developed to determine the time between collision and displacement speed of water droplets considering the forces acting on the droplets. Their study focused on the effect of operational variables such as temperature of the emulsion, strength or magnitude of electric field and the inlet rate of fresh water on desalting efficiency. Mahdi et al. [5] demonstrated the effect of process variables, namely chemical dosage of demulsifier, temperature of oil stream, wash water dilution ratio, settling time and mixing time with water stream on desalting efficiency. The efficiency of the desalting process was determined in terms of water cut and degree of salt removal. Vafajoo et al. [15] investigated the influence of concentration of chemical demulsifier, temperature and pH of the oil stream on the performance of desalting process. The two performance variables studied were salt removal efficiency and bottom sediments and water (BSW). Bansal and Ameensayal [16] designed a horizontal electrostatic desalting unit using computational fluid dynamics (CFD) to investigate the flow profile (velocity distribution) of crude oil in the system. They studied the impact of geometry on transitions of flow regime of the crude oil stream based on Reynolds number. The results explained the position of the obstruction plate with respect to the inlet header at which separation of crude oil and water is uniform and separation efficiency is optimum. Shariff and Oshinowo [17] modeled the flow of water-in-oil emulsion using a multiphase approach and showed that the presence of vortices formed by the inlet distributor reduces the separation efficiency to a large extent. The increase in desalting with a reduction in wash water consumption through an enhanced turbulent mixing of wash water with the raw crude oil stream was investigated by Ilkhaani [18]. Xu et al. [19] investigated the desalting efficiency of an external electric field using four different electrodes: plate, folded plate, grid and copper mesh electrodes. They showed that the desalting efficiency of a flat plate electrode is lower than the other configurations, owing to the non-uniformity of electric field distribution.

This study is divided into two main parts. First, a simplified collision model between two droplets in water-in-oil (W/O) emulsions was developed, based on the pioneering work of Bresciani et al. [7]. Then, this model was employed to perform a complete process analysis determining the effect of the main process variables on the collision frequency. The objective of this part is to provide an empirical correlation of the collision frequency to be used in CFD simulations shown in the second part of the study. The ultimate objective is to predict the separation process in a complex desalting unit with realistic assumptions by coupling an analytical analysis of collision between droplets into a CFD simulation in order to include physical phenomena that otherwise would be neglected if only the CFD tool is used which would be scientifically incorrect.

Table 1 summarizes the highlights from research on modeling of the desalting process of crude oil, along with those briefly discussed above. It can be seen from the table that no CFD model has been published so far about this complex phenomenon of collision between droplets, so the effect of the electrostatic field is not yet accounted in the simulation of desalting. This research will lead to understanding of and ultimately optimizing the desalting of crude oil reducing risks of corrosion in the refinery, improving process safety, and in general providing a more environmentally friendly process.

Table 1. Summary of proposed models for desalting of crude oil.

Reference	Correlation/Ranges	Notable Results
Bresciani et al. [7] Bresciani et al. [9]	$tc = 3{:}57\, l\mu_o^{-2}a^{-5}$ $l_0 = (1.613x^{-1/3} - 2)\,a$ $E[0\ 0.1-3\,\text{kV/cm}], x[0.02-0.08], a[1-100\ \mu\text{m}]$	Increase in operational variables promotes desalting
Abdul-Wahab et al. [10]	$\eta_1 = \beta_0 + \beta_1 X_1 + \beta_2 X_2 + \beta_3 X_3 + \beta_4 X_4 + \beta_5 X_5 + \beta_{12} X_1 X_2 + \beta_{13} X_1 X_3 + \beta_{14} X_1 X_4 + \beta_{15} X_1 X_5 + \beta_{23} X_2 X_3 + \beta_{24} X_2 X_4 + \beta_{25} X_2 X_5 + \beta_{34} X_3 X_4 + \beta_{35} X_3 X_5 + \beta_{45} X_4 X_5 + \beta_{1212} X_1 X_2 (X_1 - X_2) + \beta_{1313} X_1 X_3 (X_1 - X_3) + \beta_{1414} X_1 X_4 (X_1 - X_4) + \beta_{1515} X_1 X_5 (X_1 - X_5) + \beta_{2323} X_2 X_3 (X_2 - X_3) + \beta_{2424} X_2 X_4 (X_2 - X_4) + \beta_{2525} X_2 X_5 (X_2 - X_5) + \beta_{3434} X_3 X_4 (X_3 - X_4) + \beta_{3535} X_3 X_5 (X_3 - X_5) + \beta_{123} X_1 X_2 X_3$ $\eta_1 = 50.277 + 8.040 X_2 + 1.863 X_3 + 0.679 X_5$ $(R^2 = 0.7)$ $\mu_o[0.017-0.048\,\text{Pa s}], X_1[55-70\,^\circ\text{C}], X_2[1-3\,\text{min}], X_3[1-9\,\text{min}], X_4[1-15\,\text{mg/kg}], X_5[1-10\%]$	Depends on settling time, mixing time and rate of dilution
Al-Otaibi et al. [2]	N/A	Increase in operational variables promotes desalting
Aryafard et al. [12]	No correlation $\Delta p[137.895-206.843\,\text{Pa}]$ $E[1-2\,\text{kV/cm}]$ $x[0.03-0.06]$	Decrease in pressure drop, increase in electric field and inlet rate of fresh water promote desalting
Mohammadi [20]	$E[1.4-2.8\,\text{kV/cm}], a[600\ \mu\text{m}], \mu_o[0.015\,\text{Pa s}]$	Strong electric field Small distance of separation of droplets
Kakhki et al. [14]	No correlation $T[50\,^\circ\text{C}], x[0.03-0.05], f[50-400\,\text{Hz}]$	Increase in frequency promotes desalting
Alnaimat et al. [21]	No correlation, no ranges given	Low flow and charged particle velocities High electric and magnetic fields

Operational variables: temperature, T; oil viscosity, μ_o; strength of electric field, E; volume fraction of water, x; collision time, tc; initial separation between droplets, l_0; droplet size, a; temperature, X_1, T; settling time, X_2; mixing time, X_3; chemical dosage, X_4; dilution rate, X_5; efficiency of salt separation, η_1; pressure drop, Δp; current frequency, f; parameters used for calculating of salt separation, $\beta_0-\beta_5$, $\beta_{12}-\beta_{15}$, $\beta_{23}-\beta_{25}$, $\beta_{34}-\beta_{35}$, β_{45}, $\beta_{1212}-\beta_{1515}$, $\beta_{2323}-\beta_{2525}$, $\beta_{3434}-\beta_{3535}$, β_{123}.

2. Materials and Methods

For a basic understanding of the effect on the main process variables on the desalting efficiency, the analysis was done in a simplification of the real geometry of the system as shown in Figure 1a [9]. The analysis takes place in a space bounded by the inlet of emulsion and by the position of upper and lower electrodes. The separation process involving the coalescence of droplets to increase their size and facilitate the separation from the oil by gravity is oversimplified in this study to understand the time to obtain first collision that eventually determines the real frequency of collisions, which in turn indirectly determines the separation process rate and efficiency. Besides, instead of analyzing the actual process where a population of many droplets of different sizes is present in a water in oil emulsion entering into the electric field zone of the desalting unit, only two droplets are analyzed to set up a model that predicts, from fundamental principles, the rate at which the droplets collide. Figure 1b shows the schematic representation of the two droplets analyzed in this study which experience several forces that determine the time for them to collide.

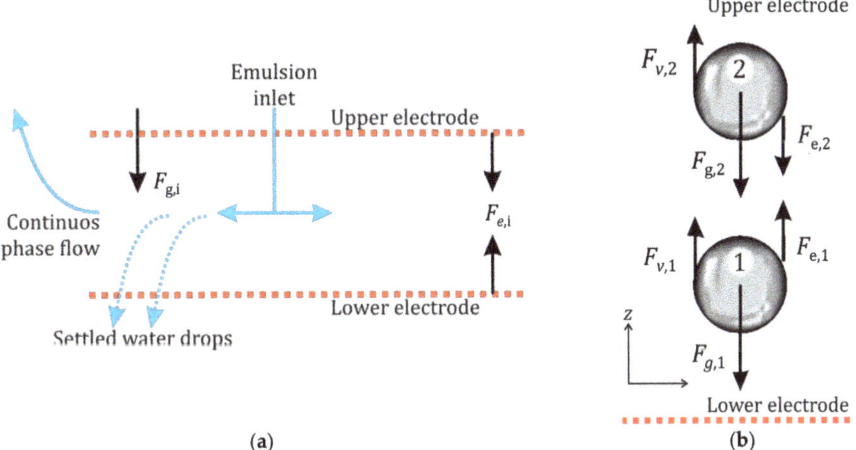

Figure 1. (**a**) Schematic representation of the desalting system, including the inlet of the emulsion, the electric field set up by the upper and lower electrodes and the flow of the continuous and droplets. (**b**) Schematic representation of all forces acting on two droplets of water, one on top of the other and both being in between the two electrodes. Forces include the electrostatic attraction (F_e), drag (F_v), gravity and buoyancy forces (indicated as F_g together).

2.1. Force Balance

The forces acting on both the droplets are indicated in Figure 1b and are, namely, gravitational, buoyant, drag and electrostatic attraction forces. Several simplifying assumptions besides the consideration of a simplified geometry and considering only two droplets one on the top of the other, are stated in this formulation, which is based on the work by Bresciani et al. [7]. These assumptions are as follows: (i) the study does not intend to define the physics of coalescence, but only the process leading to the collision of two droplets; (ii) breakup of droplets is neglected; (iii) system is isothermal so the physical properties are constant; (iv) drag due to the main fluid flow in the desalting unit is neglected and only the movement of the droplets due to the other forces is considered; (v) interfacial tension forces are neglected; (vi) spherical droplets of constant size and evenly distributed at the beginning of the process are considered; (vii) some other forces like the added mass, lift and electrophoresis forces are also neglected; (vii) water-in-oil (W/O) emulsions are only considered in this work. The sum of forces for *i*th drop, F_i is:

$$F_i = F_{v,i} + F_{g,i} + F_{e,i} \tag{1}$$

where $F_{v,i}$, $F_{g,i}$, and $F_{e,i}$ are the drag, gravity plus buoyant and electrostatic forces of the *i*-th droplet. Newton's second law of movement is the mass of a droplet of size a_i times its acceleration and it can be expressed as:

$$F_i = \frac{4}{3}\pi a_i^3 \rho_w \frac{\mathrm{d}v_i}{\mathrm{d}t} \tag{2}$$

where ρ_w and $\frac{\mathrm{d}v_i}{\mathrm{d}t}$ are the density and acceleration of the droplet represented as the time derivative of the droplet velocity, v_i. The drag force, $F_{v,i}$, is represented by:

$$F_{v,i} = -6\pi\mu_o a_i v_i \tag{3}$$

where μ_o is the viscosity of the continuous phase (oil). The gravitational and buoyant forces of the droplet are represented as:

$$F_{g,i} = \frac{4}{3}\pi a_i^3 (\rho_w - \rho_o)g \tag{4}$$

where ρ_o and g are the oil density and gravitational constant, respectively. Finally, the electrical forces for droplet 1 and droplet 2 can be estimated as:

$$F_{e,1} = -6KE^2 (a_2)^2 \left(\frac{a_1}{l}\right)^4 \tag{5}$$

$$F_{e,2} = 6KE^2 (a_1)^2 \left(\frac{a_2}{l}\right)^4 \tag{6}$$

where K, E and l are the dielectric constant of oil, the electric field and the distance between droplets, respectively. Subscripts 1 and 2 stand for droplet 1 (lower) and 2 (upper) as represented in Figure 2.

By substituting Equations (2)–(6) into Equation (1) for droplets 1 and 2, the following system of three ordinary differential equations arises:

$$\frac{\mathrm{d}v_1}{\mathrm{d}t} = \frac{(\rho_w - \rho_o)g}{\rho_w} - \frac{9KE^2 a_1 (a_2)^2}{2\pi\rho_w l^4} - \frac{9\mu_o}{2a_1^2\rho_w}v_1 \tag{7}$$

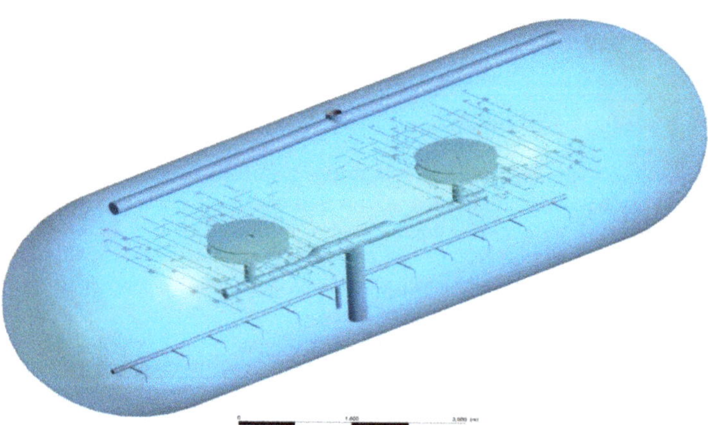

Figure 2. *Cont.*

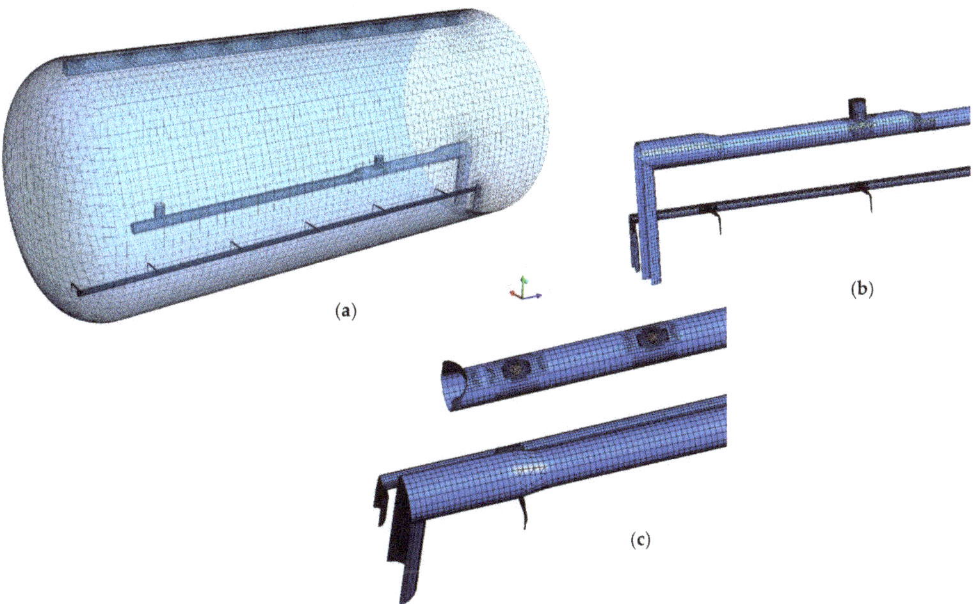

Figure 2. Schematic of the whole desalter unit indicating (**a**) general grid of the whole desalter without showing the electrodes, (**b**) inlet of emulsion and water outlet and (**c**) oil outlet.

$$\frac{dv_2}{dt} = \frac{(\rho_w - \rho_o)g}{\rho_w} + \frac{9KE^2 a_2 (a_1)^2}{2\pi\rho_w l^4} - \frac{9\mu_o}{2a_2^2\rho_w} v_2 \tag{8}$$

$$\frac{dl}{dt} = v_1 - v_2 \tag{9}$$

Equations (7) and (8) come out by applying Newton's second law of motion for droplet 1 Equation (7) and for droplet 2 Equation (8), while the third equation comes out by updating the distance between the droplets. Initial conditions for the velocities of both droplets assume static conditions i.e., $v_1 = v_2 = 0$, while the initial distance of the droplets l_0 takes into account the amount of water expressed in volume fraction x and the size of the two droplets a_1 and a_2 in a cubic volume according to:

$$l_0 = \left(\frac{2\pi \left(a_1^3 + a_2^3 \right)}{3x} \right)^{1/3} - (a_1 + a_2) \tag{10}$$

A system of 3 differential equations was numerically and simultaneously solved by employing the iterative Runge–Kutta 4th order method that provides a very low total error if a small step is employed [22]. In this case, an optimal time step was used for each case Δt between 1×10^{-8} and 1×10^{-6} s. A Python® code version 3.6 was written to solve the problem for $v_1^{t+\Delta t}$, $v_2^{t+\Delta t}$, $l^{t+\Delta t}$ i.e., the velocities of the droplets 1 and 2 and the distance between droplets at time step $t + \Delta t$. The simulation ends when the distance between the droplets approaches zero. The code is simulated in a PC with 32 Gb RAM, process Intel Xeon Silver 4110 CPU 2.1 GHz. Cases with smallest droplet size took until 3.51×10^9 iterations to converge and computational time was about 55 h each one.

Table 2 lists the physical properties of the droplets (water) and of the continuous phase (oil) as well as typical operating variables including electric field, droplet sizes, temperature (through the oil viscosity) and water volume fraction of the emulsion as standard conditions in the present study.

Table 2. Operating conditions of a case with standard conditions.

Process Variable	Quantity
Electric field	1 kV/cm
Lower droplet radius	10 μm
Upper droplet radius	10 μm
Water density	943 kg/m^3
Oil density	892 kg/m^3
Gravitational constant	9.81 m/s^2
Oil viscosity	0.044 Pa s
Water volume fraction	0.07
Oil dielectric constant	2.2 ε_0

The model was used to perform a process analysis to elucidate the effect of the main process analysis such as the water content in the emulsion, the strength of the electric field, the oil viscosity and the size of the two droplets for the case where both are the same size or they are not. Table 3 shows the values employed for each variable.

Table 3. Levels of the variables considered in this work to perform the process analysis.

Process Variable	Quantity	Units
Water volume fraction	0.03, 0.05, 0.07, 0.10, 0.12	-
Electric field	0.1, 0.5, 1, 2, 3	kV/cm
Oil viscosity	0.017, 0.027, 0.044, 0.071	Pa s
Lower droplet and upper droplet let radii	1, 5, 10, 15, 20	μm

2.2. Mathematical Modeling

The previous process analysis provides a theoretical collision frequency correlation between droplet lets coming out from first principles and including the main process parameters. This correlation may be used to develop a robust 3D mathematical model on the multiphase fluid flow of the emulsion in a real geometry of a desalting unit through a computational fluid dynamics (CFD) simulation. The fact that the model includes the collision frequency correlation developed through the model presented in the previous section, is necessary to predict and understand the effect of the presence or absence of the electric field in the separation kinetics in an industrial desalting unit. The mathematical model, including simplifying assumptions, governing equations and boundary conditions, is presented below.

2.2.1. Assumptions

The assumptions to model the separation of water-in-oil emulsions are listed below:

- Isothermal system, i.e., there are no gradients of temperature.
- Constant physical properties, each fluid phase is Newtonian and incompressible.
- Non-slip and impermeable conditions were applied for all the internal walls and standard wall functions were used.
- Oil liquid phase is considered as a continuous primary phase and water is considered as the dispersed secondary phase.
- The multiphase system is modeled with a mixture approach which solves a single set of equations: the continuity, the turbulent Navier–Stokes equations and the $k - \varepsilon$ realizable turbulence model, which showed the best convergence behavior and that is being used recently in many complex flow systems.

- A conservation equation for interfacial area concentration is used as an approximation to take into account the coalescence and break up of droplets due to the electrostatic and turbulence effects to compute a droplet t diameter distribution.
- Coalescence terms depend on the frequency of collision, which is taken as the inverse of the time between collisions, which in turn is obtained from the previous droplet collision study.

2.2.2. Governing Equations

The mixture model uses the volume fraction composition of the mixture contained in every cell to compute the density, the viscosity, and the velocity of the mixture. These properties are calculated for N phases according to:

$$\rho_m = \sum_{i=1}^{N} \alpha_i \rho_i \tag{11}$$

$$\mu_m = \sum_{i=1}^{N} \alpha_i \mu_i \tag{12}$$

$$v_m = \frac{\sum_{i=1}^{N} \alpha_i \rho_i v_i}{\rho_m} \tag{13}$$

where ρ_m, μ_m, and v_m are the density, viscosity, and velocity for the mixture; ρ_i, μ_i, and v_i are the same properties for each phase. α_i is the volume phase fraction, and i subscript denotes the i-th phase. For this case, subscripts w and o are used to denote water (secondary, dispersed phase) and oil (primary, continuous phase), respectively.

The continuity and momentum conservation equations are formulated for mixture of phases and taking properties of the mixture in a regular way [23], including in the convective terms the drift velocity $v_{dr,w}$, which specifies the slip velocity between the dispersed phase (water) with respect to the continuous phase (oil) as:

$$v_{dr,w} = v_{wo} - v_m \tag{14}$$

$$v_{wo} = \frac{(\rho_w - \rho_m)d_w^2}{18\mu_o f_{drag}} \vec{a} - \frac{\eta_t}{Pr_t}\left(\frac{\nabla \alpha_w}{\alpha_w} - \frac{\nabla \alpha_o}{\alpha_o}\right) \tag{15}$$

where $\vec{U}_{wo}$ is the slip velocity modeled by an algebraic slip formulation according to Manninen et al. [23], d_w the water droplet diameter (defined later), $\vec{a}$ is the acceleration given by gravity and centrifugal forces. The second term in the right side of (15) contains the dispersion due to the turbulence, where η_t is the turbulent diffusivity calculated from a continuous-dispersed fluctuation velocity correlation, Pr_t is the Prandtl turbulent number fixed to a value of 0.75, and f_{drag} is the drag coefficient, calculated by the Schiller–Nauman correlation as:

$$f_{drag} = \begin{cases} 1 + 0.15Re^{0.687} & Re \leq 1000 \\ 0.018Re & Re > 1000 \end{cases} \tag{16}$$

where Re is the relative Reynolds number. The volume fraction conservation equation is derived from the continuity equation of the secondary dispersed phase:

$$\nabla \cdot (\rho_w v_m \alpha_w) = -\nabla \cdot (\rho_w v_{dr,w} \alpha_w) \tag{17}$$

2.2.3. Interfacial Area Concentration Model

To include the emulsion breakup, a single transport equation is added to solve the interfacial area concentration which is defined as the interfacial area formed between the dispersed phase (water) and the continuous (oil) phase per unit volume of mixture. This equation represents a population balance which considers the droplets volume fraction as a function of droplet size distribution resulting when it breaks up or two or more droplets coalesce.

$$\nabla \cdot (\rho_w \boldsymbol{v}_w \chi_w) = \rho_w (S_{RC} + S_{TI}) \tag{18}$$

where χ_w is the interfacial area concentration (m^2/m^3), and the right-hand side terms in Equation (18) account for the coalescence and breakup of the dispersed phase. S_{RC} is the coalescence source term due to random collisions, and S_{TI} is the breakup source term due to turbulence. In this study, the coalescence term was considered as a modified Hibiki–Ishii model [24] as follows:

$$S_{RC} = -\frac{1}{108\pi} \left(\frac{\alpha_w}{\chi_w} \right)^2 n_c f_c \lambda_c \tag{19}$$

$$n_c = 108\pi \frac{\alpha_w}{d_w^3} \tag{20}$$

$$f_c = \frac{1}{22469.1829\alpha_w^{-3.294} \mu_o^{0.968} E_0^{-2.007} d_w^{-0.013}} \tag{21}$$

$$\lambda_c = 1 \tag{22}$$

where λ_c is a total coalescence probability after a collision, n_c is the number of droplets per volume of mixture, and f_c is the frequency of collisions, which is proposed in the first part of this study Equation (28). The breakup source term by turbulent impacts is calculated by the Hibiki–Ishii model:

$$S_{TI} = \frac{1}{108\pi} \left(\frac{\alpha_w}{\chi_w} \right)^2 n_b f_b \lambda_b \tag{23}$$

$$n_b = 108\pi \frac{(1 - \alpha_w)}{d_w^3} \tag{24}$$

$$f_b = \frac{0.264\alpha_w \varepsilon^{1/3}}{d_b^{2/3}(\alpha_{w,max} - \alpha_w)} \tag{25}$$

$$\lambda_b = \exp\left(-1.37 \frac{\sigma}{\rho_o d_b^{5/6} \varepsilon^{2/3}} \right) \tag{26}$$

where n_b is the number of eddies per volume of the two-phase mixture according to Azbel and Athanasios [25], f_b is the frequency of collision due turbulence, and λ_b is the breakup efficiency.

Additionally, the water droplet diameter is calculated as:

$$d_w = 6\frac{\alpha_w}{\chi_w} \tag{27}$$

2.2.4. Realizable $k - \varepsilon$ Turbulence Model

As the mixture approach is used, a single set of equations were used to compute the turbulent kinetic energy and its dissipation rate (k and ε) using the mixture shared properties. Conservation equations of k and ε are solved using realizable $k - \varepsilon$ turbulence model that can be found elsewhere (Shih et al. [26]). This turbulence model mainly gives a superior ability to capture the mean flow of complex structures and for flows involving rotation in closed domain as in the case of crude oil electrostatic desalting unit. Dutta et al. [27] used realizable $k - \varepsilon$ model to account for the effect of turbulence in a closed domain flow similar to the desalter and Chen et al. [28] adopted realized $k - \varepsilon$ and Eulerian models to evaluate the separation efficiency in an oil/water separator validating their numerical results with experiments. Furthermore, Mouketou and Kolesnikov [29] compared multiphase flow simulations with experimental data applying the same modeling approach and included a Lagrangian model for solid particles applicable to their oil–water process. Using the above reasoning, the flow conditions found in the device are found to be compatible with realizable $k - \varepsilon$ model and with non-equilibrium wall function.

2.2.5. Boundary Conditions and Solution

The boundary conditions include non-slip and impermeability at all the walls of the system while standard wall functions accounted for non-equilibrium conditions are used to connect the turbulent core with the laminar flow field near the static walls. At the inlet, a fixed uniform mixture velocity is set and at the outlets, the gauge pressure is set to zero. Finally, the model was implemented in the CFD code Fluent Ansys version 19.0. It ran in a PC with 32 Gb RAM and processor Intel Xeon Silver 4110 CPU 2.10 GHz and a clock speed of 2095 Mhz and it took around 6 h of simulation in steady state to get convergence simulating with 10,000 iterations for a mesh with around 1,500,000 elements, resulting after performing a grid sensitivity study, where this mesh is the optimum in reducing the computational time with good enough accuracy. For the grid sensitivity study, three more grids were simulated with around 0.5, 1 and 3 million mesh elements. The average (mixture) velocity was then calculated for each of these mesh elements and compared. Increasing from around 1,500,000 to around 3,000,000 elements brought only 0.51% (that is less than 1%) change in the average mixture velocity which was considered a marginal improvement compared to the increase in computational time (6 h of simulation to 13 h to obtain convergence). As such, it was decided to continue further simulation with around 1,500,000 elements.

In Table 4, the geometry of the desalting unit together with operating parameters are shown, while Figure 2 shows a detail of the unit including the position of inlets, outlets and electrodes. After a sensitivity study of the grid, a mesh of 1,500,000 elements was carefully designed to warrant results independent of the grid and convergence, including finer elements next to inlets, outlet and internal walls than in the zones where there are no such elements, was obtained as can be seen in Figure 2a–c.

Table 4. Operating parameters of an industrial desalting unit.

Process Variable	Quantity
Length	15.24 m
Internal diameter	3.66 m
Water volume fraction	0.10
Emulsion flow at inlet	0.126 m^3/s
Electric field	1 kV/cm
Oil viscosity	0.044 Pa s
Interfacial tension	4 mN/m

3. Results and Discussion

Figure 3 shows the plots of the time-evolution (in logarithmic scale) of the distance between droplets, which represents the process analysis performed in this study to reveal the effects of water content (Figure 3a), oil viscosity (Figure 3b), size of the droplets assuming same size (Figure 3c), electric field (Figure 3d), on the time to get a collision between the two droplets. The initial distance between droplets depends only on the water content and droplet size as expressed by Equation (10), so this is why the distances between droplets are different at initial time for Figure 3a,c (effect of water content and droplet size), while for the rest of variables the initial distances are the same in all cases (the simulations used constant droplet size of 10 μm and constant water volume fraction of 0.07). The levels of each variable analyzed are depicted in the legend of each plot and they correspond to those shown in Table 3. The rest of the operating conditions in each simulation correspond to the so-called standard conditions presented in Table 2. Figure 3a shows that as the volume fraction of water increases, both the initial distance between droplets and the time for collision decrease. When more water is present in the emulsion of evenly distributed droplets of constant size, these droplets get closer to each other and the electric field forces act inversely proportional to the distance between droplets to get the collision at shorter times (<1 s). On the contrary, when the water content is the lowest, the initial distance between droplets is three times larger than the largest water content and consequently, the electric field forces are delayed (>20 s) to bring both droplets into contact. The other forces, such as the gravitational forces (weight and buoyant) have the same significance at constant droplet size. Figure 3b shows the effect of oil viscosity on the collision time, where it can be noted that when the viscosity of the oil increases four times (temperature decreases) the time for the first collision increases as well from 0.7 to 3 s (i.e., more than four times). The oil viscosity increment increases the drag force that prevents in a certain way the motion of the droplets delaying the collision between droplets. Figure 3c shows that there is no effect of the droplet size on the time for the first collision. Actually, the only effect of reducing the droplet size is reducing the initial size between droplets for constant water content, i.e., as the size of droplet increases, they are more separated from each other. All forces acting on the droplets depend on the size of droplets, but apparently, their opposite effects in the different forces cancel out for determining the time for the first collision keeping constant the electric field, oil viscosity and water content in the emulsion. This conclusion is important but must be taken cautiously since in real emulsion the size of droplets is not constant but it presents a population of sizes that makes it complex to analyze this variable. Furthermore, in this study, the effect of the size with different droplet sizes will be analyzed. Figure 3d presents the effect of the electric field strength on the time for the first collision between two droplets of the same size. The effect of this variable is significant, since reducing the electric field from 3 to 0.1 kV/m increases the collision time from less than 1 s to approximately 175 s (keeping the other variables fixed). The electrostatic attractive force created by the dipolar effect on the droplets, attract two near droplets and dominate the force balance if the field is big enough to overcome the drag and gravitational forces. Then the desalting unit needs the electric field to achieve the separation.

These results of collision time were correlated with the four variables explored to obtain the following equation that shows a correlation applicable only for the case when the two neighboring droplets have the same size:

$$t = 22\,469.183 x^{-3.294} \mu_0^{0.968} E^{-2.007} a^{-0.013} \, , R^2 = 99.738\% \tag{28}$$

The sign and the magnitude of the exponent of each variable indicate the dependence between the time for the first collision and that variable. The more significant variables in order of decreasing importance are water content, electric field, oil viscosity, and droplet size and the sign indicates that an increment in both the water content and electric field will increase the probability of fast collisions causing eventually coalescence of droplets and therefore efficient water and salt separation from oil. On the contrary, the increment in oil viscosity (decrement in temperature) will increase the collision time, decreasing the collision frequency and reducing the rate of separation and the efficiency of the process. The size of the droplets has practically no effect on the collision time since the exponent is nearly zero. Figure 4 shows the performance of the correlation in predicting the values obtained by the model. The agreement is excellent, so the correlation represents a good quantitative summary of the effect of the variables explored on the collision time and indirectly in the efficiency of the desalting process.

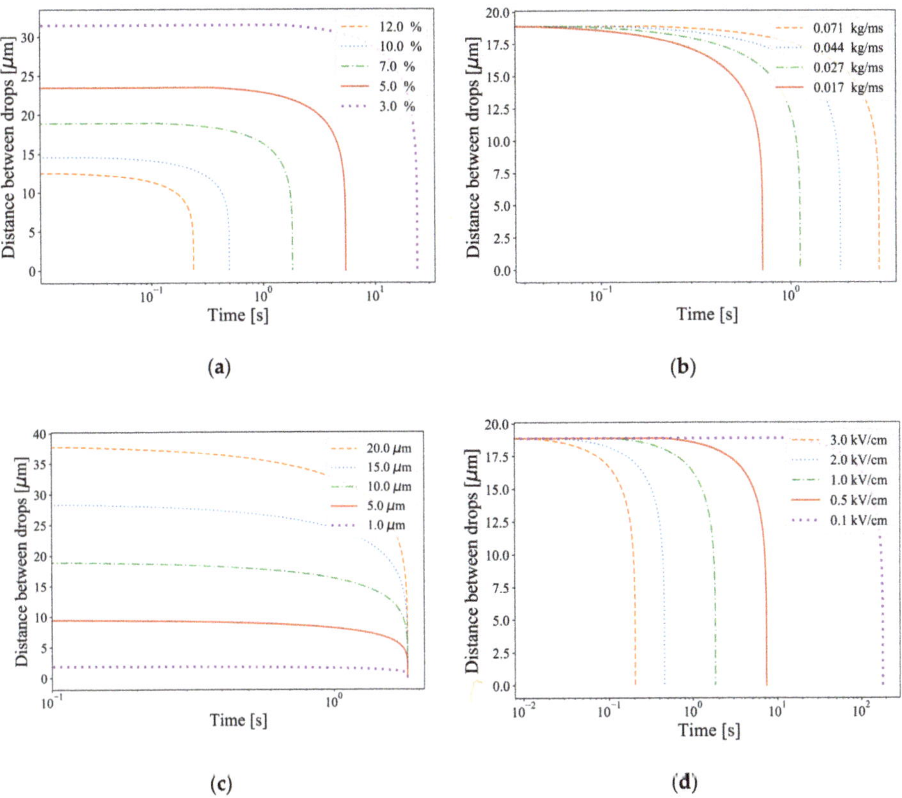

Figure 3. Effect of the (**a**) water content x, (**b**) oil viscosity μ_o (temperature), (**c**) size of the droplets (assuming two droplets of the same size), (**d**) electric field strength E. In all cases, the rest of the operating conditions correspond to the standard conditions mentioned in Table 2.

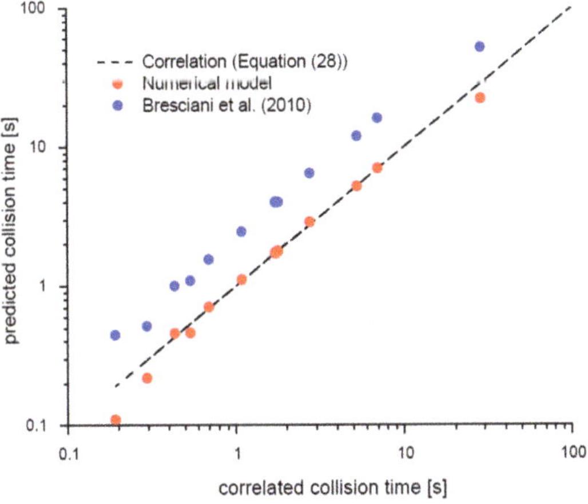

Figure 4. Predicted collision time by the correlation Equation (11) versus collision time in logarithmic scale.

In Figure 5, the time for the first collision between two droplets is plotted as a function of each variable explored in this study, such as the water content, droplet size, oil viscosity (temperature) and intensity of electric field. It can be noted that the most significant variables in order of descending importance are the electric field, water content of the emulsion, oil viscosity and the droplet size; the final variable not having a visible effect on the time for the first collision between the two droplets. The droplet Reynolds number ranges from 0.002 to 0.82 considering velocity ranges between 0.1 and 2 m/s and with a range of sizes of 1–20 μm. All cases studied fall under the Stokes regime and maybe this is why the droplet size does not influence the collision time. The effect of electric field and water content is inverse with collision time, i.e., the higher the electric field and water content, the lower the collision time, while the oil viscosity has a moderate proportional relationship with the collision time. It is necessary to note that this result was not shown by Bresciani et al. [7]. Additionally, the study of the effect of water content and oil viscosity on collision time was also not carried out by them. These results have economic and environmental implications, and these are subjected to an optimization analysis to minimize the operational costs and reduce corrosion in the refinery. In this regard, an increase in both the water content and electric field will signify increase in the freshwater consumption and the use of a bigger electric field with an associated investment and operational cost as well as environmental issues, so there must be a compromise between separation efficiency and operational (more electric energy) and investment cost (more robust equipment) in the electrostatic desalting unit. Additionally, in the case of the water content, an excessive increase in this variable may result in reverse of the emulsion from water-in-oil (W/O) to an oil-in-water (O/W) more difficult to separate and the waste of fresh water is negative to the environment and to the population. Besides, the results are just indirectly related to the separation process since the collision is not sufficient for coalescence as the presence of emulsifying substances at the oil–water interface may prevent coalescence [4].

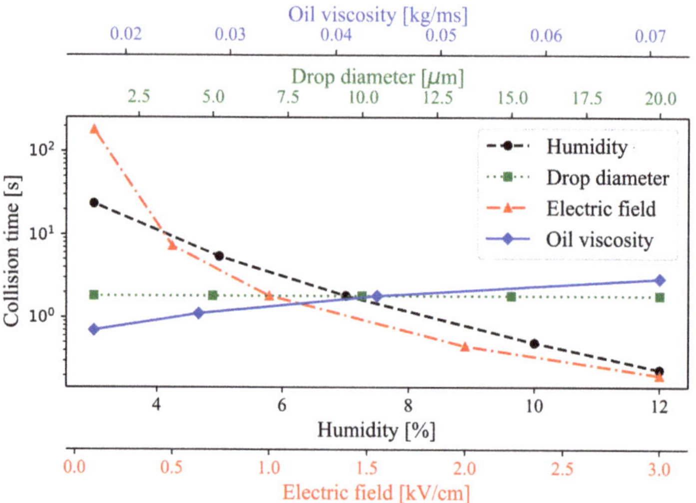

Figure 5. Collision time as a function of the main process variables explored in this study: water content, droplet diameter, electric field intensity, oil viscosity.

Finally, if the two droplets have different sizes, the analysis gets more complex. Figure 6 presents this effect when the upper droplet is 20 μm (Figure 6a), 15 μm (Figure 6b), 10 μm (Figure 6c), 5 μm (Figure 6d) and 1 μm (Figure 6e), and the lower droplet sizes vary from 5 to 20 μm while keeping the rest of variables with values mentioned in Table 2 (standard conditions). For big upper droplets of 20, 15 and 10 μm (Figure 6a–c, respectively), the time for the first collision is around 2 s when the lower droplet is 20 μm and around 10 s when the lower droplet is 5 μm. However, when the upper droplet is 5 μm (Figure 6d) and the lower droplet is also small (5 or 10 μm), the droplets do not collide but rather they separate more as time increases. This effect is even more evident when the upper droplet is 1 μm (Figure 6e) where only the lower droplet of the same size (1 μm) will eventually collide, while with the rest of lower droplet sizes no collision occurs. Then, if the lower droplet is bigger than the upper droplet, the former will sink by gravity at a higher rate than the upper small droplet, so the distance between droplets will increase. In the actual process with much more than two droplets, this effect of no collision would not likely happen since the upper droplet will sooner or later meet another droplet. However, these results indicate that neighboring droplets of similar size would enhance more the separation process than a droplet population with a broader distribution of size. It is well known that inducing a high shear stabilizes the emulsion and also adding fresh water with a big pressure drop at the mixer valve reduces the size of the droplets with a narrow distribution of sizes. Besides, the reader should take into account that the drag force included in the previous analysis is valid for droplets moving in a static fluid but in the real desalter unit there is a fluid flow that promotes a different drag regime, so caution is advised in using these results and extrapolating them into the industrial unit. Additionally, in the current model, the DLVO theory was not taken into account, which is an oversimplification, and consequently, the electric double layer and Van der Waals forces are not considered, which would include a repulsion force between the droplets as they approach closer one to the other before collision takes place and would increase in some extent the estimated collision times slowing down the coalescence process. Additionally, the application of the electric field and the presence of impurities in the crude oil affects the stability and size distribution of droplets.

For the sake of gaining quantitative knowledge on the significance of each parameter, a full factorial design at two levels was performed, and the statistical analysis outcome is presented in Figure 7 and Table 5, which present a variance analysis to determine which variable or interaction between the variables are the most significant on the collision time. Figure 7 presents a Pareto chart with the mean effects of the parameters and Table 5 has more detailed information on the same effects. It is worth noting that the interaction E^*x has a positive significant effect on the collision time, which suggests that increasing both variables at the same time may be detrimental in the separation process.

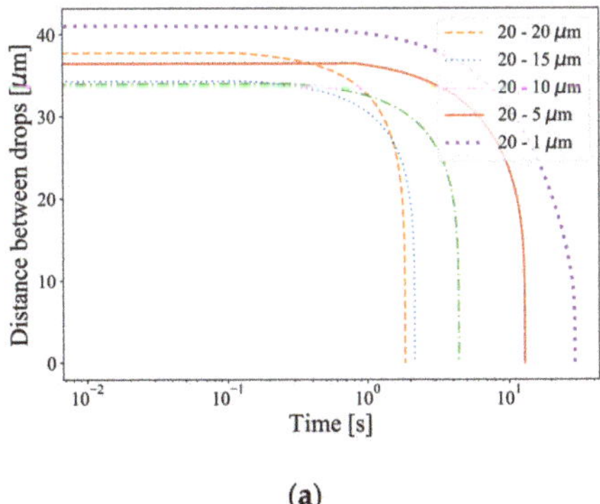

(a)

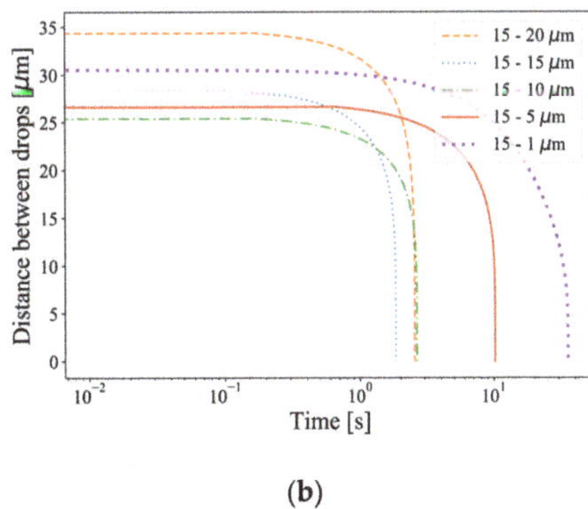

(b)

Figure 6. *Cont.*

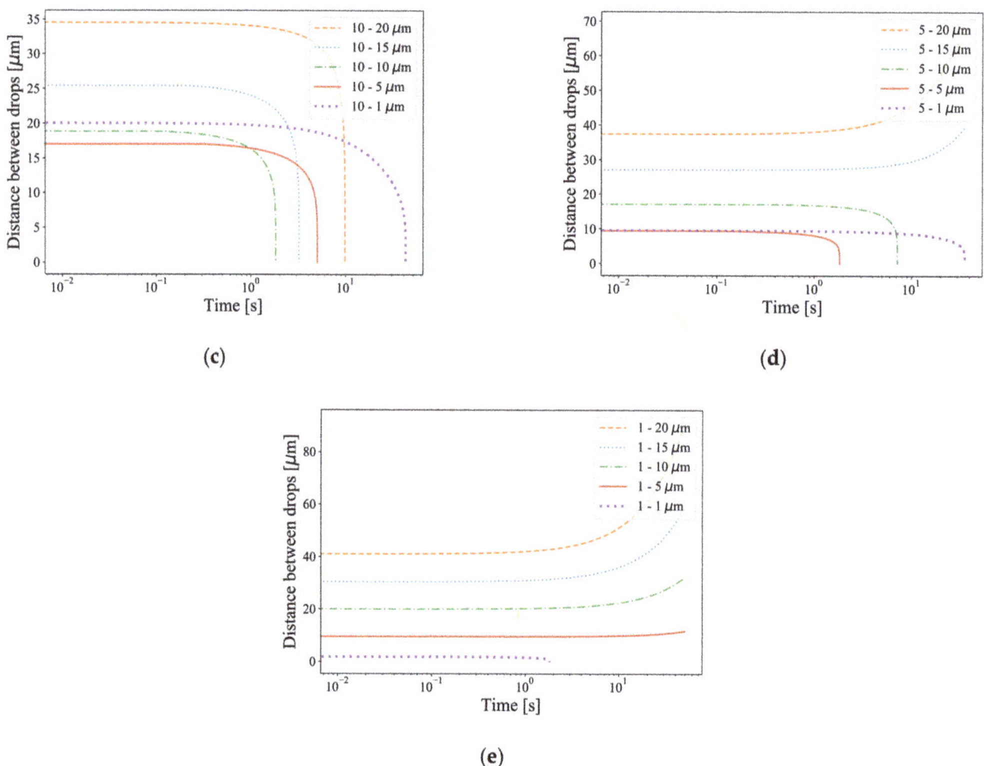

Figure 6. Effect of different droplet sizes of 5, 10, 15 and 20 μm for (**a**) upper droplet of 20 μm, (**b**) upper droplet of 15 μm, (**c**) upper droplet of 10 μm and (**d**) upper droplet of 5 μm. (**e**) upper droplet of 1 μm.

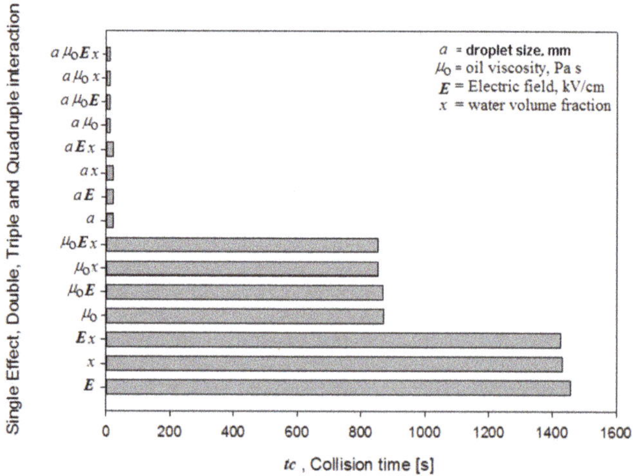

Figure 7. Pareto chart of the single, double, triple and quadruple interaction effects on the collision time, t (s), the significance level is $\alpha = 0.05$.

Table 5. Mean effect of the variables and their interactions.

Term	Mean Effect on Time, t
Droplet size, a	-29.250 ± 733.2
Viscosity of oil, μ_o	878.900 ± 733.2
Electric field, E	-1463.200 ± 733.2
Water content, x	-1436.200 ± 733.2
Double interaction, $a\mu_o$	-17.528 ± 733.2
Double interaction, aE	29.180 ± 733.2
Double interaction, ax	28.640 ± 733.2
Double interaction, $\mu_o E$	-877.000 ± 733.2
Double interaction, $\mu_o x$	-860.800 ± 733.2
Double interaction, Ex	1433.100 ± 733.2
Triple interaction, $a\mu_o E$	17.490 ± 733.2
Triple interaction, $a\mu_o x$	17.168 ± 733.2
Triple interaction, aEx	-28.580 ± 733.2
Triple interaction, $\mu_o Ex$	858.9 ± 733.2
Quadruple interaction, $a\mu_o Ex$	-17.130 ± 733.2

Correlation Equation (28) was cast into Matlab® to minimize the collision time using the optimization genetic algorithm. Bound values for each variables are x[7–12%], μ_o[0.017–0.071 Pa s], E [0.1–3 kV/cm], a[1–20 μm], and the optimal point obtained is $x = 11.998\%$, $E = 3$ kV/cm, $\mu_o = 0.017$ Pa s and $5 \leq a \leq 16$, giving an optima with a value of tc around 0.013 s.

Some basic understanding is gained with the use of such a simplified model. However, it is the intention of this study to provide a less simplified description of the desalting unit operation through a computational fluid dynamics (CFD) model. Such a model has to be considered to be a new approach with new tools in its first stages of development, which is prepared to predict breakage emulsion for 3D complex geometry with actual operating conditions of an industrial electrostatic desalting unit to see the effect of the electric field in the water–oil separation. A detailed process analysis with this CFD model is beyond the scope of the current study, and in this study only the effect of the presence of the magnetic field is explored as a closure of the methodology proposed which included the frequency of collisions correlation cast into the CFD model. Figure 7 presents the oil volume fraction contours at the middle plane in an industrial desalting unit with a presence of an electric field of 1 kV/cm (Figure 8a) and without electric field (Figure 8b). The rest of the variables are, namely, 0.044 Pa s of oil viscosity, variable droplet size from 1 to 20 μm (due to coalescence), and 10% water content. The model includes the frequency correlation as a function of the process variables, and it can clearly be seen that the electric field allows separation of water and that the absence of such a field preserves the emulsion. It has been confirmed with these results, that the development of a collision model from the present study, to predict coalescence, is a key feature of any subsequent study allowing a proper prediction of the industrial electrostatic desalting unit.

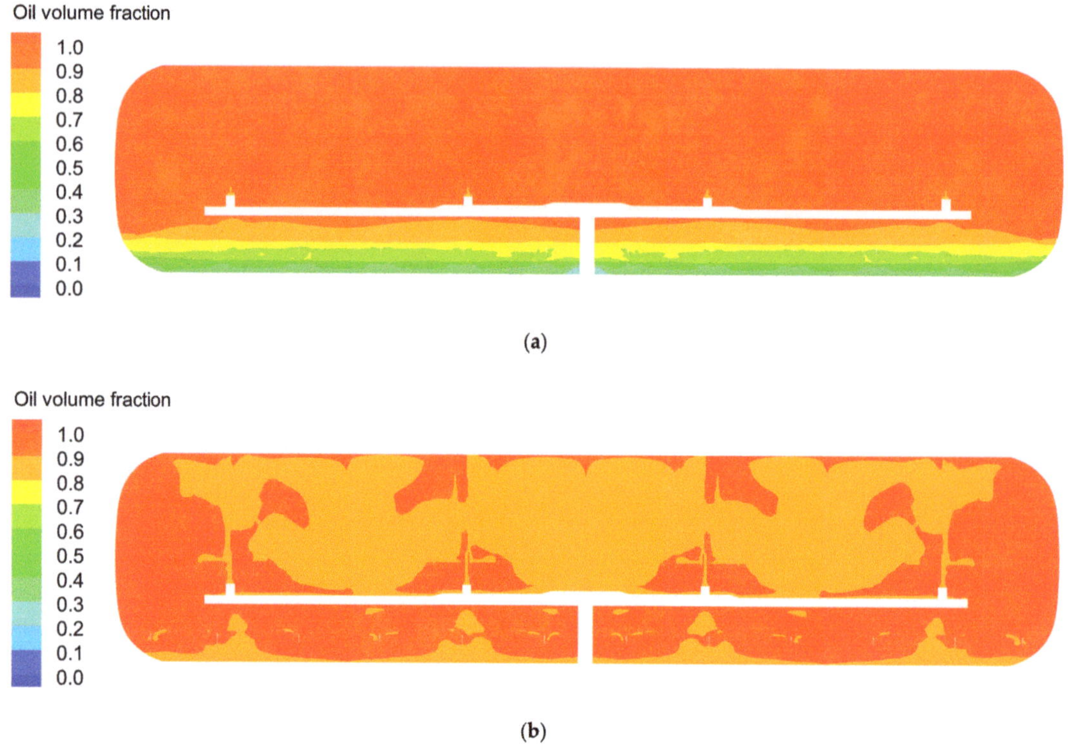

Figure 8. Oil volume fraction for (**a**) with a constant electric field of 1 kV/cm and (**b**) without an electric field. The remaining variables and properties correspond to the standard conditions (see Table 1).

4. Conclusions

Time for collision between two droplets is related to the rate of coalescence which influences the rate separation between water and oil in an electrostatic desalting unit. A process analysis was performed to determine the effect of the electrostatic field, water content, oil viscosity (temperature) and droplet size on the collision time. The analysis gave the following outcomes:

1. The time of collision varies with an inverse relationship to the third power of the water content, which indicated to be the most significant variable, valid in the range of water content of 3–12%.
2. The time of collision varies inversely to the square of the electrostatic field $(0.1 < E < 3 \text{ kV/cm})$, while it has a proportional nearly linear relationship to the oil viscosity $(0.017 < \mu_o < 0.071 \text{ Pa s})$.
3. The droplet size has no influence on the time of collision within the range of droplet sizes analyzed in this work (1 μm up to 20 μm).

If the upper droplet is smaller than the lower droplet in the unit, for emulsions with a wide distribution of sizes, collision may be delayed or even not occur since gravitational force dominates over the electrostatic force on the bottom heavier droplet. The time between collision (frequency of collision) is useful to simulate coalescence phenomena in complex multiphase fluid flow simulations in industrial desalting units performed using numerical modeling, which represents a new tool and hence a new approach (balance of forces model to get a correlation for the time between collisions of a couple of droplets cast into a CFD multiphase fluid flow model) to explore the performance and optimization of electrostatic desalting units. In this study, the effect of the electric field was shown to be a key parameter that has to be included in the description of the separation phenomenon, if the quantitative prediction of the separation kinetics is the goal of any future CFD analysis. This would help to ultimately optimize the desalting of crude oil reducing risks of corrosion in the refinery, improving process safety, and in general providing a more environmentally friendly process.

Author Contributions: Conceptualization, M.A.R.-A., J.G.-F. and A.D.; methodology D.A.-L.; software, D.A.-L. and A.D.; validation, D.A.-L. and J.G.-F.; formal analysis, M.A.R.-A.; investigation, J.G.-F. and A.D.; resources, J.G.-F. and M.A.R.-A.; writing—original draft preparation, A.D.; writing—review and editing, M.A.R.-A. and A.D.; supervision, A.D.; funding acquisition, J.G.-F. and M.A.R.-A. All authors have read and agreed to the published version of the manuscript.

Funding: This research was funded by SENER—CONACYT 144156, Alternativas tecnológicas para mejorar el sistema de desalado de crudo pesado en las refinerías.

Institutional Review Board Statement: Not applicable.

Informed Consent Statement: Not applicable.

Data Availability Statement: Data collected belongs to the property of PEMEX.

Conflicts of Interest: The authors declare no conflict of interest.

References

1. Abdel-Aal, H.K.; Aggour, M.; Fahim, M. *Petroleum and Gas Field Processing*, 1st ed.; Marcel Dekker, Inc.: New York, NY, USA, 2003.
2. Al-Otaibi, M.B.; Elkamel, A.; Nassehi, V.; Abdul-Wahab, S.A. A Computational Intelligence Based Approach for the Analysis and Optimization of a Crude Oil Desalting and Dehydration Process. *Energy Fuels* **2005**, *19*, 2526–2534. [CrossRef]
3. Check, G.R.; Mowla, D. Theoretical and experimental investigation of desalting and dehydration of crude oil by assistance of ultrasonic irradiation. *Ultrason. Sonochem.* **2013**, *20*, 378–385. [CrossRef] [PubMed]
4. Daniel-David, D.; Pezron, I.; Dalmazzone, C.; Noïk, C.; Clausse, D.; Komunjer, L. Elastic properties of crude oil/water interface in presence of polymeric emulsion breakers. *Colloids Surf. A Physicochem. Eng. Asp.* **2005**, *270–271*, 257–262. [CrossRef]
5. Mahdi, K.; Gheshlaghi, R.; Zahedi, G.; Lohi, A. Characterization and modeling of a crude oil desalting plant by a statistically designed approach. *J. Pet. Sci. Eng.* **2008**, *61*, 116–123. [CrossRef]
6. Lake, L.W.; Fanchi, J.R.; Society of Petroleum Engineers (U.S.). *Petroleum Engineering Handbook*; Society of Petroleum Engineers: Richardson, TX, USA, 2006.
7. Bresciani, A.E.; Alves, R.M.B.; Nascimento, C.A.O. Coalescence of Water Droplets in Crude Oil Emulsions: Analytical Solution. *Chem. Eng. Technol.* **2010**, *33*, 237–243. [CrossRef]
8. Sams, G.; Zaouk, M. Emulsion Resolution in Electrostatic Processes. *Energy Fuels* **2000**, *14*, 31–37. [CrossRef]
9. Bresciani, A.E.; Mendonça, C.F.X.; Alves, R.M.B.; Nascimento, C.A.O. Modeling the kinetics of the coalescence of water droplets in crude oil emulsions subject to an electric field, with the cellular automata technique. *Comput. Chem. Eng.* **2010**, *34*, 1962–1968. [CrossRef]
10. Abdul-Wahab, S.; Elkamel, A.; Madhuranthakam, C.R.; Al-Otaibi, M.B. Building Inferential estimators for modeling product quality in a crude oil desalting and dehydration process. *Chem. Eng. Process. Process Intensif.* **2006**, *45*, 568–577. [CrossRef]
11. Abdel-Aal, H.K.; Zohdy, K.; Abdelkreem, M. Waste Management in Crude Oil Processing: Crude Oil Dehydration and Desalting. *Int. J. Waste Resour.* **2018**, *8*, 326. [CrossRef]
12. Aryafard, E.; Farsi, M.; Rahimpour, M.R. Modeling and simulation of crude oil desalting in an industrial plant considering mixing valve and electrostatic drum. *Chem. Eng. Process. Process Intensif.* **2015**, *95*, 383–389. [CrossRef]
13. Aryafard, E.; Farsi, M.; Rahimpour, M.R.; Raeissi, S. Modeling electrostatic separation for dehydration and desalination of crude oil in an industrial two-stage desalting plant. *J. Taiwan Inst. Chem. Eng.* **2016**, *58*, 141–147. [CrossRef]
14. Kakhki, N.A.; Farsi, M.; Rahimpour, M.R. Effect of current frequency on crude oil dehydration in an industrial electrostatic coalesce. *J. Taiwan Inst. Chem. Eng.* **2016**, *67*, 1–10. [CrossRef]
15. Vafajoo, L.; Ganjian, K.; Fattahi, M. Influence of key parameters on crude oil desalting: An experimental and theoretical study. *J. Pet. Sci. Eng.* **2012**, *90–91*, 107–111. [CrossRef]
16. Bansal, H.; Ameensayal. CFD analysis of horizontal electrostatic desalter influence of header obstruction plate design on crude-water separation. In Proceedings of the International Conference on Research in Electrical, Electronics & Mechanical Engineering, Dehradun, India, 26 April 2014.
17. Shariff, M.M.; Oshinowo, L.M. Debottlenecking water-oil separation with increasing water flow rates in mature oil fields. In Proceedings of the 5th Water Arabia 2017 Conference and Exhibition, Al-Khobar, Saudi, 17–19 October 2017.
18. Ilkhaani, S. Modeling and Optimization of Crude Oil Desalting. Master's Thesis, University of Waterloo, Waterloo, ON, Canada, 2009.

19. Xu, X.; Yang, J.; Gao, J. Effects of Demulsifier Structure on Desalting Efficiency of Crude Oils. *Pet. Sci. Technol.* **2006**, *24*, 673–688. [CrossRef]
20. Mohammadi, M. Numerical and experimental study on electric field driven coalescence of binary falling droplets in oil. *Sep. Purif. Technol.* **2017**, *176*, 262–276. [CrossRef]
21. Alnaimat, F.; Alhseinat, E.; Banat, F.; Mittal, V. Electromagnetic–mechanical desalination: Mathematical modeling. *Desalination* **2016**, *380*, 75–84. [CrossRef]
22. Ascher, U.M.; Petzold, L.R. *Computer Methods for Ordinary Differential Equations and Differential-Algebraic Equations*, 1st ed.; SIAM: Philadelphia, PA, USA, 1 August 1998; ISBN 0898714125.
23. Manninen, M.; Taivassalo, V.; Kallio, S. On the mixture model for multiphase flow. *Tech. Res. Cent. Finl.* **1996**, *288*, 67.
24. Hibiki, T.; Ishii, M. One-group Interfacial Area Transport of Bubbly Flows in Vertical Round Tubes. *Int. J. Heat Mass Transf.* **2000**, *43*, 2711–2726. [CrossRef]
25. Azbel, D.; Athanasios, I.L. A mechanism of liquid entrainment. In *Handbook of Fluids in Motion*, 1st ed.; Cheremisino, N., Ed.; Ann Arbor Science Publishers: Ann Arbor, MI, USA, 1983.
26. Shih, T.H.; Liou, W.W.; Shabbir, A.; Yang, Z.; Zhu, J. A New K-epsilon Eddy-Viscosity Model for High Reynolds Number Turbulent Flows—Model Development and Validation. *Comput. Fluids* **1995**, *24*, 227–238. [CrossRef]
27. Dutta, A.; Ekatpure, R.; Heynderickx, G.; de Broqueville, A.; Marin, G. Rotating fluidized bed with a static geometry: Guidelines for design and operating conditions. *Chem. Eng. Sci.* **2010**, *65*, 1678–1693. [CrossRef]
28. Chen, L.; Wu, S.; Lu, H.; Huang, K.; Zhao, L. Numerical Simulation and Structural Optimization of the Inclined Oil/Water Separator. *PLoS ONE* **2015**, *10*, e0124095. [CrossRef] [PubMed]
29. Mouketou, F.; Kolesnikov, A. Modelling and Simulation of Multiphase Flow Applicable to Processes in Oil and Gas Industry. *Chem. Prod. Process Model.* **2018**, *14*, 1–16. [CrossRef]

processes

MDPI

Article

A Study of Linkage Effects and Environmental Impacts on Information and Communications Technology Industry between South Korea and USA: 2006–2015

Junhwan Mun [1], Eungyeong Yun [2] and Hangsok Choi [3,*]

[1] Department of Convergence Program for Social Innovation, Sungkyunkwan University, Seoul 03063, Korea; mjhpioneer@skku.edu
[2] Department of Business Consulting, Meta Credit Group Consulting, Seoul 07299, Korea; egyun@mcgcorp.co.kr
[3] Department of Business Administration, Dongduk Women's University, Seoul 02748, Korea
* Correspondence: hschoi918@dongduk.ac.kr

Abstract: This study examined the relationship among carbon dioxide emissions and linkage effects using Input–Output (IO) data of the information and communications technology (ICT) industry between South Korea and the USA. As we wanted to find out if the ICT industry, which the world is passionate about, is a sustainable industry. The linkage effects are analyzed to determine the impact of ICT industry on the national economy, and CO_2 emissions of the industry are analyzed to determine how much influence it has on air pollution. In addition, we classify ICT industry by ICT service and manufacturing industries as the key industries in Korea and the US. Data were collected from OECD ranging from 2006 to 2015 in order to quantitatively estimate backward linkage, forward linkage effect, and carbon dioxide emissions. The results indicated that ICT manufacturing industry in Korea has high backward and forward linkage effects. CO_2 emissions from ICT service is more than from ICT manufacturing in both Korea and the US. We wanted to find out if the ICT industry, which the world is passionate about, is a sustainable industry. As a contribution, ICT manufacturing and service industries in Korea and the United States are directly compared, and CO_2 emissions over 10 years are analyzed in a time series.

Keywords: IO analysis; CO_2 emission; ICT service; ICT manufacturing

Citation: Mun, J.; Yun, E.; Choi, H. A Study of Linkage Effects and Environmental Impacts on Information and Communications Technology Industry between South Korea and USA: 2006–2015. *Processes* **2021**, *9*, 1043. https://doi.org/10.3390/pr9061043

Received: 1 April 2021
Accepted: 9 June 2021
Published: 15 June 2021

Publisher's Note: MDPI stays neutral with regard to jurisdictional claims in published maps and institutional affiliations.

1. Introduction

Global warming is the most serious problem facing humankind today. It is because, due to human activities, greenhouse gases (GHG) are increasing at an unprecedented rate and are accumulating in the atmosphere. Such global warming raises the global temperature, leading to abnormal weather conditions and destruction of ecosystems, and it will soon affect a wide range of areas beyond them, from energy supply to human health. The necessity of reducing GHG for the survival of humankind has become an important agenda of the international community. Korea is a country that emits many greenhouse gases. However, in order to become a country responsible for reducing greenhouse gas emissions and responding to climate change in the international community, Korea has presented a challenging goal of reducing business as usual (BAU) by 37% by 2030 [1]. Low carbon and GHG reduction are becoming a target not only for Korea, but also for countries around the world. The key strategies for low carbon in major countries are summarized in energy efficiency improvement, energy conversion, and resource circulation. [2]. First, they set up a strategy to reduce production costs and reduce greenhouse gas emissions by increasing energy efficiency and reducing energy consumption. In addition, they seek sustainable growth by replacing final energy consumption in industries with clean energy such as fossils, electricity, and renewable energy, and they strive to build a resource recycling economy system that reuses and recycles resources in industrial processes that emit large amounts of greenhouse gases. The Korean government creates the foundation necessary for low-carbon and green growth for the harmonious development of the economy and environment. They attempt to promote the national economy development by using green technology and green industry as new growth engines. Furthermore, Korea aims to improve the quality of life of people by building low-carbon society and to leap forward as a mature advanced country that fulfills its responsibilities in the international community [3]. To this end, the first step begins with reducing greenhouse gas emissions. Additionally, it is necessary to focus on securing innovative technologies such as developing low-carbon fuel processes and air pollution reduction facilities, while preparing a specific roadmap for the stable supply of green hydrogen and green energy and formation of an appropriate price. However, since Korea has an industrial structure different from those of countries that are strongly pursuing low carbon, their domestic reality should be considered.

Table 1 shows share of added value in ICT industry in 4 countries and the average of OECD. Korea has the world's best network infrastructure built and distributed and has a favorable environment for developing the Fourth Industry such as Internet of Things, autonomous vehicles, and Big Data analysis. However, the information and communication technology (ICT) industry in Korea is more concentrated in the ICT manufacturing sector than in other countries [4]. Thus, software and ICT service sector is inferior to countries such as the United States and Japan. Despite the high share of added value in ICT manufacturing industry, the share of the ICT service industry is below the OECD average [5].

Table 1. Share of added value to GDP in the ICT industry (2015).

Country	ICT Manufacturing	Communication	Software and Service	Sum of ICT Industrial
Korea	5.6%	1.0%	1.6%	8.2%
USA	1.5%	1.5%	2.1%	5.1%
EU	0.8%	1.2%	2.0%	4.0%
Japan	1.7%	1.9%	2.4%	6.1%
Average of OECD	1.0%	1.2%	1.9%	4.1%

Source: Authors' own calculations on the data form [5], 'OECD digital outlook 2017', based on 2015 data; comparison of EU (23 country's) and OECD (33 country's).

However, the growth of ICT manufacturing industry in Korea, which is so strong, has slowed over the past five years. Its growth rate fell sharply from 50% in 2010 to −9% in 2015, but in the United States it fell from 10% in 2010 to 5% in 2014 [6]. Moreover, Korea is the second country in the world with the highest proportion of manufacturing to GDP after China [7]. On the other hand, the proportion of the service industry is lower than that of the US, the UK, and France [7]. The growth rate of ICT service industry in Korea was around 5% in 2011, but it has since declined. The growth rate of ICT service industry in the US fell sharply from 10% in 2012 to −5% in 2013. However, in 2014, it turned to a positive growth rate of 4% [6]. As such, it is appropriate to analyze ICT industry by categorizing it into manufacturing and service sectors because the growth rate and major powerhouses differ according to the type of final product. Additionally, this also helps in establishing a low-carbon strategy.

CO_2 emissions in ICT industry are included not only from the electricity consumed in the use of products and devices, but also from the energy used in the products. Ericsson [8] largely categorizes CO_2 emissions of ICT industry into user devices, networks, and data centers in consideration of these points. Since CO_2 emissions are different for each sector, applying low-carbon strategy of countries with a high proportion of the service industry may be a strategy that does not take into account the specificity of the industrial structure of a country that is strong in the manufacturing industry. The active national-level response to overcome the limit of reducing CO_2 emissions should be implemented in a flexible environment according to the industrial sector.

According to a study by Belkhir and Elmeligi [9], by 2040 if the carbon emission of the ICT industry remains as it is, it will account for up to 14% of the global carbon emission. In addition, countries around the world are actively working to strengthen industrial competitiveness using ICT technology. Therefore, we want to understand how much ICT industry has an impact on CO_2 emissions in a country in this study as it is becoming more and more important.

Although it is focusing on the development of industries related to the fourth industry such as Artificial Intelligence, Big Data, and autonomous vehicles, it is questionable whether these industries can be said to be eco-friendly and sustainable industries that can grow with other industries together. In other words, the research questions of this study were:

1. We confirmed that ICT industry in the US and Korea was a sustainable industry.
2. We checked whether there was a difference in the influence of the manufacturing sector and the service sector of ICT industry on the national economy.

We selected input–output analysis as the theoretical method, and compared changes in ICT manufacturing and ICT service industries in the US and Korea. We also compared CO_2 emissions of the two countries. To be a sustainable industry, we needed to make sure that the industry pollutes the air less. The United States has the world's highest sales and industry share of ICT industry and has the largest number of companies in ICT service among the global representative digital companies [10]. We think it is important to compare and analyze the connection between CO_2 emissions and linkage effects in ICT manufacturing industry in Korea and ICT service industry in the United States. This is because the world wants to foster industries with low carbon dioxide emissions and high productivity. In order to discover a sustainable industry in this future as we have mentioned, we will derive CO_2 emissions from ICT manufacturing and ICT service industries in Korea and the US from 2006 to 2015, and analyze backward and forward linkage effects. After analysis, we will present the direction of low-carbon development that suits the conditions of each country.

This study is organized as follows. Section 2 introduces the importance of CO_2 emissions research and previous studies related to ICT manufacturing and ICT service and establishes hypotheses. Then, Section 3 explains research methods and data. In Section 4, we describe the results of the analysis. Finally, in Section 5, policy suggestions are made through analysis of the results

2. Literature Review and Hypotheses Development

According to the OECD, ICT (Information and Communications Technology) industry is defined as the production of goods and services products that are primarily used to represent information processing and communication and transmission by electronic means. ICT industry can be divided into ICT manufacturing and ICT service business, and in detail, it can be classified into computer parts, electronic and electrical equipment manufacturing, information and communication related business, and software business [11]. Table 2 shows the definition of each ICT industry. In 1998, OECD member countries agreed to define the ICT sector as a combination of manufacturing and services industries that capture, transmit, and display data and information electronically [12].

Table 2. Definition of ICT manufacturing and service industries.

Sector	The Principles Underlying the Definition
ICT manufacturing Industry	• Must be intended to fulfil the function of information processing and communication including transmission and display. • Must use electronic processing to detect, measure, and/or record physical phenomena or control a physical process.
ICT service Industry	• Must be intended to enable the function of information processing and communication by electronic means.

However, there was a debate about the principle of selecting the ICT industry and the interpretation of the principle in 1998. As a result, the definition of ICT industry classified ICT industry by adding a section called ICT trade. Table 3 shows a list of industries belonging to ICT industry according to the 4th International Standard Industry Classification (ISIC).

Table 3. The list of ICT industries (ISIC Rev. 4).

Sector		ISIC Rev. 4 (2007)
ICT Manufacturing	2610	Manufacture of electronic components and boards
	2620	Manufacture of computers and peripheral equipment
	2630	Manufacture of communication equipment
	2640	Manufacture of consumer electronics
	2680	Manufacture of magnetic and optical media
ICT Trade	4651	Wholesale of computers, computer peripheral equipment, and software
	4652	Wholesale of electronic and telecommunications equipment and parts
ICT Services	5820	Software publishing
	61	Telecommunications
	6110	Wired telecommunications activities
	6120	Wireless telecommunications activities
	6130	Satellite telecommunications activities
	6190	Other telecommunications activities

Meanwhile, the OECD continued to work on the definition and classification system for the content media industry based on the prospect that structural changes that would occur in the content production and distribution industry as the number of content users increases due to the spread of ICT technology. Additionally, this was reflected in ISIC revision 4. When transitioning from ISIC revision 3 to ISIC revision 4, they included the production and distribution of information and cultural products, information technology and data processing, and information service activities. ISIC revision 4 classifies the information and communication service sector into one section and classifies the ICT industry as Table 4 [13].

Table 4. The 2006–2007 OECD ICT sector definition (based on ISIC Rev. 4).

Sector		ISIC Rev. 4 (2008)
ICT Manufacturing	2610	Manufacture of electronic components and boards
	2620	Manufacture of computers and peripheral equipment
	2630	Manufacture of communication equipment
	2640	Manufacture of consumer electronics
	2680	Manufacture of magnetic and optical media
ICT Services	5820	Software publishing
	6110	Wired telecommunications activities
	6120	Wireless telecommunications activities
	6130	Satellite telecommunications activities
	6190	Other telecommunications activities
	6201	Computer programming activities ICT
	6202	Computer consultancy and computer facilities management activities ICT
	6209	Other information technology and computer service activities
	6311	Data processing, hosting, and related activities
	6312	Web portals

There is a slight difference in whether detailed enterprise groups should be included in ICT manufacturing industry or ICT service industry by era, but as a result, it shows consistency in classifying ICT industry into ICT manufacturing industry and service industry. According to Global Industry Classification Standard, S&P capital IQ industry classification, ICT industry is classified into Semiconductors & Semiconductor Equipment industries, Technology Hardware & Equipment (ICT manufacturing), and Software & Services industries and industries related to digital technology such as IoT, fifth generation technology standard, cloud, big data, and AI software are evaluated as key drivers of the fourth industrial revolution.

Several researchers have also divided ICT industry into the manufacturing sector and the service to study their impact on the national economy as Table 5 [14–17].

Table 5. Research on ICT industry divided into manufacturing and service sectors.

Researcher	Country	Year	Aim of the Study
Xing et al. (2011)	China	2002	Analyzing the form of convergence between ICT manufacturing industry and ICT service industry, and identifying their role
Rohman (2013)	EU	1995, 2000, 2005	Analyzing the strengths of ICT industry by comparing the multiplier effects of ICT industry and non-ICT industry.
Hong, J.P., Byun, J.E., and Kim, P.R. (2016)	Korea	From 1995 to 2009	Examining structural changes and growth factors of ICT manufacturing and service industries
Abubakar, Y.A., and Mitra, J. (2010)	EU	From 2001 to 2003	Investigating factors influencing regional innovation by contrasting high-tech manufacturing (ICT manufacturing) and knowledge intensive services (ICT service)

Korea commercialized 5G, the basic infrastructure of the 4IR, for the first time in the world. Currently, 24 countries have commercialized 5G or are planning to do so, and Korea is evaluated as a world leader in 5G commercialization [18]. In addition, they have a well-distributed ICT infrastructure, excellent accessibility, and a favorable environment for responding to technology related to the fourth industry. The global cloud market is worth USD 243 trillion [19], and the top three global companies (Amazon, Microsoft, Google) account for 57% of the market [20]. American companies already dominated more than half of global ICT service market (as of 2018) [21]. By comparison, Korean Cloud market is USD 2 billion, which is only 1% of the world market. ICT industry in Korea is more concentrated in the manufacturing sector than in other countries, so the software and ICT service sectors are inferior to rival countries such as the United States and Japan [5].

Table 6 shows the R&D investment status of major countries by industry in 2019. Of the 2500 companies, there are a total of 70 Korean companies, and they invested about USD 3.67 billion in R&D budget for a year [22]. Among them, 16 ICT manufacturing industries invested about USD 24.95 billion, and five ICT service industries invested USD 846 million. This means that 70.2% of the R&D budget of Korean companies is invested in ICT manufacturing and ICT service industries. The US has the highest R&D investment ratio in ICT manufacturing and ICT service industries, and Korea has the lowest investment ratio in ICT service industry at 0.6% compared to other countries. While the US invests intensively in ICT service, ICT manufacturing, and health industries, Korea is excessively focused on ICT manufacturing. Therefore, it is necessary to promote the future development of Korea by expanding investment in ICT service industry as the ICT service industry is an important sector that connects various industries in the new era [23].

Table 6. R&D investment by industry and share by country (2019).

Country	Korea	EU	USA	Japan	China	Other	Total
ICT Manufacturing	277068 (10.9%)	350194 (13.8%)	1039653 (40.9%)	282468 (11.1%)	369681 (14.6%)	220856 (8.7%)	2539921 (100%)
ICT Service	9396 (0.6%)	192305 (11.6%)	1111437 (67.1%)	68112 (4.1%)	222116 (13.4%)	85530 (3.2%)	1655896 (100%)
Health	8362 (0.4%)	591757 (26.6%)	1089259 (49.0%)	172388 (7.8%)	60838 (2.7%)	300892 (13.5%)	2223497 (100%)
Manufacture of motor vehicles and transportation	56601 (3.4%)	795317 (47.7%)	240605 (14.4%)	424664 (25.5%)	110945 (6.7%)	39552 (2.4%)	1667684 (100%)
(manufacturing) industry	30992 (4.0%)	206451 (26.9%)	188392 (24.6%)	144571 (18.9%)	149339 (19.5%)	46685 (6.1%)	766430 (100%)
Chemical industry	3687 (1.3%)	68402 (23.3%)	65787 (22.4%)	99087 (33.8%)	19540 (6.7%)	36859 (12.6%)	293362 (100%)
Aerospace and Defense industry	4074 (1.5%)	120838 (45.8%)	109282 (41.5%)	- (0.0%)	4331 (1.6%)	25049 (9.5%)	263574 (100%)
Other	17839 (1.3%)	393693 (29.5%)	233973 (17.5%)	236951 (17.7%)	321055 (24.0%)	131807 (9.9%)	1335318 (100%)
Total	408019 (3.8%)	2718957 (25.3%)	4078388 (38.0%)	1428242 (13.3%)	1257847 (11.7%)	854230 (7.9%)	10745683 (100%)

Source: Authors' own computations on the data collected from [22,24].

As is well known, ICT industry is the future food of not only Korea and the United States, but also major countries. Moreover, low carbon and greenhouse gas emission reduction policies are also being implemented for the future. The US announced 2013 Climate Action Plan, reducing carbon emissions and leading international community's efforts to respond to climate change [25]. In China, CO_2 emissions once increased by 15–17%/year due to rapid economic growth but based on the "National Plan for Response to Climate Change (2013–2020)," they are pushing for reduction of emissions and changes in energy mix [3].

In order to simultaneously foster sustainable development and the development of the future food, the ICT industry, CO_2 emissions of ICT industry should be lower than that of other industries, and economic linkage effect should be high. After the collapse of the dot-com bubble in 2000, research in the ICT industry shifted to industrial analysis. As a result, it has been expanded to the academic doctrine that ICT industry serves as a driving force for innovation and a catalyst for innovation [26,27].

ICT industry has the power to induce innovation and speed up the innovation of companies. It can also penetrate the economy as a whole, affect all industries, and cause a technology shock [27]. Mas, de Guevara [28] find that all of them experienced growth in the ICT sector. In OECD countries, the ICT sector's share in GDP has remained relatively constant since 1995, implying a growth in total value added as GDP has also grown in that time [5,11,29]. Mattioli and Lamonica [30] classified linkage effects of the ICT industry in the overall economy, and said that ICT industry had a low backward linkage effect and a high forward linkage effect, resulting in a strong supply side to the industry as a whole. As such, it is the ICT industry that has a large linkage effect on the national economy [16,30–34].

However, if the ICT industry is the industry that causes environmental pollution due to high CO_2 emissions, then the ICT industry cannot be said to be a sustainable industry. Therefore, many researchers are continuing research to measure CO_2 emissions of each country and CO_2 emissions of major industries in order to find industries with less environmental pollution [33,35–39]. Loefgren and Muller [35] studied the relationship between energy use efficiency and carbon emissions and suggested a plan to improve carbon emission efficiency as an effective method to reduce carbon emission. Zhou et al. [36] used DEA to estimate the CO_2 emission efficiency of OECD countries. Jaeger et al. [39] concluded that low-carbon-based growth is optimistic for economic growth and job creation in their research Lee et al. [38] conducted a study to measure the productivity of the Korean manufacturing industry based on CO_2 emissions. Moon et al. [37] analyzed linkage effects of all Korean industries from 2005 to 2015 in a time series to derive eco-friendly industries with high linkage effects.

As ICT industry is highly competitive, air pollution reduction measures such as carbon emissions management are needed to become a sustainable industry. Therefore, we would like to examine the relationship between linkage effect and CO_2 emissions of ICT industry. In addition to service-based intangible industries such as big data and artificial intelligence, the ICT industry also includes hardware manufacturing industries such as semiconductors and automobiles. In addition to service-based intangible industries such as big data and artificial intelligence, the ICT industry also includes hardware manufacturing industries such as semiconductors and automobiles. There are a number of studies comparing the impact of the ICT industry on the national economy and the productivity performance of the ICT industry [27,31,32,40,41], but few studies have classified the ICT industry into manufacturing and service industries [34,42]. There are not many studies comparing the ICT industry between Korea and the United States [43,44], either. However, depending on the shape of final product, the influence on other industries or the influence from other industries may differ. Therefore, we will analyze linkage effect and CO_2 emissions of each industry by dividing ICT industry into ICT-Service and ICT-Manufacturing sector according to International Standard Industry Classification from OECD. Additionally, to test the difference, we developed hypotheses as Table 7.

Table 7. Statistical Hypotheses of the Regression Model.

	Hypotheses
H1	*The linkage effect between Korean ICT manufacturing industry and the US one is different.*
H1a	*The backward linkage effect between the Korean ICT manufacturing industry and the US one is different.*
H1b	*The forward linkage effect between the Korean ICT manufacturing industry and the US one is different.*
H2	*The linkage effect between Korean ICT service industry and the US one is different.*
H2a	*The backward linkage effect between the Korean ICT service industry and the US one is different.*
H2b	*The forward linkage effect between the Korean ICT service industry and the US one is different.*

3. Materials and Methods

This study uses IO tables from Korea and the United States provided by the OECD for 10 times from 2006 to 2015 [45]. ICT industry was divided into ICT manufacturing and service industry according to the fourth International Standard Industry Classification. CO_2 emissions are based on CO_2 emissions embodied in domestic final demand provided by the OECD from 2006 to 2015 [45]. This is because when analyzing linkage effects, it is analyzed as domestic production, so that CO_2 emissions must also be analyzed as occurring in the domestic production process to enable an accurate comparison [37]. When analyzing CO_2 emissions of ICT industry, the same industry classification standard as in the analysis of the linkage effects is applied to determine whether there is a difference in CO_2 emissions depending on the final output type.

Linkage effects suggested by Hirschman [46] are to derive the production induction coefficient using the IO table, and to indicate the degree of industrial activation through the derived production induction coefficient. Linkage effects of Hirschman [46] is that an industry induces production directly or indirectly to all industries within a country, and the degree is expressed as production induction coefficient. It can be said that the greater the production inducement coefficient, the greater the power to revitalize the entire industry. The linkage effects can be divided into backward linkage effect and forward linkage effect.

Backward linkage effect can be expressed as the pulling power that induces the input of intermediate goods in the process of an industry producing final goods [47]. In other words, the backward linkage effect is the power of dispersion because one industry induces production to other industries through demand for intermediate goods in the process of producing final goods, and the production inducement continuously promotes production to others [48,49]. The high coefficient means that it has strengths as a "demander" in the national economy [27].

Forward linkage effect is a production inducing effect that occurs when the final goods of one industry are input as intermediate goods of other industries. Forward linkage effect is explained as an interaction that directly triggers the structural relationship between supply and demand for intermediate goods across the industry as a business client inputs a product [47]. It is also explained by the reaction that occurs when the entire national industry is activated, the sensitivity of dispersion, and the effect of inducing supply [30,48–50]. The high value of this coefficient can be said to be a strength as a "supplier" in the national economy [27].

To explain these effects with a formula, we use the production inducement effect matrix, which is created based on the IO table that presents the transaction figures between industries in a certain form is used. IO table shows the interdependence of goods and services between industries, the input of production factors, and the sales process according to the final demand of products. The horizontal direction represents the sales breakdown of products in each industry and consists of intermediate demand sold as intermediate goods and final demand sold as consumer goods, capital goods, and exports. Additionally, here, excluding income becomes the total output. The vertical direction shows the input structure of each industry, and it can be divided into intermediate inputs of raw materials and value-added inputs such as labor and capital, and the total is the total input amount.

The effect on production inducement is the direct or indirect production fluctuations caused by the final demand through the input coefficient.

The concept of the effect on production inducement was proposed by Leontief [51] based on Keynesian multiplier theory. The effect on production inducement by the direct production factor can be expressed as follows.

$$(1-a)^{-1} = \frac{1}{1-a} = 1 + a + a^2 + a^3 + \dots \tag{1}$$

where 1 is the direct production factor, a is the primary effect on production inducement, and a^2 is the secondary effect on production inducement. Therefore, the effect on production inducement is expressed as the sum of infinite geometric series of $(1-a)^{-1}$ when a is $0 < a < 1$. With this logic, the production induction coefficient can be obtained through the inverse matrix $(I-A)^{-1}$ of A, which is the matrix of a_{ij}, and this is called Leontief inverse matrix.

$$(I-A)^{-1} = I + A + A^2 + A^3 + A^4 + \dots \dots \tag{2}$$

I on the right side represents the direct production effect of each industry to satisfy the final demand for each industry's product of by one unit. A is the primary production inducement effect, which is the input unit of intermediate goods necessary for the production of final goods in each industry. A^2, the secondary production inducement effect, is the unit of input of intermediate goods required for the production of each industry as a result of the primary production inducement effect and A^3, A^4, $\dots \dots$ are the 3rd, 4th, $\dots \dots$ production inducement effect. Therefore, $(I-A)^{-1}$ means the production inducement coefficient, which is the sum of the direct and indirect production inducement effects caused by an increase in final demand by one unit. Since the production inducement coefficient has the property of a multiplier representing the inducement effect derived from the final demand for an industry.

The linkage effect is the theory suggested by Rasmussen [52] and Hirschman [46]. If the element of the production inducement matrix is r_{ij} and the Leontief inverse matrix, $(I-A)^{-1}$, is changed to $\sum_{i} B_{ij}$, the backward linkage effect and forward linkage effect can be expressed as follows [53].

$$FL_i = \frac{\sum_{j=1}^{n} r_{ij}}{\frac{1}{n} \sum_{i=1}^{n} r_{ij} \sum_{j=1}^{n} r_{ij}} \tag{3}$$

$$BL_i = \frac{\sum_{i=1}^{n} r_{ij}}{\frac{1}{n} \sum_{i=1}^{n} r_{ij} \sum_{j=1}^{n} r_{ij}} \tag{4}$$

FL_i = Sensibility Index of Dispersion;
BL_i = Power Index of dispersion;
r_{ij} = The factor of Leontief Inverse Matrix;
$\sum_{j=1}^{n} r_{ij}$ = The sum of columns of Leontief Inverse Matrix;
$\sum_{i=1}^{n} r_{ij}$ = The sum of rows of Leontief Inverse Matrix.

4. Results

Table 8 shows the FL effects of the ICT manufacturing industry between Korea and USA during the entire study period from 2006 to 2015.

Table 8 shows the results of linkage effects analysis of ICT manufacturing industry in Korea and the United States. Backward linkage effect of Korean ICT manufacturing industry is 1.24 on average over 10 years, which has a large influence on other industries. On the other hand, that of the US has a small influence at 0.85. Backward linkage effect represents the influence of an industry on all industries when demand for an industry increase. If backward linkage effect of the ICT manufacturing industry is large, it means that the impact on all industries is large [54,55]. Therefore, it can be said that while Korean

ICT manufacturing industry has a great influence on all industries in the country, in the US it does not.

Table 8. Comparison of linkage effects of ICT manufacturing between Korea and USA.

Year	Forward Linkage of ICT Manufacturing		Backward Linkage of ICT Manufacturing	
	Korea	USA	Korea	USA
2006	1.719	0.777	1.262	0.946
2007	1.604	0.728	1.245	0.966
2008	1.528	0.807	1.241	0.923
2009	1.823	0.862	1.281	0.867
2010	1.54	0.806	1.253	0.832
2011	1.791	0.819	1.282	0.835
2012	1.644	0.825	1.239	0.816
2013	1.586	0.915	1.2	0.803
2014	1.573	0.889	1.213	0.781
2015	1.716	1.005	1.177	0.771
Avg.	1.652	0.843	1.239	0.854

The forward linkage effect of the Korean ICT manufacturing industry is 1.65 on average over 10 years, which has a great influence on other industries. On the other hand, that of the US has a small influence at 0.84. The large forward linkage effect means that when the demand for products of all industries increases, the influence that an industry will receive is large [54,55]. In other words, the large forward linkage effect of ICT manufacturing industry means that the output of the industry is used as intermediate goods in the production process of other industries. Since this indicator is analyzed from the standpoint of intermediate goods used during the production process of output, it also means dependence on other industries. Therefore, we can say that Korean ICT manufacturing industry plays a large role as a factor of production when producing other industries, but that of the United States is not sufficiently performing such a role. Table 9 shows the results of linkage effects analysis of ICT service industry in Korea and the United States.

Table 9. Comparison of linkage effects of ICT Service between Korea and USA.

Year	Forward Linkage of ICT Service		Backward Linkage of ICT Service	
	Korea	USA	Korea	USA
2006	1.028	0.894	0.891	0.874
2007	1.013	0.887	0.907	0.87
2008	0.939	0.897	0.897	0.86
2009	0.949	0.992	0.922	0.923
2010	0.938	0.957	0.919	0.93
2011	0.917	0.929	0.933	0.936
2012	0.905	0.926	0.919	0.945
2013	0.927	0.932	0.91	0.928
2014	0.947	0.933	0.918	0.94
2015	0.92	0.972	0.932	0.939
Avg.	0.948	0.932	0.915	0.915

Backward linkage effect of Korean ICT service industry is 0.91 on average over 10 years, which has a small influence on other industries. That of the United States is also 0.91, which has a small influence. ICT service industries of the two countries have small backward linkage effects, so their influence on other industries is small.

Forward linkage effect of the Korean ICT service industry is 0.95 on average over 10 years, with a small influence on other industries. That of the US is also 0.93, which has a small influence. If the forward linkage effect of ICT service industry is small, it means that even if the demand for other industries increases, the demand for that industry also does not increase. In other words, since the dependence on other industries is low, even if the demand of other industries increases, there is little possibility that the relevant industries will be put into the process as a production factor.

Table 10 presents a summary of the statistical test results of the two hypotheses and four sub hypotheses in this study. This study compared ICT manufacturing and service sectors between Korea and USA to the relationship linkage effects. The hypothesis H1 that linkage effects of Korean ICT manufacturing industry and the US are different was accepted. ($p < 0.000$). On the other hand, the hypothesis H2 that the linkage effects of the ICT service industry in Korea and the US are different was rejected, and the pattern of the linkage effect of ICT service industry between the two countries is drawn as shown in (Figure 1). The result of this study show that ICT manufacturing sectors of Korea and USA have very different linkage effects.

Table 10. Results of hypothesis testing.

	Hypotheses	*p*-Value	Results
H1	*The linkage effect between the Korean ICT manufacturing industry and the US is different.*	-	Accept
H1a	*The backward linkage effect between the Korean ICT manufacturing industry and the US is different.*	0.000	Accept
H1b	*The forward linkage effect between the Korean ICT manufacturing industry and the US is different.*	0.000	Accept
H2	*The linkage effect between the Korean ICT service industry and the US is different.*	-	Reject
H2a	*The backward linkage effect between the Korean ICT service industry and the US is different.*	0.980	Reject
H2b	*The forward linkage effect between the Korean ICT service industry and the US is different.*	0.445	Reject

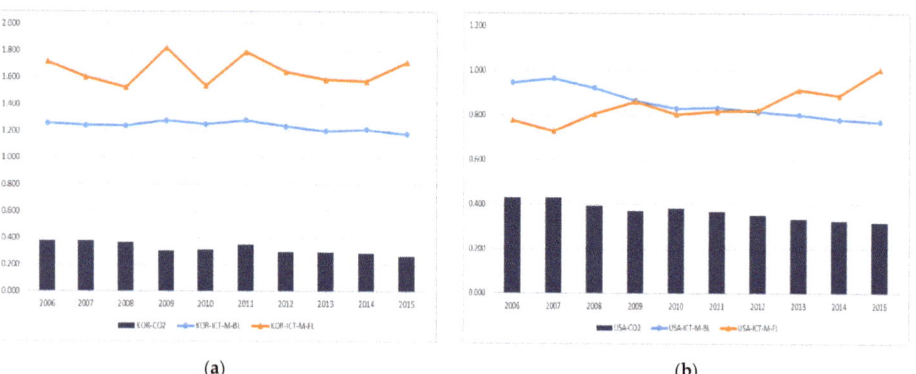

Figure 1. Trends of linkage effect and CO_2 emission of ICT manufacturing between Korea and USA: (**a**) Korea (2006–2015); (**b**) USA (2006–2015). Source: authors' own computations on the data collected from OECD stat [45].

CO_2 emissions were analyzed along with the linkage effect of ICT industry in Korea and the United States in accordance with the low carbonization trend of major countries in the world.

From 2006 to 2015, ICT manufacturing industry in Korea has an average of 1.9 t of CO_2 emissions, accounting for 0.33% of the total CO_2 emissions [45]. As shown in Table 11, CO_2 emissions are gradually decreasing from 0.379 in 2006 to 0.267 in 2015. Considering linkage effect in Figure 1, ICT manufacturing industry in Korea has a high relationship with other industries and CO_2 emissions are gradually decreasing, so it is the industry suitable for realizing the goal of low carbonization and sustainable economic development. From 2006 to 2015, CO_2 emissions in the US are 23.1 t, accounting for 0.37%

of the total CO_2 emissions [45]. In addition, compared to the entire industry, CO_2 emissions of ICT manufacturing industry in the US are gradually decreasing from 0.430 in 2006 to 0.319 in 2015. Considering the linkage effect, ICT manufacturing industry in the US is the industry that has a low relationship with other industries and CO_2 emissions are gradually decreasing. However, since the linkage effect of ICT manufacturing industry in the US are gradually increasing, it is the industry that can be continuously developed.

Table 11. Comparison of CO_2 emission of ICT industry between Korea and USA.

Year	CO_2 Emission of ICT Service			
	Korea	USA	Korea	USA
2006	0.379	0.430	0.490	0.417
2007	0.379	0.429	0.458	0.425
2008	0.366	0.394	0.432	0.436
2009	0.304	0.370	0.471	0.488
2010	0.314	0.382	0.456	0.472
2011	0.353	0.367	0.412	0.469
2012	0.298	0.352	0.417	0.466
2013	0.291	0.336	0.430	0.49
2014	0.286	0.326	0.400	0.511
2015	0.267	0.319	0.420	0.518
Avg.	0.324	0.370	0.439	0.469

From 2006 to 2015, CO_2 emissions from the ICT service industry in Korea average 2.6 t, accounting for 0.45% of the total CO_2 emissions [45]. As shown in Table 11, CO_2 emissions are gradually decreasing from 0.490 in 2006 to 0.420 in 2015. Considering the linkage effect shown in Figure 2, the ICT service industry in Korea has a small influence on other industries and CO_2 emissions are gradually decreasing. Compared with ICT manufacturing industry, ICT service industry has lower linkage effects and higher absolute CO_2 emissions than the manufacturing industry. CO_2 emissions in the US from 2006 to 2015 are 28.6 tons, accounting for 0.47% of total CO_2 emissions [45]. CO_2 emissions of ICT service industry in the US is gradually increasing from 0.417 in 2006 to 0.518 in 2015. Considering the linkage effect in Figure 2, ICT service industry in the US has a low relationship with other industries, but it is the industry that CO_2 emissions are gradually increasing. ICT service industry in the US has similar backward and forward linkage effects to the manufacturing industry, but CO_2 emissions from its service industry are higher than those from its manufacturing, so efforts to reduce CO_2 emissions are needed.

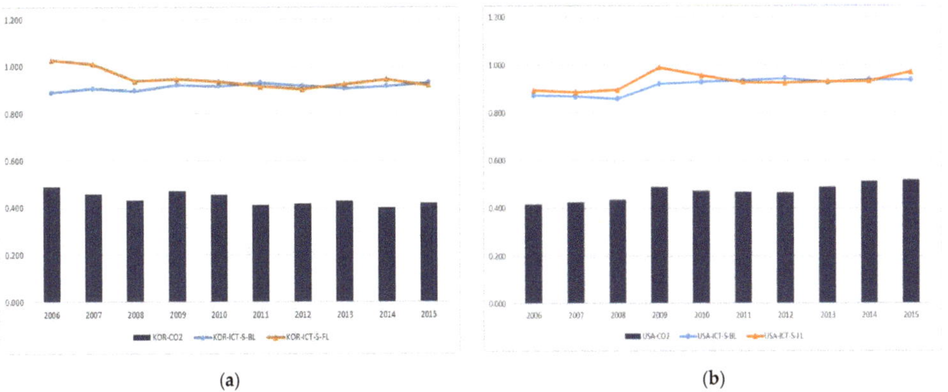

(a)   (b)

Figure 2. Trends of linkage effect and CO_2 emission of ICT service between Korea and USA: (**a**) Korea (2006–2015); (**b**) USA (2006–2015). Source: Authors' own computations on the data collected from OECD stat [45].

5. Discussion and Conclusions

In this study, we compared ICT manufacturing and service sectors between Korea and USA to analyses the relationship among linkage effects and CO_2 emissions. The ICT industry is good for reaching environmentally sound practices because it maintains high productivity while improving overall energy productivity in the national economy. At the same time, due to the ICT, all economic sectors can become more energy efficient. By increasing the energy efficiency, it reduces the environmental impacts of other sectors because ICT allows existing processes to be optimized or enables entirely new, more energy efficient processes [56]. We use data from OECD for the period between 2006 and 2015 and utilized IO analysis and Leontief's Matrix to calculate the backward and forward linkage effects. Additionally, in order to analyze whether CO_2 emissions vary according to the final output type, the same industry classification standard used in linkage effects analysis was applied. The results of this study show that the ICT industries sectors of Korea and USA have a few different effects on each national economy.

The finding of this study has important implications for both theory and practice. First, the ICT industry is classified into ICT service and ICT manufacturing industry, and the relationship between linkage effects and CO_2 emissions of the two industries is examined in this study, in a complex manner. Existing studies have conducted comparative analysis of linkage effects in all industries [48,57,58] or limited to one country [37,59,60]. However, ICT industry has different growth rates and linkage effects depending on the type of final product. Additionally, CO_2 emissions are different. Its CO_2 emissions can be divided into ICT manufacturing industry, which consumes energy to produce electronic devices, and ICT service industry, which consumes energy to utilize the produced electronic devices. Based on these results, this study is meaningful in analyzing the effects of each industry by dividing it into manufacturing and service sectors according to the type of final product.

Second, ICT service industries in Korea and the US show higher CO_2 emissions than their ICT manufacturing industries. This means that the transition from manufacturing to service industry does not unconditionally contribute to the reduction of domestic CO_2 emissions. It is thought that the power consumption in ICT industry is relatively small compared to other industries. However, due to the spread of PCs and the increase in the amount of Internet communication, standby power has increased, leading to an increase in power consumption. The ICT industry, which people thought to be the most environmentally friendly, is changing into the industry where CO_2 emissions are gradually increasing. However, because ICT technology is also connected with new ones that can quickly solve these environmental problems, the world is investing intensively in ICT. The ICT service industry is expected to be the one that can deplete CO_2 emissions by expanding non-face-to-face services and utilizing rapidly developing IT technology [61], but it is also true that its influence on air pollution is gradually increasing. In the case of linkage effects of the service industry, forward and backward linkage effects in both countries are about 0.9, and the influences on other industries are not significant. Though, in the ICT service industry in the US, CO_2 emissions are gradually increasing. In other words, the degree of development with other industries is insignificant, but caution is needed because the amount of ICT industry in the country's total CO_2 emissions is increasing. In addition, it is necessary to change the industrial structure through the development of ICT technology so that it can develop together with other industries.

Third, CO_2 emissions of ICT manufacturing industries in Korea and the United States are on the decline. This result implies that an increase in manufacturing output by itself does not increase CO_2 emissions. It was confirmed that CO_2 emissions of ICT manufacturing industry in Korea and the US are on decreasing trends. In the case of Korea, the proportion of ICT manufacturing industry is increasing, but CO_2 emissions decrease, indicating that the increase in the manufacturing industry does not affect the increase in CO_2 emissions. Rather, it can be regarded as a change in the energy-intensive nature of the manufacturing industry. In other words, if the manufacturing industry continues to pursue low-carbon emissions and high added value, it will become a sustainable industry. Therefore, government support should be expanded so that the ICT manufacturing industry makes GREEN production processes, and the ICT manufacturing industry should also make continuous efforts to create eco-friendly construction methods. As a result of linkage effect analysis, it was found that both BL and FL in Korea exceeded 1 and had a lot of influence on other industries, but in the US, both BL and FL were analyzed to be less than 1. That is, the ICT manufacturing industry in Korea has a significant influence on both forward and backward industries, but the US does not. The ICT manufacturing industry in Korea is their flagship industry, accounting for about 10% of total production (2018), 6.7% of GDP ('2019), and exports of USD 94 billion (2019) [62]. ICT manufacturing industry in Korea continues to create investment and employment as the industry leading the national economy despite the recent decline in competitiveness of traditional manufacturing industries such as steel industry and shipbuilding.

Finally, when examining ICT industries of Korea and the US by dividing them into manufacturing and service industries in this study, it is found that that CO_2 emissions of all Korean industries increases but decreases in the case of ICT industry. Additionally, as its linkage effects are large, it can be seen that sustainable development based on low carbon is possible. The high linkage effects mean that ICT industry carries out production activities and buys or sells a lot of goods and services from other industries. In this way, the ICT industry in Korea can be said to be the central industry of the national economy because it arouses the supply and demand of other industries and leads to the production activities of that industry. Continued development of this industry will entail the development of new technologies that can reduce the ever-increasing CO_2 emissions through increasing government investment. In the case of the United States, CO_2 emissions from ICT manufacturing industry are on the decline, but ICT service is on the rise. Additionally, linkage effects are less than 1, indicating that the influence on other industries was insignificant. However, being low does not mean that it has a small share of the total output in the national economy. The linkage effects are only a measure of their influence on other industries, not how much of an industry's output accounts for a country's total output. Actually, the United States is the global number one powerhouse in the ICT industry, has a great added value creation effect, and is easy to converge and combine with other industries. These results suggest that the industry should be accompanied by efforts to reduce CO_2 emissions when it develops. Unlike other industries, ICT industry is widely applied in other fields and has the characteristics that it can improve the efficiency of industries and technologies. In the short term, the development of ICT industry can increase energy consumption but in the long term, it is possible to control energy consumption and reduce CO_2 emissions by applying ICT technology to energy demand management monitoring and low-carbon optimization. The US government should invest in technology development in ICT industry for it. This is because the development of ICT industry is that of technologies and devices for other industries.

ICT industry is a representative industry that has a comparative advantage over other countries in both Korea and the United States and has a large growth effect compared to investment [57,63,64]. When the growth of the ICT industry is combined with the effect of converting energy consumption, systematic support from government for the ICT industry should continue because the effect of reducing CO_2 emissions is large.

Despite these implications, this study has the following limitations. First, this study compares and analyzes only two countries in ICT industry, Korea and the United States. Future research should analyze countries that are importantly fostering the ICT industry. Second, this study only analyzed a part of the ICT sectors in Korea and USA because of limitation of publicly available data. Lastly, only the backward linkage effect and forward linkage effect were considered in the analysis of linkage effect without net multipliers. In future studies, additional analysis indicators such as employment inducement effect and value-added inducement effect including net multipliers should be examined together.

Author Contributions: Conceptualization, J.M. and E.Y.; methodology, J.M. and E.Y.; software, J.M.; validation, E.Y.; formal analysis, J.M.; investigation, H.C.; writing—original draft preparation, J.M.; writing—review and editing, J.M. and E.Y.; supervision, H.C. All authors have read and agreed to the published version of the manuscript.

Funding: This research received no external funding.

Institutional Review Board Statement: Not applicable.

Informed Consent Statement: Not applicable.

Data Availability Statement: Not applicable.

Conflicts of Interest: The authors declare no conflict of interest.

References

1. Ministry of Environment. Available online: http://www.me.go.kr/home/web/board/read.do?menuId=286&boardMasterId=1&boardCategoryId=39&boardId=1062250 (accessed on 2 March 2021).
2. European Commission. *Communication from the Commission: The European Green Deal*; European Commission: Brussels, Belgium, 2019.
3. Eunjoo, H.; Yun Hyeok, C.; Jong Dae, K. Study on Priorities of Regional Climate Change Policy. *J. Environ. Sci. Int.* **2016**, *25*, 589–601.
4. Hyun, K.; Sang Gun, L. A comparative study on production inducement effects in key industries of Korea and the Netherlands. *Glob. Bus. Financ. Rev.* **2020**, *25*, 13–32.
5. OECD. *OECD Digital Economy Outlook 2017*; OECD: Paris, France, 2017.
6. OECD. *OECD Digital Economy Outlook 2015*; OECD: Paris, France, 2015.
7. World Bank. *World Bank Annual Report 2017*; World Bank Group: Washington, DC, USA, 2017.
8. Ericsson. *A Quick Guide to Your Digital Carbon Footprint*; Ericsson: Stockholm, Sweden, 2020.
9. Belkhir, L.; Elmeligi, A. Assessing ICT global emissions footprint: Trends to 2040 & recommendations. *J. Clean. Prod.* **2018**, *177*, 448–463.
10. Forbes. Top 100 Digital Companies. 2019. Available online: https://www.forbes.com/top-digital-companies/list/#tab:rank (accessed on 15 February 2021).
11. OECD. *Measuring the Information Economy. E-Commerce*; OECD: Paris, France, 2002.
12. OECD. *Information Economy—Sector Definitions Based on the International Standard Industry Classification (ISIC 4)*; OECD: Paris, France, 2007.
13. OECD. *OECD Guide to Measuring the Information Society 2011*; OECD: Paris, France, 2011.
14. Xing, W.; Ye, X.; Kui, L. Measuring convergence of China's ICT industry: An input–output analysis. *Telecommun. Policy* **2011**, *35*, 301–313. [CrossRef]
15. Rohman, I.K. The globalization and stagnation of the ICT sectors in European countries: An input-output analysis. *Telecommun. Policy* **2013**, *37*, 387–399. [CrossRef]
16. Hong, J.P.; Byun, J.E.; Kim, P.R. Structural changes and growth factors of the ICT industry in Korea: 1995–2009. *Telecommun. Policy* **2016**, *40*, 502–513. [CrossRef]
17. Abubakar, Y.A.; Mitra, J. Innovation performance in European regions: Comparing manufacturing and services ICT subsectors. *Int. J. Entrep. Innov. Manag.* **2010**, *11*, 156–177. [CrossRef]
18. GSM Association. *The Mobile Economy 2020*; GSM Association: London, UK, 2020.
19. Gartner. *Gartner Forecasts Worldwide Public Cloud Revenue to Grow 17% in 2020*; Gartner: Stanford, CT, USA, 2019.
20. Canalys. *Cloud Infrastructure Spend Reaches US$20 Billion in Q2 2018, with Hybrid IT Approach Dominant*; Canalys: Singapore, 2018.
21. IDC. *Semiannual Public Cloud Services Tracker*; IDC: Needham, MA, USA, 2018.
22. European Commission. *The 2019 EU Industrial R&D Investment Scoreboard*; European Commission: Luxembourg, 2019.
23. Bieser, J.; Hilty, L. Assessing Indirect Environmental Effects of Information and Communication Technology (ICT): A Systematic Literature Review. *Sustainability* **2018**, *10*, 2662. [CrossRef]
24. European Commission. *The 2018 EU Industrial R&D Investment Scoreboard*; European Commission: Luxembourg, 2018.
25. Executive Office of the President. *The President's Climate Action Plan*; Executive Office of the President: Washington, DC, USA, 2013.
26. Oliner, S.D.; Sichel, D.E. The resurgence of growth in the late 1990s: Is information technology the story? *J. Econ. Perspect.* **2000**, *14*, 3–22. [CrossRef]
27. García-Muñiz, A.S.; Vicente, M.R. ICT technologies in Europe: A study of technological diffusion and economic growth under network theory. *Telecommun. Policy* **2014**, *38*, 360–370. [CrossRef]
28. Fernandez-deGuevara, J.; Lopez-Cobo, M.; Mas, M. *The 2017 PREDICT Key Facts Report. An Analysis of ICT R&D in the EU and beyond*; Joint Research Centre (Seville Site): Luxembourg, 2017.
29. OECD. *Measuring the Digital Transformation: A Roadmap for the Future*; OECD: Paris, France, 2019.

30. Mattioli, E.; Lamonica, G.R. The ICT role in the world economy: An input-output analysis. *J. World Econ. Res.* **2013**, *2*, 20–25. [CrossRef]
31. Inklaar, R.; O'Mahony, M.; Timmer, M. ICT and Europe's productivity performance: Industry-level growth account comparisons with the United States. *Rev. Income Wealth* **2005**, *51*, 505–536. [CrossRef]
32. Irawan, T. ICT and economic development: Comparing ASEAN member states. *Int. Econ. Policy* **2014**, *11*, 97–114. [CrossRef]
33. Malmodin, J.; Lundén, D. The energy and carbon footprint of the global ICT and E&M sectors 2010–2015. *Sustainability* **2018**, *10*, 3027.
34. Talib, F.; Rahman, Z.; Akhtar, A. An instrument for measuring the key practices of total quality management in ICT industry: An empirical study in India. *Serv. Bus.* **2013**, *7*, 275–306. [CrossRef]
35. Aasa, L.; Adrian, M. *The Effect of Energy Efficiency on Swedish Carbon Dioxide Emissions 1993–2004*; Gothenburg University: Goeteborg, Sweden, 2008.
36. Zhou, P.; Ang, B.W.; Poh, K.L. Slacks-based efficiency measures for modeling environmental performance. *Ecol. Econ.* **2006**, *60*, 111–118. [CrossRef]
37. Moon, J.; Yun, E.; Lee, J. Identifying the Sustainable Industry by Input–Output Analysis Combined with CO_2 Emissions: A Time Series Study from 2005 to 2015 in South Korea. *Sustainability* **2020**, *12*, 6043. [CrossRef]
38. Lee, S.; Noh, D.-W.; Oh, D.-H. Characterizing the difference between indirect and direct CO_2 emissions: Evidence from Korean manufacturing industries, 2004–2010. *Sustainability* **2018**, *10*, 2711. [CrossRef]
39. Jaeger, C.C.; Paroussos, L.; Kupers, R.T.L.; Mangalagiu, D. *A New Growth Path for Europe: Generating Prosperity and Jobs in the Low-Carbon Economy*; Synthesis Report PIK; University of Oxford: Oxford, UK, 2011.
40. Chaminade, C.; Plechero, M. Do regions make a difference? Regional innovation systems and global innovation networks in the ICT industry. *Eur. Plan. Stud.* **2015**, *23*, 215–237. [CrossRef]
41. Halkos, G.E.; Tzeremes, N.G. International competitiveness in the ICT industry: Evaluating the performance of the top 50 companies. *Glob. Econ. Rev.* **2007**, *36*, 167–182. [CrossRef]
42. Lee, D.H.; Hong, G.Y.; Lee, S.-G. The relationship among competitive advantage, catch-up, and linkage effects: A comparative study on ICT industry between South Korea and India. *Serv. Bus.* **2019**, *13*, 603–624. [CrossRef]
43. Lechman, E.; Marszk, A. ICT technologies and financial innovations: The case of exchange traded funds in Brazil, Japan, Mexico, South Korea and the United States. *Technol. Forecast. Soc. Chang.* **2015**, *99*, 355–376. [CrossRef]
44. Rhee, K.H.; Pyo, H.K. Aggregate Total Factor Productivity and Resource Reallocation Effect of ICT Sectors in Korea: A Comparison with the USA, Japan and EU7. *Korean Econ. Rev.* **2012**, *28*, 189–219.
45. OECD. *OECD Stat*; OECD: Paris, France, 2020.
46. Hirschman, A.O. *The Strategy of Economic Development*; Yale University Press: New Haven, CT, USA, 1958.
47. Ye, Z.P.; Yin, Y.P. Economic Linkages and Comparative Advantage of the UK Creative Sector. University of Hertfordshire Business School Working Paper No. UHBS, 2. Available online: https://papers.ssrn.com/sol3/papers.cfm?abstract_id=1310948 (accessed on 16 February 2021).
48. Chiu, R.-H.; Lin, Y.-C. Applying input-output model to investigate the inter-industrial linkage of transportation industry in Taiwan. *J. Mar. Sci. Technol.* **2012**, *20*, 173.
49. Sari, K.; Arifin, M. The linkage among technology-intensive manufacture industries in east java by input-output analysis approach. *J. ST Policy RD Manag.* **2014**, *12*, 45–54.
50. Morrissey, K.; O'Donoghue, C. The role of the marine sector in the Irish national economy: An input–output analysis. *Mar. Policy* **2013**, *37*, 230–238. [CrossRef]
51. Leontiew, W. The structure of American economy, 1919–1929. In *An Empirical Application of Equilibrium Analysis*; Harvard University Press: Cambridge, MA, USA, 1941.
52. Rasmussen, P.N. *Studies in Inter-Sectoral Relations*; Einar Harcks: København, Denmark, 1956.
53. Lin, S.J.; Chang, Y.F. Linkage effects and environmental impacts from oil consumption industries in Taiwan. *J. Environ. Manag.* **1997**, *49*, 393–411. [CrossRef]
54. Chang, Y.-T.; Shin, S.-H.; Lee, P.T.-W. Economic impact of port sectors on South African economy: An input–output analysis. *Transp. Policy* **2014**, *35*, 333–340. [CrossRef]
55. Kwak, S.-J.; Yoo, S.-H.; Chang, J.-I. The role of the maritime industry in the Korean national economy: An input–output analysis. *Mar. Policy* **2005**, *29*, 371–383. [CrossRef]
56. Ozturk, A.; Umit, K.; Medeni, I.T.; Ucuncu, B.; Caylan, M.; Akba, F.; Medeni, T.D. Green ICT (Information and Communication Technologies): A review of academic and practitioner perspectives. *Int. J. eBus. eGov. Stud.* **2011**, *3*, 1–16.
57. Han, I.; Byun, S.-Y.; Shin, W.S. A comparative study of factors associated with technology-enabled learning between the United States and South Korea. *Educ. Technol. Res. Dev.* **2018**, *66*, 1303–1320. [CrossRef]
58. Jeong, K.; Hong, T.; Kim, J. Development of a CO_2 emission benchmark for achieving the national CO_2 emission reduction target by 2030. *Energy Build.* **2018**, *158*, 86–94. [CrossRef]
59. Hwang, W.-S.; Shin, J. ICT-specific technological change and economic growth in Korea. *Telecommun. Policy* **2017**, *41*, 282–294. [CrossRef]
60. Zheng, H.; Fang, Q.; Wang, C.; Wang, H.; Ren, R. China's carbon footprint based on input-output table series: 1992–2020. *Sustainability* **2017**, *9*, 387. [CrossRef]

61. Faisal, F.; Turgut Tursoy, A.; Pervaiz, R. Does ICT lessen CO_2 emissions for fast-emerging economies? An application of the heterogeneous panel estimations. *Environ. Sci. Pollut. Res* **2020**, *27*, 10778–10789. [CrossRef] [PubMed]
62. KOSIS. Kostat. 2019. Available online: http://kostat.go.kr/portal/korea/index.action (accessed on 28 December 2020).
63. Kim, J.; Eunnyeong, H. Effect of ICT capital on the demands for labor and energy in major industries of Korea, US, and UK. *Environ. Resour. Econ. Rev.* **2014**, *23*, 91–132. [CrossRef]
64. Min, Y.-K.; Lee, S.-G.; Aoshima, Y. A comparative study on industrial spillover effects among Korea, China, the USA, Germany and Japan. *Ind. Manag. Data Syst.* **2019**, *119*, 454–472. [CrossRef]

 *processes*

Article

Sustainable Development in EU Countries in the Framework of the Europe 2020 Strategy

Elena Širá, Rastislav Kotulič *, Ivana Kravčáková Vozárová and Monika Daňová

Faculty of Management, University of Prešov in Prešov, Konštantínova 16, 080 01 Prešov, Slovakia; elena.sira@unipo.sk (E.Š.); ivana.kravcakova.vozarova@unipo.sk (I.K.V.); monika.danova@unipo.sk (M.D.)
* Correspondence: rastislav.kotulic@unipo.sk; Tel.: +421-51-4880-590

Abstract: The Europe 2020 Strategy was proposed with a long-term vision to ensure prosperity, development, and competitiveness for the member countries. This strategy is divided into three main areas named "growth". One of these is sustainable growth. This is an area of sustainability, where the partial targets are referred to as the "20-20-20 approach", and includes a reduction of greenhouse gas emissions, an increase in energy efficiency, and the sharing of renewable energy sources. However, questions arise, including: How do member states meet these targets? Which countries are leaders in this area? According to these stated questions, the aim of this article is to assess how EU countries are meeting the set targets for sustainable growth resulting from the Europe 2020 strategy and to identify the countries with the best results in this area. We looked for answers to these questions in the analysis of sustainable indicators, which were transformed into a synthetic measure for comparability of the resulting values. Finally, we identified the Baltic states, Nordic countries (European Union members), Romania, and Croatia as the best countries in fulfilling the sustainable growth aims. As sustainable development and resource efficiency are crucial areas for the future, it is important to consider these issues.

Keywords: Europe 2020; sustainable growth; resource efficiency; sustainable development

Citation: Širá, E.; Kotulič, R.; Kravčáková Vozárová, I.; Daňová, M. Sustainable Development in EU Countries in the Framework of the Europe 2020 Strategy. *Processes* **2021**, *9*, 443. https://doi.org/10.3390/pr9030443

Academic Editor: Peter Glavič

Received: 4 February 2021
Accepted: 25 February 2021
Published: 28 February 2021

1. Introduction

The Lisbon Strategy was launched in Lisbon in March 2000, formulated and agreed to by the heads of state and governments of all European Union (EU) member states. It is a 10-year strategy with an ambitious goal of growth and jobs, at the end of which Europe would become the world's most dynamically knowledge-based economy capable of sustainable growth and able to compete worldwide [1,2]. In 2005, due to the unsatisfactory results of the previous period, this strategy was renamed and innovated as the Growth and Jobs Strategy, or the renewed Lisbon Strategy. This strategy enables the weakest countries to catch up with the living standards of the most developed countries in the EU. Only through rapid and long-term economic growth can such a goal be achieved [3].

The main problems leading to failure included many set goals and insufficient focus on them, the inconsistent pursuit of these goals, the reluctance of some states to meet these goals (as, in some cases, they required unpopular measures), and the fact that not all goals suited all countries, as each country has its own level of economy and reformation. The strategy was better fulfilled by states that already had reforms in place within their country [4]. A major problem that created significant regional disparities at the national level was the ability to adopt new technologies.

This problem is especially true for the new EU member states and lagging regions. The creation of regional disparities hampers the development of the EU as a whole; therefore, it is important to set up the system in such a way that it mainly supports the development of the new EU member states. The ability to adopt new technologies is followed by policies such as science, research, and innovation policies. There is a real concern that more developed regions, which have better and quantitatively more room to adapt to change, will benefit from the promotion of innovation, the knowledge economy, and electronics.

While less developed countries are devoting resources to attracting investment in new technologies, developed countries will move forward, thus widening the regional disparities. Unfortunately, European policies do not address the specific problems of lagging regions or the possible effects of increasing or decreasing regional disparities [5]. "Following the more or less unsuccessful completion of the Lisbon Strategy, the EU adopted a new strategy in 2010 that aims to improve the European competitiveness of the Union and its Member States" [6].

The new strategy, called the Europe 2020 Strategy, aims to eliminate disparities between member countries. Much research is dedicated to this area and evaluates the achieved results of countries based on individual indicators, such as [1,7–12]. The Europe 2020 Strategy contains several areas for which objectives are specified for improving the current situation in EU member states. But only a few studies, e.g., [13–17], are dedicated to the field of sustainable growth. Therefore, we are interested in how individual countries are meeting their goals in this very specific area, which is linked to many areas of economics and human life. So, our goal is to assess how each country is achieving its goals as the 10-year period is slowly coming to an end and the results should be clear.

In the field of sustainability, the Europe 2020 Strategy has specified an area called 20-20-20, which sets out the priorities and goals that EU countries have to meet by 2020 [18]. They are:

- To increase the share of renewable energy sources in final energy consumption to 20%,
- To reduce greenhouse gas emissions by at least 20% compared to 1990 levels, and
- A 20% increase in energy efficiency, associated with a reduction in energy consumption.

2. Literature Review

The current reform agenda, focusing mainly on growth and jobs, is the Europe 2020 Strategy [9]. This global policy strategy, proposed with a long-term vision, was proposed by the European Commission and subsequently endorsed by the European Council in June 2010. The year 2010 was a new beginning for the European Union. Following the previous Lisbon Strategy and the consequences of the economic crisis, Europe wanted to emerge from this crisis, both financially and economically stronger. This is why the European Union decided to launch a new Europe 2020 Strategy, the main aim of which was to build a highly innovative and competitive economy throughout the EU. The Europe 2020 Strategy was launched by the European Commission as a 10-year plan, the short-term priority of which was to emerge from the crisis and, at the same time, achieve a sustainable future with an ever-evolving world economy. In designing the Europe 2020 Strategy, the lessons learned from the implementation of the Lisbon Strategy were taken into account, inspired by its strengths but also by its weaknesses [7].

This strategy aimed to help Europe recover from the economic crisis and transform it into a smart, sustainable, and inclusive economy with high levels of employment, productivity, and social cohesion, and to re-strengthen the EU's position as a major player in global governance [10]. Sustainable growth consists of efforts to improve countries in the areas of renewable energy, CO_2 emissions and energy consumption [8]. The basic goal of the strategy was to increase the sustainable, inclusive, and smart growth of the European Union; therefore, the European Commission proposed to set five measurable and ambitious goals, which have gradually become national goals, in the following areas: employment, research and innovation, climate change and energy, education, and poverty and social exclusion.

Sustainable growth is made up of three main objectives, as mentioned above. One is to increase the share of renewable energy sources in final energy consumption to 20%. To achieve this goal, certain underlying issues in related areas need to be taken into account, such as objectives in transport, temperature, and electricity regulation; energy policy measures to combine different types of renewable technologies; and the use of business mechanisms to promote joint support schemes and joint projects between member states [19].

According to Gogkoz and Guvercin [20], geographical mismatches between the main centers of energy production and consumption lead to higher vulnerability of the energy exporting and importing countries to various risks. Wars, political tensions, and technical failures can reduce energy supplies. In this respect, renewable energy is the key to ensuring energy security. Renewable energy is typically provided by local natural resources, which are constantly replenished and environmentally friendly. As a result, renewable energy is available for each region and can take into account its geographical and natural specifics and use them in this area.

Renewable energy sources can be divided into two groups, namely common and emerging energy technologies. Common energy technologies include hydro and wind energy, solar energy, biomass, biofuels, and geothermal energy. The group of emerging energy technologies includes marine energy, concentrated solar photovoltaics, improved geothermal energy, cellulose ethanol, and artificial photosynthesis [21]. In order for renewables to be sustainable, they should be unlimited; however, at the same time it is good to keep in mind that they should not harm the environment. Another typical attribute of renewable energy sources is the price, and so this energy should be cheap in the long run.

Global warming is one of the most significant changes in the world today. Global warming is caused by the increased concentration of greenhouse gases in the atmosphere and leads to a phenomenon commonly known as the "greenhouse effect" [22]. Anthropogenic greenhouse gas emissions disrupt the radioactive balance of the atmosphere, resulting in a change in climatic patterns. Carbon dioxide (CO_2) emissions produce changes that are visible in the natural environment, e.g., they change the alternation and course of individual seasons, and they disrupt the habitats of marine ecosystems.

In addition, we can see the effects of increasing CO_2 emissions on the climate, even with the increase and yield of rainfall. Current policies in this area aim to reduce extreme weather events, and thus avoid the danger of climate change. As climate change caused by carbon dioxide has been predicted to be irreversible for approximately 1000 years, efforts are being made to reduce the concentration of these emissions in the atmosphere [23]. Many countries are taking action against climate change. There are currently approximately 1800 laws and regulations in the world addressing the need to protect the environment. No country in the world does not have at least one law on environmental protection [24]. The Europe 2020 Strategy sets out the priority of reducing greenhouse gas emissions by at least 20% compared to 1990 levels.

Several authors have stated that climate change is occurring with increasing energy consumption. Energy-related activities are the main source of greenhouse gas emissions [25]. Studies have confirmed that two-thirds of greenhouse gases generally come from fossil-energy-related greenhouse gas emissions, and that these emissions are on the rise. Therefore, the first step in reducing greenhouse gas emissions is to reduce energy consumption or increase energy efficiency [26,27]. Various types of primary energy include oil, coal, gas, hydropower, and other renewable energy sources [28].

The authors Bekun, Alola, and Sarkodie [29] suggested an interrelationship between carbon dioxide emissions, economic growth, natural resource rent, renewable energy, and non-renewable energy consumption. The links between these individual elements can be seen in Figure 1. The authors stated that the consumption of non-renewable energy and economic growth increased carbon emissions, while the consumption of energy from renewable sources decreased. Positive effects can be deduced from ecological energy sources, especially considering the goals of sustainable development.

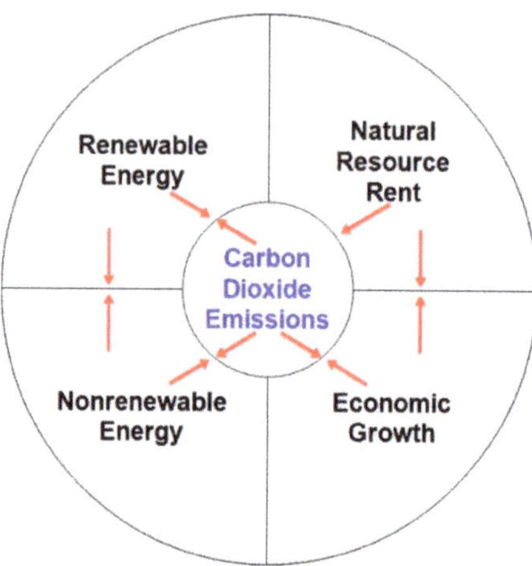

Figure 1. Factors influencing carbon dioxide emissions [29].

Addressing climate and environmental issues is a crucial challenge today with a vision for the years to come. This will be a challenging period that requires interventions in the functioning of an energy self-sufficiency economy. These changes affect all sectors. At the same time, this offers an opportunity to modernize and streamline the economy [18]. The Europe 2020 Strategy is a strategy for smart, sustainable, and inclusive growth that distinguishes the following major factors that contribute to the strengthening of the economy [30,31]. Sustainable growth represents the promotion of a more resource-efficient, greener, and more competitive economy. The specific objectives of this strategy in the area of sustainability are to reduce greenhouse gas emissions by at least 20% compared to 1990 levels; increase the share of renewable energy sources in final energy consumption to 20%; and increase energy efficiency by 20% [18].

The EU's strategic document Europe 2030 follows Europe 2020 for the next decade. According to the nature of the 2030 Agenda issued by the European Commission, creativity and innovation in particular are considered to be key elements of development. They are the driving forces behind the personal, social, and economic development of Europe. The terms "creative solutions" and "innovative practices" are considered to be catalysts for the growth and prosperity of the regions of the European Union.

Agenda 2030 is a plan designed to ensure prosperity for the people of the EU. In today's world of many economic, political, and social challenges, the program supports world peace and also helps to eradicate poverty in all its forms. Poverty is one of the biggest challenges in the world today. Agenda 2030 can be described as a comprehensive approach that fills a gap with a multilateral approach, and is also an important frame of reference for the protection of values, nature, humanity, and human rights. In the current context of Agenda 2030, indicators to monitor the achievement of the sustainable development goals are becoming essential [32].

Innovation is needed to achieve the goals of sustainable development. Green knowledge management processes have a special role to play in sustainable development, especially in the field of the creation, acquisition, exchange, and use of knowledge, as well as its impact on green technologies, eco-innovation, and the socio-economic dimension of sustainable development [30]. Over the last 15 years, the field of sustainability has aroused interest among many authors. Many sustainability concepts are based on the three pillars, namely, environmental, economic, and social [33]. Many authors have different approaches with defining the terms of sustainable development. Jabareen [34] stated that the definitions of sustainable development are vague and that there is a lack operative definition of this term. Daly [35] pointed out the utility for future generations. Development might be better defined as utility per throughput of GDP growth. We can find the ethical concept of this term in the work of Ebner and Baumgartner [36]. Sustainable development is explained as a balance between the pillars of sustainable development, which are the environmental, social, and economic pillars; however, some authors also add a fourth cultural pillar, or a pillar of good governance [37].

To ensure the benefits of sustainable growth, the entire world economy must grow. This is mainly due to the fact that if rich countries do not grow rapidly, they will not have a surplus to invest in poor countries, nor any additional income that can be used to buy exports from poor countries [35].

Every economy that wants to be successful and competitive must pay attention to its performance. The country's competitive position in a sustainable environment is a key element. Competitiveness in today's world is an important aspect of every country. In the case of long-term periods, it is important to build a sustainable economy. Based on this, we set out the main aim of the article as follows: to assess how EU countries are meeting the set targets for sustainable growth resulting from the Europe 2020 Strategy and to identify the countries with the best results in this area.

3. Materials and Methods

The sustainability of the overall development of the EU community is emphasized by the Europe 2020 Strategy. This strategy has set priorities and targets that EU countries are scheduled to meet by 2020. One of them is sustainable growth. This is a summary term for three targets, named 20-20-20 [18].

We divided these goals into two groups according to the nature of their consequences, where the first group is formed by the stimulating factor (share of renewable energy) and the second group is formed by non-stimulating factors (the remaining two goals).

As a sample of countries, we chose the countries forming the EU. The selection was also conditioned by the commitment of the EU member states to the Europe 2020 Strategy. The current contribution also includes the United Kingdom within the EU, as we evaluated a period of 14 years, i.e., from 2005 to 2018. During this period the UK was a full member of the EU and had the resulting obligations.

Methods of multicriteria analysis originated in the 1960s. They were used as a decision tool. Today, these methods are becoming more popular [38]. Multicriteria analysis provides a systematic approach to combining different information inputs to rank project alternatives. It is also used to quantify decision makers to compare alternatives. There are many approaches that fall under the multicriterial decision analysis, each of which includes different structures for their representation, algorithms for their combination, and processes for interpretation [14,39].

We have a set of data related to our research problem. We analyzed data on Europe 2020 indicators from the field of climate change and energy. For the needs of our analyses, we worked with this data: share of renewable energy [40], greenhouse gas emissions [41], and primary energy consumption [42]. This data was obtained mainly from Eurostat databases. In the first step, we ordered the data into a matrix, where we have m values in n objects [43–45].

$$X = \begin{bmatrix} x_{11} & x_{12} & \cdots & x_{1m} \\ x_{21} & x_{22} & \cdots & x_{2m} \\ \cdots & \cdots & \cdots & \cdots \\ x_{n1} & x_{n2} & \cdots & x_{nm} \end{bmatrix} \tag{1}$$

The indicators represent the different units; therefore, it is a problem to compare them in various units, and they should be normalized into standardized values (SV). We chose different formulas for stimulating and non-stimulating factors [46], which take into account their specificities.

$$SV_{ij} = \frac{x_{ij} - \min\{x_{ij}\}}{\max\{x_{ij}\} - \min\{x_{ij}\}} \quad \text{for stimulating factors} \tag{2}$$

$$SV_{ij} = \frac{\max\{x_{ij}\} - x_{ij}}{\max\{x_{ij}\} - \min\{x_{ij}\}} \text{ for non} - \text{stimulating factors} \tag{3}$$

where SV_{ij} represents the standardized value of x_{ij}; x_{ij} represents the actual value of the ith value in jth country; and max and min refer to the maximum and minimum value [47].

The evaluation and comparison of various indicators respecting countries to each other is an important issue in recent years [48]. Regions are compared according to several criteria to determine the most successful one. This topic will continue to be an important challenge for the coming years. In this context, the calculation of the synthetic measure is one of the offered options. The advantage is that this is one of the most recognized alternatives, because the output is very simple and clear [49,50] and ultimately supports the comparability of the examined objects. This index is based on the taxonomic distances of a chosen indicator from the best object, where the best object has the highest parameters according to the character of the indicator [51,52].

The aim of the research of the synthetic measure is to measure complex groups of variables [53] that are determined to describe the state from several perspectives [54]. This method has recently gained popularity, especially in applied sciences such as economics. Synthetic measures use a statistical methodology equipped with the necessary mathematical formalization to define a synthetic measure from a group of ordinal and dichotomous variables [49].

Based on the above knowledge of the benefits of this approach, we also decided to use this synthetic measure for our research. The construction of the synthetic measure (SM) was as follows:

$$SM_i = \frac{1}{m} \sum_{j=1}^{m} SV_{ij} \tag{4}$$

Finally, the results of the synthetic measure are comparable and complex, because they create a picture of the country on the basis of a summary of several facts or indicators. They represent the summary data of the country in a given year.

According to obtained values in the synthetic measure, the four groups were created. The individual groups were formed with respect to the methodology used by Stec and Grzebyk [46]. They analyzed, in their study, the fulfillment of all five fundamental objectives of the Europe 2020 Strategy in a period of six years, specifically 2009–2014. Molina, Fernández, and Martín [55] divided the countries into three groups, according to results of their synthetic measure. However, we determined that a four-group division was more effective. It contains several groups that can take into account the fine specifics in the obtained results, and thus more accurately divide the analyzed countries. Based on Stec and Grzebyk [46], the groups are as follows:

$$\text{Group I: } \mathbf{SM} \geq \text{oSM} + \text{STD} \tag{5}$$

$$\text{Group II: oSM} + \text{STD} > \mathbf{SM} \geq \text{oSM} \tag{6}$$

$$\text{Group III: oSM} > \mathbf{SM} \geq \text{oSM} - \text{STD} \tag{7}$$

$$\text{Group IV: oSM} - \text{STD} > \mathbf{SM} \tag{8}$$

where **SM** is the synthetic measure obtained by countries, oSM is the overall synthetic measure, and STD is the standard deviation.

The clarity, ease of processing, and classification of the results, and thus the countries, and last but not least the use in research similar to ours, have convinced us to use this approach rather than a cluster analysis. We also did not a use cluster analysis due to its negatives, which are mentioned by several authors, such as indistinctness, measurability [56], a large diversity of clustering algorithms [57,58], and the fact that clustering always provides groups, even if there is no group structure [59]. When using clusters, groups are assumed to exist. But it is this assumption that may be wrong or weak. Now, we will analyze the development of countries based on sustainable growth indicators. In conclusion, based on the recalculations and synthesis of the obtained data into a synthetic measure, we shall evaluate how the countries have done in this area. Finally, we identify the countries that are the strongest EU economies in this area and investigate how their positions have changed over time.

The share of renewable energy is a stimulating factor. It is desirable for EU countries to obtain the highest values in this indicator. In the table below (Table 1) [40], the highlighted values in bold and italics represent the status in which the countries have met the target. There are, of course, differences between countries in the target values. The lowest target value was set in Luxembourg, in the amount of 11%, and the highest was in Sweden, at about 49%. If we average the target values of the individual countries, we have 21%, which is slightly above the level of the set target of this indicator in the Europe 2020 Strategy. We can see that certain countries obtained the target value only in the last years (Greece, Cyprus, and Latvia). Croatia was the only country that obtained the target values in the whole analyzed period. The target value for this country was 20%. The remaining countries that met the target values did so mostly in the period 2011–2018.

Table 1. Share of renewable energy in gross final energy consumption.

	2005	2006	2007	2008	2009	2010	2011	2012	2013	2014	2015	2016	2017	2018
BG	9.173	9.415	9.098	10.345	12.005	13.927	14.152	15.837	*18.898*	*18.5*	*18.261*	*18.76*	*18.701*	*20.592*
CZ	7.115	7.363	7.895	8.674	9.978	10.514	10.945	12.813	*13.926*	*15.073*	*15.067*	*14.924*	*14.796*	*15.138*
DK	15.956	16.334	17.748	18.544	19.949	21.889	23.39	25.466	27.174	29.323	*30.866*	*32.052*	*34.677*	*35.413*
EE	17.429	15.972	17.056	18.672	22.941	24.599	*25.347*	*25.521*	*25.321*	*26.141*	*28.528*	*28.715*	*29.168*	*29.993*
EL	7.277	7.458	8.249	8.183	8.731	10.077	11.153	13.741	15.326	15.683	15.69	15.39	17.3	*18.051*
HR	*23.691*	*22.668*	*22.161*	*21.986*	*23.596*	*25.103*	*25.389*	*26.757*	*28.4*	*27.817*	*28.969*	*28.267*	*27.28*	*28.047*
IT	7.549	8.328	9.807	11.492	12.775	13.023	12.881	15.441	16.741	*17.082*	*17.525*	*17.415*	*18.267*	*17.775*
CY	3.131	3.263	4.004	5.134	5.925	6.173	6.261	7.137	8.456	9.172	9.929	9.859	10.503	*13.898*
LV	32.264	31.141	29.615	29.811	34.317	30.375	33.478	35.709	37.037	38.629	37.538	37.138	39.019	*40.029*
LT	16.77	16.889	16.482	17.825	19.798	19.64	19.944	21.438	22.69	*23.594*	*25.75*	*25.613*	*26.039*	*24.695*
HU	6.931	7.433	8.575	8.564	11.673	12.742	*13.972*	*15.53*	*16.205*	*14.618*	*14.495*	*14.377*	*13.543*	*12.535*
RO	17.571	17.096	18.195	20.204	22.157	22.834	21.186	22.825	*23.886*	*24.845*	*24.785*	*25.032*	*24.454*	23.875
FI	28.814	30.043	29.561	31.235	31.198	32.294	32.664	34.341	36.728	*38.78*	*39.321*	*39.013*	*40.917*	*41.16*
SE	40.265	42.04	43.551	44.288	47.476	46.595	*48.135*	*50.027*	*50.792*	*51.817*	*52.947*	*53.328*	*54.157*	*54.651*
BE	2.332	2.633	3.101	3.59	4.715	6.002	6.275	7.088	7.65	8.043	8.026	8.752	9.113	9.478
DE	7.167	8.466	10.039	10.072	10.851	11.667	12.453	13.543	13.76	14.385	14.906	14.889	15.476	16.673
IE	2.822	3.073	3.519	3.992	5.246	5.781	6.57	7.006	7.582	8.568	9.044	9.165	10.465	10.888
ES	8.442	9.155	9.669	10.749	12.978	13.831	13.247	14.314	15.347	16.156	16.259	17.423	17.563	17.454
FR	9.599	9.337	10.242	11.189	12.216	12.672	10.858	13.274	13.908	14.422	14.861	15.501	15.904	16.444
LU	1.402	1.469	2.725	2.809	2.929	2.85	2.856	3.114	3.499	4.469	4.987	5.361	6.198	8.973
MT	0.123	0.149	0.177	0.195	0.221	0.979	1.85	2.862	3.76	4.744	5.119	6.208	7.219	7.968
NL	2.478	2.778	3.298	3.596	4.266	3.917	4.524	4.659	4.691	5.415	5.668	5.802	6.456	7.34
AT	24.355	26.277	28.145	28.79	31.041	31.207	31.553	32.736	32.666	33.553	33.502	33.374	33.141	33.806
PL	6.9	6.888	6.93	7.711	8.699	9.3	10.354	10.97	11.463	11.614	11.888	11.4	11.117	11.477
PT	19.526	20.794	21.914	22.934	24.411	24.155	24.607	24.578	25.703	29.511	30.518	30.868	30.614	30.206
SI	19.773	18.369	19.615	18.569	20.661	20.95	20.773	21.304	22.855	22.125	22.428	21.454	21.105	20.912
SK	6.36	6.584	7.766	7.723	9.368	9.099	10.348	10.453	10.133	11.713	12.882	12.029	11.465	11.896
UK	1.281	1.488	1.734	2.814	3.448	3.862	4.392	4.461	5.524	6.737	8.385	9.032	9.858	11.138

Where BG—Bulgaria, CZ—Czech Republic, DK—Denmark, EE—Estonia, EL—Greece, HR—Croatia, IT—Italy, CY—Cyprus, LV—Latvia, LT—Lithuania, HU—Hungary, RO—Romania, FI—Finland, SE—Sweden, BE—Belgium, DE—Germany, IE—Ireland, ES—Spain, FR—France, LU—Luxembourg, MT—Malta, NL—Netherlands, AT—Austria, PL—Poland, PT—Portugal, SI—Slovenia, SK—Slovakia, UK—United Kingdom.

Now, we focus on the situation in the EU. Data regarding greenhouse gas emissions [41] are presented in Table 2. Highlighted values in bold and italics represent the status in which the countries have met the target. In the case of this factor, all EU member states set the same value. It is highly desirable for each EU country to reduce greenhouse gas emissions by 20% to 80% compared to 1990. This aim is equal for all countries. Compared to the development of the previous indicator, we can see that the countries have been meeting this target for several years. There were no exceptions for countries reaching the required value in the whole monitored period. We can see this fact in Bulgaria, Czech Republic, Estonia, Latvia, Lithuania, Romania, and Slovakia. Even large economies, such as Germany, the UK, and Sweden have met the required values of the indicator for several years.

Table 2. Greenhouse gas emissions, base year 1990.

	2005	2006	2007	2008	2009	2010	2011	2012	2013	2014	2015	2016	2017	2018
BE	100.19	98.38	96.17	96.33	87.61	92.63	85.76	83.61	83.48	*79.77*	82.81	81.96	82.14	82.67
BG	*63.1*	*63.63*	*67.42*	*66.12*	*57.18*	*59.74*	*64.88*	*59.93*	*54.78*	*57.74*	*60.93*	*58.52*	*60.88*	*57.16*
CZ	*75.13*	*75.76*	*76.68*	*74.28*	*69.75*	*71.07*	*70.28*	*68.14*	*65.46*	*64.41*	*65.13*	*66.06*	*65.56*	*64.82*
DK	95.48	106.07	99.77	94.84	90.58	90.89	83.78	*77.52*	*79.88*	*74.35*	*70.66*	*73.74*	*70.67*	*70.69*
DE	80.56	81.24	*79.2*	*79.44*	*74.05*	*76.64*	*74.73*	*75.25*	*76.68*	*73.49*	*73.8*	*74.17*	*73.22*	*70.44*
EE	*47.42*	*45.67*	*54.98*	*49.72*	*41.18*	*52.3*	*52.48*	*49.89*	*54.46*	*52.5*	*45.22*	*48.98*	*52.26*	*49.98*
HR	93.22	94.41	98.85	96.19	88.77	87.51	86.55	80.99	*76.7*	*74.34*	*75.62*	*76.15*	*78.71*	*75.23*
LV	*43.7*	*45.6*	*47.48*	*45.89*	*43.38*	*47.6*	*44.62*	*43.97*	*43.76*	*43.26*	*43.41*	*43.55*	*43.94*	*45.95*
LT	*47.38*	*48.03*	*52.63*	*50.82*	*41.68*	*43.44*	*44.67*	*44.59*	*42.09*	*42.02*	*42.64*	*42.77*	*43.24*	*42.64*
HU	80.69	*79.38*	*77.63*	*75.56*	*69.04*	*69.43*	*67.72*	*63.62*	*60.64*	*61.33*	*64.95*	*65.48*	*68.25*	*67.82*
RO	*61*	*61.33*	*62.32*	*60.42*	*51.64*	*50.11*	*52.03*	*50.66*	*46.83*	*46.96*	*47.08*	*46.29*	*47.39*	*46.84*
SK	*69.87*	*69.76*	*67.4*	*68.13*	*62.24*	*63.25*	*62.3*	*58.85*	*58.39*	*55.6*	*57.04*	*57.72*	*59.31*	*59.16*
FI	98.59	114.53	112.53	101.51	96.18	107.11	96.8	89.12	89.8	84.01	*79.06*	83.16	*79.59*	81.41
SE	94.55	94.13	92.85	89.94	83.56	91.82	86.07	82.01	*79.79*	*77.4*	*77.11*	*76.99*	*76.52*	*75.28*
UK	89.73	88.89	87.38	84.8	*77.67*	*79.33*	*73.68*	*75.63*	*73.97*	*69.03*	*66.9*	*63.76*	*62.7*	*61.59*
IE	127.69	126.82	125.74	124.38	113.15	112.47	104.76	105.23	105.41	105.34	109.56	113.34	113.29	113.6
EL	131.44	127.92	130.56	127.41	120.39	114.48	111.82	108.44	99.42	96.53	92.98	89.74	93.62	90.84
ES	154.68	152.43	156.4	144.77	130.96	126.21	126.38	123.77	114.87	115.77	119.83	116.51	121.49	119.74
FR	102.48	100.44	98.67	97.32	93.64	94.78	89.76	89.78	89.96	84.41	85.28	85.44	86.35	83.1
IT	114.36	112.59	111.37	108.88	98.19	100.44	98.12	94.43	87.63	83.71	86.3	85.8	85.05	84.41
CY	159.6	162.99	168.15	171	166.42	161.52	156.65	147.73	135.68	141.56	142.01	151	155.75	153.81
LU	108.82	106.9	103.13	102.29	97.71	102.38	100.82	98.04	93.98	91.21	88.67	87.98	90.88	94.16
MT	117.04	119.32	123.35	121.73	114.59	118.84	119.43	126.67	115.91	118	94.18	83.84	93.45	96.14
NL	99.72	97.55	96.93	96.71	93.84	99	92.95	91.04	90.96	87.73	91.63	91.57	90.78	88.58
AT	118.92	116	112.78	112.05	103.39	109.19	106.41	102.82	103.25	98.68	101.6	103.05	106.17	102.66
PL	85.21	88.49	88.45	87.17	83.13	87.1	86.9	85.35	84.64	82.02	82.73	84.56	87.7	87.42
PT	146.22	138.51	134.95	131.1	125.57	118.96	116.78	113.69	110.69	110.82	118.01	115.32	123.78	118.9
SI	109.96	110.92	111.92	115.86	105.09	105.2	105.23	102.21	98.51	89.22	90.18	94.69	93.47	94.35

In Table 3, the highlighted values in bold and italics represent the status in which the countries have met the target. It is desirable for all EU countries to reduce their consumption [42]. In percentage terms compared to 2005, the average values should decrease to 80%. In Table 3, there are set data in million tonnes of oil equivalent. Three countries, namely, Estonia, Croatia, and Latvia, met the specified values during the whole monitored period. The Czech Republic, Denmark, Cyprus, Malta, Austria, Sweden, and the UK met the specified values only in one or two years. The remaining countries managed to reach the target in recent years. The Europe 2020 Strategy has set target values to be met by 2020. Therefore, it is very desirable to identify the countries that in recent years have grown closer to meeting the specified values. Almost all achieved values in EU countries had a declining trend. If we look at the aggregate values of the indicators for all 28 countries, the values already achieved in 2018 were close to the set target amount.

Table 3. Primary energy consumption in million tonnes of oil equivalent.

	2005	2006	2007	2008	2009	2010	2011	2012	2013	2014	2015	2016	2017	2018
CZ	42.51	43.49	43.65	42.5	40.14	42.66	41.03	40.59	40.94	39.16	39.74	40.04	40.35	40.39
DK	19.45	20.84	20.37	19.91	18.91	20.2	18.52	17.82	17.84	16.93	16.92	17.57	17.85	17.96
EE	5.5	4.93	5.54	5.37	4.74	5.58	5.6	5.42	5.98	5.7	5.33	5.9	5.65	6.17
IE	14.95	15.12	15.98	15.65	14.9	14.7	13.53	13.7	13.8	13.24	13.92	14.61	14.39	14.54
EL	30.17	30.14	30.18	30.35	29.32	27.11	26.55	26.39	23.28	23.14	23.23	22.9	23.12	22.42
ES	136.56	136.74	139.35	134.44	123.38	123.34	122.98	123.41	116.06	114.2	118.6	119.29	125.79	124.63
HR	9.14	9.11	9.44	9.2	8.95	8.86	8.65	8.18	8	7.6	7.96	8.5	8.33	8.18
IT	180.83	178.95	178.67	176.12	164.08	167.28	162	156.56	152.05	142.66	149.12	147.97	148.95	147.24
CY	2.48	2.57	2.7	2.85	2.77	2.68	2.65	2.5	2.18	2.22	2.28	2.43	2.53	2.55
LV	4.49	4.66	4.77	4.58	4.43	4.56	4.28	4.44	4.36	4.36	4.27	4.29	4.47	4.69
LT	8.5	7.89	8.1	8.26	7.82	6.17	5.91	5.98	5.8	5.75	5.79	6.4	6.16	6.33
LU	4.77	4.69	4.61	4.61	4.34	4.61	4.53	4.42	4.3	4.19	4.14	4.15	4.29	4.46
HU	26.35	25.99	25.39	25.16	23.95	24.62	24.39	23.13	22.41	21.99	23.3	23.74	24.5	24.49
MT	0.92	0.92	0.95	0.96	0.88	0.93	0.93	0.97	0.87	0.88	0.75	0.71	0.81	0.82
AT	32.71	32.62	32.18	32.47	30.64	32.86	31.97	31.64	32.11	30.8	31.62	31.9	32.81	31.8
PL	87.96	92.35	91.9	93.09	89.53	96.56	96.55	93.1	93.53	89.49	90.06	94.83	99.16	101.06
PT	24.85	24.4	23.85	23.59	23.62	22.64	22	21.4	21.3	20.68	21.64	21.76	22.82	22.64
RO	36.01	37.53	37.44	37.32	32.66	32.97	33.55	33.26	30.41	30.5	30.73	30.62	32.37	32.48
SI	7.1	7.2	7.3	7.49	6.8	7	7.8	6.81	6.63	6.37	6.32	6.54	6.73	6.67
SK	17.41	17.24	16.43	16.98	15.52	16.66	15.97	15.59	15.69	14.83	15.22	15.37	16.15	15.79
FI	33.56	36.67	36.03	34.57	32.39	35.5	34.25	33.02	32.04	32.7	31.15	32.43	32.09	32.99
SE	49.26	47.96	47.65	47.49	43.33	48.59	47.62	47.59	46.44	46.03	44.32	45.41	46.45	46.78
UK	223.48	220.41	214.5	211.8	195.99	205.09	190.09	195.15	191.63	180.72	183.11	179.01	176.87	176.27
BE	51.56	51.41	50.32	51.13	50.47	54.14	50.52	47.78	49.34	45.7	46.06	49.18	49.09	46.84
BG	19.22	19.85	19.51	19.02	16.91	17.4	18.57	17.84	16.51	17.27	17.96	17.68	18.34	18.36
DE	321.62	332.75	315.79	320.76	299.92	315.15	297.8	301.12	308.29	293.6	295.93	297.63	298.12	291.75
FR	260.92	256.17	252.66	255.39	246.32	254.45	249.19	249.15	250.37	239.77	244.4	240.11	239.15	238.91
NL	70.11	69.52	69.37	69.88	67.63	71.72	67.05	66.75	66.21	62.32	63.74	64.77	65.08	64.71

4. Results and Discussion

4.1. Evaluation of the Performance of EU Countries in the Context of Sustainable Indicators

The share of renewable energy is a stimulating factor. Renewable energy sources are currently unevenly and insufficiently exploited in the European Union [60]. Energy is crucial for economic progress. The current growth of the world's population requires more energy. The transition to renewable energy produced from renewable natural sources is an opportunity to meet growing demand, promote energy security, and contribute to tackling global warming and climate change [21].

It is desirable for all countries to obtain the highest values of this indicator. The target for all EU countries is to reach 20%. In addition to this goal, all member states have set their own specific goal in this area. The aim is higher than the overall set value of 20%, in the case of 10 countries.

Overall, we can say that the values of the share of renewable energy in gross final energy consumption are increasing every year. This is true in the case of countries that already meet the set values and in the case of countries that still lack a few percentage points to reach them. If we examine the average values for all EU countries, there is also an increase in the shares of renewable energy in gross final energy consumption. In the last analyzed year, the value of the EU average was over 18%, which is close to the set target value of 20%.

Many scientists assume that increasing the share of renewable energy will lead to a reduction of CO_2 emissions [61]. The current situation in the area of greenhouse gas emissions is as follows. The six largest producers of greenhouse gases in the world together account for 62% of total production. These are China (26%), the United States (13%), the European Union (about 9%), India (7%), the Russian Federation (5%), and Japan (almost 3%). Three of them showed a decrease in greenhouse gas emissions in 2019, namely the European Union (-3.0%), the USA (-1.7%), and Japan (-1.6%) [62].

In the case of this factor, all EU member states set the same value. It is highly desirable for each EU country to reduce greenhouse gas emissions by 20%. This aim is equal for all countries. Compared to the development of the previous indicator, we can see in Table 2 that 15 countries have been meeting this target for several years.

Primary energy consumption was a non-stimulating factor. For this reason, it is desirable for each country to reduce the achieved values of this indicator to at least the set target value, possibly even below this level. Even for this indicator, each country has its own target value, taking into account its specificities and capabilities. We can notice that in the case of this indicator, compared to the previous ones, most of the EU countries met the set values. The lowest volume of primary energy consumption was given for Malta, while the highest was for Germany.

In Tables 1–3 we marked the names of the selected countries in colour. In this case, these are countries that reached the target values in at least one year in all three analyzed sustainable development indicators. These countries are the Czech Republic, Denmark, Estonia, Croatia, Latvia, Lithuania, Hungary, Romania, Finland, and Sweden.

4.2. Evaluation of the Results of EU Countries

As the data on individual indicators are of a different nature (stimulating and non-stimulating factors), and the countries reached different values in the observed period, it is problematic to compare countries on the basis of such data. In addition, for some indicators, countries had individual targets [13,63]. Thus, the evaluation of absolute values does not allow us comparability. To resolve this issue, we had to convert this data to a standardized value for each indicator.

The question is how to compare countries with each other if we consider all the analyzed indicators. We used a synthetic measure to solve this problem. It is calculated as follows:

$$SM_i = \frac{1}{3}\sum_{j=1}^{3} SV_{ij} \tag{9}$$

This synthetic measure can reach the values [0,1]. The results obtained with the synthetic measure in the analyzed countries are below, in Table 4.

Table 4. Synthetic measure.

	2005	2006	2007	2008	2009	2010	2011	2012	2013	2014	2015	2016	2017	2018
BE	0.4776	0.4842	0.4939	0.4956	0.5256	0.5169	0.5398	0.5530	0.5552	0.5708	0.5626	0.5660	0.5679	0.5710
BG	0.6471	0.6466	0.6353	0.6467	0.6820	0.6866	0.6736	0.6974	0.7307	0.7171	0.7095	0.7190	0.7120	0.7330
CZ	0.5803	0.5792	0.5799	0.5920	0.6140	0.6113	0.6176	0.6350	0.6483	0.6598	0.6573	0.6538	0.6540	0.6579
DK	0.6052	0.5789	0.6042	0.6222	0.6428	0.6527	0.6816	0.7111	0.7155	0.7437	0.7626	0.7613	0.7850	0.7893
DE	0.2865	0.2815	0.3134	0.3080	0.3475	0.3305	0.3576	0.3596	0.3501	0.3769	0.3769	0.3741	0.3797	0.4005
EE	0.7521	0.7478	0.7299	0.7535	0.8021	0.7828	0.7869	0.7948	0.7813	0.7916	0.8253	0.8162	0.8108	0.8212
IE	0.4467	0.4503	0.4550	0.4617	0.4989	0.5042	0.5300	0.5312	0.5349	0.5410	0.5324	0.5227	0.5310	0.5326
EL	0.4491	0.4592	0.4573	0.4648	0.4872	0.5128	0.5268	0.5514	0.5874	0.5971	0.6062	0.6130	0.6145	0.6270
ES	0.2897	0.2997	0.2900	0.3314	0.3916	0.4090	0.4054	0.4182	0.4547	0.4592	0.4450	0.4600	0.4415	0.4465
FR	0.3060	0.3144	0.3280	0.3345	0.3593	0.3510	0.3581	0.3729	0.3751	0.4031	0.3989	0.4067	0.4078	0.4197
HR	0.6687	0.6594	0.6445	0.6505	0.6797	0.6922	0.6967	0.7198	0.7388	0.7439	0.7473	0.7416	0.7287	0.7424
IT	0.3433	0.3545	0.3670	0.3862	0.4336	0.4261	0.4365	0.4671	0.4971	0.5186	0.5082	0.5100	0.5161	0.5165
CY	0.3792	0.3712	0.3624	0.3618	0.3785	0.3927	0.4058	0.4342	0.4735	0.4627	0.4661	0.4425	0.4341	0.4598
LV	0.8529	0.8410	0.8267	0.8322	0.8663	0.8312	0.8582	0.8733	0.8820	0.8931	0.8861	0.8833	0.8936	0.8944
LT	0.7451	0.7444	0.7299	0.7425	0.7785	0.7747	0.7737	0.7829	0.7972	0.8029	0.8145	0.8131	0.8143	0.8075
LU	0.4967	0.5022	0.5196	0.5223	0.5350	0.5223	0.5264	0.5352	0.5481	0.5613	0.5710	0.5751	0.5726	0.5810
HU	0.5811	0.5879	0.6000	0.6054	0.6424	0.6473	0.6594	0.6807	0.6932	0.6822	0.6708	0.6683	0.6553	0.6503
MT	0.4717	0.4660	0.4558	0.4600	0.4786	0.4723	0.4761	0.4636	0.4969	0.4975	0.5611	0.5943	0.5757	0.5734
NL	0.4611	0.4691	0.4740	0.4759	0.4896	0.4701	0.4940	0.5001	0.5010	0.5176	0.5077	0.5077	0.5134	0.5248
AT	0.5831	0.6024	0.6225	0.6281	0.6659	0.6498	0.6599	0.6767	0.6747	0.6932	0.6845	0.6798	0.6694	0.6835
PL	0.5075	0.4945	0.4954	0.5022	0.5222	0.5086	0.5156	0.5268	0.5312	0.5429	0.5422	0.5297	0.5156	0.5166
PT	0.4913	0.5197	0.5359	0.5523	0.5755	0.5918	0.6009	0.6096	0.6242	0.6475	0.6342	0.6431	0.6188	0.6290
RO	0.6870	0.6817	0.6860	0.7033	0.7424	0.7502	0.7346	0.7484	0.7676	0.7735	0.7721	0.7758	0.7677	0.7654
SI	0.6039	0.5928	0.5978	0.5809	0.6220	0.6233	0.6220	0.6333	0.6525	0.6721	0.6716	0.6538	0.6546	0.6512
SK	0.6144	0.6162	0.6303	0.6276	0.6542	0.6489	0.6596	0.6695	0.6686	0.6863	0.6894	0.6823	0.673 9	0.6773
FI	0.6617	0.6251	0.6280	0.6680	0.6836	0.6591	0.6891	0.7203	0.7341	0.7609	0.7785	0.7648	0.7859	0.7818
SE	0.7263	0.7395	0.7524	0.7645	0.8045	0.7727	0.7978	0.8198	0.8314	0.8442	0.8535	0.8551	0.8603	0.8662
UK	0.3254	0.3320	0.3433	0.3592	0.3973	0.3864	0.4192	0.4095	0.4238	0.4549	0.4680	0.4842	0.4941	0.5054

where BE—Belgium, BG—Bulgaria, CZ—Czechia, DK—Denmark, DE—Germany, EE—Estonia, IE—Ireland, EL—Greece, —Spain, FR—France, HR—Croatia, IT—Italy, CY—Cyprus, LV—Latvia, LT—Lithuania, LU—Luxembourg, HU—Hungary, MT—Malta, NL—Netherlands, AT—Austria, PL—Poland, PT—Portugal, RO—Romania, SI—Slovenia, SK—Slovakia, FI—Finland, SE—Sweden, UK—United Kingdom.

Based on the calculated synthetic measure (SM), we already had comparable data available for the analyzed countries. As we analyzed 28 EU member states over a period of 14 years, we decided to classify the calculated values into groups. We determined the groups with regard to the achieved results of the country. We created four groups, where group I represents the formation of the most developed countries, which concerns the indicators we examined. Conversely, group IV was formed of the weakest countries, or countries that had the greatest difficulty in meeting the targets and individual targets for sustainable growth set out in the Europe 2020 Strategy.

For better comparability, with respect to sketching the progress between the individual countries, we classified the countries into the four groups mentioned in 2005, 2010, and 2018. That is, in the first analyzed year (2005), as a reference to the starting point situation, in the year 2010, as the first year of implementation of the Europe 2020 Strategy, and in the last analyzed year (2018), where we wanted to point out the progress of some countries. Countries that improved their positions in 2010 and 2018 by moving to a higher group are highlighted.

The individual groups were formed with respect to the methodology used by Stec and Grzebyk [46]. This four-group division contains several groups that can take into account the fine specifics in the obtained results, and thus more accurately divide the analyzed countries. Based on Stec and Grzebyk [46] and the obtained results from our research, the groups are as follows:

$$\text{Group I: } \mathbf{SM} \geq 0.7386 \tag{10}$$

$$\text{Group II: } 0.7386 > \mathbf{SM} \geq 0.5926 \tag{11}$$

$$\text{Group III: } 0.5926 > \mathbf{SM} \geq 0.4467 \tag{12}$$

$$\text{Group IV: } 0.4467 > \mathbf{SM} \tag{13}$$

where **SM** is the synthetic measure obtained by countries.

Table 5 provides us with a clear order of countries according to defined groups, namely in 2005 as the first analyzed year, in 2010 as the first year of implementation of the Europe 2020 Strategy, and in 2018 as the last analyzed year.

Table 5. Distribution of countries into individual groups according to the performance of the synthetic measure.

Group	2005	2010	2018
I. **High level**	Estonia Lithuania Latvia	Estonia Lithuania Latvia **Romania** **Sweden**	Estonia Lithuania Latvia Romania Sweden **Croatia** **Finland** **Denmark**
II. **Medium-high level**	Bulgaria Denmark Croatia Romania Slovenia Slovakia Finland Sweden	Bulgaria Denmark Croatia Slovenia Slovakia Finland **Czech Republic** **Hungary** **Austria**	Bulgaria Slovenia Slovakia Czech Republic Hungary Austria **Greece** **Portugal**
III. **Medium-low level**	Belgium Ireland Greece Portugal Luxembourg Hungary Malta Netherlands Austria Poland Czech Republic	Belgium Ireland Greece Portugal Luxembourg Malta Netherlands Poland	Belgium Ireland Luxembourg Malta Netherlands Poland **Cyprus** **Italy** **United Kingdom**
IV. **Low level**	Germany Spain France Italy Cyprus United Kingdom	Germany Spain France Italy Cyprus United Kingdom	Germany Spain France

In 2005, there were only three Baltic countries in group I of the high level countries. After 14 years, this group grew by five other countries, namely the Nordic countries, Croatia, and Romania. The last two countries may come as a surprise for many readers, as they are not significantly strong EU economies. However, these countries deserved their position, because they met the required values in the monitored indicators in the area of sustainable growth. In addition, their individual targets were set at approximately the same levels as the EU average targets. Romania entered group I after five years, together with Sweden, while Croatia achieved this in 2018.

Group II, the medium-high level countries, consisted of the same number of countries in 2005 and 2018. However, there were changes in the representation of countries within this group. Some countries that were included in this group in 2005 improved their position in 2010 and 2018 and moved to group I. They were Romania and Sweden in 2010, and Denmark, Croatia, Finland in 2018.

The representation of the largest number of countries is in group III. In the years 2005, 2010, and 2018, Belgium, the Netherlands, Luxembourg, Malta, Ireland, and Poland entered this group. These are countries with both smaller and larger populations and, therefore, different goals. However, these countries met the set targets only in some years for selected indicators, and mostly had difficulty meeting the set quotas of the Europe 2020 Strategy.

Finally, in the last analyzed year, group IV was reduced to three countries, compared to six in 2005 and 2010. This is a good development. There were the same countries in 2005 and 2010. We can divide this group into two parts in the years 2005 and 2010. First, these countries include some of the EU's largest economies, which also have targets set at higher levels than other countries. On the other side are the weakest economies, including countries with more problems and lower economic performance.

The trend of the decreasing number of countries in the weakest group is desirable. It provides a positive report in moving of individual countries in terms of fulfilling and ensuring sustainable growth. In this regard, we can see the considerable progress that countries have made in the 14 years. Not only was there a decrease in the number of countries belonging to the weakest group in the last year, but there was also a consistent shift upward between the individual groups to higher groups. A shift in the opposite direction was not found.

5. Conclusions

The area of sustainable growth is a very important issue that has been addressed intensively for several years. This is an area that has an impact on the future, whose challenges have recently become very urgent. We can usually observe changes in the climate that will certainly have an effect on virtually everyone. Therefore, it is very important that countries also work together as a whole to minimize the negative impacts and try to pay attention and take measures to improve the future situation.

When we started the analyses, we were concerned about how the countries would cope with the ambitious goals arising from the Europe 2020 Strategy. Over a period of 14 years, we can see a significant improvement in individual countries, which also contributes to meeting the goals of this strategy. It is advisable to focus on a longer period of time, because the processes are demanding, and we cannot expect to see results after a year or two. We can see that countries are attempting to take these concerns seriously. Many countries can be admired for meeting the strategy's stated objectives in recent years only on the basis of a systematic approach over several years.

In the analysis of individual indicators, we identified countries that met the set criteria throughout the period. There were countries that met the required targets only in some years. In the end, there were countries that did not meet the set values for a single year, although they were constantly approaching them. These countries had another two years to meet the set targets, at least at the end of the set period, in 2020.

For some targets (e.g., primary energy consumption), the required values were reached by more countries than other targets. Setting goals is also important. In the case of greenhouse gas emissions, there was a single goal for all countries, while in other cases, the goals were adapted to the conditions of a given country, of course with regard to the fulfillment of the overall EU goal. The countries that reached the target values in at least one year in all three analyzed sustainable development indicators were the Czech Republic, Denmark, Estonia, Croatia, Latvia, Lithuania, Hungary, Romania, Finland, and Sweden.

Based on the synthetic measure, we comprehensively evaluated the resulting performances of the countries in the entire period. After the subsequent division of the countries into development stages, we identified the most successful countries in the area of strategy implementation. In the last year they were Latvia, Lithuania, Estonia, Denmark, Finland, Sweden, Croatia, and Romania. Except for the Czech Republic and Hungary, these were all countries that reached the target values in all three indicators simultaneously. As far as individual groups are concerned, a shift can be seen from 2018 compared to 2005.

Significantly more countries reached higher groups, and at the same time, the number of countries in the weakest group decreased. This trend of a decreasing number of countries in the weakest group is desirable. It provides a positive report in the moving of individual countries in terms of fulfilling and ensuring sustainable growth. In this regard, we can see the considerable progress that countries have made in 14 years. Not only did the number of countries belonging to the weakest group in the last year decrease, but there was also a consistent shift upward between the individual groups to higher groups. A shift in the opposite direction was not found.

An important area for further research in continuing this work may be analyses of changes in the Europe 2030 Strategy in the area of sustainable growth. What are the other specific targets for this area for the next 10 years? What are the other, new targets in this area? How will individual countries continue to succeed in meeting the set goals and improving the overall situation in this area? How will these improvements help companies and individuals, and how this will affect a country's overall performance? All these suggestions are possible directions for future research in this area. It is still necessary to examine this area and pay close attention, as this will provide the foundation for the future of humanity and future generations.

Author Contributions: Conceptualization, E.Š., I.K.V., M.D., and R.K.; methodology, E.Š. and R.K.; software, E.Š.; validation, E.Š.; formal analysis E.Š.; investigation, E.Š., R.K., and I.K.V.; resources, E.Š., M.D., R.K. and, I.K.V.; data curation, E.Š.; writing—original draft preparation, E.Š., I.K.V., M.D., and R.K.; writing—review and editing, E.Š., I.K.V., M.D., and R.K.; visualization, E.Š. and R.K.; supervision, E.Š., and R.K.; project administration, R.K.; funding acquisition, R.K. All authors have read and agreed to the published version of the manuscript.

Funding: This research was funded by the Cultural and Educational Grant Agency of the Ministry of Education, Science, Research and Sport of the Slovak Republic, grant numbers KEGA 011PU-4/2019 and KEGA 024PU-4/2020; by the Scientific Grant Agency of the Ministry of Education, Science, Research, and Sport of the Slovak Republic and the Slovak Academy of Sciences, grant number VEGA 1/0648/21.

Institutional Review Board Statement: "Not applicable" for studies not involving humans or animals.

Informed Consent Statement: "Not applicable" for studies not involving humans.

Data Availability Statement: The data presented in this study are openly available in [Eurostat], reference number [40–42].

Acknowledgments: The authors also thank the journal editor and anonymous reviewers for their guidance and constructive suggestions.

Conflicts of Interest: The authors declare no conflict of interest. The funders had no role in the design of the study; in the collection, analyses, or interpretation of data; in the writing of the manuscript, or in the decision to publish the results.

References

1. Dul'ováSpišáková, E.; Mura, L.; Gontkovičová, B.; Hajduova, Z. R&D in the Context of Europe 2020 in Selected Countries. *Econ. Comput. Econ. Cybern. Studies Res.* **2017**, *51*, 243–244.
2. Pereira, J. *Europe 2020–The European Strategy for Sustainable Growth. What Does it Look Like from Outside*; Friedrich Ebert Stiftung: Bonn, Germany, 2011; ISBN 978-3-86872-886-6.
3. Mikloš, I. Stratégia konkurencieschopnosti Slovenska do roku 2010 (Competitoveness strategy of the Slovakia till 2010). Available online: http://www.vuvh.sk/download/VaV/Vyznamne%20dokumenty%20EU/Narod_Lisabonska_strategia.pdf (accessed on 3 January 2021).
4. Veselovski, A.; Vaňová, L. Implementácia Lisabonskej stratégie v podmienkach SR (Implementation of Lisbon strategy in Slovakia). pp. 312–315. Available online: http://www.agris.cz/Content/files/main_files/72/150765/57Veselovski.pdf (accessed on 10 November 2020).
5. Világi, A.; Bilčík, V.; Klamár, R.; Benč, V. *Kohézna politika EÚ na roky 2007–2013 a Lisabonská stratégia*; (EU Cohesion Policy 2007–2013 and the Lisbon Strategy); Výskumné centrum SFPS: Bratislava, Slovakia, 2007; p. 39.
6. Gabrielová, H. *Stratégia Európa 2020 a Podmienky na Formovanie štruktúry Slovenskej Ekonomiky (Europe 2020 Strategy and Conditions for Dorminh the Structure of Slovakia)*; Working Papers; Ekonomický ústav SAV: Bratislava, Slovakia, 2014; ISSN 1337-5598. Available online: http://www.ekonom.sav.sk/uploads/journals/271_wp66_gabrielova.pdf (accessed on 18 December 2020).
7. Staníčková, M. Can the implementation of the Europe 2020 Strategy goals be efficient? The challenge for achieving social equality in the European Union. *Equilib. Q. J. Econ. Econ. Policy* **2017**, *12*, 383–398.
8. Tural, K.; Abuzarli, U.; Laman, M. Climate change and energy–Europe 2020 Strategy. *Qual. Access Success* **2018**, *19*, 552–555.
9. Radulescu, M.; Fedajev, A.; Sinisi, C.I.; Popescu, C.; Iacob, S.E. Europe 2020 implementation as driver of economic performance and competitiveness. Panel analysis of CEE countries. *Sustainability* **2018**, *10*, 566. [CrossRef]
10. Ayllón, S.; Gábos, A. The interrelationships between the Europe 2020 poverty and social exclusion indicators. *Soc. Indic. Res.* **2017**, *130*, 1025–1049. [CrossRef]
11. Šírá, E. Fulfillment of strategy Europe 2020 on the example of the Slovak Republic. In *Management Mechanisms and Development Strategies of Economic Entities in Conditions of Institutional Transformations of the Global Environment*; ISMA University: Riga, Latvia, 2019; pp. 313–321.
12. Bley, S.J.; Hametner, M.; Dimitrova, A.; Ruech, R.; De Rocchi, A.; Gschwend, E.; Umpfenbach, K. *Smarter, Greener, More Inclusive? Indicators to Support the Europe 2020 strategy-2017 Edition*; Publications Office of the European Union: Luxembourg, 2017; ISBN 978-92-79-70105-4.
13. Vavrek, R.; Chovancová, J. Assessment of economic and environmental energy performance of EU countries using CV-TOPSIS technique. *Ecol. Indic.* **2019**, *106*, 105519. [CrossRef]
14. Pili, S.; Grigoriadis, E.; Carlucci, M.; Clemente, M.; Salvati, L. Towards sustainable growth? A multi-criteria assessment of (changing) urban forms. *Ecol. Indic.* **2017**, *76*, 71–80. [CrossRef]
15. Štreimikienė, D.; Strielkowski, W.; Bilan, Y.; Mikalauskas, I. Energy dependency and sustainable regional development in the Baltic states: A review. *Geogr. Pannonica* **2016**, *20*, 79–87. [CrossRef]
16. Onofrei, M.; Cigu, E. Regional economic sustainable development in EU: Trends and selected issues. In Proceedings of the EURINT 2015, Regional Development and Integration: New Challenges for the EU, Iasi, Romania, 22–23 May 2015; pp. 268–280.
17. Lipińska, D. Resource-Efficcent Growth In The Eu's Sustainable Development–A Comparative Analysis Based On Selected Indicators. *Comp. Econ. Res.* **2016**, *19*, 101–117. [CrossRef]
18. Barroso, J.M. Europe 2020-A Strategy for Smart, Sustainable and Inclusive Growth. 2010. Available online: https://ec.europa.eu/eu2020/pdf/COMPLET%20EN%20BARROSO%20%20%20007%20-%20Europe%202020%20-%20EN%20version.pdf (accessed on 17 January 2020).
19. Saint Akadiri, S.; Alola, A.A.; Akadiri, A.C.; Alola, U.V. Renewable energy consumption in EU-28 countries: Policy toward pollution mitigation and economic sustainability. *Energy Policy* **2019**, *132*, 803–810. [CrossRef]
20. Gökgöz, F.; Güvercin, M.T. Energy security and renewable energy efficiency in EU. *Renew. Sustain. Energy Rev.* **2018**, *96*, 226–239. [CrossRef]
21. Armeanu, D.Ş.; Vintilă, G.; Gherghina, Ş.C. Does renewable energy drive sustainable economic growth? multivariate panel data evidence for EU-28 countries. *Energies* **2017**, *10*, 381. [CrossRef]
22. Hussain, S.; Peng, S.; Fahad, S.; Khaliq, A.; Huang, J.; Cui, K.; Nie, L. Rice management interventions to mitigate greenhouse gas emissions: A review. *Environ. Sci. Pollut. Res.* **2015**, *22*, 3342–3360. [CrossRef] [PubMed]
23. Malik, A.; Lan, J.; Lenzen, M. Trends in global greenhouse gas emissions from 1990 to 2010. *Environ. Sci. Technol.* **2016**, *50*, 4722–4730. [CrossRef] [PubMed]
24. Eskander, S.M.; Fankhauser, S. Reduction in greenhouse gas emissions from national climate legislation. *Nat. Clim. Chang.* **2020**, *10*, 750–756. [CrossRef]
25. Adamišin, P.; Huttmanová, E. The analysis of the energy intensity of economies by selected indicators of sustainability (Rio+ 20). *J. Econ. Dev. Environ. People* **2013**, *2*, 7–18. [CrossRef]
26. Liobikienė, G.; Butkus, M. The European Union possibilities to achieve targets of Europe 2020 and Paris agreement climate policy. *Renew. Energy* **2017**, *106*, 298–309. [CrossRef]

27. Cao, X.; Dai, X.; Liu, J. Building energy-consumption status worldwide and the state-of-the-art technologies for zero-energy buildings during the past decade. *Energy Build.* **2016**, *128*, 198–213. [CrossRef]

28. Valadkhani, A.; Smyth, R.; Nguyen, J. Effects of primary energy consumption on CO2 emissions under optimal thresholds: Evidence from sixty countries over the last half century. *Energy Econ.* **2019**, *80*, 680–690. [CrossRef]

29. Bekun, F.V.; Alola, A.A.; Sarkodie, S.A. Toward a sustainable environment: Nexus between CO2 emissions, resource rent, renewable and nonrenewable energy in 16-EU countries. *Sci. Total Environ.* **2019**, *657*, 1023–1029. [CrossRef]

30. Guo, M.; Nowakowska-Grunt, J.; Gorbanyov, V.; Egorova, M. Green Technology and Sustainable Development: Assessment and Green Growth Frameworks. *Sustainability* **2020**, *12*, 6571. [CrossRef]

31. Adamisin, P.; Pukala, R.; Chovancova, J.; Novakova, M.; Bak, T. Fullfilment of environmental goal of the strategy Europe 2020. Is it realistic? In Proceedings of the 3rd International Multidisciplinary Scientific Conference on Social Sciences and Arts, SGEM 2016, Vienna, Austria, 22–31 August 2016; pp. 181–188.

32. Firoiu, D.; Ionescu, G.H.; Băndoi, A.; Florea, N.M.; Jianu, E. Achieving sustainable development goals (SDG): Implementation of the 2030 Agenda in Romania. *Sustainability* **2019**, *11*, 2156. [CrossRef]

33. Paolotti, L.; Gomis, F.D.C.; Torres, A.A.; Massei, G.; Boggia, A. Territorial sustainability evaluation for policy management: The case study of Italy and Spain. *Environ. Sci. Policy* **2019**, *92*, 207–219. [CrossRef]

34. Jabareen, Y. A new conceptual framework for sustainable development. *Environ. Dev. Sustain.* **2008**, *10*, 179–192. [CrossRef]

35. Daly, H.E. Sustainable development—definitions, principles, policies. In *The Future of Sustainability*; Springer: Dordrecht, The Netherlands, 2006; pp. 39–53.

36. Ebner, D.; Baumgartner, R.J. The relationship between sustainable development and corporate social responsibility. In *Corporate Responsibility Research Conference*; Queens University: Belfast, Northern Ireland, 2006; Volume 4.

37. Adamišin, P.; Huttmanová, E.; Chovancová, J. Evaluation of sustainable development in EU countries using selected indicators. *Reg. Dev. Cent. East. Eur. Ctries.* **2015**, *2*, 188.

38. Borza, S.; Inta, M.; Serbu, R.; Marza, B. Multi-criteria analysis of pollution caused by auto traffic in a geographical area limited to applicability for an eco-economy environment. *Sustainability* **2018**, *10*, 4240. [CrossRef]

39. Huang, I.B.; Keisler, J.; Linkov, I. Multi-criteria decision analysis in environmental sciences: Ten years of applications and trends. *Sci. Total Environ.* **2011**, *409*, 3578–3594. [CrossRef]

40. Share of Renewable Energy. Available online: https://ec.europa.eu/eurostat/databrowser/view/t2020_31/default/table?lang=en (accessed on 27 December 2020).

41. Greenhouse Gas Emissions. Available online: https://ec.europa.eu/eurostat/databrowser/view/t2020_30/default/table?lang=en (accessed on 27 December 2020).

42. Primary Energy Consumption in Million Tonnes of Oil Equivalent. Available online: https://ec.europa.eu/eurotat/databrowser/view/t2020_33/default/table?lang=en (accessed on 27 December 2020).

43. Hindls, R.; Hronová, S.; Seger, J.; Fisher, J. *Statistika Pro Ekonomy (Statistics for Eoconomists)*; Professional Publishing: Praha, Czech Republic, 2017; ISBN 978-80-86946-43-6.

44. Kukuła, K. *Metoda Unitaryzacji Zerowanej*; PWN: Warsaw, Poland, 2000.

45. Rimarčík, M. Štatistika Pre Prax (Statistic for Practice). Košice, Slovakia. 2007. Available online: https://books.google.sk/books?hl=sk&lr=&id=n9a86Nb9EZ8C&oi=fnd&pg=PA45&dq=rimarcik&ots=11arS0uhjO&sig=xnvVa-r0FmGtIxmXwaxb9PWDRn0&redir_esc=y#v=onepage&q=rimarcik&f=false (accessed on 4 January 2021).

46. Stec, M.; Grzebyk, M. The implementation of the Strategy Europe 2020 objectives in European Union countries: The concept analysis and statistical evaluation. *Qual. Quant.* **2018**, *52*, 119–133. [CrossRef]

47. Lu, C.; Yang, J.; Li, H.; Jin, S.; Pang, M.; Lu, C. Research on the Spatial–Temporal Synthetic Measurement of the Coordinated Development of Population-Economy-Society-Resource-Environment (PESRE) Systems in China Based on Geographic Information Systems (GIS). *Sustainability* **2019**, *11*, 2877. [CrossRef]

48. Vavrek, R. Evaluation of the Impact of Selected Weighting Methods on the Results of the TOPSIS Technique. *Int. J. Inf. Technol. Decis. Mak.* **2019**, *18*, 1821–1843. [CrossRef]

49. Boccuzzo, G.; Caperna, G. Evaluation of life satisfaction in Italy: Proposal of a synthetic measure based on poset theory. In *Complexity in Society: From Indicators Construction to Their Synthesis*; Springer: Cham, Switzerland, 2017; pp. 291–321.

50. Vaňková, I.; Vavrek, R. Evaluation of local accessibility of homes for seniors using multi-criteria approach–Evidence from the Czech Republic. *Health Soc. Care Community* **2020**. [CrossRef] [PubMed]

51. Siedlecki, R.; Papla, D. Conditional correlation coefficient as a tool for analysis of contagion in financial markets and real economy indexes based on the synthetic ratio. *Procedia-Soc. Behav. Sci.* **2016**, *220*, 452–461. [CrossRef]

52. Vavrek, R.; Bečica, J. Capital City as a Factor of Multi-Criteria Decision Analysis—Application on Transport Companies in the Czech Republic. *Mathematics* **2020**, *8*, 1765. [CrossRef]

53. Godin, B. The Knowledge Economy: Fritz Machlup's Construction of a Synthetic Concept. In *The Capitalization of Knowledge: A Triple Helix of University-Industry-Government*; 2010; p. 261. Available online: http://www.chairefernanddumont.ucs.inrs.ca/wp-content/uploads/2013/01/Godin_37.pdf (accessed on 10 January 2021).

54. Kononova, O.; Prokudin, D. Synthetic Method in Interdisciplinary Terminological Landscape Research of Digital Economy. In Proceedings of the SHS Web of Conferences, Samara, Russia, 26–27 November 2018; Volume 50, p. 01082.

55. Molina, M.D.M.H.; Fernández, J.A.S.; Martín, J.A.R. A synthetic indicator to measure the economic and social cohesion of the regions of Spain and Portugal. *Revista Economía Mundial* **2015**, *39*, 223–239.
56. Tvaronaviciene, M.; Razminiene, K.; Piccinetti, L. Aproaches towards cluster analysis. *Econ. Sociol.* **2015**, *8*, 19. [CrossRef] [PubMed]
57. Xu, D.; Tian, Y. A comprehensive survey of clustering algorithms. *Ann. Data Sci.* **2015**, *2*, 165–193. [CrossRef]
58. Lukač, J.; Freňaková, M.; Manová, E.; Simonidesová, J.; Daneshjo, N. Use of Statistical Methods as an Educational Tool in the Financial Management of Enterprises in the Implementation of International Financial Reporting Standards. *TEM J.* **2019**, *8*, 819.
59. Madhulatha, T.S. An overview on clustering methods. *J. Eng.* **2012**, *2*, 719–725. [CrossRef]
60. Menegaki, A.N. Growth and renewable energy in Europe: Benchmarking with data envelopment analysis. *Renew. Energy* **2013**, *60*, 363–369. [CrossRef]
61. Bilan, Y.; Streimikiene, D.; Vasylieva, T.; Lyulyov, O.; Pimonenko, T.; Pavlyk, A. Linking between renewable energy, CO2 emissions, and economic growth: Challenges for candidates and potential candidates for the EU membership. *Sustainability* **2019**, *11*, 1528. [CrossRef]
62. Olivier, J.G.; Peters, J.A.H.W. Trends in global CO2 and total greenhouse gas emissions: 2020 Report. *PBL Neth. Environ. Assess. Agency* **2017**, *5*, 2020.
63. Širá, E.; Vavrek, R.; Kravčáková Vozárová, I.; Kotulič, R. Knowledge economy indicators and their impact on the sustainable competitiveness of the EU countries. *Sustainability* **2020**, *12*, 4172. [CrossRef]

Article

Optimal Cleaning Cycle Scheduling under Uncertain Conditions: A Flexibility Analysis on Heat Exchanger Fouling

Alessandro Di Pretoro, Francesco D'Iglio and Flavio Manenti *

Politecnico di Milano, Department of Chemistry, Materials and Chemical Engineering Giulio Natta, Piazza Leonardo da Vinci 32, 20133 Milan, Italy; alessandro.dipretoro@polimi.it (A.D.P.); francesco.diglio@mail.polimi.it (F.D.)
* Correspondence: flavio.manenti@polimi.it; Tel.: +39-02-2399-3273

Abstract: Fouling is a substantial economic, energy, and safety issue for all the process industry applications, heat transfer units in particular. Although this phenomenon can be mitigated, it cannot be avoided and proper cleaning cycle scheduling is the best way to deal with it. After thorough literature research about the most reliable fouling model description, cleaning procedures have been optimized by minimizing the Time Average Losses (TAL) under nominal operating conditions according to the well-established procedure. For this purpose, different cleaning actions, namely chemical and mechanical, have been accounted for. However, this procedure is strictly related to nominal operating conditions therefore perturbations, when present, could considerably compromise the process profitability due to unexpected shutdown or extraordinary maintenance operations. After a preliminary sensitivity analysis, the uncertain variables and the corresponding disturbance likelihood were estimated. Hence, cleaning cycles were rescheduled on the basis of a stochastic flexibility index for different probability distributions to show how the uncertainty characterization affects the optimal time and economic losses. A decisional algorithm was finally conceived in order to assess the best number of chemical cleaning cycles included in a cleaning supercycle. In conclusion, this study highlights how optimal scheduling is affected by external perturbations and provides an important tool to the decision-maker in order to make a more conscious design choice based on a robust multi-criteria optimization.

Keywords: maintenance; scheduling; fouling; flexibility; heat exchanger

Citation: Di Pretoro, A.; D'Iglio, F.; Manenti, F. Optimal Cleaning Cycle Scheduling under Uncertain Conditions: A Flexibility Analysis on Heat Exchanger Fouling. *Processes* **2021**, *9*, 93. https://doi.org/10.3390/pr9010093

Received: 11 December 2020
Accepted: 30 December 2020
Published: 4 January 2021

Publisher's Note: MDPI stays neutral with regard to jurisdictional claims in published maps and institutional affiliations.

1. Introduction

Due to the higher CAPital EXpenses (CAPEX) related to equipment oversizing and OPerating EXpenses (OPEX) related to energy and production losses as well as frequent maintenance, fouling still represents, nowadays, a relevant issue for the process industry.

The total heat exchanger fouling costs for highly industrialized countries were about 0.25% of the countries' Gross National Product (GNP) in 1992 [1,2]. Table 1 shows the annual costs of fouling in some different countries based on the actualization of the money value of the 1992 estimation, considering 2018 GNP.

Table 1. Fouling vs. Gross National Product (GNP) (2018).

Country	Costs (M\$/a)	GNP 2018 (M\$/a)	Costs/GNP (%)
US	14,175	20,891,000	0.13%
Germany	4875	4,356,353	0.21%
France	2400	2,962,799	0.15%
Japan	10,000	5,594,452	0.33%
Australia	463	1,318,153	0.06%
New Zealand	64.5	197,827	0.06%

As it can be noticed, additional energy duties required to compensate for an ineffective heat transfer and frequent unit cleaning and maintenance result in highly relevant expenses. A considerable part of this work is then focused on the OPEX optimization for already designed heat exchanger systems.

The overall fouling process is the net result of two simultaneous sub-processes, namely a deposition and a removal process [3,4]. The combination of these basic phenomena affects the growth of the deposit on the surface, mathematically defined as the rate of deposit growth (fouling resistance or fouling factor, R_f).

The two best ways to describe fouling in a model suitable way have been found in literature as the "Two-layer model" and "Distributed model". The latter still does not result as reliable as the first one for optimization purposes, due to its computational intensiveness [5].

In order to keep the energetic and economic efficiencies of the whole heat transfer process acceptable, heat exchangers cleaning should be often performed. Cleaning methodologies can be classified into two main groups, namely chemical cleaning and mechanical cleaning.

Mechanical cleaning methods completely restore the heat transfer surface from fouling but may damage the equipment. On the contrary, chemical cleaning techniques, which are not actually able to completely remove the fouling layers, do not cause stress or damage to the heat exchanger internals. As will be later discussed, in the developed model mechanical cleaning removes both the layers, while chemical cleaning only removes the one exposed to the fluid.

This physical description of the fouling process is useful to assess the system heat transfer performances and thus to find the optimal operation scheduling. Batch process scheduling is indeed a major research topic in process system engineering and its optimization algorithms were widely studied during the last decades [6–8].

With the advent of major concerns related to the sustainability issue, the environmental considerations concerning energy and waste analysis of batch processes have become part of the scheduling optimization domain thanks to several studies carried out by the most influential exponents of the topic [9,10].

Until the end of the 20th century, the vast majority of the studies concerning batch processes design were mainly referred to as nominal operating conditions, i.e., no uncertainty was accounted for. Moreover, the scheduling optimization and the control design were performed in two different steps of the process design procedure even though they substantially affect each other.

With the advances in design under uncertainty methodologies and the spread of its applications to thermodynamics [11], unit operations [12–16], reacting systems [17], and other fields of process engineering, the way process scheduling was conceived has started changing even if—differently from process units—a considerably smaller amount of publications about this topic is available in the literature. A pioneering work under this perspective was performed by Balasubramanian and Grossmann that analyzed the scheduling optimization under uncertain processing times with a branch and bound [18] and with a fuzzy programming [19] approach. Bonfill et al. (2005) [20] later discussed the scheduling optimization accounting for variable market demand with a two-stage stochastic optimization strategy.

Two papers, with similar research topics to that presented in this publication, were proposed by Smaïli et al. (1999) [21] and Diaby et al. (2016) [22]. The first one deals with the optimal scheduling of cleaning cycles in heat exchanger networks subject to fouling and the uncertainty analyzed refers to the measurements used to derive the fouling model. On the other hand, Diaby et al. propose the use of the genetic algorithm in order to optimize heat exchanger networks as well.

Uncertainty in batch process scheduling may refer as well to the control system performances to optimize the operation times. Further details concerning the current state of the art and the new challenges of simultaneous scheduling design and control under uncertainties can be found in the interesting literature review recently carried out by Dias and Ierapetritou (2016) [23].

The main innovative aspect of this article with respect to the literature presented here above is that, given a heat exchanger unit undergoing fouling, the optimal cleaning scheduling is thoroughly analyzed and optimized accounting for uncertain operating conditions, i.e., the so-called aleatory uncertainty related to the process input parameters is discussed under a flexibility point of view. The energy analysis of the system is coupled indeed with the economic one in order to provide a reliable estimation of each operating cycle time as well as of the associated cleaning costs.

In order to have a clear insight into this research study, the case study followed by further details and hypotheses concerning the fouling process, the flexibility assessment, and the optimization algorithm is presented in the following sections.

2. The Heat Exchanger Case Study

This research work aims at the application of the newly proposed procedure starting from the most elementary unit in order to set the basis for the possible system scale up to the several possible arrangements of multiple heat exchanger to form a more complex network. Therefore, the selected case study is a simple heat exchanger unit undergoing fouling as shown in Figure 1. The purpose of this unit is the effective heat recovery between hot and cold hydrocarbon process streams. The former passes on the tube sides while the latter on the shell side of the unit.

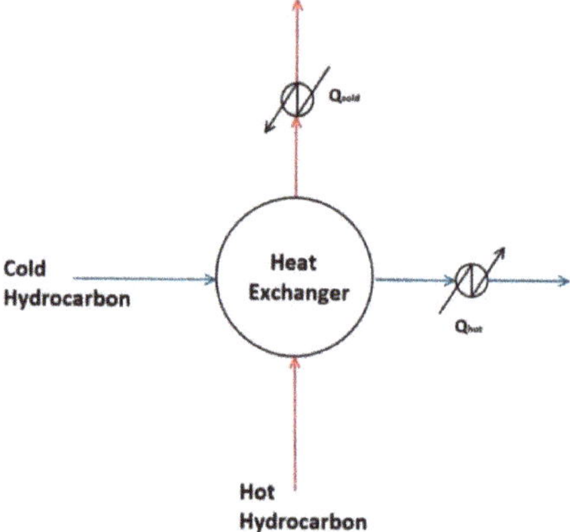

Figure 1. Heat exchanger system layout.

Two additional external duty exchangers are present to make the streams achieve the desired temperature specifications in case the heat recovery was not sufficient. The value of these heat duties expressed in Watts are defined as Q_{cold} and Q_{hot} respectively.

The process parameters for this case study were accurately selected in order to be compliant with the industrial practice. Based on the same case study proposed by Ishiyama et al. 2011 [24] and later used by Pogiatzis et al. (2012) [5], some process parameters, in particular specific heat capacities and heat transfer coefficient, were adjusted to more reliable values for the given mixtures according to the most reputable literature in the process engineering domain [25–28].

The equipment sizing was performed as well, in the case of maximum heat recovery for a fixed minimum temperature approach under nominal operating conditions and a heat transfer surface area equal to 140 m^2 was found.

The complete list of the obtained process parameters used for this research work is reported in Table 2. The details concerning the cleaning cycle scheduling and optimization procedure as well as the flexibility analysis applied to the case study here above are presented and discussed in detail in the following section.

Table 2. Process parameters.

	Symbols	Quantities	Value	Unit
Heat exchanger	U	Overall heat transfer coefficient	350	W/(m^2 K)
	A	Surface area	140	m^2
Cold stream	$q_{m,c}$	Mass flow rate	135	kg/s
	T_c^{in}	Inlet temperature	525	K
	c_P^c	Specific heat capacity	3125	J/(kg K)
Hot stream	$q_{m,h}$	Mass flow rate	68	kg/s
	T_h^{in}	Inlet temperature	575	K
	c_P^h	Specific heat capacity	2200	J/(kg K)
Fouling	K_g	Deposition rate	5×10^{-6}	(m^2 K)/(W d)
	K_c	Ageing rate	2.5×10^{-7}	(m^2 K)/(W d)
	λ_g	Gel thermal conductivity	0.1	W/(m K)
	λ_c	Coke thermal conductivity	0.8	W/(m K)

3. Methodology

The methodology section deals with the three main aspects related to cleaning cycle scheduling under uncertain operating conditions.

The first part to be defined concerns the fouling kinetic model used to describe the fouling physical phenomenon that will affect heat transfer effectiveness dynamics and will be used to assess the optimal cycle duration. Moreover, the two different cleaning techniques are presented in order to highlight the effect of the cleaning process on the fouling parameters.

After that, the definition of uncertain operating conditions via the flexibility index is carried out. The analysis is focused both on deterministic and stochastic indexes in order to provide a complete overview of flexibility and the way it can be quantified and associated with process variables.

Finally, the scheduling optimization and the corresponding decisional algorithm are discussed and compared to those already employed in the available literature. The definition of the algorithm is indeed required to outline a thorough and general procedure to be used no matter the case study under analysis.

3.1. Fouling Kinetics and Cleaning Techniques

3.1.1. Fouling Kinetics

The fouling model employed in this work is the so-called "two-layer model" proposed in similar publications by Ishiyama et al. (2011) [24] and Pogiatzis et al. (2012) [5].

As stated by its name, this model assumes a fouling deposition distributed over two layers characterized by different physical properties and defined as:

- Gel layer, located at the interface between the hard solid deposition and the process fluid;
- Coke layer, formed between the Gel layer and the exchanger heat transfer surface.

Please note that "coke" as defined here above is used only for fouling formed starting from an organic compound, otherwise, it should be referred to as "crust".

The zeroth-order kinetic model assumed to describe the behavior of the growing layers are:

- Coke layer:

$$\begin{cases} \frac{d\delta_c}{dt} = \lambda_c \cdot K_c, & \delta_g > 0 \\ \frac{d\delta_c}{dt} = 0, & \delta_g = 0 \end{cases} \tag{1}$$

- Gel layer:

$$\frac{d\delta_g}{dt} = \lambda_g \cdot K_g - \frac{d\delta_c}{dt} \tag{2}$$

where δ_g and δ_c are the thicknesses and where λ_g and λ_c are the thermal conductivities of the gel and coke layer, respectively.

The sum of these basic components represents the deposit growth on the heat exchanger surface. The thermal fouling resistance (or fouling factor) of the fouling layer R_f, also defined as the rate of deposit growth, is evaluated by treating the layers as a pair of thin slabs of insulating material and mathematically described as:

$$R_f = \Phi_d - \Phi_r = \frac{\delta_g}{\lambda_g} + \frac{\delta_c}{\lambda_c} \tag{3}$$

where Φ_d and Φ_r are the rates of deposition and removal respectively. R_f, as well as the deposition and the removal rate, can be expressed in the units of thermal resistance as $[(m^2\,K)/W]$ or in the units of the rate of thickness change as $[m/s]$ or units of mass change as $[kg/(m^2\,s)]$ [3].

Finally, in the fouling model adopted, the overall heat transfer coefficient U is given by:

$$\frac{1}{U} = \frac{1}{U_{clean}} + R_f \Leftrightarrow U = \frac{U_{clean}}{1 + U_{clean} \cdot R_f} \tag{4}$$

where U_{clean} indicates its value at $t = 0$.

3.1.2. Cleaning Techniques

Both the two different cleaning techniques presented by Ishiyama et al. (2011) [24] have been studied in this research work. They can be classified and qualitatively described as follows:

- Chemical Cleaning: the heat exchanger is shut down in order to perform the cleaning. Only the gel layer is removed with this technique by means of a proper solvent. This cleaning procedure is generally shorter and has lower fixed costs with respect to mechanical cleaning;
- Mechanical Cleaning: the heat exchanger is shut down in order to perform the cleaning procedure. Deposits are completely removed by means of mechanical strength and the heat exchanger recovers its original heat transfer capacity. This cleaning procedure is generally longer and has higher fixed costs with respect to chemical cleaning.

The cost related to the two techniques are those suggested in the referenced publications. Although the results are affected by those values, the procedure and the related code are of general validity whatever the cost function is.

The effect of these cleaning techniques on the model equations concerns the terms δ_g and δ_c. In particular, the chemical cleaning process resets the gel layer δ_g to zero, restoring the heat transfer coefficient defined by the Equation (4) to the value:

$$\frac{1}{U} = \frac{1}{U_{clean}} + \frac{\delta_c}{\lambda_c} \Leftrightarrow U = \frac{U_{clean} \cdot \lambda_c}{\lambda_c + U_{clean} \cdot \delta_c} \tag{5}$$

It is worth remarking that, once the chemical cleaning is performed, the coke layer keeps on increasing starting from the thickness achieved before the unit shut down.

On the other hand, the mechanical cleaning process is able to remove the entire fouling layer so that the heat transfer coefficient is equal to U_{clean} once the unit is restarted.

The analysis of more cleaning cycle techniques with respect to the usual scheduling based on the Mechanical Cleaning only implies the possibility of multiple cleaning cycles configurations obtained by combining them. The implications related to what will be called "supercycle" and the way we should treat it during the design phase are better discussed in Section 3.3.

3.2. Flexibility Indices

Flexibility, defined as the ability of a process to accommodate a set of uncertain parameters [29], can be quantified by the mean of several indices proposed in the literature. As already observed by Di Pretoro et al. (2019) [13], flexibility indices can be classified into deterministic and stochastic according to the way uncertainty is treated. In this research work, both a deterministic index and a stochastic one have been employed in order to compare their behavior on the applied case study.

The main definitions and mathematical formulations are discussed here below in order to provide a complete overview before their application to the cleaning cycle scheduling problem.

The first to be defined and the most widespread flexibility index is the Swaney and Grossmann (1985) [30] one.

Let us define the general hyperrectangle:

$$T(\delta) = \left\{ \theta : \theta^N - \delta \cdot \Delta\theta^- < \theta < \theta^N + \delta \cdot \Delta\theta^+ \right\} \tag{6}$$

With δ non-negative scalar variable for which

$$\begin{cases} \delta < 1, \ T(\delta) \in T \\ \delta > 1, \ T(\delta) \ni T \end{cases} \tag{7}$$

The flexibility index F_{SG}, proposed by Swaney and Grossmann, is then the solution to the constrained optimization flexibility problem:

$$F_{SG} = \max \delta \tag{8}$$

$$\text{s.t.} \ \max_{\theta \in T(\delta)} \ \min_z \ \max_{j \in J} f_j(d, z, \theta) \leq 0 \tag{9}$$

where d and z represent the design and the manipulated variables respectively. By referring to Figure 2, F_{SG} graphically represents the maximum scale factor so that the corresponding hyperrectangle is bounded by the feasible zone. Moreover, for constraints jointly quasi-convex in z and 1D quasi-convex in θ, the solution lies at a vertex of the hyperrectangle allowing to solve the optimization problem by evaluating the feasibility of the design at each vertex. On the contrary, certain types of non-convex domains may lead to nonvertex solutions.

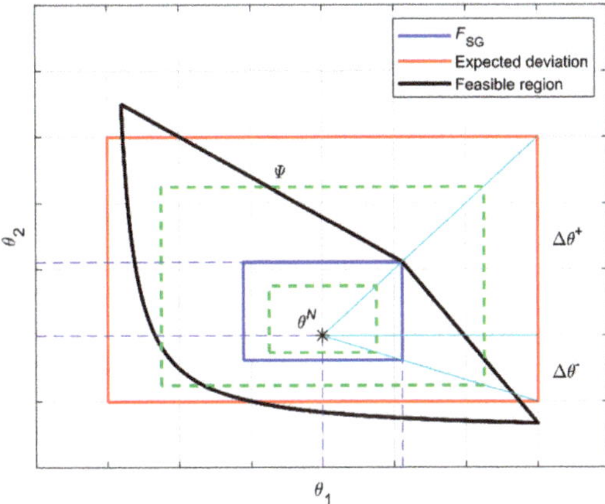

Figure 2. Flexibility representation on the uncertain domain.

The index presented here above can be referred to as "deterministic" since it is based on the assessment of a "perturbation magnitude" in the uncertain domain without taking into account the perturbation likelihood. Thus, this index definition results in a rather conservative estimation of the system flexibility.

For these reasons, a few years later, Pistikopoulos and Mazzuchi (1990) [31] introduced a stochastic flexibility index based on the perturbation likelihood. Given the uncertain parameters Probability Distribution Function (hereafter PDF) $P(\theta)$ and the feasible region

$$\Psi(d, z, \theta) \leq 0 \tag{10}$$

the stochastic flexibility index SF can be defined as:

$$SF = \int_{\Psi} P(\theta) \cdot d\theta \tag{11}$$

Unlike the deterministic indices, the stochastic flexibility one does not quantify a perturbation magnitude but assesses the percentage of the possible operating conditions for which the system keeps being feasible. However, the higher cost to pay for such an accurate result does not lie on the higher need for data only. As shown in Figure 3 indeed, more than one subregion in the uncertain domain corresponding to the same value of integral (11) could exist.

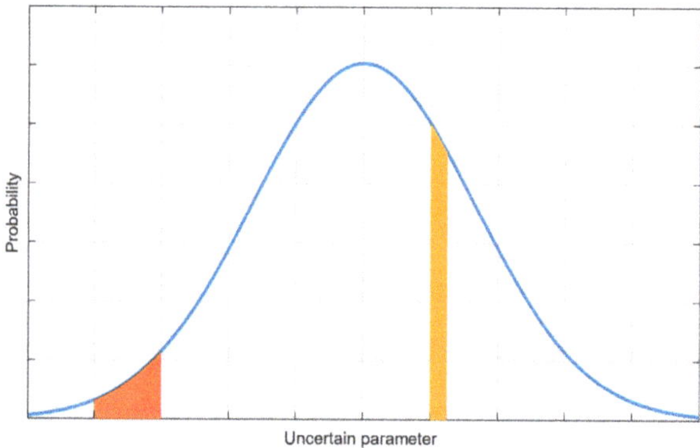

Figure 3. Stochastic flexibility index SF representation on a Normal Probability Distribution Function (PDF).

Therefore, the design choice and the related Time Average Losses (TAL, hereafter denoted by Θ) associated with each value of SF is the result of an economic optimization problem stated as:

$$\Theta(SF) = \max_{D} \Theta \tag{12}$$

where

$$D = D^k \Big|_{} \max_{D^k} \Theta \leq \max_{D^j} \Theta \; \forall j : \int_{D^k} P(\theta) \cdot d\theta = \int_{D^j} P(\theta) \cdot d\theta = SF \tag{13}$$

This formulation states that, for each value SF of the stochastic flexibility index, the associated subregion D, represented by the maximum value of Θ in it, is the cheapest one among all the subregions satisfying the Equation (11). Thus, for each step of the stochastic flexibility analysis an additional optimization problem should be solved, considerably increasing the required computational effort.

The particular definition of this index makes it different from the deterministic ones by two main properties as pointed out by Di Pretoro et al. (2019) [13]:

- The stochastic flexibility index SF has always a value between 0 and 1;
- Under nominal operating conditions, SF can be higher than 0.

The importance of including flexibility when assessing operating costs has already been proved by Di Pretoro et al. (2020) [32]. However, in order to do that the decisional algorithm for the operation scheduling to which the flexibility indices will be applied should be thoroughly defined. For this reason, a detailed explanation will be discussed in the following section.

3.3. Decisional Algorithm for Operation Scheduling

Given the heat exchanger case study under analysis with the related cleaning techniques and the associated fouling kinetic model described in Sections 2 and 3.1 respectively, the decisional algorithm for the cleaning operation schedule should give the answers to two main questions:

1. WHEN the cleaning operation should be performed?
2. WHICH cleaning operation should be performed?

In order to do that, some cost functions are required. The fact that the cycle is optimized with respect to an economic criterion implies that, once questions 1 and 2 are answered, a third question "what are the expected operating costs?" is answered as well.

The first cost function to use is called the "Energy Loss" function and it is defined as:

$$EL = C_E \cdot \int_0^t (Q_{cl} - Q) \cdot dt \tag{14}$$

where C_E is the cost per energy [€/J], t is the operating time, Q_{cl} is the heat exchanged under clean conditions, i.e., at $t = 0$, and Q is the actual heat recovery at $t = t_{op}$. In particular, it can be observed that the integrated function represents the external heat duty demand, i.e., the sum between Q_{cold} and Q_{hot}. This cost function describes operating costs to be afforded when the heat recovery is still ongoing.

Two other cost functions should be defined in order to quantify the expenses incurring during the unit shutdown. The first one refers to the cost of the energy to be provided by the external duty when no heat recovery is performed. This value is equal to:

$$EL_C = C_E \cdot Q_{cl} \cdot \tau_C \tag{15}$$

for solvent cleaning or

$$EL_M = C_E \cdot Q_{cl} \cdot \tau_M \tag{16}$$

for mechanical cleaning where τ_i is the time required for the corresponding cleaning operation.

Finally, the last cost item is represented by the cost of the cleaning operation itself, abbreviated with C_C^{Cl} and C_M^{Cl} for the chemical and mechanical cleaning techniques respectively. The costs of cleaning have been assumed as constant in time. Although the code does not account for inflation and market trends' influence, it is still possible to implement within it these aspects for real applications. However, this is not the purpose of this work.

Once all these cost items are calculated, the optimization problem concerning the optimal cycle duration needs to be solved. Its general formulation is:

$$\min_{t,\ \tau_M, \tau_C} \Theta(t, \tau_M, \tau_C) \tag{17}$$

$$\Theta(t, \tau_M, \tau_C) = \Theta_M(t, \tau_M) + \Theta_C(t, \tau_C) \tag{18}$$

This formulation is used to solve the so-called supercycle, i.e., the sequence of multiple chemical cleaning operations (N_C) required before performing the mechanical one. In particular, whether a cleaning operation is required, i.e., when the energy losses are higher than the cost of cleaning, the minimum value between Θ_M and Θ_C determines which kind of cleaning technique should be employed. In fact, each time the solvent cleaning is preferred, the coke layer keeps increasing until the chemical operation becomes more expensive than the mechanical one and the so-called supercycle is closed.

In the case where only mechanical cleaning is considered, the optimization problem is simplified by removing all the chemical cleaning related terms.

The decisional algorithm described here above has been graphically summarized in the flowchart reported in Figure 4.

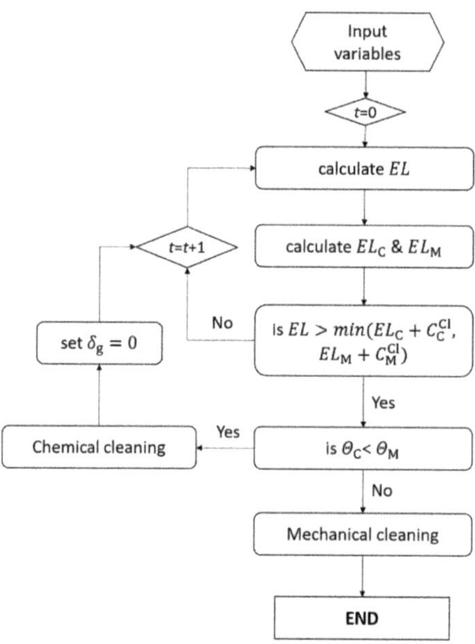

Figure 4. Decisional algorithm.

The following section presents the obtained results both for the conventional procedure and for the stochastic flexibility based one. In the latter case, the economic optimization carried out to select the optimal scheduling has been coupled with the optimization required by the stochastic flexibility assessment. This innovative procedure allows outlining the trend of the cycle duration and the associated costs obtained from the first optimization as a function of the flexibility index resulting from the second one.

The heat exchanger and fouling models, the flexibility analysis, and the decisional algorithm for scheduling optimization calculations were all performed by means of MatLab® codes.

4. Results

The following sections present in detail the results obtained. In particular, they have been classified into three main parts. The first one shows the results of the scheduling algorithm applied to our case study under nominal operating conditions both in case of a single cleaning process and of the combination of the two.

A sensitivity analysis with respect to two possible uncertain variables will follow in order to detect the most critical one and use it to perform the flexibility assessment.

Finally, the third part presents and discusses in detail the outcome of the stochastic flexibility assessment for the single heat exchanger case study by coupling economic and operational aspects.

4.1. Nominal Operating Conditions

This section presents the results obtained with the process parameters discussed in Section 2 accounting both for the mechanical cleaning methodology only and for both chemical and mechanical ones.

4.1.1. Mechanical Cleaning Only

The scheduling optimization has been performed first according to the conventional methodology accounting for mechanical cleaning only. The time average losses have been calculated for a range of cycle time and the t_{op} corresponding to its minimum value has been detected.

The overall Energy Losses and the TAL trend as a function of the operation time are shown in Figure 5a,b respectively. As it can be noticed, for low t_{op} values, the TAL are very high since the cleaning costs are much higher than the cost of the external duty required to compensate for the lower performance of the heat exchanger due to the fouling phenomenon. This value decreases until a minimum of 306.16 $/d in correspondence of 195 days and starts increasing again for the longer t_{op} because of the higher impact of the *EL* with respect to the cleaning expenses.

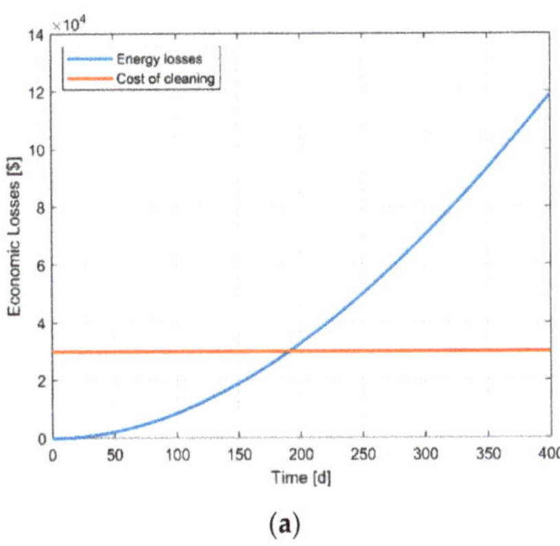

(a)

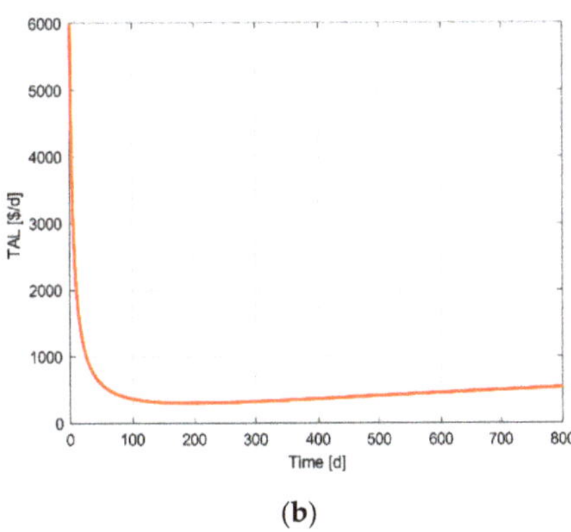

(b)

Figure 5. Mechanical cleaning. (**a**) Energy Losses and Cleaning Costs vs. cycle time; (**b**) Time Average Losses (TAL) vs. cycle time.

4.1.2. Supercycle

The decisional algorithm for scheduling optimization has been then applied accounting for both the gel and the coke layers according to the methodology presented in Section 3.3. The TAL cost function was evaluated for different numbers of chemical cleaning cycles N_C before performing the mechanical operation.

Figure 6a shows the optimal TAL value for N_C ranging from 0 to 5. The function in 0 represents the result obtained when no chemical cleaning is performed before the mechanical one. This value is exactly the one obtained for a simple cycle scheduling in the previous section, i.e., 306.16 $/d. The TAL trend shows a minimum of 234 $/d in correspondence of N_C equal to 2. This means that the optimal supercycle consists of two chemical cleanings and a mechanical one. For lower values, i.e., 1 and 0, the energy losses due to the lower efficiency of the heat recovery do not justify the cost of a mechanical operation. On the other hand, for higher values, the impact of the increasing coke layer on the heat transfer implies substantial energy losses that would compromise the profitability of the operation more than the cost of complete cleaning.

In order to ease the understanding of this phenomenon, the gel and coke layer thicknesses as a function of time have been represented in Figure 6b for the optimal operation scheduling.

The time interval when the layer thickness has a stationary trend refers to cleaning operation time, i.e., the unit maintenance shutdown time. In particular, it can be noticed that the last shutdown required for the mechanical cleaning is longer than the previous ones (5 days vs. 1 day) due to the higher complexity related to the coke layer removal.

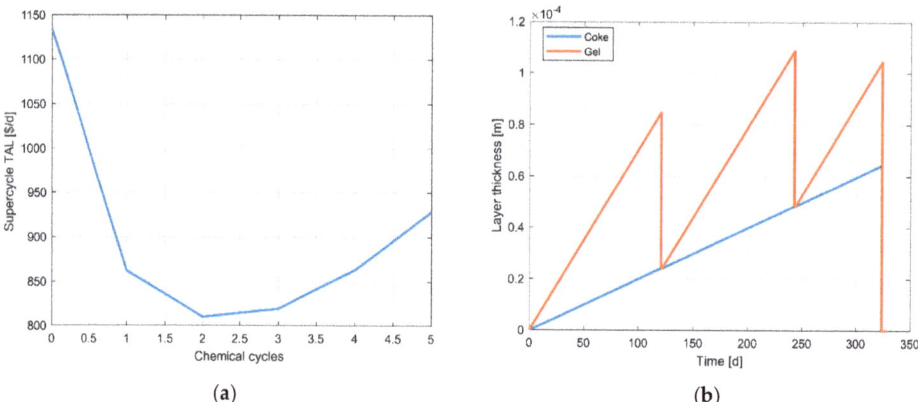

(a)   **(b)**

Figure 6. Supercycle optimal scheduling: (**a**) TAL vs. number of solvent cleaning N_C; (**b**) Supercycle layer thickness trend.

The most relevant outcome of this section is that the combination of two different cleaning strategies allows to save up to 24% of the time average losses thanks to the lower complexity and duration of the solvent cleaning subcycle.

4.2. Sensitivity Analysis

The sensitivity analysis was performed on the hot side inlet temperature T_h^{in} and the heat transfer surface area A. The choice of perturbing both an input parameter and a design variable has the purpose to verify which one of them has a higher impact on the operation scheduling. The most critical one indeed will be selected as an uncertain parameter to perform the stochastic flexibility assessment.

Figure 7 shows the results of the sensitivity analysis for wide ranges of the perturbed variables. In particular, Figure 7a refers to the optimal cycle TAL while Figure 7b shows the corresponding optimal cycle time.

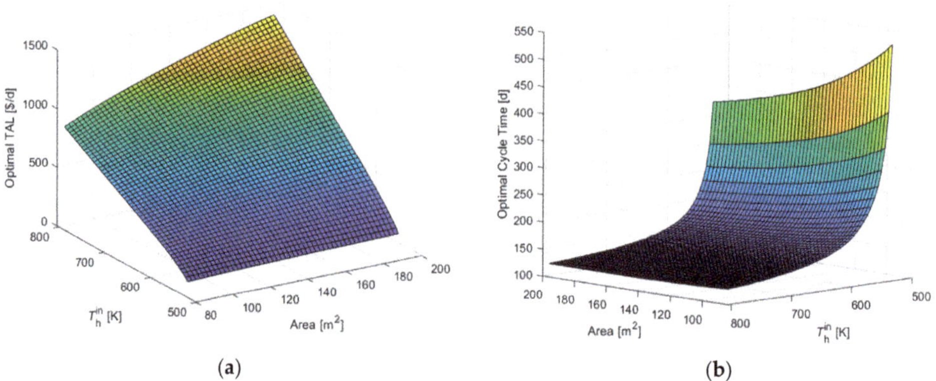

(a)   **(b)**

Figure 7. Sensitivity analysis results: (**a**) Optimal cycle TAL; (**b**) Optimal cycle time.

On the one hand, in the former Figure, an almost linear trend of the Time Average Losses as a function of the two perturbed variables can be noticed. More precisely, for the same deviation in terms of percentage, the TAL shows a slightly higher sensitivity with respect to T_h^{in} than to A. In particular, the sensitivity of the cost function with respect to one of the two variables becomes more relevant as the value of the other one gets higher.

On the other hand, in the latter Figure, the higher sensitivity of the optimal cycle time with respect to the temperature results to be much more relevant. The perturbation of t_{op} with respect to the heat transfer surface area is almost negligible for low T_h^{in} values and shows a moderate response for very high-temperature perturbations. On the contrary, the optimal cycle time exhibits an almost exponential growth with respect to the inlet temperature for any value of the heat transfer surface area. This behavior can be reconducted to the fact that this parameter directly acts on the external duty demand required to achieve the hot side specification while the heat exchanger was already properly sized and affects the optimal scheduling solution only in case of very ineffective heat transfer.

Thus, in the light of these results, the intermediate value $A = 140$ m^2 was finally kept as the nominal one while the hot side inlet temperature has been selected as uncertain parameters for the flexibility analysis presented in detail in the next section.

4.3. Flexibility Assessment

As already discussed in detail by previous studies dealing with the aleatory uncertainty and the stochastic flexibility index [13,31,32], the uncertainty characterization, i.e., the probability distribution function used to describe the uncertain parameter likelihood, plays a main role in the final result. For this reason, in order to have a more complete overview, two PDFs, namely the Gaussian and the Beta distributions, are employed in this study. The former describes an uncertain variable perturbation with a symmetric behavior with respect to the central point (i.e., its mode) usually corresponding to the nominal operating conditions. On the other hand, the Beta distribution defines an aleatory parameter for which either the positive or the negative deviation is more likely to occur. Further details about their mathematical formulation are provided in the Appendix A.

Both the Gaussian (or Normal) and the Beta distributions belong to the class of two-parameter PDFs, therefore two conditions should be imposed in order to uniquely determine their trends. In the case of real applications, the distribution of the uncertain parameter can be outlined according to the available frequency data or accounting for its expected behavior. For this paper, since the study is not related to any existing unit, the maximum likelihood of the PDF was set equal to the nominal operating conditions point and the variance was selected so that the entire uncertain space is covered with a residual probability lower than the 0.01%.

The trends resulting from these hypotheses have been outlined in Figure 8. As it can be noticed, the Gaussian distribution results particularly narrow with respect to the entire temperature range since there is a minimum inlet temperature below which the heat transfer is not feasible due to a minimum temperature approach lower than 10 °C. This value was then set as the lower boundary of the uncertain domain.

Once the uncertain parameter PDF has been properly defined, the stochastic flexibility assessment can be finally carried out according to the methodology presented in Section 3.2. Each probability distribution function was then integrated over the uncertain domain and the maximum TAL and cycle time have been assessed at each value of the SF index. This procedure allowed to estimate the expected costs and operation time to be respectively afforded and waited for each simple cycle.

The cost trends obtained this way are shown in Figure 9. As it can be noticed, the expected TAL grows fast for very low flexibility index values until a lower slope increased is attained over a wide range of the uncertain parameter. As the residual probability gets lower, much higher losses should be expected in order to almost entirely cover the uncertain domain.

The interval width reflects the PDF shape as expected and has already been discussed in the previous sections. In particular, the "steady" region is much wider in the case of the Gaussian one. The TAL for lower values resulted higher for the Beta PDF while the cost increase in the residual part of the stochastic flexibility index is more relevant when using the Gaussian one.

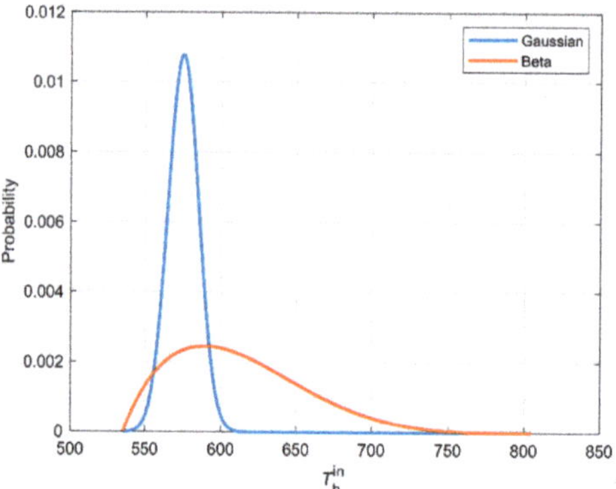

Figure 8. Hot inlet temperature probability distribution.

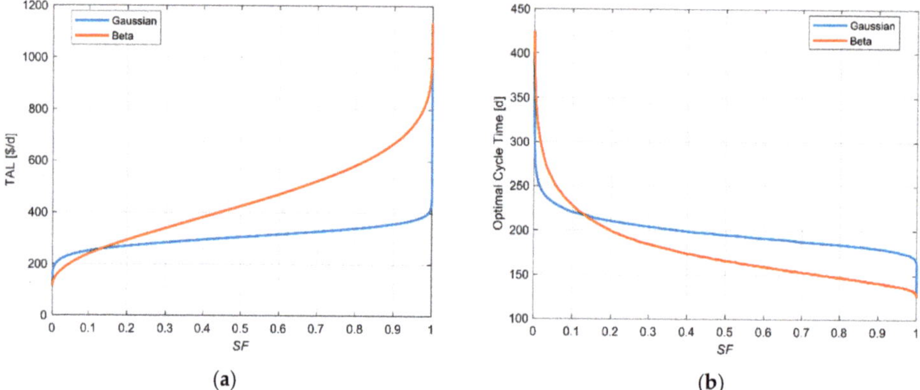

(a)   (b)

Figure 9. Flexibility analysis results—mechanical cleaning: (**a**) Optimal cycle TAL vs. *SF*; (**b**) Optimal cycle time vs. *SF*.

In regards to the optimal cycle time, the obtained curves show an opposite trend with respect to the corresponding TAL functions. This behavior is due to the fact that higher inlet temperature requires cleaning cycles to be more frequently performed as already pointed out by the sensitivity analysis. Analogously to the economic remarks discussed in Section 4.1.2 about the costs, the use of a combined chemical and mechanical cleaning methodology allows extending the operating time before a complete shutdown is required.

In any case, the two trends are qualitatively similar. It is worth remarking that the presence of some edges in these trends is due to the discretization of the uncertain domain that was not kept too dense in order to avoid excessive computational time.

The same procedure was then carried out for the supercycle accounting for both chemical and mechanical cleaning operations.

The optimal number of solvent cleaning cycles over the uncertain domain always resulted as 2. However, the corresponding costs and optimal cycle time considerably vary over the inlet temperature interval as shown by the plots in Figure 10a,b for the two probability distributions. Given that the optimal supercycle configuration does nott change, relatively smooth trends can be observed for both the output variables and both the PDFs without any particular discontinuous point. As for the simple cycle, a fast increase in the cost function can be observed for very low and very high *SF* values. However, for the supercycle, the Beta distribution exhibits a higher slope in the intermediate range with respect to the Gaussian one.

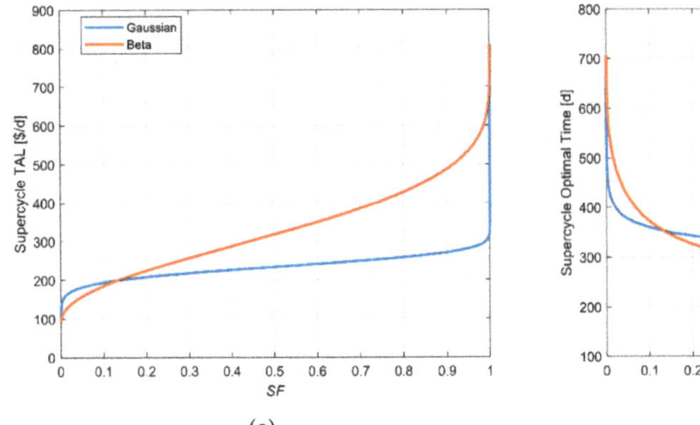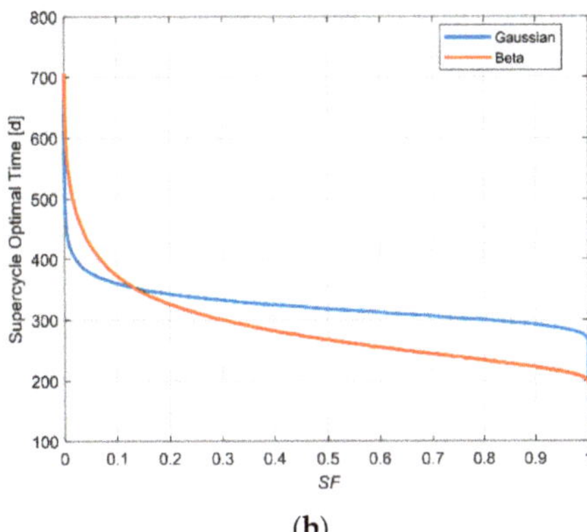

Figure 10. Flexibility analysis results—supercycle: (**a**) Optimal supercycle TAL vs. *SF*; (**b**) Optimal supercycle time vs. *SF*.

Even in this case, an opposite trend can be detected for the optimal cycle time as for the previous analysis. In general, higher costs, and thus lower optimal cycle time, can be pointed out for the Beta distribution due to its higher variance.

Therefore, in the light of these results, it can be also concluded that, in the case of a narrow probability distribution, the scheduling parameters obtained accounting for nominal operating conditions only are much closer to the real estimation. This is due to the fact that the trends TAL and t_{op} vs. *SF* exhibit a lower slope over a wider flexibility range and just the very last part of residual probability significantly affects their values.

5. Conclusions

The purpose of this research work was to include the uncertainty of the operating condition in the heat cleaning cycles scheduling optimization algorithm. The outcome of the study was successful and allowed for the definition of a thorough procedure to be applied in the case of heat exchanger cleaning cycle optimal scheduling.

The general validity of the innovative algorithm with respect to the uncertainty characterization was proved by testing both a symmetric (i.e., Gaussian) and a skewed (i.e., Beta) probability distribution function. Moreover, the procedure does not depend on the fouling kinetic model that is used to describe the physical phenomena related to fouling.

Furthermore, the employment of a preliminary sensitivity analysis is proposed in this study with the purpose of restraining the number of uncertain parameters to be accounted for by identifying the most critical ones from a flexibility perspective and thus decreasing the dimension of the uncertain domain.

A further result obtained during this study was the algorithm aimed at the optimally combined solvent and mechanical cleaning cycle under uncertainty. As for nominal operating conditions, the possibility to exploit different cleaning operations allows reducing the costs related to fouling removal. However, the optimization of the so-called "supercycle" implies the introduction of an additional decisional variable, i.e., chemical vs. mechanical, requiring a non-negligible computational effort already in the case of a single unit.

In conclusion, a general procedure for operation scheduling under uncertainty was outlined by means of a stochastic flexibility indicator and, in particular, its application to a single heat exchanger cleaning cycle was analyzed in deep. The aleatory uncertainty characterization by means of a probability distribution function combined with the proposed algorithm provides a tool to the decision-maker to have a more reliable cost estimation and to define the best scheduling solution according to the specific operating conditions.

This work sets the basis for future applications on more complex heat exchanger networks as well as additional perspectives on other kinds of operations different than unit maintenance. Moreover, besides the use of the methodology proposed in this research study to deal with an undesired disturbance, i.e., fouling, the scheduling algorithm could be also actively exploited to take advantage of certain operational parameters in order to reduce operating costs and enhance the process performances.

Author Contributions: Conceptualization, A.D.P.; Data curation, F.D.; Methodology, A.D.P. and F.D.; Supervision, F.M.; Writing—original draft, A.D.P. and F.D. All authors have read and agreed to the published version of the manuscript.

Funding: This research received no external funding.

Institutional Review Board Statement: Not applicable.

Informed Consent Statement: Not applicable.

Data Availability Statement: Data sharing not applicable.

Conflicts of Interest: The authors declare no conflict of interest.

Glossary

Symbol	Definition	Unit
A	Heat transfer surface area	m^2
CAPEX	CAPital EXpenses	$/a
c_P^i	Specific heat at constant pressure	$J/(kg\,K)$
C_i^{Cl}	Cleaning costs	$
C_E	Energy costs	$/J
D	Stochastic Flexibility index domain	n-D space
d	Design variable	variable
EL	Energy Losses	$
F_{SG}	Swaney & Grossmann flexibility index	1
GNP	Gross National Product	acronym
K_i	Fouling kinetic constant	$(m^2\,K)/J$
N_C	Number of chemical cleaning cycles	1
OPEX	OPerating EXpenses	$/a
PDF	Probability Distribution Function	function
$P(\theta)$	Probability function	function
$q_{m,i}$	Mass flow rate	kg/s
Q_i	Heat duty	W
R_f	Fouling heat transfer resistance	$(m^2\,K)/W$
SF	Stochastic Flexibility index	1
t	Time	d
t_{op}	Operation time	d
T	Flexibility hyper-rectangle	function
TAL	Time Average Losses	acronym
T^{in}	Inlet temperature	K
U	Heat transfer coefficient	$W/(m^2\,K)$
z	Control variable	variable
Greek letters		
α	Beta distribution shape parameter	1
β	Beta distribution scale parameter	1
δ	Flexibility index scale factor	1
δ_i	Fouling layer	m
Θ	Time Average Losses	$/d
θ	Uncertain variable	various
λ_i	Thermal conductivity	$W/(m^2\,K)$
μ	Mean	various
σ	Variance	various
τ_i	Cleaning time	d
Φ_i	Fouling deposition/removal rate	$(m^2\,K)/W$
Ψ	Feasible domain	n-D space

Appendix A

The formulation of the probability distribution functions employed to describe the uncertain variable deviation likelihood will be discussed in detail here below in order to provide a better understanding of the associated mathematical properties. For further details please refer to Severini (2011) [33].

Appendix A.1. Normal Probability Distribution Function

As already mentioned, the condition of "general validity" is represented by the Gaussian or normal probability distribution function. It is symmetrical with respect to its mean and the 99.73% of cumulative probability falls in the range ±3 times the variance $[-3\sigma, +3\sigma]$.

Different Gaussian distributions are plotted in Figure A1 for different values of the characteristic parameters μ and σ.

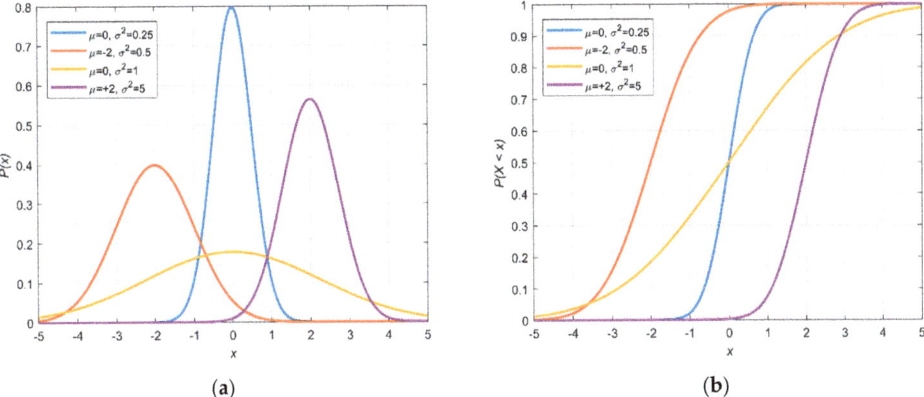

Figure A1. Normal (Gaussian) distribution: (**a**) Probability & (**b**) Cumulative Distribution Functions.

The single variable Normal PDF mathematical formulation states as:

$$P(x) = \frac{1}{\sigma \cdot \sqrt{2 \cdot \pi}} \cdot e^{-\frac{(x-\mu)^2}{2 \cdot \sigma^2}} \tag{A1}$$

where μ is the mean and σ is the variance.

For a general number of variables the n-variables normal PDF with can be defined as:

$$P(x) = \left(\frac{1}{2 \cdot \pi}\right)^{\frac{n}{2}} \cdot |\Sigma|^{-\frac{1}{2}} \cdot e^{-\frac{1}{2} \cdot (x-\mu)' \cdot \Sigma^{-1} \cdot (x-\mu)} \tag{A2}$$

where Σ is the variance-covariance matrix, Σ^{-1} its inverse and $|\Sigma|$ its determinant.

Moreover we can standardize, i.e., reconduct to a 0 mean value and variance equal to 1 (variance-covariance matrix equal to the identity matrix), the normal distribution by mean of the independent variable substitution:

$$z = \frac{x - \mu}{\sigma} \tag{A3}$$

obtaining:

$$P(z) = \left(\frac{1}{2 \cdot \pi}\right)^{\frac{n}{2}} \cdot |I|^{-\frac{1}{2}} \cdot e^{-\frac{1}{2} \cdot z' \cdot I \cdot z} \tag{A4}$$

for a general n variables standard normal probability distribution.

This transformation besides making the calculations easier allows to compare variables with different dimensions, e.g., temperature vs. flowrate vs. velocity etc.

The boundaries of the feasibility domain, if analytically available, have then to be rewritten as functions of the new variable z by inverting the Equation (A3).

Appendix A.2. Beta Probability Distribution Function

The second distribution function used in this paper to describe the uncertain variable deviation likelihood is the so called Beta PDF. It is a two-parameter continuous probability distribution defined for positive values of the independent variable $x \in [0, \infty]$.

Beta distribution is defined according to two parameters; the parametrization with a shape parameter α and a scale parameter β is used here below. Different Gaussian distributions are plotted in Figure A2 for different values of the characteristic parameters.

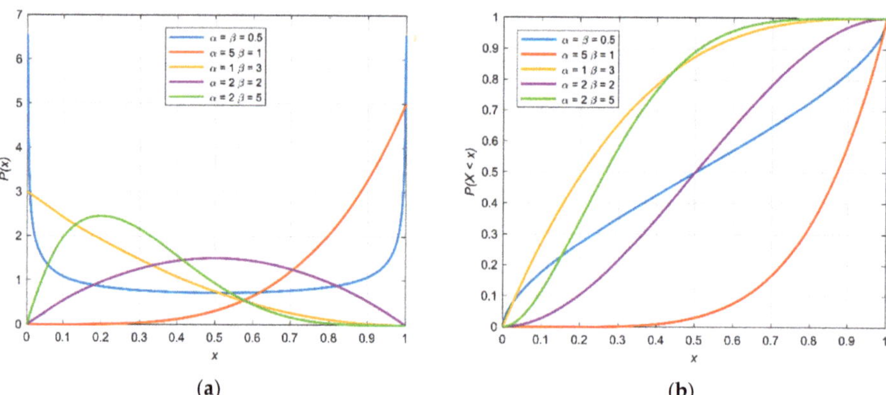

Figure A2. Beta distribution: (**a**) Probability & (**b**) Cumulative Distribution Functions.

The analytical expression of the single variable Beta PDF is formulated as:

$$P(x) = x^{\alpha-1} \cdot (1-x)^{\beta-1} \cdot \frac{\Gamma(\alpha+\beta)}{\Gamma(\alpha)+\Gamma(\beta)} \tag{A5}$$

The mode with $\alpha, \beta > 1$ assumes the meaning of the most likely value of the distribution, it graphically corresponds to the peak in the PDF and is mathematically formulated as:

$$mode = \frac{\alpha-1}{\alpha+\beta-2} \tag{A6}$$

For $\alpha, \beta < 1$ the behavior is opposite and is defined as the anti-mode, or the lowest point of the probability density curve.

For $\alpha = \beta$, the expression for the mode simplifies to $1/2$, showing that for $\alpha = \beta > 1$ the mode (respectively anti-mode when $\alpha, \beta < 1$), is at the center of the distribution: it is symmetric in those cases as can be seen in the figure.

References

1. Pritchard, A.M. The Economics of Fouling. In *Fouling Science and Technology*; Melo, L.F., Bott, T.R., Bernardo, C.A., Eds.; NATO ASI Series; Springer: Dordrecht, The Netherlands, 1988; pp. 31–45. [CrossRef]
2. Steinhagen, R.; Müller-Steinhagen, H.; Maani, K. Problems and Costs due to Heat Exchanger Fouling in New Zealand Industries. *Heat Transf. Eng.* **1993**, *14*, 19–30. [CrossRef]
3. Awad, M.M. Fouling of Heat Transfer Surfaces. *Chem. Eng. Technol.* **2011**. [CrossRef]
4. Bott, T.R. *Fouling of Heat Exchangers*, 1st ed.; Elsevier Science: Amsterdam, The Netherlands; New York, NY, USA, 1995.
5. Pogiatzis, T.; Ishiyama, E.M.; Paterson, W.R.; Vassiliadis, V.S.; Wilson, D.I. Identifying optimal cleaning cycles for heat exchangers subject to fouling and ageing. *Appl. Energy* **2012**, *89*, 60–66. [CrossRef]
6. Méndez, C.A.; Cerdá, J.; Grossmann, I.E.; Harjunkoski, I.; Fahl, M. State-of-the-art review of optimization methods for short-term scheduling of batch processes. *Comput. Chem. Eng.* **2006**, *30*, 913–946. [CrossRef]
7. Mockus, L.; Reklaitis, G.V. A New Global Optimization Algorithm for Batch Process Scheduling. In *State of the Art in Global Optimization*; Nonconvex Optimization and Its Applications; Floudas, C.A., Pardalos, P.M., Eds.; Springer: Boston, MA, USA, 1996; Volume 7, pp. 521–538.

8.   Xueya, Z.; Sargent, R.W.H. A new unified formulation for process scheduling. In Proceedings of the AIChE Annual Meeting, Paper No. 144c, St. Louis, MO, USA, 7–12 November 1993.
9.   Stefanis, S.; Livingston, A.; Pistikopoulos, E. Environmental impact considerations in the optimal design and scheduling of batch processes. *Comput. Chem. Eng.* **1997**, *21*, 1073–1094. [CrossRef]
10. Grau, R.; Graells, M.; Corominas, J.; Espuña, A.; Puigjaner, L. Global strategy for energy and waste analysis in scheduling and planning of multiproduct batch chemical processes. *Comput. Chem. Eng.* **1996**, *20*, 853–868. [CrossRef]
11. Di Pretoro, A.; Montastruc, L.; Manenti, F.; Joulia, X. Exploiting Residue Curve Maps to Assess Thermodynamic Feasibility Boundaries under Uncertain Operating Conditions. *Ind. Eng. Chem. Res.* **2020**, *59*, 16004–16016. [CrossRef]
12. Hoch, P. Evaluation of design flexibility in distillation columns using rigorous models. *Comput. Chem. Eng.* **1995**, *19*, S669–S674. [CrossRef]
13. Di Pretoro, A.; Montastruc, L.; Manenti, F.; Joulia, X. Flexibility analysis of a distillation column: Indexes comparison and economic assessment. *Comput. Chem. Eng.* **2019**, *124*, 93–108. [CrossRef]
14. Sudhoff, D.; Leimbrink, M.; Schleinitz, M.; Gorak, A.; Lutze, P. Modelling, design and flexibility analysis of rotating packed beds for distillation. *Chem. Eng. Res. Des.* **2015**, *94*, 72–89. [CrossRef]
15. Adams, T.A.; Thatho, T.; Le Feuvre, M.C.; Swartz, C.L.E. The Optimal Design of a Distillation System for the Flexible Polygeneration of Dimethyl Ether and Methanol under Uncertainty. *Front. Energy Res.* **2018**, *6*, 41. [CrossRef]
16. Floudas, C.A.; Gümüş, A.Z.H. Global Optimization in Design under Uncertainty: Feasibility Test and Flexibility Index Problems. *Ind. Eng. Chem. Res.* **2001**, *40*, 4267–4282. [CrossRef]
17. Huang, W.; Yang, G.; Qian, Y.; Jiao, Z.; Dang, H.; Qiu, Y.; Cheng, Z.; Fan, H. Operation Feasibility Analysis of Chemical Batch Reaction Systems with Production Quality Consideration under Uncertainty. *J. Chem. Eng. Jpn.* **2017**, *50*, 45–52. [CrossRef]
18. Balasubramanian, J.; Grossmann, I.E. A novel branch and bound algorithm for scheduling flow shop plants with uncertain processing times. *Comput. Chem. Eng.* **2002**, *26*, 41–57. [CrossRef]
19. Balasubramanian, J.; Grossmann, I.E. Scheduling optimization under uncertainty—An alternative approach. *Comput. Chem. Eng.* **2002**, *27*, 469–490. [CrossRef]
20. Bonfill, A.; Espuña, A.; Puigjaner, L. Addressing Robustness in Scheduling Batch Processes with Uncertain Operation Times. *Ind. Eng. Chem. Res.* **2005**, *44*, 1524–1534. [CrossRef]
21. Smaili, F.; Angadi, D.; Hatch, C.; Herbert, O.; Vassiliadis, V.; Wilson, D. Optimization of Scheduling of Cleaning in Heat Exchanger Networks Subject to Fouling. *Food Bioprod. Process.* **1999**, *77*, 159–164. [CrossRef]
22. Diaby, A.L.; Miklavcic, S.; Addai-Mensah, J. Optimization of scheduled cleaning of fouled heat exchanger network under ageing using genetic algorithm. *Chem. Eng. Res. Des.* **2016**, *113*, 223–240. [CrossRef]
23. Dias, L.S.; Ierapetritou, M.G. Integration of scheduling and control under uncertainties: Review and challenges. *Chem. Eng. Res. Des.* **2016**, *116*, 98–113. [CrossRef]
24. Ishiyama, E.; Paterson, W.R.; Wilson, D.I. Optimum cleaning cycles for heat transfer equipment undergoing fouling and ageing. *Chem. Eng. Sci.* **2011**, *66*, 604–612. [CrossRef]
25. Sinnott, R.K.; Towler, G. *Chemical Engineering Design*, 6th ed.; Butterworth-Heinemann: Waltham, MA, USA, 2019.
26. Wenner, R.R. *Thermochemical Calculations*, 1st ed.; McGraw-Hill Book Co.: New York, NY, USA; London, UK, 1941.
27. Green, D.; Perry, R.; Southard, M.Z. *Perry's Chemical Engineers' Handbook*, 9th ed.; McGraw-Hill Education: New York, NY, USA, 2019.
28. Hall, S.M. *Rules of Thumb for Chemical Engineers*, 4th ed.; Gulf Professional Publishing: Houston, TX, USA, 2011.
29. Hoch, P.; Eliceche, A.M. Flexibility analysis leads to a sizing strategy in distillation columns. *Comput. Chem. Eng.* **1996**, *20*, S139–S144. [CrossRef]
30. Swaney, R.E., Grossmann, I.E. An index for operational flexibility in chemical process design. Part I: Formulation and theory *AIChE J.* **1985**, *31*, 621–630. [CrossRef]
31. Pistikopoulos, E.; Mazzuchi, T. A novel flexibility analysis approach for processes with stochastic parameters. *Comput. Chem. Eng.* **1990**, *14*, 991–1000. [CrossRef]
32. Di Pretoro, A.; Montastruc, L.; Manenti, F.; Joulia, X. Flexibility assessment of a biorefinery distillation train: Optimal design under uncertain conditions. *Comput. Chem. Eng.* **2020**, *138*, 106831. [CrossRef]
33. Severini, T.A. *Elements of Distribution Theory*; Cambridge University Press (CUP): Cambridge, UK, 2005.

Article

Research on Optimization of Coal Slime Fluidized Bed Boiler Desulfurization Cooperative Operation

Yangjian Xiao [1], Yudong Xia [1,*], Aipeng Jiang [1], Xiaofang Lv [2], Yamei Lin [1] and Hanyu Zhang [1]

[1] School of Automation, Hangzhou Dianzi University, Hangzhou 310018, China; yjxiao@hdu.edu.cn (Y.X.); jiangaipeng@hdu.edu.cn (A.J.); linyamei@hdu.edu.cn (Y.L.); xiaoyu77@hdu.edu.cn (H.Z.)
[2] Zhejiang Supcon Technology Co., Ltd., Hangzhou 310053, China; lvxf@supcon.com
* Correspondence: ydxia@hdu.edu.cn

Abstract: The semi-dry desulfurization of slime fluidized bed boilers (FBB) has been widely used due to its advantages of low cost and high desulfurization efficiency. In this paper, the cooperative optimization of a two-stage desulfurization processes in the slime fluidized bed boiler was studied, and a model-based optimization strategy was proposed to minimize the operational cost of the desulfurization system. Firstly, a mathematical model for the FBB with a two-stage desulfurization process was established. The influences of coal slime elements on combustion flue gas and the factors that may affect the thermal efficiency of the boiler were then analyzed. Then, on the basis of the developed model, a number of parameters affecting the SO_2 concentration at the outlet of the slime fluidized bed boiler were simulated and deeply analyzed. In addition, the effects of the sulfur content of coal slime, excess air coefficient, and calcium to sulfur ratio were also discussed. Finally, according to the current SO_2 emission standard, the optimization operation problems under different sulfur contents were studied with the goal of minimizing the total desulfurization cost. The results showed that under the same sulfur content, the optimized operation was able to significantly reduce the total desulfurization cost by 9%, consequently improving the thermal efficiency of the boiler, ensuring the stable and up-to-standard emission of flue gas SO_2, and thus achieving sustainable development.

Keywords: slime fluidized bed; mechanism model; boiler thermal efficiency; simulation; cooperative optimization

Citation: Xiao, Y.; Xia, Y.; Jiang, A.; Lv, X.; Lin, Y.; Zhang, H. Research on Optimization of Coal Slime Fluidized Bed Boiler Desulfurization Cooperative Operation. *Processes* **2021**, *9*, 75. https://doi.org/10.3390/pr9010075

Received: 2 December 2020
Accepted: 28 December 2020
Published: 31 December 2020

Publisher's Note: MDPI stays neutral with regard to jurisdictional claims in published maps and institutional affiliations.

1. Introduction

China is one of the countries with the most serious air pollution. It is reported that more than 80% of the total coal consumption is used for direct combustion, and the total SO_2 emissions of thermal power plants account for 51% of the total SO_2 emissions of the country [1]. Coal slime is a by-product in the coal washing process. It is a viscous substance composed of fine coal, weathered stone, and water, which may eventually cause environmental pollution in the case of accumulation. It is the most effective way to burn it for electricity generation. However, its combustion operation produces a large amount of SO_2, resulting in environment pollution, and finally causing immeasurable losses to the country's social and economic development.

The flue gas desulfurization technology mainly uses absorbents to remove SO_2 in the flue gas and converts it into stable sulfur compounds or sulfur [2]. Generally, flue gas desulfurization methods can be divided into three categories: dry, semi-dry, and wet. Wet desulphurization technology is widely used in newly built large power plant boilers; this method has many advantages, such as low operating cost, abundant raw materials, and high stability. However, there are many disadvantages of wet desulphurization technology, such as serious blockage and wear, the byproduct occupying land problem, large water consumption, and air pollution [3]. For small and medium power plants, dry or semi-dry desulfurization is generally adopted. Compared with wet desulfurization, dry desulfurization is more advantageous in terms of its cost, as it does not require water and reheat energy [4,5]. However, it has not been widely used for its high cost of desulfurizer and low sulfur dioxide removal rate [6]. Semi-dry flue gas desulfurization avoids the shortcomings of wet and dry desulfurization and is widely used in desulfurization systems for its low operating cost and high desulfurization efficiency [7,8]. Therefore, it is of great significance to study the way in which to reduce the operating cost of the semi-dry desulfurization system with the slime fluidized bed boiler and to study the influence of various parameters on the performance of the semi-dry desulfurization.

In order to investigate the factors that affecting the desulfurization efficiency, a detailed mathematical model is required. Over the years, a number of researchers have been dedicated to developing a desulfurization model. For example, Zheng et al. [9] proposed a simplified desulfurization model in the fluidized bed boiler. Through the model, it was found that the activity of limestone and calcium/sulfur ratio were two main factors affecting desulfurization process, providing a promising strategy for desulfurization control. Neathery [10] established a mathematical model on the basis of a large number of desulfurization experiments and used the model to simulate the influence of operating parameters on the desulfurization efficiency. Due to the low high temperature desulfurization efficiency in the slime fluidized bed boiler furnace, and in order to further improve the desulfurization efficiency of the boiler, a combined desulfurization technology with low cost and high desulfurization efficiency is required. Tampella Power of Finland [11] initially proposed the "limestone injection into the furnace and activation of unreacted calcium" (LIFAC) technology. In the LIFAC, a humidification activation chamber was installed in the flue at the end, and the unreacted CaO in the furnace was activated by humidification water. This method could effectively improve the calcium utilization rate of the desulfurizer and save the cost of desulfurization. However, the desulfurization rate in the furnace and the utilization rate of calcium base are still not high enough, and thus its economic performance is limited. Therefore, the operation of desulfurization should be further optimized. Cai Yi [12] studied the operation strategy and parameter optimization of calcium spray in the furnace and limestone wet desulfurization system in the furnace, proposing a comprehensive optimization method based on environmental protection standards and technical and economic indicators. Through combination with field trials, the accuracy of the optimization method was verified. However, the proposed operation strategy was based on wet desulfurization process and thus was only suitable for large power plants. Du Zhao [13] established a series of simulation mathematical models by analyzing various factors that affect desulfurization under different working conditions, and on the basis of these models, developed a set of simulation software for fluidized bed boiler desulfurization, which provided a basis for the later collaborative desulfurization. In summary, while the desulfurization technology has been extensively studied, all these reported studies focus on the wet desulfurization applied in large power plants. Therefore, the small- and medium-sized power plants urgently need a combined semi-dry desulfurization model with high desulfurization efficiency and economic performance.

On the other hand, the two-stage desulfurization technology is a promising method to improve the desulfurization efficiency. The most widely adopted method is optimized each stage separately in order to obtain the best operational conditions. For example, desulphurization is achieved in the furnace to a certain extent, and the concentration of SO_2 at the furnace outlet is reduced to a reasonable range. Then, the flue gas semi-dry desulfurization system is used to remove more sulfur dioxide. However, due to the coupling and a certain timing relationship between the two processes, it is necessary to consider the two processes together to realize cooperative optimization to reduce cost. Therefore, the way in which to coordinate the relationship between the two processes and rationally allocate the share of the two-stage desulfurization requires consideration of various factors [14]. In this case, the SO_2 emission concentration can reach the standard operation, and thus the overall material consumption and energy consumption costs can be realized at their corresponding lowest levels. The collaborative optimization approach was initially proposed by Kroo et al. [15] and Balling and Sobieski [16] in 1994. Its idea is to coordinate the coupling relationship between the various systems and obtain the optimal solution, and it has been successfully employed in a variety of applications. In the field of denitrification in power plants, Yu Han et al. [17] believed that a reasonable increase in the temperature of Selective catalytic reduction (SCR) denitrification flue gas would bring about better de-NOx performance in the cold end of the coal-fired boiler, and proposed a collaborative optimization strategy for energy conversion and NOx removal in the cold end of the boiler. Results showed that great energy-saving, NOx removal, and economic performances could be achieved using the proposed collaborative optimization strategy. Temple et al. [18] used the collaborative optimization method to optimize the operation and maintenance cost of naval warships by comparing different aspects of the total cost of ownership of a ship. Chen et al. [19] considered the problem of electric vehicle capacity allocation and economic dispatch, and proposed the use of a collaborative optimization method to reduce the total cost of pure electric vehicle owners. In order to increase the high efficiency range of electric vehicles and improve the cruising range, Zhao et al. [20] first proposed a collaborative optimization control strategy. In the field of power grid, Tian et al. [21] proposed the collaborative optimization allocation of voltage-detection active power filters (VDAPFs) and static var generators (SVGs) for simultaneous mitigation of voltage harmonic and deviation in distribution networks by using the improved particle swarm optimization algorithm, wherein the total cost (including investment and operating cost) was minimized and optimized. Therefore, collaborative optimization method is a powerful technic to obtain the optimum operation of a complicated process. However, no studies on the collaborative optimization of the coal slime fluidized bed boiler desulfurization system can be identified.

Therefore, in this paper, cooperation optimization of the desulfurization system combined with dry and semi-dry desulfurization processes was studied for operational cost saving. In this study, the boiler's thermal efficiency loss and the influence of different coal slime components on the flue gas volume and the added limestone on the boiler efficiency were considered. On the basis of the calcium injection in the furnace, we coordinated the semi-dry desulfurization to obtain the best operating conditions and realize the economic operation of the entire desulfurization system. The organization of the paper is as follows. Firstly, the descriptions of the two-stage desulfurization process are presented. Then, the developed model for the desulfurization process is detailed. This is followed by presenting the optimization method and results. Finally, conclusions are given.

2. Cooperative Mechanism of Desulfurization

Almost all of the SO_2 generated in the slime fluidized bed boiler comes from the sulfur content in the coal. Therefore, changes in the amount of coal or the sulfur content in the coal directly affect the amount of SO_2 generated. When the amount of SO_2 produced changes, firstly, the transportation amount of limestone should be adjusted accordingly, and secondly, the amount of slaked lime in the semi-dry method should be adjusted according to the SO_2 value after desulfurization in the furnace. The desulfurization synergistic process diagram of the slime fluidized bed boiler is shown in Figure 1 below.

As shown in Figure 1, the slime fluidized bed boiler firstly adjusts its calcium/sulfur ratio according to the slime control signal, and then adjusts the limestone input to achieve the purpose of controlling the first stage of desulfurization. The incompletely absorbed SO_2 in the boiler reaches the absorption tower directly through the superheater, economizer, and air preheater, which reduces the flue gas temperature, recovers the heat in the flue gas, and improves the thermal efficiency of the boiler. The slaked lime enters the absorption tower after being atomized by the atomizing fan from the lower part of the absorption tower. In the absorption tower, the slaked lime atomized into fine droplets is mixed and contacted with the flue gas, and chemically reacts with the SO_2 in the flue gas to form $CaSO_3$, and the SO_2 in the flue gas is removed. The final SO_2 concentration at the exit of the chimney is controlled by adjusting the amount of slaked lime to make it reach the standard.

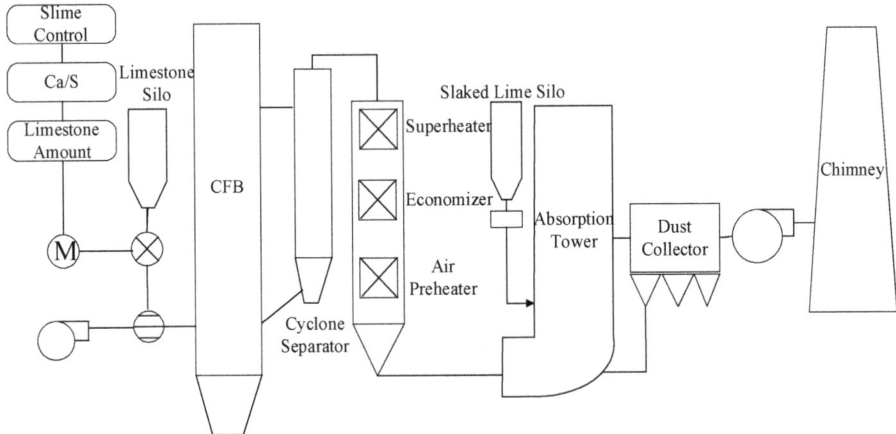

Figure 1. Two-stage desulfurization process diagram of a slime fluidized bed boiler.

3. Mathematical Model of Desulfurization

On the basis of the above process of coal slime fluidized bed boiler synergistic desulfurization operation, this paper mainly studied the calculation that is included in the composition of coal slime combustion flue gas, consumption of the material and the energy of two-stage desulfurization, and the thermal efficiency of the boiler by adding limestone.

3.1. The Flue Gas from Coal Slime Combustion

During the combustion of coal slime, some of the oxygen will be consumed [22], and thus it will additionally increase the requirement for air volume and also cause changes in flue gas volume. Through the analysis of the coal composition, the smoke composition model is established [23]; the amount of oxygen required is as follows:

$$n(O_2) = 0.8333\gamma(C_{ar}) + 2.5\gamma(H_{ar}) + 0.3125\gamma(S_{ar}) - 0.3125\gamma(O_{ar}) \tag{1}$$

The volume fraction of oxygen in the air is 21%, and thus the theoretical air volume required for the desulfurization reaction is expressed as

$$\begin{aligned} V'_{air} &= 4.78 \times n(O_2) \times 22.4/1000 \\ &= 0.089\gamma(C_{ar}) + 0.265\gamma(H_{ar}) + 0.0333\gamma(S_{ar}) - 0.0333\gamma(O_{ar}) \end{aligned} \tag{2}$$

The theoretical amount of flue gas generated after coal slime combustion is expressed as

$$\begin{aligned} V'_{sm} &= ((n(O_2) + 3.78 \times (n(O_2))) + \gamma(M_{ar})) \times 22.4/1000 \\ &= 0.089\gamma(C_{ar}) + 0.324\gamma(H_{ar}) + 0.0333\gamma(S_{ar}) - 0.026\gamma(O_{ar}) \\ &\quad + 0.0224\gamma(M_{ar}) \end{aligned} \tag{3}$$

The boiler combustion is carried out under the excess air coefficient α, and thus the actual flue gas volume is calculated as

$$\begin{aligned} V_{sm} &= V'_{sm} + V'_{air} \times (\alpha - 1) \\ &= 0.089\gamma(C_{ar}) + (0.056 + 0.2675\alpha)\gamma(H_{ar}) + 0.0333\gamma(S_{ar})\alpha \\ &\quad + (0.0074 - 0.0333\alpha)\gamma(O_{ar}) + 0.0224\gamma(M_{ar}) \end{aligned} \tag{4}$$

In summary, the theoretical SO_2 concentration can be calculated as

$$N_{SO_2} = \gamma(S_{ar})/100 \times 1000 \times 2/V_{sm} \times 1000 \tag{5}$$

where $n(O_2)$ is the amount of oxygen required, and the unit is mol. $\gamma(C_{ar})$, $\gamma(H_{ar})$, $\gamma(S_{ar})$, $\gamma(O_{ar})$, and $\gamma(M_{ar})$ are the contents of carbon, hydrogen, sulfur, oxygen, and moisture in the slime fuel, respectively. V'_{air} is the theoretical air volume required for the slime fuel, V'_{sm} is the theoretical amount of smoke, and V_{sm} is the actual amount of smoke, and their units are all m^3. N_{SO_2} is the theoretical SO_2 concentration, and the unit is mg/m^3. The theoretical SO_2 concentration corresponding to different coal slime element content, is shown in Table 1.

Table 1. Theoretical SO_2 concentration corresponding to different coal slime element content.

γ (C_{ar})/%	γ (H_{ar})/%	γ (O_{ar})/%	γ (N_{ar})/%	γ (S_{ar})/%	γ (A_{ar})/%	γ (M_{ar})/%	Theory SO_2 (mg/m^3)
34.51	2.31	7.29	0.61	0.5	29.49	25.29	1870.66
33.51	3.31	6.29	0.66	0.6	30.34	25.29	2107.98
33.68	2.38	6.68	0.72	0.7	31.34	24.50	2643.61
33.76	2.56	5.85	0.75	0.8	31.39	24.89	2949.14
33.82	2.67	6.35	0.76	0.9	31.12	24.38	3297.77

From Table 1, we can see that the amount of oxygen required for slime combustion is directly affected by the composition of the slime. In the case of a certain excess air coefficient, the theoretical SO_2 concentration generation also changes, and the SO_2 generation basically comes from the sulfur content in the slime.

In order to fully burn coal slime fuel, more air must be supplied. Under different excess air coefficients, the theoretically generated SO_2 concentration will also change. The change chart is shown below:

It can be seen from Figure 2 that as the excess air coefficient increases, the theoretical SO_2 concentration gradually decreases. However, the excess air coefficient is too large, and the boiler's air supply is too much, which will not only reduce the furnace temperature and worsen the combustion, but also increase the amount of flue gas, which will increase the heat loss of the boiler exhaust gas and reduce the thermal efficiency of the boiler. Therefore, in the case of ensuring complete combustion, the excess air coefficient should be minimized [24,25].

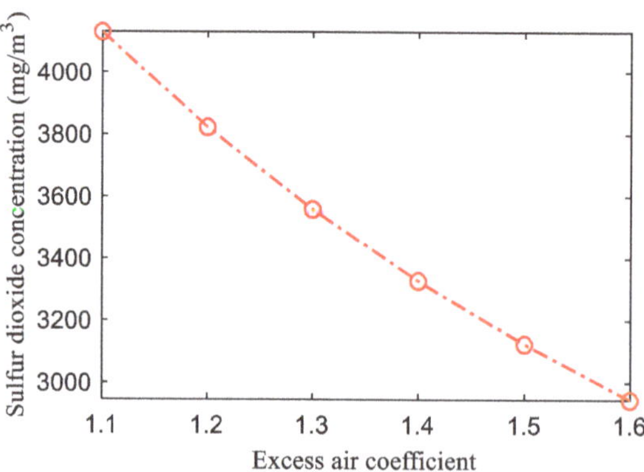

Figure 2. The influence of the theoretical SO_2 concentration of excess air coefficient.

3.2. The Heat Loss of Boiler

In order to understand the various losses of the boiler, researchers generally use the reverse balance method in the boiler performance calculations. In the slime fluidized bed boiler, its various losses include the following: mechanical incomplete combustion loss q_4, exhaust heat loss q_2, heat dissipation loss q_5, ash heat loss q_6, and chemical incomplete combustion loss q_3, with q_i being the effective utilization of heat or the percentage of heat loss to the input heat.

Mechanical incomplete combustion loss is caused by unburned or incompletely burned carbon in the fuel:

$$q_4 = \frac{32,866\gamma(A_{ar})}{Q_r}\left(\frac{\alpha_{fh}\varphi_{fh}}{100 - \varphi_{fh}} + \frac{\alpha_{hz}\varphi_{hz}}{100 - \varphi_{hz}}\right) \tag{6}$$

where $\gamma(A_{ar})$ is the total ash mass fraction in percent after desulfurization reaction of unit slime combustion. α_{fh} and α_{hz} are, respectively, the proportion of the amount of ash in fly ash and the proportion of the amount of ash in ash in the furnace. φ_{fh} and φ_{hz} are the percentage of the combustible content of fly ash and ash slag in the amount of fly ash and ash slag, respectively. Q_r is the heat brought into the furnace per kilogram of fuel, and its unit is kJ.

The heat loss due to exhaust gas is caused by the fact that the exhaust gas temperature of the boiler is higher than the ambient temperature, which is equal to the difference between the exhaust gas enthalpy and the furnace air enthalpy. The exhaust gas enthalpy includes the theoretical flue gas enthalpy, excess air enthalpy, and fly ash enthalpy. The heat loss due to exhaust gas is

$$q_2 = \frac{H_{y0} + (\alpha - 1)H_{k0} + H_{fh} - \alpha H_{kl}}{Q_r}(100 - q_4) \tag{7}$$

In the formula, H_{y0}, H_{k0}, H_{fh}, and H_{kl} are the flue gas enthalpy, exhaust air enthalpy, fly ash enthalpy, and cold air enthalpy, respectively, and their units are kJ.

Among them, the flue gas enthalpy value is

$$H_{y0} = (V_{RO_2}C_{RO_2} + V_{N_2}C_{N_2} + V_{H_2O}C_{H_2O})t_y \tag{8}$$

In the formula, V_{RO_2}, V_{N_2}, and V_{H_2O} are the volume of triatomic gas, nitrogen, and water vapor in theoretical flue gas, respectively, in m^3; C_{RO_2}, C_{N_2}, and C_{H_2O} are the volumetric heat capacity of triatomic gas, the volumetric heat capacity of nitrogen, and the volumetric heat capacity of water vapor under standard conditions, respectively, in $kJ/(m^3 \cdot K)$; and t_y is the exhaust gas temperature in K.

The air enthalpy value in the exhaust smoke is

$$H_{k0} = C_{k0}(V_{gk}^0)^c t_y \tag{9}$$

In the formula, C_{k0} is the volumetric heat capacity of air in the flue gas under standard conditions in $kJ/(m^3 \cdot K)$, and $(V_{gk}^0)^c$ is the theoretical air volume in m^3. The fly ash enthalpy is calculated as

$$H_{fh} = \frac{\gamma(A_{ar})\varphi_{fh}}{100 \times 100}C_{fh}t_y \tag{10}$$

Among them, C_{fh} is the volumetric heat capacity of fly ash under standard conditions in $kJ/(m^3 \cdot K)$, and the enthalpy of cold air is

$$H_{kl} = C_{kl}(V_{gk}^0)^c t_{kl} \tag{11}$$

In the formula, C_{kl} is the volumetric heat capacity of cold air under standard conditions in $kJ/(m^3 \cdot K)$, and t_{kl} is the temperature of the cold air in K.

The heat loss of the boiler is proportional to the load of the boiler:

$$q_5 = k_{q5}\frac{D_e}{D_{re}} \tag{12}$$

In the formula, k_{q5} is the coefficient of heat dissipation loss, and D_e and D_{re} are the rated evaporation capacity of boiler and actual evaporation capacity of the boiler, respectively, in t/h.

The heat loss of ash and slag is the loss caused by the discharge of high-temperature ash from the furnace:

$$q_6 = \frac{\gamma(A_{ar})\alpha_{hz}C_{hz}}{Q_r}t_{hz} \tag{13}$$

In the formula, C_{hz} is the volumetric heat capacity of the slag in the standard conditions in $kJ/(m^3 \cdot K)$. t_{hz} is the ash temperature, and the unit is K.

Because q_3 is difficult to measure and calculate, it is generally necessary to consider the concentration of incomplete combustion such as CO. For slime fluidized bed boilers, the loss is very small. According to experience, its value generally does not exceed 0.5%.

In summary, the thermal efficiency of the boiler can be seen as

$$\eta_{gl} = 100 - q_2 - q_3 - q_4 - q_5 - q_6 \tag{14}$$

3.3. Model of Cooperative Desulfurization

In-furnace calcium injection and semi-dry method for synergistic desulfurization is used in the entire coal slime fluidized bed boiler desulfurization system. In-furnace calcium injection is widely used in industrial processes due to its low investment and operating costs. Due to the general in-furnace calcium injection desulfurization's efficiency being between 50% and 60%, as the calcium/sulfur ratio increases, its efficiency is significantly reduced, and it is affected by the cost of the desulfurizer and the amount of NOx generated [26], and thus it is necessary to add the semi-dry flue gas method for synergistic desulfurization on the original basis so that it meets the national emission standards.

3.3.1. Model of Desulfurization in Furnace

After limestone is added to the coal slime fluidized bed boiler, it will be decomposed into calcium oxide and carbon dioxide. Furthermore, calcium oxide reacts with sulfur dioxide released by combustion in the furnace to form calcium sulfate, which is discharged together with the slag [27]. The chemical reaction heat is shown in Figure 3 below:

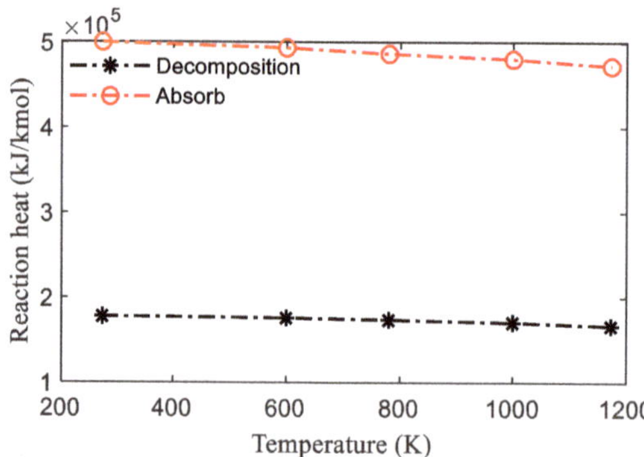

Figure 3. Chemical reaction heat of calcium injection desulfurization in the furnace.

The chemical equation of the reaction of limestone in the furnace is as follows [28], which can be divided into the following two processes:

$$CaCO_3 \rightarrow CaO + CO_2 - 177,900 \ kJ \cdot kmol^{-1} \tag{15}$$

$$CaO + SO_2 + 1/2O_2 \rightarrow CaSO_4 + 500,600 \ kJ \cdot kmol^{-1} \tag{16}$$

Since many parameters are non-linear in the production process, mechanism modeling and data analysis modeling can be carried out on the desulfurization scheme. Considering that the slime fluidized bed boiler desulfurization system is a nonlinear fast time-varying system, some parameters are set as follows.

Suppose that in the desulphurization system of a slime fluidized bed boiler, the amount of slime fed is $M(t/h)$, the mass fraction of sulfur in coal slime fuel is $\gamma(S_{ar})(\%)$, the purity of calcium carbonate contained in the desulfurizer is $\lambda(\%)$, and the transport volume of limestone powder of the desulfurizer is $T_s(t/h)$, then, the expression of calcium/sulfur ratio can be derived as

$$R = \frac{n_{Ca}}{n_S} = \frac{\frac{\lambda T_s}{100}}{\frac{M\gamma(S_{ar})}{32}} = \frac{8\lambda T_s}{25M\gamma(S_{ar})} \tag{17}$$

For the desulfurization process in the slime fluidized bed furnace, natural limestone is used as the desulfurizer, and its main chemical composition is $CaCO_3$. Its consumption is mainly related to the content of SO_2 gas generated during combustion, and thus it can be derived as

$$T_s = M \times \frac{\gamma(S_{ar})}{100} \times \frac{100}{32} \times R \times \frac{100}{\lambda} = \frac{25}{8\lambda}MR\gamma(S_{ar}) \tag{18}$$

It can be seen from Figure 4 that with the increase of calcium/sulfur ratio, the desulfurization efficiency of flue gas shows an upward trend. When R reaches a certain value (about 3), it has little effect on the desulfurization efficiency [29], and then we need to be considered the influence of other factors on the desulfurization efficiency. The calcium-sulfur ratio reflects the utilization rate of the desulfurizer, and to a certain extent, it can also reflect the working efficiency and operating cost of the device.

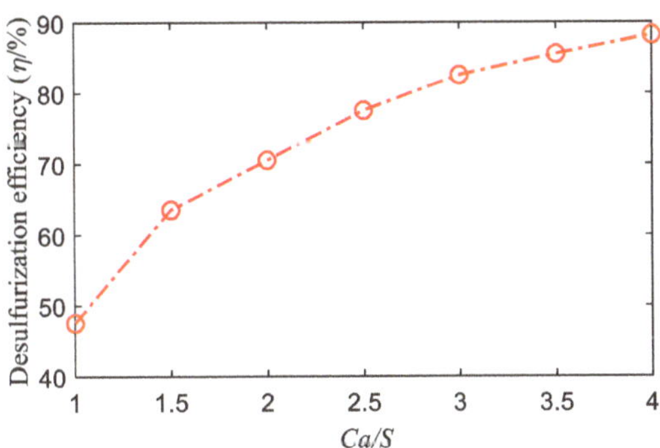

Figure 4. The effect of calcium to sulfur ratio on desulfurization efficiency.

The relationship between calcium/sulfur ratio and limestone desulfurization efficiency can be approximately expressed as [30]

$$\eta'_{SO_2} = 1 - \exp(-mR) \tag{19}$$

In the above formula, m is an influencing parameter, and $m > 0$. In this study, m was taken as 0.6 based on actual data.

3.3.2. Model of Semi-Dry Desulfurization

As SO_2 cannot be eliminated ideally in the actual flue gas, it is necessary to add an appropriate amount of slaked lime to achieve the effect of the second stage of desulfurization. The unreacted SO_2 in the boiler flows into the absorption tower along with the flue gas. The semi-dry desulfurization reaction is completed in the absorption tower. The chemical equation is as follows:

$$SO_2 + Ca(OH)_2 \rightarrow CaSO_3 + H_2O \tag{20}$$

$$CaSO_3 + 1/2H_2O \rightarrow CaSO_3 \cdot 1/2H_2O \tag{21}$$

$$CaSO_3 + 2H_2O + 1/2O_2 \rightarrow CaSO_4 \cdot 2H_2O \tag{22}$$

The main products produced by the reaction in the desulfurization tower are calcium sulfite hemihydrate, calcium sulfate dihydrate, and unreacted slaked lime.

By fitting the relevant experimental data, we can obtain the relationship between the SO_2 concentration and the consumption of slaked lime in the semi-dry flue gas desulfurization system as follows:

$$N'_{SO_2} - N''_{SO_2} = -2400T_x^2 + 2200T_x + 240 \tag{23}$$

where N'_{SO_2} and N''_{SO_2} are the SO_2 concentration at the outlet of the slime fluidized bed boiler and the semi-dry flue outlet, respectively, in mg/m^3, and T_x is the consumption of slaked lime in t/h. Finally, the SO_2 concentration at the exit of the semi-dry flue can be obtained as

$$N''_{SO_2} = N_{SO_2}e^{-0.6R} + 2400T_x^2 - 2200T_x - 240 \tag{24}$$

From Equation (24) above, it can be seen that in order to make the SO_2 concentration at the final semi-dry flue outlet reach the standard and minimum the material and energy consumption, it is necessary to coordinate the relationship between the calcium/sulfur ratio and the consumption of slaked lime.

3.4. Loss of Boiler Heat Efficiency by Adding Limestone

Although adding limestone to the slime fluidized bed boiler effectively reduces SO_2 emissions, it will affect the calculation of system combustion products, ash balance, and boiler thermal efficiency [31,32]. In the coal slurry fluidized bed boiler, the combustion temperature is generally about 850–950 °C. When limestone is thermally decomposed at high temperature, a part of the heat needs to be absorbed from the furnace. The heat absorbed during thermal decomposition of limestone is not equal to that released during the absorption of sulfur dioxide. In other words, it will also affect the heat balance calculation of the boiler.

3.4.1. The Heat Loss of Chemical Reaction

According to the chemical reaction process of desulfurization, the amount of calcium carbonate in the desulfurizer required for every kilogram of coal slime (kmol)

$$n_{CaCO_3} = \frac{R\gamma(S_{ar})}{100 \times 32} \tag{25}$$

The heat $Q_x(kJ)$ required by calcium carbonate in the calcination process can be calculated by the following formula:

$$Q_x = \frac{n_{CaCO_3}d_rQ_1}{100} = \frac{Rd_rQ_1\gamma(S_{ar})}{3.2 \times 10^5} \tag{26}$$

In the formula, Q_1 is the heat absorbed by calcium carbonate during calcination, the unit is kJ, and $d_r(\%)$ is the decomposition rate of limestone. The amount of sulfur dioxide that reacts with the desulfurizer (kmol) is

$$n(SO_2) = \frac{\gamma(S_{ar})}{100 \times 32} \times \eta'_{SO_2} \tag{27}$$

In the process of calcium oxide reacting with sulfur dioxide, the heat released is

$$Q_f = n(SO_2) \times Q_2 = \frac{Q_2\eta'_{SO_2}\gamma(S_{ar})}{3.2 \times 10^3} \tag{28}$$

In the formula, Q_f is the heat released by the reaction between calcium oxide and sulfur dioxide, in kJ, and Q_2 is the heat released by the reaction between calcium oxide and sulfur dioxide, in kJ, and thus the total heat absorption is

$$
\begin{aligned}
Q_h &= Q_x - Q_f \\
&= \frac{Rd_r Q_1 \gamma(S_{ar})}{3.2 \times 10^5} - \frac{Q_2 \eta'_{SO_2} \gamma(S_{ar})}{3.2 \times 10^3}
\end{aligned}
\tag{29}
$$

It can be seen from the above formula that in the desulfurization process, only when $Q_x = Q_f$ will the heat balance calculation of the boiler not be affected by the heat absorbed by limestone decomposition and the heat released by sulfur dioxide.

Therefore, the influence of the heat loss of the boiler chemical reaction on the boiler thermal efficiency is

$$
q_h = \frac{Q_x - Q_f}{Q_y} = \frac{Rd_r Q_1 \gamma(S_{ar})}{3.2 \times 10^5 \times Q_y} - \frac{Q_2 \eta'_{SO_2} \gamma(S_{ar})}{3.2 \times 10^3 \times Q_y}
\tag{30}
$$

where Q_y is the heat per kilogram of coal slime fuel substituted into the furnace, and the unit is kJ.

3.4.2. The Heat Loss of Physical

For the unreacted calcium oxide, calcium sulfate produced by the reaction and various impurities in limestone during the desulfurization process are removed from the furnace with ash, or it escapes with the flue gas in the form of fly ash. If the mass of ash produced when adding limestone is m_h (t/h), $\beta\%$ of the ash leaves the furnace in the form of slagging, and $(100 - \beta)\%$ of the ash leaves the furnace in the form of fly ash. The specific heat of the ash is c_h (J/(kg·K)), the exhaust gas temperature is t_y (K), the slag discharge temperature is t_z (K), and the ambient temperature is t_0 (K). Therefore, the heat loss caused by adding desulfurizer limestone is

$$
q_w = \frac{m_h c_h}{Q_y} \left[\frac{100 - \beta}{100}(t_y - t_0) + \frac{\beta}{100}(t_z - t_0) \right]
\tag{31}
$$

The relationship between the total ash ratio of slag discharge and the physical heat loss is shown in Figure 5 below. It can be seen that as the total ratio of slag discharge increases, the physical heat loss also increases, and thus the amount of solid slag discharge must be minimized.

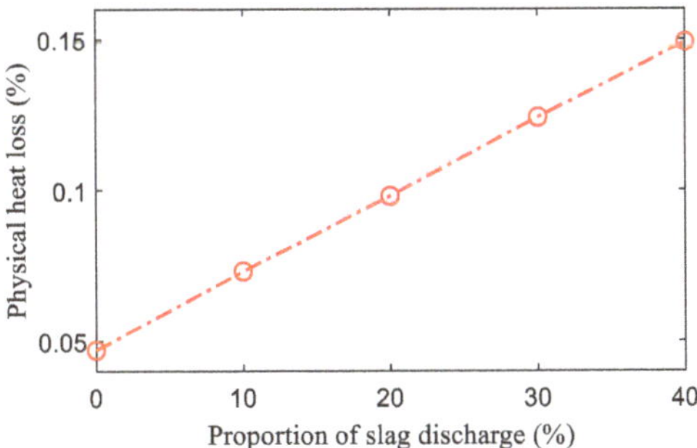

Figure 5. The relationship between proportion of slag discharge and the physical heat loss.

3.4.3. The Heat Loss of Exhaust

During the calcination process of limestone, the moisture will increase, and the desulfurization reaction needs to increase the air and at the same time cause more flue gas to be generated in the device, as well as have a certain amount of influence on the flue gas, which will increase the heat loss of the flue gas [33,34]. The volume of a certain amount of carbon dioxide produced during the calcination of limestone is

$$V_{CO_2} = \frac{R\gamma(S_{ar})}{32 \times 100} \times \frac{d_r}{100} \times 22.4 = 7 \times 10^{-5} R d_r \gamma(S_{ar}) \tag{32}$$

Among them, the volume of sulfur dioxide absorbed and oxygen consumed by calcium oxide in the absorption process are, respectively,

$$V_{SO_2} = \frac{\eta'_{SO_2}\gamma(S_{ar})}{32 \times 100} \times 22.4 = 7 \times 10^{-3} \eta'_{SO_2}\gamma(S_{ar}) \tag{33}$$

$$V_{O_2} = \frac{\eta'_{SO_2}\gamma(S_{ar})}{32 \times 100} \times 0.5 \times 22.4 = 3.5 \times 10^{-3} \eta'_{SO_2}\gamma(S_{ar}) \tag{34}$$

The moisture produced during the calcination of limestone is

$$\begin{aligned} V_{H_2O} &= \frac{R\gamma(S_{ar})}{32\times100} \times 100 \times \frac{100}{\delta_{Ca}} \times \frac{\gamma_{H_2O}}{100} \times \frac{22.4}{18} \\ &= 0.0389 R\gamma(S_{ar})\frac{\gamma_{H_2O}}{\delta_{Ca}} \end{aligned} \tag{35}$$

where $\gamma_{H_2O}(\%)$ is the amount of water contained in limestone, and $\delta_{Ca}(\%)$ is the purity of calcium carbonate contained in the desulfurizer.

The latent heat loss of evaporation Q_{H_2O} taken away by moisture is

$$\begin{aligned} Q_{H_2O} &= \frac{R\gamma(S_{ar})}{32\times100} \times 100 \times \frac{100}{\delta_{Ca}} \times \frac{\gamma_{H_2O}}{100} \times 2257 \\ &= 70.53 R\gamma(S_{ar})\frac{\gamma_{H_2O}}{\delta_{Ca}} \end{aligned} \tag{36}$$

Assuming that the respective volumetric heat capacities of the flue gas in the standard state are C_{CO_2}, C_{SO_2}, C_{O_2}, and C_{H_2O}, and their units are $kJ/(m^3 \cdot K)$, the heat loss from flue gas thus is

$$q_y = \frac{(t_y - t_0)(V_{CO_2}C_{CO_2} + V_{H_2O}C_{H_2O} - V_{SO_2}C_{SO_2} - V_{O_2}C_{O_2}) - Q_{H_2O}}{Q_y} \tag{37}$$

As the calcium/sulfur ratio increases, adding limestone to the boiler will reduce the thermal efficiency of the boiler. When the sulfur-based $\gamma(S_{ar})$ content of coal slime is 0.9, the heat loss of each part is as shown in the following Figure 6.

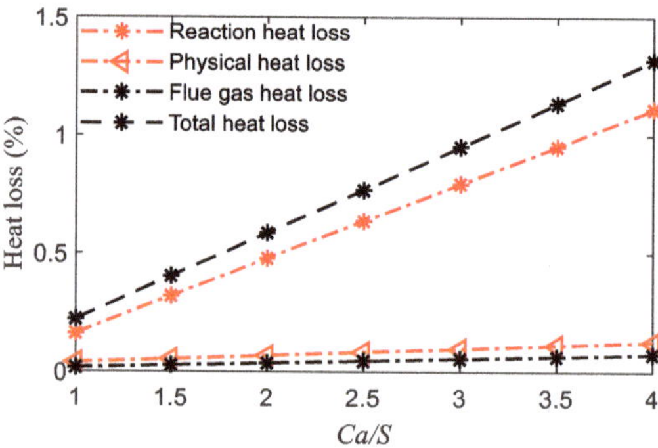

Figure 6. The relationship between the heat loss of each part and Ca/S.

It can be seen from Figure 6 that among the losses of each part, the thermal efficiency loss caused by the chemical reaction is more obvious. Effectively reducing the loss caused by the chemical reaction is beneficial to improving the boiler efficiency.

4. Cooperative Optimization of the Two-Stage Desulfurization System

4.1. Research on Optimization Based on The Lowest Total Cost

The optimization proposition of the total desulfurization cost of the slime fluidized bed boiler [35] is

$$\text{Min}\sum F(x) + P_s M_s + P_x M_x \tag{38}$$

$$\text{Total desulfurization efficiency } (\%) : \eta_{SO_2} \geq 95 \tag{39}$$

$$SO_2 \text{ concentration } \left(\text{mg/Nm}^3\right) : N_{SO_2} \leq 35 \tag{40}$$

$$\text{Calcium/sulfur ratio} : 1 \leq R \leq 5 \tag{41}$$

$$\text{Bed temperature } (^\circ C) : 850 \leq t_B \leq 1050 \tag{42}$$

$$\text{Slaked lime consumption } (t/h) : 0 < T_x < 1 \tag{43}$$

P_s and P_x are the price of limestone and slaked lime in CNY/t, respectively. M_s and M_x are the mass of limestone and slaked lime consumption, respectively, with the unit t. $\sum F(x)$ includes the power consumption of raw material machines, the power consumption of the air compressor, and the power loss of each part [36]. According to power plant operation guidance data and related materials, it is approximated as a functional relationship, and the unit is CNY. Checking the relevant experimental data of the power plant, this part of the cost accounts for about 10% of the total desulfurization cost. Taking into account the national emission requirements, the total desulfurization efficiency is greater than 95%, and the SO_2 concentration should be less than 35 mg/Nm^3. In the actual operation process, the calcium/sulfur ratio is generally between 1 and 5. Considering economic issues, the consumption of slaked lime will be controlled between 0 and 1, and the unit is t/h. For slime fluidized bed boilers, the bed temperature is generally between 850 and 1050, and the unit is °C. According to the above model, the optimization proposition of the desulfurization system composition was optimized to solve [37]. Considering that the prices of limestone and slaked lime will change with the market, further analysis of them shows that the relationship between their prices and total costs is shown in Figures 7 and 8 below:

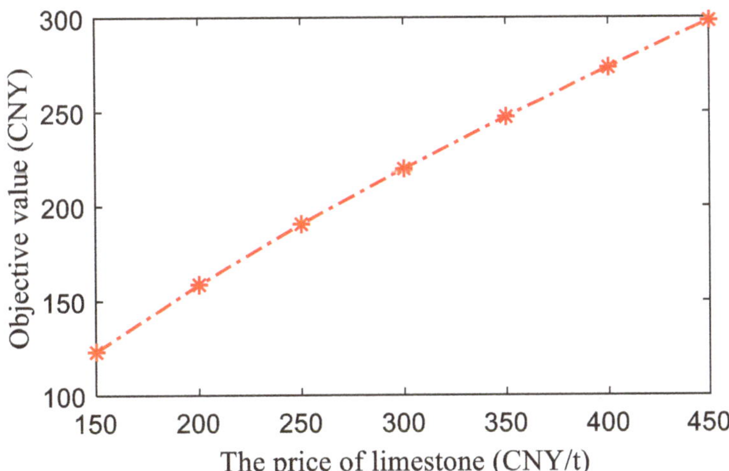

Figure 7. The relationship between limestone price and total cost.

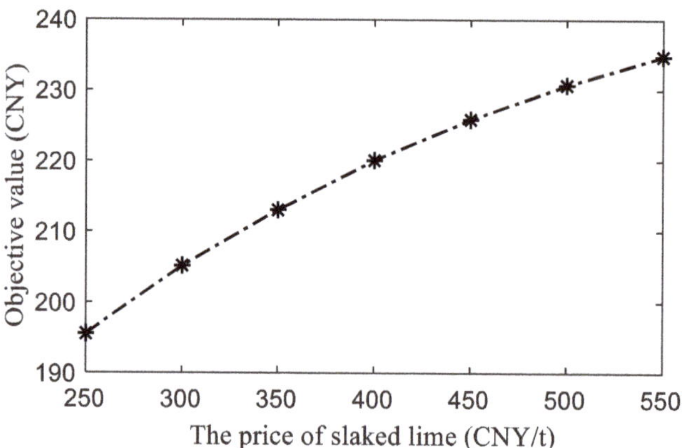

Figure 8. The relationship between slaked lime price and total cost.

According to the industrial market, the purity of limestone in the furnace is 90%, and the price is 300 CNY/t; the purity of slaked lime in the desulfurization tower is 95%, and its price is 400 CNY/t. Considering that there are not many variables involved in this article, and regarding strong nonlinear optimization, sequence quadratic program (SQP) algorithm has better advantages than other optimization algorithms in solving nonlinear optimization problems, and is mainly used to optimize the hybrid optimization model under the MATLAB platform. We analyzed its best operating conditions under various constraints. Under certain conditions, when the influence of boiler thermal efficiency on the desulfurization efficiency of the desulfurization system was not taken into consideration, the entire desulfurization system was optimized for different sulfur content of coal slime to obtain the best calcium/sulfur molar ratio, with the goals before and after optimization the value comparison being shown in Figure 9 below.

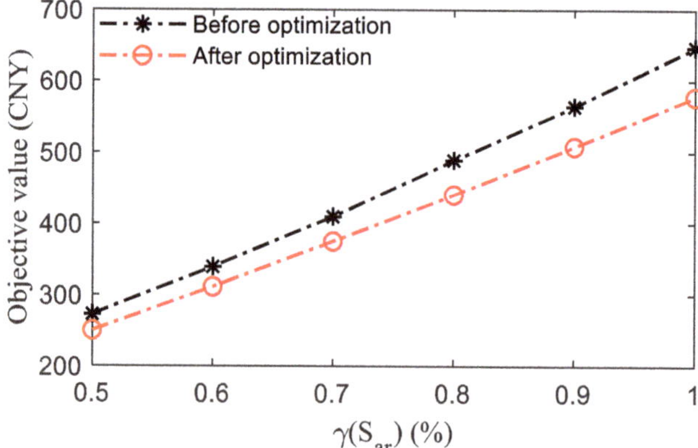

Figure 9. Comparison of objective value before and after optimization.

It can be seen from Figure 9 that the object value after optimization was reduced by 9% compared with before optimization. For different sulfur content, the best calcium/sulfur molar ratio was obtained, and the calculation results are shown in Table 2.

Table 2. The best operation for different sulfur bases.

$\gamma\,(S_{ar})/\%$	Objective Value after Optimization/CNY	Objective Value before Optimization/CNY	Limestone Consumption/t	Hydrated Lime Consumption/t	Optimal Ca/S	Excess Air Coefficient α
0.5	250.08	272.56	0.566	0.119	2.17	1.45
0.6	311.18	339.28	0.728	0.150	2.32	1.43
0.7	374.61	409.52	0.902	0.176	2.47	1.46
0.8	440.23	488.84	1.089	0.199	2.61	1.52
0.9	507.96	564.12	1.287	0.218	2.74	1.46
1.0	577.66	647	1.495	0.236	2.87	1.53

It can be seen from Table 2, considering that the excess air coefficient α was too large, the heat loss of the boiler increased, and thus the edge processing was performed on it. With the continuous increase of the sulfur-based content in the coal slime, the calcium/sulfur ratio and the total cost of desulfurization were also continuously increasing, and the optimal calcium/sulfur ratio was obtained through optimization calculation. Comparing the values before and after optimization, we found that the coordinated two-stage desulfurization could significantly reduce the desulfurization cost. When the sulfur-based content in the slime reached 1%, the calcium/sulfur ratio reached 2.87, and the cost could be reduced by 69.34 CNY. Considering that the addition of limestone had a certain impact on the thermal efficiency loss of the boiler, the calcium/sulfur ratio was thus not as high as possible.

4.2. Research on Optimization of Boiler Thermal Efficiency Loss by Adding Limestone

If only limestone is added for desulfurization, although the desulfurization effect can be achieved, it greatly reduces the thermal efficiency of the boiler and increases the total cost. Considering that the addition of limestone has an impact on the thermal efficiency loss of the boiler, then the effect of limestone thermal efficiency loss is added to the objective function and optimized to solve it. The mathematical model of the optimization proposition is as follows:

$$\text{Min} \sum F(x) + G(x) + P_s M_s + P_x M_x \tag{44}$$

$$\text{Total desulfurization efficiency (\%) : } \eta_{SO_2} \geq 95 \tag{45}$$

$$\text{SO}_2 \text{ concentration } \left(\text{mg/Nm}^3\right) : N_{SO_2} \leq 35 \tag{46}$$

$$\text{Calcium/sulfur ratio : } 1 \leq R \leq 5 \tag{47}$$

$$\text{Bed temperature } (^\circ\text{C}) : 850 \leq t_B \leq 1050 \tag{48}$$

$$\text{Slaked lime consumption (t/h) : } 0 < T_x < 1 \tag{49}$$

Among them, $G(x)$ is the cost of heat loss caused by adding limestone, and the unit is CNY. The Ca/S comparison before and after optimization is shown in Figure 10 below:

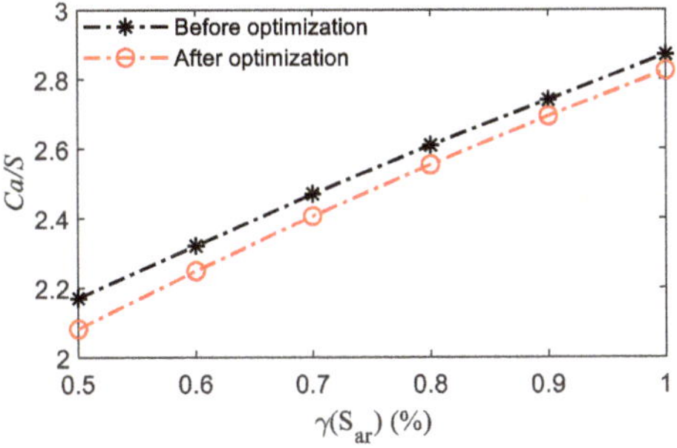

Figure 10. Ca/S comparison before and after optimization.

It can be seen from Table 3 that with the increase of the sulfur content in coal slime, the loss of boiler thermal efficiency also increased, which indirectly affected the total cost of the entire desulfurization system. The total cost in Table 3 includes the cost of limestone and slaked lime for desulfurization and the additional cost of slime caused by boiler heat loss. In comparison with Table 2, it can be seen that with the same sulfur content of slime, when adding the influence of boiler thermal efficiency loss on the total desulfurization cost, the calcium/sulfur ratio was also reduced accordingly. When the sulfur content was 0.8%, the calcium/sulfur ratio dropped from 2.61 to 2.55. Obviously, it was not a case of the higher the calcium/sulfur ratio, the better, which is consistent with the actual situation. It should be pointed out that the feed conditions including the composition, temperature, and pressure would impact the performance of desulfurization. In the current study, for simplifying this optimization problem, only the bed boiler conditions that would directly influence the desulfurization process were considered. The feed conditions will be further studied in future work.

Table 3. The effect of adding limestone on the heat loss of the boiler of the desulfurization system.

$\gamma\ (S_{ar})/\%$	Objective Value/CNY	Limestone Consumption/t	Hydrated Lime Consumption/t	Heat Loss by Adding Limestone/%	Bed Temperature/$^\circ$C	Optimal Ca/S	Excess Air Coefficient α
0.5	265.89	0.542	0.138	0.342	874.5	2.082	1.43
0.6	331.91	0.703	0.168	0.451	876.2	2.248	1.51
0.7	400.74	0.877	0.195	0.571	878.5	2.406	1.43
0.8	472.21	1.064	0.217	0.701	882.6	2.554	1.52
0.9	546.16	1.262	0.237	0.839	889.3	2.693	1.51
1.0	622.44	1.471	0.253	0.986	894.2	2.824	1.46

5. Conclusions

As the country's requirements for flue gas pollutant emission increase, the optimized operation of coal slurry fluidized bed boiler desulfurization is urgent. It is of great significance to achieve the best thermal efficiency and desulfurization conditions of fluidized bed boilers by improving the operating conditions of the boiler. This paper firstly analyzed the elements of coal slime to obtain the theoretical amount of flue gas produced, and then calculated the theoretical sulfur dioxide concentration. By studying the process of calcium injection desulfurization and semi-dry desulfurization in the furnace, we derived the desulfurization system model. By coordinating the relationship between the two stages of desulfurization, the best desulfurization situation was obtained. Finally, the influence of limestone addition on the thermal efficiency of desulfurization was considered, and the optimal objective function of the entire desulfurization system and its corresponding constraints were determined. By solving the optimization proposition, the best operating condition was obtained. This study aimed to reach the emission standard of SO_2 concentration in China through cooperative operations of dry and semi-dry desulfurization processes, and at the same time reducing the total operational cost as much as possible. The results show that the optimized operation can significantly reduce the total cost of desulfurization by 9%, improve the thermal efficiency of the boiler, reduce the calcium-sulfur ratio, and ensure that the SO_2 concentration reaches the national emission standard. This research provides an important basis for improving the desulfurization efficiency, reducing the desulfurization cost of the entire slime fluidized bed boiler, and guiding the desulfurization process.

Author Contributions: Y.X. (Yangjian Xiao) and Y.X. (Yudong Xia) performed the simulations and analyzed the data, A.J. designed the process scheme and optimization of the paper, Y.L. and H.Z. wrote the paper and reviewed it, X.L. checked the results of the whole manuscript. All authors have read and agreed to the published version of the manuscript.

Funding: The work was supported by the Natural Science Foundation of Zhejiang (no. LY20F030010), the National Natural Science Foundation of China (no. 61973102), and the National Science and Technology Major Project (2018AAA0101601).

Institutional Review Board Statement: Not applicable.

Informed Consent Statement: Informed consent was obtained from all subjects involved in the study.

Data Availability Statement: Data available on request due to restrictions eg privacy or ethical. The data presented in this study are available on request from the corresponding author. The data are not publicly available due to our research group is still in the process of further research in this area.

Conflicts of Interest: The authors declare no conflict of interest.

References

1. National Bureau of Statistics of the People's Republic of China. *China Statistical Yearbook*; China Statistics Press: Beijing, China, 2019.
2. Long, N.V.D.; Lee, D.Y.; Jin, K.M.; Choongyong, K.; Mok, L.Y.; Won, L.S.; Lee, M. Advanced and intensified seawater flue gas desulfurization processes: Recent developments and improvements. *Energies* **2020**, *13*, 5917. [CrossRef]
3. Yan, M.-P.; Shi, Y.-T. Thermal and economic analysis of multi-effect concentration system by utilizing waste heat of flue gas for magnesium desulfurization wastewater. *Energies* **2020**, *13*, 5384. [CrossRef]
4. Liu, Y.; Bisson, T.M.; Yang, H.; Xu, Z. Recent developments in novel sorbents for flue gas clean up. *Fuel Process. Technol.* **2010**, *91*, 1175–1197. [CrossRef]
5. Slack, A. Removing sulfur dioxide from stack gases. *Environ. Sci. Technol.* **1973**, *7*, 110–119. [CrossRef]
6. Xu, H.; Yang, D.; Hu, G.-G.; Chen, H.-P. Analysis of the Heat Loss due to Exhaust Gas of CFB Boiler with the Limestone Desulfurization. In Proceedings of the International Conference on Energy and Environment Technology, Guilin, China, 16–18 October 2009; Volume 1, p. 4145.
7. Xie, D.; Wang, H.-M.; Chang, D.-W.; You, C.-E. Semidry desulfurization process with in situ supported sorbent preparation. *Energy Fuels* **2017**, *31*, 4211–4218. [CrossRef]
8. Zhang, X.-P.; Wang, N.-H. Effect of humidification water on semi-dry flue gas desulfurization. *Energy Procedia* **2012**, *14*, 1659–1664. [CrossRef]
9. Zheng, J.; Yates, J.G.; Rowe, P.N. A model for desulphurization with limestone in a fluidized coal combustor. *Chem. Eng. Sci.* **1982**, *37*, 167–174. [CrossRef]
10. Neathery, J.K. Model for flue-gas desulfurization in a circulating dry scrubber. *AICHE J.* **1996**, *42*, 259–268. [CrossRef]
11. Anthony, E.J.; Berry, E.E.; Blondin, J.; Bulewicz, E.M.; Burwell, S. LIFAC ash-Strategies for management. *Waste Manag.* **2005**, *25*, 265–279. [CrossRef]
12. Cai, Y. Desulfurization Atmosphere Effect in Circulating Bed Furnace and Optimization of Combined Desulfurization Operation. Ph.D. Thesis, Zhejiang University, Zhejiang, China, 2016.
13. Du, Z. Simulation Study of Circulating Fluidized Bed Boiler Desulfurization. Master's Thesis, North China Electric Power University, Beijing, China, 2004.
14. Yu, F.-X. Research on Limestone Desulfurization Technology of Large Circulating Fluidized Bed. Master's Thesis, Kunming University of Science and Technology, Kunming, China, 2014.
15. Balling, R.J.; Sobieszczanski-Sobieski, J. Optimization of coupled system: A critical overview of approaches. *AIAA J.* **1996**, *34*, 6–17. [CrossRef]
16. Balling, R.J.; Sobieszczanski-Sobieski, J. An algorithm for solving the system-level problem in multilevel optimization. In Proceedings of the 5th Symposium on Multidisciplinary Analysis and Optimization, Panama City Beach, FL, USA, 7–9 September 1994; pp. 794–809.
17. Han, Y.; Sun, Y.-Y. Collaborative optimization of energy conversion and NOx removal in boiler cold-end of coal-fired power plants based on waste heat recovery of flue gas and sensible heat utilization of extraction steam. *Energy* **2020**, *207*, 118172. [CrossRef]
18. Temple, D.; Colette, M. A goal-programming enhanced collaborative optimization approach to reducing lifecyle costs for naval vessels. *Struct. Multidiscip. Optim.* **2016**, *53*, 1–15. [CrossRef]
19. Chen, C.-S.; Chen, J.; Xiao, L.-L.; Duan, S.-X.; Chen, J. Cooperative optimization of electric vehicles in microgrids considering across-time-and-space energy transmission. *IEEE Trans. Ind. Electron.* **2019**, *66*, 1532–1542. [CrossRef]
20. Zhao, J.-F.; Hua, M.-Q.; Liu, T.-Z. Research on a sliding mode vector control system based on collaborative optimization of an axial flux permanent magnet synchronous motor for an electric vehicle. *Energies* **2018**, *11*, 3116. [CrossRef]
21. Tian, S.-Y.; Jia, Q.-Q.; Xue, S.-W.; Yu, H.; Qu, Z.-W.; Gu, T.-Y. Collaborative optimization allocation of VDAPFs and SVGs for simultaneous mitigation of voltage harmonic and deviation in distribution networks. *Int. J. Electr. Power Energy Syst.* **2020**, *120*, 106034. [CrossRef]
22. Wei, X.; Jiang, X.-H.; Jin, S.-W.; Gao, H.-P. Test of the ultra-low emission of CFB boiler of 300 MW unit by two stages combined desulfurization matching. *Therm. Power* **2017**, *46*, 107–118.
23. Chen, L. Research on modeling of circulating fluidized bed boiler and its intelligent control system. Ph.D. Thesis, Zhejiang University, Hangzhou, China, 2005.

24. Gao, L.-X.; Yuan, L.-J.; Zhou, Z.-N.; Li, C. A new method for determining the optimal excess air coefficient. *Coal Mine Mach.* **2009**, *30*, 31–33.
25. Zhao, G.-L.; Li, R.-H.; Huang, Q.-Y. Study on the influence of gas boiler excess air coefficient on boiler thermal efficiency. *Chem. Eng. Des. Commun.* **2017**, *43*, 123–124.
26. Ke, X.-W.; Cai, R.-X.; Yang, H.-R.; Zhang, M.; Zhang, H.; Wu, Y.-X.; Lu, J.-F.; Liu, Q.; Li, J.-F. NOx generation and ultra-low emission in circulating fluidized bed combustion. *Proc. Chin. Soc. Electr. Eng.* **2018**, *38*, 390–669.
27. Tan, Q.-Y.; Peng, F.; Peng, H.-W.; Tang, X.-S. Optimization design of two-stage desulfurization system for CFB boiler. *Electr. Power Constr.* **2014**, *35*, 89–94.
28. Zhu, Y.-F. Economic Analysis and Optimization Study of Two-Stage Desulfurization of 300MW Circulating Fluidized Bed Boiler. Master's Thesis, North China Electric Power University, Beijing, China, 2016.
29. Zhu, Y.-Q.; Wei, X.; Zhang, X.-Y.; Fu, Y. Experimental study on desulfurization performance of semi-dry flue gas circulating fluidized bed. *Chem. Eng. Pet. Nat. Gas* **2020**, *49*, 8–11.
30. Wang, Q.; Bai, J.-Y.; Wang, X.-F.; Wang, L. Multi-objective optimization of CFB boiler desulfurization and denitrification system. *China Electr. Power* **2017**, *50*, 109–121.
31. Wang, C.; Cheng, L.-M.; Qiu, K.-Z.; Zhou, X.-L.; Xu, M.-L.; Cheng, H. The influence of adding desulfurizer to circulating fluidized bed boiler on heat balance. *Therm. Power Gener.* **2011**, *40*, 72–77.
32. Heydar, M.; Milad, S.; Mohammad, H.A.; Ravinder, K.; Shahaboddin, S. Modeling and efficiency optimization of steam boilers by employing neural networks and response-surface method (RSM). *Mathematics* **2019**, *7*, 629.
33. Wang, H.-C.; Liu, K.; Cui, H. The influence of adding limestone to the CFB boiler's thermal calculations. *Shandong Electr. Power Technol.* **2016**, *43*, 34–37.
34. Yang, D.; Xu, H.; Hu, G.-G.; Chen, H.-P. Study on the heat loss of 300MW circulating fluidized bed boiler exhaust flue gas by adding limestone for desulfurization. *Boil. Technol.* **2011**, *42*, 32–36.
35. Jiang, A.-P.; Lin, W.-W.; Ding, Q.; Wang, J.; Jiang, Z.-S.; Huang, G.-H. Optimization of dry desulfurization operation of slime fluidized bed boiler based on hybrid modeling. *CIESC J.* **2012**, *63*, 2783–2788.
36. Cai, J.; Zhang, M.; Wang, Z.-W.; Rong, Y.-J.; Yao, X.; Wu, Y.-X. Economic analysis of the desulfurization process of circulating fluidized bed boiler. *Clean Coal Technol.* **2020**, *26*, 90–98.
37. Zhang, Z.-H. Research on Energy Saving Optimization and Control of CFB Unit Combined Desulfurization Based on Environmental Constraints. Master's Thesis, Shanxi University, Taiyuan, China, 2016.

 *processes*

Article

Responsible Design for Sustainable Innovation: Towards an Extended Design Process

Ricardo J. Hernandez [1,2,*] **and Julian Goñi** [1]

[1]   DILAB Escuela de Ingeniería, Facultad de Ingeniería, Pontificia Universidad Católica de Chile, Santiago 7820436, Chile; jvgoni@uc.cl

[2]   Escuela de Diseño, Pontificia Universidad Católica de Chile, Santiago 7820436, Chile

*   Correspondence: rhernandep@ing.puc.cl

Received: 20 October 2020; Accepted: 27 November 2020; Published: 29 November 2020

Abstract: Design as a discipline has changed a lot during the last 50 years. The boundaries have been expanded partially to address the complexity of the problems we are facing nowadays. Areas like sustainable design, inclusive design, codesign, and social design among many more have emerged in response to the failures of the production and consumption system in place. In this context, social, environmental, and cultural trends have affected the way artefacts are designed, but the design process itself remains almost unchanged. In some sense, more criteria beyond economic concerns are now taken into consideration when social and environmental objectives are pursued in the design process, but the process to reach those objectives responds to the same stages and logic as in traditional approaches motivated only by economic aims. We propose in this paper an alternative way to understand and represent the design process, especially oriented to develop innovations that are aligned with the social, environmental, and cultural demands the world is facing now and it will face in the future. A new extended design process that is responsible for the consequences produced by the artefacts designed beyond the delivery of the solutions is proposed.

Keywords: design process; innovation; responsible design; sustainable design; sustainable innovation

1. Introduction

This article presents a critical discussion on the role of design in society as well as an exploration into how the design process itself may need to be reconsidered in order to comply with the need for a more responsible vision of design for innovation. Design as a discipline has changed a lot during the last 50 years. The boundaries have been expanded partially to address the complexity of the problems we are facing nowadays [1]. To many design practitioners and scholars, design has been one of the most prominent contributors to both the ills that affect society as well as our best shot to overcome them [2–4]. There has also been institutional pressure to reframe our current understanding of design into a more progressive and active agenda. For instance, this has been the case with institutions in the UK such as the Sorrell Foundation, Royal Society of Arts, National Endowment for Science, Technology and the Arts (NESTA), and the Design Council [5].

Overall, these trends aim to vindicate design as a means to overcome the world's challenges by actively avoiding being ideologically neutral and engaging in value-driven activity. Design's areas like sustainable design [6,7], inclusive design [8,9], participatory design [10], social design [11,12], and value-sensitive design [13,14] have sought to change the role of design in society. Learning from the frameworks of responsible innovation [15], we can group all these areas under the label of "responsible designs". We define "responsible designs" as all the wide and heterogenous range of design's areas, approaches, and activities seeking to establish what design ought to be, instead of what it currently is. We define this responsibility extrapolating the work of Voegtlin and Sherer [16] as a tripartite concept: (a) the responsibility to avoid harm, (b) the responsibility to good, and (c) the responsible governance of the design process. We assert that all forms of responsible designs ultimately reflect the need to reconsider the criteria of successful design in the face of the global challenges and the responsibility of designers in achieving that success.

In this context, social, environmental, political, and cultural trends have affected the way artefacts are designed, but the design process itself remains almost unchanged. In some sense more criteria beyond economic concerns are now taken into consideration when social and environmental objectives are pursued in the design process, but the process to reach those objectives responds to the same stages and logic as in traditional approaches motivated only by economic aims. For instance, when a product is designed with the objective to be sustainable, the normal approach is to bring principles and considerations from the sustainable design area, and inject them ideally at the start of the design process. There are tools developed to help designers to consider those principles and concerns, but in general the essence of the design process remains unchanged. This situation can be extrapolated to inclusive design, participatory design, social design, and almost all the "responsible designs" we mentioned earlier.

We propose in this paper an alternative way to understand and represent the design process, especially oriented to develop innovations that are aligned with the social, environmental, political, and cultural demands the world is facing now and it will face in the future. A new extended design process that is responsible for the consequences produced by the artefacts designed beyond the delivery of the solutions. We recognize there are multiple models that represent the design process [17–19]. Some of those models have been developed by engineers [18,20,21], others by designers and architects [22,23], some by practitioners and others even by business consultancies [17]. In these models there are different sources of variability associated with the discipline of origin, their orientation to the users, and the scope of the process [19]. However, there are also some common elements that build the structure of the design process. Those elements include the iterative rationality of the process, the divergent and convergent way of thinking embedded in the stages of the process, their orientation towards actions, and the compatibility of the process with a variety of possible outcomes [19,21,24]. Between these common elements there is not any one that represents the responsibility the design process should have to assure responsible results.

Learning from the experience of the responsible innovation framework, we observe that an extended design process should include ideals of anticipation, reflexivity, inclusion, and responsiveness [15]. In turn, it implies that design professionals and institutions should develop anticipatory, participatory, and integrative capacities [25]. We argue that an extended design process should have at its very base some of those elements that include the ideals of what we called "responsible designs", not as injections of external concepts but as structural pillars of the process. With this ambition comes also the idea to develop a model that can represent that kind of extended design process.

In sum, this paper is a conceptual article [26,27] that seeks to "provide an integration of literatures, offer an integrated framework, provide value added, and highlight directions for future inquiry" [26] (p. 17) in the intersection of design and responsibility. This added value, directions and propositions are also inspired by many years supervising and guiding design processes, and observing how responsibility is enacted in practice using the most traditional view of design. The aim of this article is twofold. Firstly, to expose the need to reconsider and potentially redesign the mainstream representations of the design process. Secondly, to explore the feasibility to incorporate responsibility into the core of the more traditional view of the design process through a model involving three elements: pertinence, transparency, and distributed agencies.

2. Responsible Designs

Current societal challenges put pressure on the design discipline and design professionals to integrate goals and success indicators beyond profit making [11,28]. These challenges involve different aspects of society, for instance, the environmental crisis, rising wealth inequality, ageing and over population, human and technology tensions, diminishing quality of life, decolonization and decentralization, and so on.

To many, design has been one of the most prominent contributors of progress, but also to social harm [2–4]. In practice, this has meant that over the past 20 years, the design discipline has sought to be reformulated and updated in order to live up to their aspirations to bring about a better world [29].

Overall, these trends set out to update design as a vehicle for change through value-driven activities. For this reason, we can group all these trends under the label of "responsible designs". We define "responsible designs" as all the wide and heterogeneous range of design trends, approaches, and activities seeking to establish what design ought to be, instead of what it currently is. It is important to note that despite the fact that many of the responsible design trends encompass each other interests. For instance, socially responsible design or sustainable design may incorporate elements of codesign or universal design, but because they belong to differentiated historical roots and have their own conceptual tradition, we concluded that it was best not to give any of these design trends primacy and to categorize all of them under a more neutral label. Figure 1 describes some of the concerns and trends involved in responsible designs:

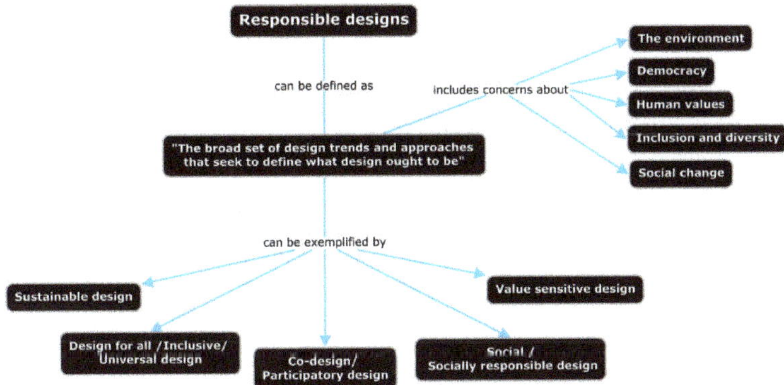

Figure 1. Responsible designs. Original work.

2.1. Inclusive/Universal Design

Inclusive design, universal design, accessible design or design for all is a group of design trends that seek to design and redesign products and services to make them accessible and usable to all or at least to a broader set of users [9]. It has mostly centered on issues of age and ability [8], but its underlying principle of universal access can be applied to all forms of individual differences. The idea of inclusive became more prominent in the 50s in the US under the label "Barrier-free design" in response to the difficulties of war veterans with injuries and physical disabilities [9]. Since then, inclusive design has been promoted under antidiscrimination laws both in Europe and the US [30].

2.2. Codesign/Participatory Design

Participatory design and codesign approaches are design trends that broadly seek to include all relevant stakeholders into the design process aspiring to equal power to all parts [10,31]. The reasoning behind the involvement of the stakeholders is both practical and political; they know what is best for them and they ought to be heard [32]. Historically, the idea of participatory design can be traced back to Scandinavia in the 70s in the context of the political demands for more democracy and collectivism, labor unions and ideals of social emancipation [33].

2.3. Social Design/Socially Responsible Design

Social design is a broad label that has been used to characterize design practices that aim at social change or social improvement instead of (solely) financial gains [34]. Social design is typically traced back to Victor Papanek, Nigel Whiteley, and Victor Margolin and particularly to Papanek's book "Design for the real world" [35]. The idea of socially responsible design takes that same principle and extends it further in the sense that it implies that designers not only should design for social change but are ultimately responsible for the societal effects of their creation [12]. Socially responsible design often seeks to integrate the principles of sustainability, universal access, and participatory design as all of those approaches are seen to be constituent of social change [11].

2.4. Sustainable Design

Sustainable design or design for sustainability is a relatively new area of study in the design discipline. Following the concerns raised in the 70s and 80s about the negative impact industry and human actions were causing on the environment [36,37], many disciplines and between them design were at the center of the discussion. While it is recognized that industry has produced important benefits for society, it has also produced negative consequences like depletion of natural resources, loss of biodiversity, pollution, social inequalities, among others. In response to these problems from a design perspective emerged approaches first to reduce those negative consequences and more recently to create value in social and environmental dimensions as it created economic value during the design process. Those approaches include cleaner production, eco-design, eco efficiency, sustainable product service systems, design for sustainability, and recently circular economy [6,38–41].

2.5. Value Sensitive Design

Value sensitive design (VDS) is a design framework developed by Batya Friedman [13], that seeks to incorporate in a systematic way human values into the development of new technologies, especially in the fields of information systems design and human–computer interactions [42]. The value sensitive design framework emphasizes methods and has since its inception been founded on a defined and formal methodology. VDS is structured as a tripartite methodology of Conceptual, Empirical, and Technical Investigations [14] and a series of design actions devoted to make visible and to manage the human values embedded in new technologies and the cultural forms of life of the stakeholders [43].

2.6. New Criteria for Successful Design

Due to all the new and renewed concerns that have been championed by design practitioners, citizens and scholars, the criteria used to define the success and failure of designed artefacts needs to be reconsidered and expanded. Through our revision of responsible design trends, we observe many central concerns that have emerged from different perspectives. As seen in Figure 1, these include: "The Environment", "Democracy", "Inclusion/Diversity", and "Social Change". All of these concerns can be interpreted as emergent criteria to define the quality of design. Melles, Vere, and Misic [11] proposed new criteria to define successful design under a socially responsible design approach that summarize these concerns:

- Need: does the user or community need this product/solution?
- Suitability: is the design culturally appropriate?
- Relative affordability: is the outcome locally and regionally affordable?
- Advancement: does it create local or regional jobs and develop new skills?
- Local control: can the solution be understood, controlled and maintained locally?
- Usability: is it flexible and adaptive to changing circumstances?
- Empowerment: does it empower the community to develop and own the solution?
- Dependency: does it add to third world dependency?

As can be observed in Melles et al.'s [11] criteria, some elements present in other responsible design traditions are missing in this set of questions. Complementary criteria can also be added to the list when considering other responsible design trends:

- Inclusivity: Will all users be able to profit from this design while respecting their diversities?
- Sustainability: Is the solution socially and environmentally sustainable?
- Value driven: Does it represent the value system of the stakeholders?

Meeting all of these standards may seem overwhelming to designers and may also make us wonder about the nature and extent of the responsibility of designers. To authors, such as Dobson [44] design responsibility should be a change in our professional attitudes towards the challenges of today's world. According to Szenasy [45] the key is understanding that all good design is responsible design, and thus, responsibility should be embedded in disciplinary rigor. According to Thorpe and Gamman [5] designers should be responsive instead of responsible for the outcomes of their work. By this, they mean that designers should accept the uncertainty embedded in complex real-world problems and avoid adopting a paternalistic stance in which the designers portrait themselves as saviors to other communities. According to Cipolla and Bartholo [46] design responsibility means acting where you are, that is, acting inside your own community in a coresponsible and dialogical manner. For Markussen [47], socially responsible designers are concerned with micro- and meso-level life improvements for a confined and marginal group or minorities.

Overall, the premise behind responsible designs is that there is a need to reconsider the ways that design incorporates the question of ethics. The philosopher Glenn Parsons [48] describes the three ways in which today's design incorporates ethics:

1. Designers have to comply with the ethical norms, rules, regulations that constrain their professional activity.
2. Designers also engage in ethics when they decide what to create.
3. The products of the design professionals also change the social landscape and thus, the concept of ethics itself.

The interesting fact about all of those ways in which ethics is embedded into the design practice, is that they act implicitly and are not part of the formal activities in the design process. In other words, the ethics and responsible concerns of design have not been designed into the design process itself. This is not to say that design teams have not incorporated some of the responsible design criteria, but rather that in some cases its incorporation has not substantially altered the overall representation of the stages involved in the design process [49–55]. More than that, these concerns have not been systematically transformed how designers outside the responsible design trends conceive and visualize the stages of the design process, at least not in all cases [56].

3. Conceptualization of the Design Process

The design process has been represented historically by different models. Some of them more normative than others, some very detailed, while others are more abstract representations of the process [17–21,57]. It is true that in recent years new ways to represent and to guide the design process have emerged in areas like speculative design, critical design, and design fictions, however, a traditional view of design still remains valid and widely used in industry and independent practice. This traditional view of the design process is usually recognized as solution-oriented, and it is in our interest to build our proposal of an extended design process on this traditional view of the design process.

These solution-oriented models of the design process are characterized by factors coming from different dimensions. One of those dimensions has to do with the discipline of origin. There are models developed by engineers that have a very comprehensive understanding of the resources and the stages involved in the conception and production of an artefact from a functional perspective [18,20,21].

There are also models developed by designers and architects that respond to other rationality [23], and even models developed by business consultancies and practitioners that address a totally different set of drivers [22,58].

However, the discipline of origin in not the only dimension affecting how the design process is represented, there is also the fact that some of those models represent processes centered in the users while others take a different approach [24,59]. The scope of the process is also a source of variability. There are models that consider the design process from the requirements gathering with the users to the final development of the solution including stages of conceptual design, detail design, and even manufacturing [21,60]. In contrast, there are more simplified models that consider the design process only as the stages that concerns giving form to a preliminary solution later to be developed by other areas in an organization.

Despite the variability between different models to represent the design process there are some common elements between them that should be remarked. First, there is the notion that the design process is iterative and it is expected to come back from later stages to previous ones in a continuous cycle of improvements [24]. This is true in models with few steps up to very complex representations with multiple stages and substages. Second, it is the divergent and convergent way of thinking embedded in the stages of the process [57]. This logic responds to complementary stages of analysis and synthesis to understand the context and the opportunities of intervention, and also to develop and select the possible solutions. A third common element between different representations of the design process corresponds to their orientation towards action [23,24,61]. This means that the models that represent the design process favor a practical approach over a theoretical one. This orientation towards action is capitalized in different moments where the design process gets closer to the user to collect information, to involve them in the process, or to test a solution. Finally, there is a fourth common element between different models of the design process. This common element is the compatibility of the model to be used in very different situations and to produce very varied outcomes. In general, a model of the design process is a representation that can be used to design a table as it can be used to design a car, or even a more complex artefact. The level of detail will be off course different in each case but the general stages that should be followed are basically the same.

Between these common elements in different models that represent the design process there is not one that addresses the responsibility the design process should have to assure responsible results. In this context, we recognize that social, environmental, political, and cultural trends have affected the way artefacts are designed, but the classical and most used design process in its core elements remains almost unchanged. In some sense, more criteria beyond economic concerns are now taken into consideration when social and environmental objectives are pursued in the design process, but the process to reach those objectives responds to the same stages and logic as in traditional approaches motivated only by economic aims. For instance, when a product is designed with the objective to be sustainable, the normal approach is to bring principles and considerations from the sustainable design area, and inject them ideally at the start of the design process. There are tools developed to help designers to consider those principles and concerns [38,62,63], but in general the essence of the design process remains unchanged. This situation can be extrapolated to inclusive design, participatory design, social design, and almost all the "responsible designs" we mentioned earlier. An extended design process that includes elements in its more basic structure that can represent the responsibility the process should have to assure responsible results is needed, especially for solution-oriented processes aiming to develop innovations aligned with social, environmental, political, and cultural objectives.

In the next section we present an analysis of the emergence and evolution of three applications of the Internet of Things (IoT), as examples of innovations developed with some additional criteria in mind that supports our idea of developing a new extended design process. In these examples we highlight some of the elements we believe should be taken into consideration in an extended design process towards responsible innovations. We do not claim these examples are cases of responsible design, not even case studies. The purpose of these examples is to illustrate how the consideration of elements such as distributed agencies, pertinence, and transparency can change the dynamics and the results of the design process favoring the reflection on responsible design in a broad sense, and to show how those elements can be manifested in practice. All three examples were selected purposively as documented examples of design within the same theme (i.e., IoT). In these examples we reconstruct the design process and characteristics of the design solution in order to analyze them through the lens of each proposed component of a responsible design process.

4. Internet of Things (IoT): An Atypical Innovation

The term Internet of Things (IoT) is relatively new. According to Coulton, Lindley, and Cooper [64], the term "The Internet of Things" was coined by Kevin Ashton. Kevin was executive director of the AutoIDCentre in the MIT in 1999 [65]. The concept of IoT combines previous interest on ubiquitous computing proposed by Mark Weiser [66], the development of sensor technology [67], and the idea of device to device (D2D) communication proposed by Bill Joy [68]. The first documented case of IoT occurred in 1984, when a coke machine was connected to the internet to report the availability and temperature of the drink [69]. Since its proliferation in the 1990s, IoT technology has mostly been used to expedite industrial logistics and since the 2010s has systematically permeated the homes of users [67]. As Sharma, Shamkuwar, and Singh [67] assert, IoT is now much more than installing sensors on objects and calling it "Smart". New IoT developments are increasingly more complex and integrated with other technological trends such as Artificial Intelligence, Big data, Cloud Computing, and Robotics among others.

There are multiple definitions of IoT that include different elements, but that converge to very similar ideas. In the Little Book of Design Fiction for the Internet of Things [64], there is one explanation that covers well the nature and purpose of what is called IoT nowadays:

"we use the term to describe any objects or things that can be interconnected via the Internet, making them to be readable, recognizable, locatable, addressable, and controllable by computers. The things themselves can be more or less anything. Later in the book we use examples such as a kettle, a door lock, an electricity meter, a toy doll, and a television, but it's important to remember that there is no limit on what could be an IoT thing. Anything that is connected to the Internet is arguably part of the IoT, including us." [64]

For the purpose of this paper, we are not interested to debate about the definitions or even the technical characteristics of the IoT. What interested us is to analyze how different artefacts using IoT have been designed, developed, and in some cases implemented or introduced to the market. Understanding artefacts in a wide vision as it is presented by [24], including products, services, and systems. In particular, we are interested in the decisions made during the design process of those artefacts. We want to use those examples to discover and to highlight the elements we believe should be incorporated in an extended design process representing the idea of responsible design absent in current design models.

The following three examples of designs and developments under the IoT umbrella are analyzed to discuss the same number of elements we believe should be included in an extended design process.

4.1. IBM's Smarter Cities Project

One of the clearest ways in which the Internet of Things is pushing for transformation in our worldview is through its potent vision of *distributed agencies*. Just as our everyday devices are becoming connected and their decisions distributed among many different information sources, our lives also seem to become more global and our decisions more dependent on the technologies that we use (our smartphones, google, etc.). However, surprisingly enough, the idea that our minds are somewhat interconnected with our environments and our technological tools is not that new.

Ever since the 1940s, cognitive psychologists and philosophers have discussed the extent to which our thinking and action is occurring "outside our heads". This is what is often called distributed cognition. According to Gavriel Salomon [70] there are two forms of distributed cognition: shared cognition and off-loading. Shared cognition occurs when a cognitive process is shared among people in a collaborative activity, while off-loading occurs when a mental process is somewhat externalized to a material object, such as a calculator, a grocery list or even, a smart device. We can see that the distribution that characterizes IoT devices also relates to the characteristics of the individual mind itself. However, how about groups of people, or communities or even cities?

Smart cities are a good example of how design projects seem increasingly distributed. Smart cities are based on the new possibilities generated by IoT to interconnect, monitor, and boost the productivity and responsiveness of all services and user contact points embedded in the city as a system [71]. One of the largest smart city designers is actually a former hardware designer: IBM.

In the 1990s and the early 2000s IBM reportedly was going through a difficult financial time [72]. This ultimately led to a strategic shift from hardware design to consultancy and software and the sale of the PC division to Lenovo by 2004. In this context, and in the midst of the financial crisis Sam Palmisano (IBM's CEO) coined the term "smart cities" in a talk entitled "A Smarter Planet: The Next Leadership Agenda". According to one of the IBM managers at the time "in 2008 the launch of Smarter Planet was not merely the announcement of a new strategy but an assertion of a new IBM's worldview, showing how the world has been changed over the previous decade" [73]. As Wiig [74] puts it, IBM's smarter city strategy is ultimately a techno-utopian vision of the new city.

IBM's proposal was ambitious, they sought to help cities redefine public services, social programs, and public infrastructure using emergent digital technologies. How could they redesign such complex systems? In truth, a closer exploration of IBM's smarter cities projects shows that this designing was a very collaborative and distributed process. IBM partnered with city governments, local universities to tackle the multiple variables involved in the challenge. Sometimes, they had to acquire technologies from smaller size firms and even produce new technologies in their Smarter Cities Technology Center in Dublin where world-class researchers were recruited [71].

IBM's design process was not without challenge. As Scuotto, Ferraris, and Bresciani [73] describe, during the design and implementation of their strategies, IBM managers face notable barriers in collaborating with governments, such as scarce management and technological skills and the overall passive role of the public sector. Another difficulty faced in the implementation of their solutions was the management of Intellectual Property Rights (IPR). IBM managers conditioned the sharing of their intellectual property with public organizations to clear contractual conditions carefully negotiated beforehand [73]. IBM tended to share as little knowledge as possible related to the core competitive capabilities in order to avoid knowledge leaks to other government's partners, but had no issue sharing knowledge in more peripheral activities [73].

Looking at IBM's vision for Smarter Cities we can extract many lessons to be incorporated in the design process. Just as our cities, our digital technology and our mind itself seems to be distributed in nature, so should perhaps be elements of the design process. In the documented cases of IBM smarter cities strategies, we can observe that the redesign of cities services involved the agency of many different stakeholders: partnerships with universities to help in the research phase, active involvement of local authorities to fund but also to place priorities, technology firms of lesser size to provide strategic services and finally, the help of top international researchers through IBM facilities worldwide.

In that sense, the design process of IBM is characterized by a sort of distributed agency—or shared cognition in the words of Salomon [70]—in which actions, burdens, benefits, and properties were negotiated among different actors. We can also learn that even if responsibilities are distributed, they also have to be clear in order facilitate collaboration. Even though the IBM design process is collaborative, it does not mean that, for instance, property rights could be diluted. On the contrary, smarter cities were more likely to succeed if the division of both labor and rights was clear and adjusted to the expectations of all involved actors.

We can also extend the argument even further and also analyze other ways in which responsibility is distributed in the design process. In some sense, users also share a responsibility in how artefacts impact society, as they hold the ultimate right to choose how to use them. In the case of Artificial Intelligence, Wong [75] argues that users should no longer be considered just passive agents. It is users that have the power to accept/reject certain artefacts that shape how technologies work with their input of data, they may benefit from others wrongdoing, and their decisions with technology can affect other people or the world. Technologies/artefacts themselves can also be regarded to have an agency.

In sum, when designing responsibility into the design process it seems that the distribution should be considered as a fundamental aspect to incorporate. The design process involves multiple agencies, from the designers themselves to institutions, to stakeholders, to end-users and to the technologies. The boundaries, actions, and properties of that distribution should be arranged at some point, or multiple points of the design process.

4.2. Smart Labels and IoT Sensing Technologies

One interesting concept that it is in some sense missing in the conceptualization of the design process is *pertinence*. It is defined by the oxford dictionary as: "the quality of being appropriate to a particular situation" [76]. Not because something can be made, it should be made. A premise usually ignored in the design and development of artefacts. There are multiple examples of products that actually nobody needs, or even worse products that might harm instead of providing a purposeful function.

Smart packaging and IoT sensing technologies are a good example to talk about *pertinence*. Smart labels are not a new concept. They have been around for at least four decades [77,78]. These labels use passive technologies like RFID to store a certain amount of information about a product that can be read by specific scanners and facilitates logistics and inventory control [79]. Initially these kinds of technologies were mainly used in warehouses and during transportation of goods to track products along the supply chain. In the earliest 2000, these smart labels were introduced in retail used for similar purposes, to have some level of control over inventories and to make the restock process more expedited. Actually, Walmart, a major US retailer started to request the use of these technologies to their suppliers, something that increased substantially the use of smart labels and decreased its price [77].

During the last two decades the applications of those smart labels have been increased and moved to new environments. In medical services pilots looking to prevent human errors for example in transfusion of blood have been put in place such as the pilot tested in the Massachusetts General in Boston [77]. Additionally, in the pharmaceutical industry they have been used in drugs to prevent counterfeit [80]. The technology has also evolved during this time, from passive labels to active tags that can be connected to the Internet and display a whole new set of capabilities.

Recently, these technologies have been considered to approach the major problems we face today as society, waste management. The amount of waste we produce every day is alarming and with it the loss of valuable resources and materials. Addressing this big issue, smart labels and sensing technology connected to the Internet have been used to design systems that can improve the sorting and disposing of waste, also to facilitate the management of bins and municipal centers of waste collection using sensors and real time data [78]. In parallel, there are people interested in using these technologies in packaging with purposes such as increasing food safety, design of sustainable biodegradable packaging, and also the development of new business models based on big data opportunities [81]. Even applications like including tags in wine bottles that can provide information about their temperature and perfect moment to be drunk. In this case of smart labels, and IoT applied to packaging, there are two issues that catch our attention. First, the fact that these technologies have been evolving from very basic applications to more sophisticated ones at low speeds. Smart labels have been around for about 40 years and yet we do not find all packages in the supermarket are connected to Internet collecting and transferring data. Why? We believe mainly because of costs and because there is not a strong case that supports the investments to use those technologies in everyday products. We hope some reasons behind this slow introduction to massive markets include also questions about the pertinence of doing so. It is not hard to justify the use of these technologies to improve the operations in a warehouse or even better to avoid human errors in medical treatments, but what about collecting commercial data from users or to provide information about a product that can be meaningless for the majority.

What is the limit? What should be the criteria to decide the application of these technologies? Do we want to have everyday products in the supermarket collecting and using information about our tastes and routines? As we mentioned earlier not because something can be made it should be made, or in this case not because something can be used it should be used. Pertinence should be one of the basic elements in the design process. It should be the element that makes designers and engineers think if what they are designing is worth it. Is it making good beyond economic considerations? Is there a possible harm involved? Who should decide if something is pertinent to be designed or not? Is it a task designers should take or should be on the market? When should it be decided if a product is pertinent or not? Questions that should be explored during the design process.

4.3. IoT Door Lock: Imagination Lancaster

Design IoT products have major issues related to security, privacy, and management of personal information between other dimensions. These issues can be encompassed in a bigger and more general concept, *transparency*. This is an important term that is usually undervalued in the design process. *Transparency* is a property of a product, a service, a process, or a system, and it can be linked to different agents. It means for example that a product can be transparent to the users in the way it works and in the negative and positive impacts it can produce [82], but it can also be transparent to investors, or to the community beyond the role of being a user or not.

A good example of how this concept of *transparency* can be treated in the design process of a product is presented in an experiment done at Lancaster University by PETRAS, an IoT Research Hub [64]. In this experiment the main question behind was how users accept agreements to use IoT products. In general, all products, IoT or not, came with a user agreement that users have to accept if they want to use the product. In the case of IoT products with multiple functionalities what can happen is that one agreement actually is related to many different functions. When a user gives its consent, that consent covers all those functionalities despite the user wanting or not to use all of them [64].

In order to explore this issue, the group of PETRAS used design fiction as a tool to picture a probable situation. They designed an IoT door lock that has multiple functionality including [64]:

- Near field communication (NFC) feature to unlock instead of a key;
- Geofencing (lock and unlock automatically using geographical location);
- Providing guest access via smartphone app;
- Voice activation using an external service like Amazon Echo;
- Connection to If This Then that IFTT to integrate with other systems.

A relevant characteristic of those functionalities is that each one requires to collect and manage different information from the user, and that information has to be stored at different places. For instance, to address the fact that a user should be able to give consent through a product agreement just to the functionalities he wants to use or to the ones he feels comfortable with, in the experiment an Orbit system was proposed [64]. In their own words:

"Orbits are circular infographics which explain, based on what data a system gathers and where it shares it, how identifiable a person might be from this data. The Orbit design uses three concentric circles, where each circle represents data that might be used to identify you. The inner circle represents data on devices you own (like your router, smartphone, or television). The middle circle represents data on servers of companies you know, but where you have no control over the servers (like the server of the lock company, Amazon's Echo servers, or your mail provider). The outer circle represents data on largely unknown 3rd party servers (this might be IFTTT or a marketing company they sold data to)" [64].

Basically, what this Orbits app gives to the user is the chance to know in advance what information will be collected by the different functionalities of the product (Door Lock), where the information will be stored, and how identifiable they will be through the information they will give [64]. With this information the users can accept the product agreement at different levels, allowing the product to perform all its functionalities or just the ones the users are comfortable with [64].

The interesting issue behind this example is the fact that artefacts should embed in their design, features that assure *transparency* to the users. This *transparency* has to be related to how the artefact works, how safe it is, what the potential impacts it can produce are, and in general all the information required to make an informed decision regarding the ownership and the use of the artefact. *Transparency* is an umbrella term part of what we recognize as responsible design and cover issues like security, privacy, safety, and ownership.

5. Extended Design Process

Between multiple models to represent the design process, especially the solution-oriented design models, the double diamond developed by the Design Council [57] is the one we chose to present the idea of an extended design process. The selection of this model to be used as the basis of our proposal lies on the fact that this model represents well the more commonly used stages in a traditional design process. It integrates the elements we identified in Section 3, elements that we consider transversal to many solution-oriented models proposed during the years. More than a referent, the double-diamond model was selected as it represents a common ground of the stages of the design process usually used in industry, taught in college, and used in many innovation processes. Rather than to propose a totally new disruptive model we aim to change current practice taking advantage of known behaviors and language.

The proposal is divided in two parts, what should happen before the delivery of the artefact (solution), and what we think should happen after the delivery during the use of the artefact. In the next section we explain in detail the implications of an extended design process in these two moments.

5.1. Before Delivery

The Double Diamond design model proposes two consecutive diamonds each one made of one divergent and one convergent stage. In this process the four stages are named as discover, define, develop, and deliver [57]. The first declaration we made, despite that it seems implicit in the model, is that the first two stages in the first diamond relates to the formulation of the design problem, and the third and fourth stages in the second diamond correspond to the development of the solution, see Figure 2.

Then, the major change is to open in the middle of each diamond a space for a new stage. These stages are neither divergent nor convergent, they are included to be evaluation and reflexion points in the design process. Instead of inserting criteria for example of sustainable or inclusive design at one of the original four stages, we propose to have dedicated moments in the design process to make a holistic evaluation of the context and the solutions in relation with the elements we believe represent well what is behind the idea of responsible design: *distributed agencies, transparency, and pertinence.*

In the first stage of reflexion and evaluation during the formulation of the problem (first diamond) the elements that should be included are transparency and pertinency. The transparency should be oriented to assure that all possible actors, stakeholders, and interactions have been taken into account to build the context in which an area of intervention (opportunity) will be defined. In the example of the Door Lock presented in Section 4 this should refer to consider all the actors that should give and access data during the design and use stages. All the possible relationships between those actors should be mapped and any potential conflict anticipated. The pertinence refers to evaluatin the purpose and relevance of the context built in relation to a set of criteria defined by the areas that delimits a responsible design like sustainability, inclusion, safety, and more depending on each situation. Referring to the example of Smart Labels, it is at this stage that the relevance of the context in which the design should take place should be evaluated according not only to economic concerns but arguments from a social and environmental perspective. Finally, distributed agencies are reflected in the strategies used to engage multiple stakeholders in the problem identification and framing. This form of distributed agency relates strongly to the advances and tools originated in codesign/participatory design.

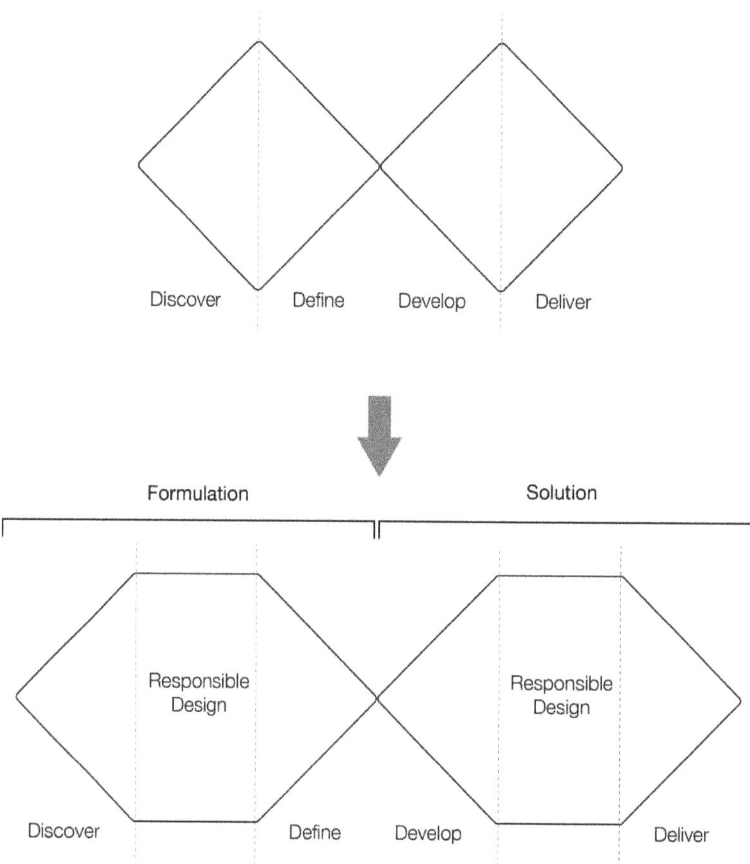

Figure 2. Responsible design before delivery. Original work adapted from [57].

In the second stage of reflection and evaluation during the development of the solution (second diamond) the elements that should be included are transparency, pertinency, and distributed agency. In this case the transparency refers to the information shared by the designers to all the actors and stakeholders that will be involved with the solution. This information includes the way the solution was made, the impacts generated during its development and the impacts it might produce during its use and final disposal, and the requirements of data to operate the solution, between others. The pertinence relates to the relevance and fulfilment of the purpose of the solution. This can be evaluated using for example the set of questions proposed by Melles, Vere, and Misic [11], regarding need, suitability, relative affordability, advancement, local control, usability, empowerment, and dependency; complemented by questions on inclusivity and sustainability. In the two examples mentioned earlier, Door Lock and Smart Labels, this stage of reflection and evaluation should focus on interrogating the solutions developed before any release to the market. This interrogation should address any potential problem during the use of the solutions, like the fact the user should have to give data they do not want to share as it was the case in the Door Lack example, or any possible misinterpretation of information given by the Smart Labels that could derive into misuses that could harm the user or others involved. Finally, a third element we believe should be included in this second stage of evaluation is the distributed agencies. It means the assignment of responsibilities between all the agents that will be involved with the solution. These agents include designers, producers, users, communities, and any other one that might be affected directly or indirectly by the solution.

5.2. Beyond Delivery

There seems to be a limit to the extent that our imagination can predict or anticipate the future impact of our design solutions. Additionally, this is not just a statement about the abilities and skills or even tools that may help professionals do a better job anticipating, but rather a statement about the nature of the design problems themselves. According to Rittel and Webber [83], designers usually have to deal with complex problems that cannot simply be addressed in a straightforward way using conventional problem-solving methods that could be applied to structured situations such as natural science experiments, finite solutions games such as chess or engineering optimization problems with a definite answer. Design problems usually resist such rationalistic approaches because they are characterized by blurred boundaries, conflicting information, and systemic complexities. This is what Rittel and Webber denominate "wicked problems" in opposition to "tame problems" that could define other disciplines such as the hard sciences.

Wicked problems are also defined by their unforeseeable impact in the systems they are implemented in. As Rittel and Webber [83] assert: "there is no immediate and no ultimate test of a solution to a wicked problem" (p. 163). Tame problems, on the other hand, have relatively predictable consequences and thus impact evaluation may be conducted. Design solutions to wicked problems produce different sorts of consequences that take place over different periods of time (short, middle, and long term). Therefore, as a final judgement would require all consequences to have played out and as there is no strict time limit for this to happen, an impact evaluation becomes a wicked problem itself. We could extract two different conclusions from this: one may think that this wicked nature of the impact of design solutions prevents us from attempting to assess its impact, but that approach would be conflicting with the adoption of a more responsible vision for design. Another way of looking at it is to deal with this challenge in a more comprehensive and designerly way to assess the impact of our design solution in recognition that they will most likely produce unexpected outcomes in their use, that are somewhat unforeseeable during the design process and that all evaluations are to be considered provisional. Our argument is that to achieve this goal, we will be required to extend the design process beyond delivery and to design in more detail how this evaluation is to be conducted.

However, what are the impacts of design? As Kiran [84] asserts, designers ultimately aim at changing behaviors. Through their solutions, designers aim at altering user behaviors through different mechanisms, such as invitation or seduction (e.g., a fly sticker at an airport bathroom that invites biological men to point correctly when they pee) or through inhibition and coercion (e.g., a mandatory seatbelt in our cars). In that sense, learning (i.e., behavioral change) is the direct impact of design.

Designs, however, not only have a direct impact on the user, but in the social and environmental system as a whole. Boenink, Swierstra, and Stemerding [85] argue that the effects of technology can be categorized under two different concepts:

1. Hard impacts: These are quantifiable impacts relating to the observable risk/benefits that agents of the system are exposed to because of the incorporation of the technology.
2. Soft impacts: These are subtle changes in the distribution of social roles and responsibilities, moral norms and values, or social identities that are produced because of the incorporation of the technology.

These two categories can also be broadly applied to design [84]. We can seek to assess how design solutions impact the direct user behavior and the hard and soft impacts they produce in the overall system. However, this impact cannot simply be forecasted during the development of the solution, as the wicked nature of the problems will unveil its consequences in use and through time.

It is for that reason that we propose that a more responsible design process should extend beyond the delivery of the solution (as most versions of the design process models do end). An extended design process would although not aim at producing a summative evaluation of the solution as this seem to be impossible, but rather, at monitoring and learning from the unfolding of the design's impact in the system as a means to decide whether it is responsible to maintain the current design, accelerate, pivot, change, fix, reconsider or even in some extreme instances, to stop it.

The criteria used to evaluate our design solutions could be essentially the same that we define as our criteria for pertinence. As we have stated, we believe that the set of questions proposed by Melles, Vere, and Misic [11] that we have expanded in the previous section could serve as a guide for this evaluation. Designers could additionally ask themselves about the unintended consequences of their solutions in practice—both positive and negative—and what they can learn and extract from observing its impact on behaviors, soft and hard transformations in the system. Transparency in this stage would also be necessary. Transparency beyond delivery allows for accountability. Designers should define and communicate how the artifact will be monitored and how that feedback will be managed. Finally, distributed agencies beyond delivery relates to who will do the monitoring. A distributed monitoring could involve individual users, communities, NGOs, Universities, and even governmental agencies depending on the complexity of the project. For instance, the impact of a project such as IBM's Smart Cities should be distributedly monitored by different actors, including the company itself but also through governmental agencies, independent evaluators and potentially the citizens that are affected by the outcomes of this design process. All these partnerships are embedded in dynamics of power and politics that add to the complexity of the design process.

We ultimately believe that the form that the monitoring and learning process beyond the delivery to take place is something that each design team will have to 'design' themselves. We do believe that regardless of the definitive form that design teams will choose, they will have to address at least these three questions:

1. How many monitoring and learning instances will be included?
2. How long into the use of the design will be considered sufficient?
3. What actors will be involved in the monitoring and learning process?

In the end, the complete proposal presented in Figure 3, shows the transversal elements we believe should be included in an extended design process oriented towards the design and development of responsible solutions. In this proposal, besides the appearance of the elements mentioned, pertinence, transparency, and distributed agencies, there are other important remarks. First, the fact that we believe the monitoring of the solutions beyond its delivery to the market should give feedback to the design process and affect any further development of the same solution or new versions of it. Second, the monitoring after the solution (artefact) is delivered should happen at different moments. It is not necessarily a continuous evaluation, neither at equal intervals of time. A major issue in this extended design process as it was mentioned earlier is to define who should do this monitoring, when, and for how long after the solution is delivered. Finally, another important feature in this proposal of an extended design process is the articulation of the two parts (before delivery and beyond delivery), the evaluation and reflection during the first part, before the delivery of the solution, should orient the monitoring during the second part, beyond the delivery of the solution.

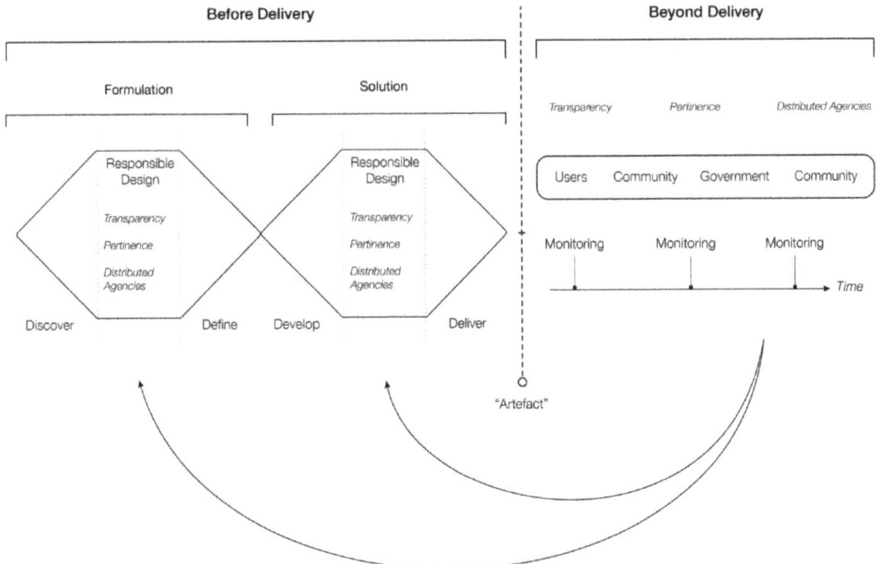

Figure 3. Responsible design before delivery. Original work adapted from [57].

6. Discussion

There are many topics we would like to discuss around this proposal of a new extended design process and its representation. Considering the scope of the paper we will summarize them in two issues. First, we would like to discuss the elements we believe represent well the ideas behind a responsible design and the fact that they are not necessarily an exhaustive list. Second, we would like to talk about an additional element we believe should be considered in an extended design process, proportionality.

Regarding the three elements we included in this proposal of an extended design process, *distributed agencies, transparency and pertinence*, we believe they represent well the main ideas behind the concept of responsible designs, but there might be others. This means that we accept that other elements can be included in the stages of reflexion and evaluation in the two diamonds of the model of the design process used to explain the additions. Any further elements that will be included in the extended design process should respond to the logic of being pillar elements and not specific topics regarding particular contexts. For example, in the field of sustainable design an important issue is to avoid the use of dangerous materials. Beyond the importance of this issue for the evaluation of a sustainable design, this is not a basic element that represents transversally the idea of responsible design, and in consequence it should not be one of the structural elements of the extended design process.

In relation with the second issue of discussion, we believe *proportionality* should be a theme that has to be taken into account in the development of an extended design process. Both before and after the delivery, we assert that there are many questions and decisions that design teams will have to answer for themselves in order to engage in a more responsible design process. It is not our ambition to provide specific answers, but rather, to suggest some key questions that we think are helpful to facilitate that decision-making process. We do believe, however, that in order to make those decisions, is it important to keep in mind a *principle of proportionality*. With proportionality we refer to the fact that now a model of the design process like the double diamond presented by the design council [57], is used indistinctly to guide the design of a chair as it is used to design a car, or a service like a financial loan. These three "artefacts" are by nature very different between them, and have different impacts in their fabrication, use, and final disposal. In this context we believe a model that wants to represent a responsible design process should consider the scale of the process and the scale of the resulting solution.

In our view, applying a principle of proportionality means that designers should consider the degree of complexity (i.e., the sensibility of the problem, the scale of the solution, the forecasted impact in the system after delivery, etc.) when seeking to design their own design process. As a rule of thumb, we propose that the more complexity, the more responsibility. Of course, as all rules of thumb, this is not an exact science and each case is different, as are the designers themselves. We are not sure yet how to include this element in our proposal, then we expose this topic as a future line of work open to receive ideas and propositions from other researchers working in the area.

As we have asserted, the impact of design is ultimately unpredictable. This should not lead to pessimism or the desire to abandon all attempts for a responsible governance of the design process. Collingridge [86] classically proposed the idea of a "dilemma of control" in technology, but it can also be applied for many of the wicked design problems. Collingridge's dilemma [86] is sustained in two premises:

1. Impacts cannot be easily predicted until the technology has been implemented and widely used.
2. Controlling or governing technology is hard when it has become entrenched in the system.

Both these premises seem to be correct and although it paints a bleak picture for design responsibility, we can look at it as a design challenge. We need to design the design process in order to improve our ability to make better judgments and predictions about the impact of our solutions and to plan how to learn and change as it is being implemented in a system.

Our argument for an extended design process also contests some of the current assumptions that including anticipation and reflexivity are good enough as the basis of responsibility in innovation [15,25]. This is not to say that they are not important, but rather, to state that in order to enact those principles, they have to be designed into the process itself. Additionally, sometimes that means defining their limits (for instance, monitoring in use cannot be replaced solely by anticipatory exercises).

In sum, the path to develop a new extended design process will need to address serious ethical and sociotechnical design challenges. For instance: (a) Dimensionality: what are the new dimensions of responsibility (e.g., pertinence and transparency) that would need to be incorporated? (b) Temporality: when does the design process end? (c) Agency: which actors need to be responsible for what elements of the extended design process? (d) Tooling: what design tools need to be created to mediate and facilitate the extension of the design process?

The projections of this work are multiple. Our next research steps involve first developing our proposals of pertinence, transparency, and distributed agencies further. This means that we will set out to operationalize them through concrete activities, tools, and indicators of success for every new step of the extended design process. Afterwards, we will seek to assess these initial ideas through the analysis and documentation of an implementation with real-life design teams. In this article we explore the issue of ethics in the design process, but as we noted throughout this exposition, there are other neighboring dimensions that could potentially be explored further, for instance, the role of power and politics, the education of future designers, and how technical knowledge relates to ethical concerns during the design process. We expect that by framing the challenges of responsible design and by attempting some initial ideas to extend the design process, we will ignite new debates and confrontations on the current state of the design process.

Finally, the limitations of this proposal must be kept in mind. Most importantly of all, as a conceptual article, we sought to frame the challenges with mainstream representations of the design process and offer a possible model that ultimately serves as proof of the possibility to reimagine the design process. As such, this article lacks the systematic empirical evidence that should support a more definitive extension of the design process. We hope that this initial exploration will ignite further explanations to show the roots and rationale of adapting this existing process, as well as systematic evidence pointing out strategic and comprehensive ways to incorporate responsibility into the core of the design process.

Author Contributions: Conceptualization, R.J.H. and J.G.; writing—original draft preparation, R.J.H. and J.G.; writing—review and editing, R.J.H. and J.G. All authors have read and agreed to the published version of the manuscript.

Funding: This research received no external funding.

Conflicts of Interest: The authors declare no conflict of interest.

References

1. Hernandez, R.J.; Cooper, R.; Jung, J. The understanding and use of design in the UK industry: Reflecting on the future of design and designing in industry and beyond. *Des. J.* **2017**, *20*, S2823–S2836. [CrossRef]
2. Margolin, V.; Margolin, S. A "Social Model" of Design: Issues of Practice and Research. *Des. Issues* **2002**, *18*, 24–30. [CrossRef]
3. Papanek, V. *Design for the Real World: Human Ecology and Social Change*, 2nd Revised ed.; Thames & Hudson: London, UK, 1991.
4. Whiteley, N. *Design for Society*; Reaktion Books: London, UK, 1993.
5. Thorpe, A.; Gamman, L. Design with society: Why socially responsive design is good enough. *CoDesign* **2011**, *7*, 217–230. [CrossRef]
6. Bhamra, T.; Lofthouse, V. *Design for Sustainability: A practical Approach*; Cooper, R., Ed.; Gower Publishing Limited: Hampshire, UK, 2007.
7. Margolin, V. Design for a Sustainable World. *Des. Issues* **1998**, *14*, 83–92. [CrossRef]
8. Bichard, J.-A.; Coleman, R.; Langdon, P. Does My Stigma Look Big in This? Considering Acceptability and Desirability in the Inclusive Design of Technology Products. In Proceedings of the International Conference on Universal Access in Human-Computer Interaction, Beijing, China, 22–27 July 2007; pp. 622–631.
9. Persson, H.; Åhman, H.; Yngling, A.A.; Gulliksen, J. Universal design, inclusive design, accessible design, design for all: Different concepts—One goal? On the concept of accessibility—Historical, methodological and philosophical aspects. *Univers. Access Inf. Soc.* **2015**, *14*, 505–526. [CrossRef]
10. Robertson, T.; Simonsen, J. Challenges and Opportunities in Contemporary Participatory Design. *Des. Issues* **2012**, *28*, 3–9. [CrossRef]
11. Melles, G.; de Vere, I.; Misic, V. Socially responsible design: Thinking beyond the triple bottom line to socially responsive and sustainable product design. *CoDesign* **2011**, *7*, 143–154. [CrossRef]
12. Cooper, R. Ethics and Altruism: What Constitutes Socially Responsible Design? *Des. Manag. Rev.* **2010**, *16*, 10–18. [CrossRef]
13. Friedman, B. *Human Values and the Design of Computer Technology*; Cambridge University Press: New York, NY, USA, 1997.
14. Friedman, B.; Hendry, D.G.; Borning, A. A Survey of Value Sensitive Design Methods. *Found. Trends Hum. Comput. Interact.* **2017**, *11*, 63–125. [CrossRef]
15. Stilgoe, J.; Owen, R.; Macnaghten, P. Developing a framework for responsible innovation. *Res. Policy* **2013**, *42*, 1568–1580. [CrossRef]
16. Voegtlin, C.; Scherer, A.G. Responsible Innovation and the Innovation of Responsibility: Governing Sustainable Development in a Globalized World. *J. Bus. Ethics* **2017**, *143*, 227–243. [CrossRef]
17. Dubberly, H. *How Do You Design? A Compendium of Models*; Dubberly Design Office: San Francisco, CA, USA, 2004.
18. Pugh, S. *Total Design: Integrated Methods for Successful Product Engineering*; Addison-Wesley: Boston, MA, USA, 1991.
19. Wynn, D.; Clarkson, J. Models of designing. In *Design Process Improvement: A Review of Current Practice*; Springer: Berlin/Heidelberg, Germany, 2005; pp. 34–59.
20. French, M. *Conceptual Design for Engineers*; Springer: Berlin/Heidelberg, Germany, 1999.
21. Ulrich, K.; Eppinger, S. *Product Design and Development*; McGraw-Hill: Boston, MA, USA, 2003.
22. Brown, T. *Change by Design (Google eBook)*; Harper Collins: New York, NY, USA, 2009; ISBN 0061937746.
23. Cross, N. *Design Thinking: Understanding How Designers Think and Work*, 1st ed.; Berg Publishers: Oxford, UK; New York, NY, USA, 2011.
24. Ulrich, K.T. *Design: Creation of Artifacts in Society*; University of Pennsylvania: Philadelphia, PA, USA, 2011; ISBN 978-0-9836487-0-3.

25. Fisher, E. Reinventing responsible innovation. *J. Responsible Innov.* **2020**, *7*, 1–5. [CrossRef]
26. Gilson, L.L.; Goldberg, C.B. Editors' Comment: So, What Is a Conceptual Paper? *Gr. Organ. Manag.* **2015**, *40*, 127–130. [CrossRef]
27. Hulland, J. Conceptual review papers: Revisiting existing research to develop and refine theory. *AMS Rev.* **2020**, *10*, 27–35. [CrossRef]
28. Stevenson, N. *A Better World by Design? An Investigation Into Industrial Design Consultants Undertaking Responsible Design Within Their Commercial Remits*; Loughborough University: Loughborough, UK, 2013.
29. Margolin, V. Social design: From utopia to the good society. In *Design for the Good Society*; Bruinsma, M., van Zijl, I., Eds.; Stichting Utrecht Biennale: Utrecht, The Netherlands, 2015; pp. 28–42.
30. John Clarkson, P.; Coleman, R. History of Inclusive Design in the UK. *Appl. Ergon.* **2015**, *46*, 235–247. [CrossRef]
31. Hartson, R.; Pyla, P. Background: Design. In *The UX Book*; Elsevier: Amsterdam, The Netherlands, 2019; pp. 397–401.
32. Simonsen, J.; Robertson, T. (Eds.) *Routledge International Handbook of Participatory Design*; Routledge: London, UK, 2012.
33. Huybrechts, L.; Benesch, H.; Geib, J. Institutioning: Participatory Design, Co-Design and the public realm. *CoDesign* **2017**, *13*, 148–159. [CrossRef]
34. Koskinen, I.; Hush, G. Utopian, molecular and sociological social design. *Int. J. Des.* **2016**, *10*, 65–71.
35. Papanek, V. *Design for the Real World*; Van Nostrand Reinhold: New York, NY, USA, 1984.
36. Meadows, D.; Meadows, D.; Randers, J.; Behrens, W. *The Limits to Growth: A Report for the Club of Rome's Project on the Predicament of Mankind*; Pan Books Ltd: London, UK, 1972.
37. WCED. *Our Common Future*; Oxford University Press: Oxford, UK, 1987.
38. Bhamra, T.; Hernandez, R.J.; Mawle, R. Sustainability: Methods and Practices. In *The Handbook of Design for Sustainability*; Bloomsbury Publishing Plc: London, UK, 2013; pp. 106–120.
39. Brezet, H.; Van Hemel, C. *Ecodesign: A Promising Approach to Sustainable Production and Consumption*; United Nations Environmental Programme (UNEP): Nairobi, Kenya, 1997.
40. Hernandez, R.J. Sustainable Product-Service Systems and Circular Economies. *Sustainability* **2019**, *11*, 5383. [CrossRef]
41. Hernandez, R.J.; Bhamra, T.; Bhamra, R. Sustainable Product Service Systems in Small and Medium Enterprises (SMEs): Opportunities in the Leather Manufacturing Industry. *Sustainability* **2012**, *4*, 175–192. [CrossRef]
42. Friedman, B.; Kahn, P.H.; Borning, A.; Huldtgren, A. Value Sensitive Design and Information Systems. In *Early Engagement and New Technologies: Opening up the Laboratory*; Doorn, N., Schuurbiers, D., van de Poel, I., Gorman, M.E., Eds.; Springer Netherlands: Dordrecht, The Netherlands, 2013; pp. 55–95.
43. Winkler, T.; Spiekermann, S. Twenty years of value sensitive design: A review of methodological practices in VSD projects. *Ethics Inf. Technol.* **2018**. [CrossRef]
44. Dobson, A. Environmental citizenship: Towards sustainable development. *Sustain. Dev.* **2007**, *15*, 276–285. [CrossRef]
45. Szenasy, S. Ethical Design Education: Confessions of a Sixties Idealist. In *Citizens Designer. Perspectives on Design Responsibility*; Heller, S., Vienne, V., Eds.; Allworth Press: New York, NY, USA, 2003; pp. 20–24.
46. Cipolla, C.; Bartholo, R. Empathy or Inclusion: A Dialogical Approach to Socially Responsible Design. *Int. J. Des.* **2014**, *8*, 87–100.
47. Markussen, T. Disentangling 'the social' in social design's engagement with the public realm. *CoDesign* **2017**, *13*, 160–174. [CrossRef]
48. Parsons, G. *The Philosophy of Design*; Polity Press: Malden, MA, USA, 2016.
49. Barbera, E.; Garcia, I.; Fuertes-Alpiste, M. A Co-Design Process Microanalysis: Stages and Facilitators of an Inquiry-Based and Technology-Enhanced Learning Scenario. *Int. Rev. Res. Open Distrib. Learn.* **2017**, *18*. [CrossRef]
50. Lucero, A.; Vaajakallio, K.; Dalsgaard, P. The dialogue-labs method: Process, space and materials as structuring elements to spark dialogue in co-design events. *CoDesign* **2012**, *8*, 1–23. [CrossRef]
51. Jarke, M.; Klamma, R.; Pohl, K.; Sikora, E. Requirements Engineering in Complex Domains. In *Graph Transformations and Model-Driven Engineering*; Springer: Berlin/Heidelberg, Germany, 2010; pp. 602–620.

52. Li, W.-T.; Ho, M.-C.; Yang, C. A Design Thinking-Based Study of the Prospect of the Sustainable Development of Traditional Handicrafts. *Sustainability* **2019**, *11*, 4823. [CrossRef]

53. Wang, S.H.; Chang, S.-P.; Williams, P.; Koo, B.; Qu, Y.-R. Using Balanced Scorecard for Sustainable Design-centered Manufacturing. *Procedia Manuf.* **2015**, *1*, 181–192. [CrossRef]

54. Sya'bana, Y.M.K.; Sanjaya, K.H.; Muharam, A. Ergonomie assessment on charging station touch screen based on task performance measurement. In Proceedings of the 2017 International Conference on Sustainable Energy Engineering and Application (ICSEEA), Jakarta, Indonesia, 23–24 October 2017; IEEE: Piscataway, NJ, USA, 2017; pp. 8–13.

55. Schulz, T.; Fuglerud, K.S.; Arfwedson, H.; Busch, M. A Case Study for Universal Design in the Internet of Things. In *Universal Design 2014*; IOS-Press: Lund, Sweden, 2014; pp. 45–54.

56. Waidelich, L.; Richter, A.; Kolmel, B.; Bulander, R. Design Thinking Process Model Review. In Proceedings of the 2018 IEEE International Conference on Engineering, Technology and Innovation (ICE/ITMC), Stuttgart, Germany, 17–20 June 2018; IEEE: Piscataway, NJ, USA, 2018; pp. 1–9.

57. Design Council. *Design for Innovation: Facts, Figures and Practical Plans for Growth*; Design Council: London, UK, 2011.

58. Brown, T. Design Thinking. *Harv. Bus. Rev.* **2008**, *86*, 84–92.

59. Cross, N. *Engineering Design Methods: Strategies for Product Design*, 2nd ed.; Wiley: Chichester, UK; New York, NY, USA, 1994; ISBN 978-0471942283.

60. Cooper, R.G. Stage-Gate Systems: A New Tool for Managing New Products. *Bus. Horiz.* **1990**, *33*, 44–54. [CrossRef]

61. Cross, N. The nature and nurture of design ability. *Des. Stud.* **1990**, *11*, 127–140. [CrossRef]

62. Brezet, H. Dynamics in ecodesign practice. *Ind. Environ.* **1997**, *20*, 21–24.

63. Hernandez, R.J.; Brissaud, D.; Mathieux, F.; Zwolinski, P. Contribution to the characterisation of eco-design projects. *Int. J. Sustain. Eng.* **2011**, *4*, 301–312. [CrossRef]

64. Coulton, P.; Lindley, J.; Cooper, R. *The Little Book of Design Fiction for the Internet of Things*; Lancaster University: Lancaster, UK, 2018.

65. Ashton, K. That 'Internet of Things' Thing. *RFID J.* **2009**, *22*, 97–114.

66. Weiser, M. The Computer for the 21st Century. *Sci. Am.* **1991**, *265*, 94–104. [CrossRef]

67. Sharma, N.; Shamkuwar, M.; Singh, I. The History, Present and Future with IoT. In *Internet of Things and Big Data Analytics for Smart Generation*; Balas, V., Solanki, V., Kumar, R., Khari, M., Eds.; Springer: Cham, Switzerland, 2019; pp. 27–51.

68. Pontin, J. *Bill Joy's Six Webs*; MIT Technology Review: Massachusetts, MA, USA, 2005.

69. Department, T.C.M.U.C.S. The "Only" Coke Machine on the Internet. Available online: https://www.cs.cmu.edu/~{}coke/history_long.txt (accessed on 1 October 2020).

70. Salomon, G. No Distribution without Individual's Cognition: A dynamic Interactional View. In *Distributed Cognitions: Psychological and Educational Considerations*; Salomon, G., Ed.; Cambridge University Press: Cambridge, UK, 1993; pp. 111–139.

71. McNeill, D. Global firms and smart technologies: IBM and the reduction of cities. *Trans. Inst. Br. Geogr.* **2015**, *40*, 562–574. [CrossRef]

72. Söderström, O.; Paasche, T.; Klauser, F. Smart cities as corporate storytelling. *City Anal. Urban Trends Cult. Theory Policy Action* **2014**, *18*, 307–320. [CrossRef]

73. Scuotto, V.; Ferraris, A.; Bresciani, S. Internet of Things: Applications and challenges in smart cities. A case study of IBM smart city projects. *Bus. Process Manag. J.* **2016**, *22*, 357–367. [CrossRef]

74. Wiig, A. IBM's smart city as techno-utopian policy mobility. *City* **2015**, *19*, 258–273. [CrossRef]

75. Wong, P.-H. Three arguments for "responsible users". AI ethics for ordinary people. In Proceedings of the ETHICOMP 2020 Paradigm Shifts in ICT Ethics, Logroño, La Rioja, Spain, 15 June–6 July 2020; Pelegrín-Borondo, J., Arias-Oliva, M., Murata, K., Lara, A.M., Eds.; Universidad de la Rioja: Logroño, La Rioja, Spain, 2020.

76. Oxford English Dictionary. Oxford English Dictionary Online. Available online: https://www.oed.com/ (accessed on 1 October 2020).

77. Butalla, L. The hottest trends in smart labels & packaging. *Convert. Mag.* **2004**, *22*, 46–51.

78. Ellen MacArthur Foundation. *Intelligent Assets: Unlocking the Circular Economy Potential*; Ellen MacArthur Foundation: Cowes, UK, 2016.

79. Wang, J.; Lim, M.K.; Zhan, Y.; Wang, X. An intelligent logistics service system for enhancing dispatching operations in an IoT environment. *Transp. Res. Part E Logist. Transp. Rev.* **2020**, *135*, 101886. [CrossRef]
80. Safkhani, M.; Rostampour, S.; Bendavid, Y.; Bagheri, N. IoT in medical & pharmaceutical: Designing lightweight RFID security protocols for ensuring supply chain integrity. *Comput. Netw.* **2020**, *181*, 107558. [CrossRef]
81. Schaefer, D.; Cheung, W.M. Smart Packaging: Opportunities and Challenges. *Procedia CIRP* **2018**, *72*, 1022–1027. [CrossRef]
82. Brunswicker, S.; Jensen, B.; Song, Z.; Majchrzak, A. Transparency as design choice of open data contests. *J. Assoc. Inf. Sci. Technol.* **2018**, *69*, 1205–1222. [CrossRef]
83. Rittel, H.W.J.; Webber, M.M. Dilemmas in a general theory of planning. *Policy Sci.* **1973**. [CrossRef]
84. Kiran, A.H. Responsible Design. A Conceptual Look at Interdependent Design-Use Dynamics. *Philos. Technol.* **2012**, *25*, 179–198. [CrossRef]
85. Boenink, M.; Swierstra, T.; Stemerding, D. Anticipating the Interaction between Technology and Morality: A Scenario Study of Experimenting with Humans in Bionanotechnology. *Stud. Ethics. Law. Technol.* **2010**, *4*. [CrossRef]
86. Collingridge, D. *The Social Control of Technology*; Open University: Milton Keynes, UK, 1980.

Publisher's Note: MDPI stays neutral with regard to jurisdictional claims in published maps and institutional affiliations.

Article

Thermal Assessment of a Micro Fibrous Fischer Tropsch Fixed Bed Reactor Using Computational Fluid Dynamics

Aya E. Abusrafa [1,2,†], Mohamed S. Challiwala [1,2,3,†], Benjamin A. Wilhite [2] and Nimir O. Elbashir [1,3,*]

[1] Chemical Engineering & Petroleum Engineering Program, Texas A&M University at Qatar, Education City, Doha 23874, Qatar; aya.abusrafa@qatar.tamu.edu (A.E.A.); m.challiwala@tamu.edu (M.S.C.)
[2] Artie McFerrin Department of Chemical Engineering, Texas A&M University, College Station, TX 77840, USA; benjaminwilhite@exchange.tamu.edu
[3] Gas and Fuels Research Center, Texas A&M Engineering Experiment Station, College Station, TX 77843, USA
* Correspondence: nelbashir@tamu.edu
† First and Second authors have contributed equally.

Received: 24 August 2020; Accepted: 24 September 2020; Published: 27 September 2020

Abstract: A two-dimensional (2D) Computational Fluid Dynamics (CFD) scale-up model of the Fischer Tropsch reactor was developed to thermally compare the Microfibrous-Entrapped-Cobalt-Catalyst (MFECC) and the conventional Packed Bed Reactor (PBR). The model implements an advanced predictive detailed kinetic model to study the effect of a thermal runaway on C_{5+} hydrocarbon product selectivity. Results demonstrate the superior capability of the MFECC bed in mitigating hotspot formation due to its ultra-high thermal conductivity. Furthermore, a process intensification study for radial scale-up of the reactor bed from 15 mm internal diameter (ID) to 102 mm ID demonstrated that large tube diameters in PBR lead to temperature runaway >200 K corresponding to >90% CO conversion at 100% methane selectivity, which is highly undesirable. While the MFECC bed hotspot temperature corresponded to <10 K at >30% CO conversion, attributing to significantly high thermal conductivity of the MFECC bed. Moreover, a noticeable improvement in C_{5+} hydrocarbon selectivity >70% was observed in the MFECC bed in contrast to a significantly low number for the PBR (<5%).

Keywords: Fischer Tropsch; syngas; CFD; entrapped cobalt catalyst; thermal management

---

1. Introduction

Fischer Tropsch (FT) synthesis, central to many gas-to-liquid (GTL) processes, is a process in which synthesis gas (or syngas, i.e., a mixture of H_2 and CO) is converted to a variety of hydrocarbon products including paraffin, olefins, and value-added chemicals [1–6]. FT synthesis is a highly exothermic reaction (the total heat released per mole of CO consumed is from 140 kJ/mol to 160 kJ/mol), and therefore, efficient heat removal is one of the main considerations while designing commercial-scale FT reactors [7,8]. Uncontrollable temperature gradients lead to the formation of local hotspots, and in some cases, unstable temperature runaways, which promote methane formation, lower the selectivity of the desired hydrocarbon products and lead to fast catalyst deactivation [2,6,9–16].

Currently, there are two modes for FT synthesis operation; High-Temperature Fischer Tropsch (HTFT) and Low-Temperature Fischer Tropsch (LTFT) [17,18]. The LTFT process is a three-phase process (gas-liquid-solid) and typically operates at temperatures ranging from 473 K to 513 K and utilizes cobalt-based catalysts to produce heavy hydrocarbons such as diesel and wax [4,19,20]. On the other hand, the HTFT process mainly involves two phases (gas-solid) [21], which operates at temperatures from 593 K to 623 K, and utilizes fused iron-based catalysts to produce lighter hydrocarbons such as olefins, oxygenates and gasoline [22]. HTFT operation is mainly conducted in fluidized bed reactors (FBR), while LTFT is applied in multi-tubular fixed bed (or packed bed reactors (PBR)), and Slurry Bubble Column reactors (SR). PBR has several advantages and is most often used in commercial applications [21]. This is due to the simple operation, high catalyst holdup, easy scale-up from a single tube to an industrial size multi-tubular reactor and shutdown robustness of the PBR compared to SBR and FBR [23,24]. Moreover, the separation of the Catalyst from the liquid product is not required in PBR. The liquid products in PBR trickle down through the reactor bed and are separated from the exit gas using a knock out vessel [7]. This imposes significant reductions in the operational costs of the process. On the other hand, due to the pressure drop limitations in PBR, particles relatively larger than 1 mm are utilized. These requirements lead to mass transfer limitations, as well as hotspot formations, that can negatively affect the product selectivity [25]. Therefore, to achieve higher productivity within the reactor bed, more active catalysts with higher thermal conductivity are needed, hence increasing the amount of heat released during the reaction. This would result in high radial and axial temperature gradients along the catalytic bed. This in return will lead to temperature runaways and loss of selectivity due to the poor effective thermal conductivity of the PBR [26,27]. For this reason, such types of reactors utilize several hundred to around ten thousand small diameter tubes (from about 20 mm to 50 mm) to facilitate heat removal [21]. The short distance between the catalyst particles and the tube walls provides more efficient heat transfer from the catalytic bed to the cooling medium. Additionally, the single-pass conversion in PBR is typically kept at 50% or lower to avoid temperature runaways and hotspot formation [28]. Therefore, a key factor to optimal FT performance in PBR is the efficient heat removal from the reactor bed to the cooling media.

To overcome the heat transfer limitations associated with PBR, a novel catalytic configuration consisting of thin copper microfibrous structures that provide a holder for the catalyst particles has been developed by Tatarchuk et al. [29,30]. Microfibers catalysts (MFEC) made of sintered micron-sized metal, glass or polymer fibers with small catalyst particles entrapped inside have also been reported [29,31]. These catalysts have been utilized in several studies to mitigate bed channeling, improve electrical conductivity in fuel cells and remove harmful airborne contaminants in air filtration systems [29,32–34]. The Microfibrous Entrapped Cobalt-based Catalytic Structure (MFECC) is produced by entrapping small cobalt particles in a porous metal sheet (made of copper) of interlocking microfibers [9]. Due to the micron-sized fibers of the microfibrous Catalyst, a large geometric surface area is offered, which dramatically enhances the thermal conductivity of the reactor bed compared to conventional PBR [29]. The high thermal conductivity and low thermal resistance of this catalytic matrix provide a significant improvement in temperature control compared to conventional PBR. Improved temperature control leads to longer catalyst lifetime and activity. The improved heat characteristics provided using MFECC structures allows the use of smaller catalyst particles with diameters ranging from 0.01 mm to 0.1 mm to eliminate mass transport resistances [29]. This provides better utilization of the Catalyst, and thus, higher productivity is achieved. Furthermore, MFECC provides a higher void fraction that results in a reduction of pressure drops compared to conventional PBR [29]. An experimental study of thermal conductivity of a conventional PBR catalyst and MFECC catalyst was conducted by Sheng et al. [29]. The study revealed that the effective radial thermal conductivity of MFECC was 56 times higher than that of PBR diluted with fresh alumina, while the wall heat transfer coefficient was ten times higher than that of the alumina PBR. Another study done by Kalluri et al. investigated the effect of bed porosity on the transport resistances for MFEC structures and diluted PBR [32]. They found that decreasing the bed void of the PBR only improved the flow disturbances and radial dispersion to a small extent. In contrast, the high void of the microfibrous structures promoted radial dispersion, which in turn led to more uniform radial concentration profiles and reduced flow disturbances.

To date, the implementation of this novel FT reactor technology has been limited to the laboratory scale. Several aspects still need to be addressed before the commercialization of MFECC reactor beds. The most important aspect is the scale-up to larger sized reactors to study the hydrodynamics and reactor performance under conditions used in the industry, which cannot be achieved experimentally.

The latter, however, can be done by applying Computational Fluid Dynamics (CFD) to represent the fluid behavior inside the reactor bed accurately. In our previous work, a two-dimensional (2D) model was developed to study the effect of using a non-conventional Supercritical Fluid (SCF) reaction media on the heat management characteristics of the bed [3,5]. Moreover, Challiwala et al. developed a 2D pseudo-homogenous model of an MFECC bed and conventional PBR using a simple kinetic model for a cobalt-based catalyst [9]. They analyzed the performance of the MFECC bed and PBR in terms of heat management and CO conversion at the various process and design parameters (inlet temperature, inlet flowrate, tube diameter). This study aims to extend on their works [3,5,9] and develop a 2D model of an FT fixed bed reactor (FB) in COMSOL® Multiphysics v5.3a utilizing a detailed kinetic model for a cobalt-based catalyst for two systems; non-conventional MFECC bed and conventional PBR, both operating under gas-phase conditions (GP-FT). The model is used to study the effect of the thermal performance of both the reactor systems on conversion levels and hydrocarbon product selectivity. Moreover, the potential of scaling-up the reactor tube diameter above typical industrial diameter (more than 102 mm internal diameter (ID)) was studied as a starting point to improve the performance of conventional FT technologies.

2. Materials and Methods

The fixed bed reactor was developed in 2D axisymmetric space via COMSOL® Multiphysics v5.3a [Multiphysics, 1998 #508]. The model geometry comprises three zones of pre-packing, catalytic bed and post-packing, respectively, as shown in Figure 1. For model validation purposes, the reactor dimensions in the CFD model were specified based on the geometry of the reactor used to conduct the FT experiments. Two-dimensional correlations were used in this modeling study for momentum, heat and mass transfer to account for the variation of concentration and temperature in the radial and axial directions. This is because, for larger tube diameters (scaling-up) [35], higher radial temperature gradients are expected, which will have a direct effect on the overall performance of the reactor bed. Since the main goal of this modeling work is to study the effect of heat generation on the reactor bed performance, including conversion and hydrocarbon product selectivity at larger tube diameters (up to 102 mm), a 2D modeling approach was chosen [35].

The fixed bed reactor developed in this work was modeled as a pseudo-homogeneous model. This assumption indicates that the interfacial mass and heat resistances occurring between the solid phase and the fluid phase are neglected. This implies that the catalyst effectiveness factor (the ratio of the overall reaction rate in the pellet to the surface reaction rate for a specific component) is equal to 1. Sheng et al. [29,30] calculated the effectiveness factor for 0.162 mm particle size to study the effect of intraparticle diffusion. Results showed that the effectiveness factor was relatively high (0.87); therefore, in this modeling study the internal diffusion effects were neglected. The fluid flow was assumed to be a single-phase flow; however, the presence of the liquid was considered by calculating the liquid physical properties (heat capacity, thermal conductivity, viscosity, diffusivity, density) of the hydrocarbon cuts which exist in the liquid and gas phase (GP).

Modeling and simulation of the FT reactor bed require a simultaneous solution of momentum, mass and energy balance equations in the three zones specified earlier: pre-packing, catalytic bed and post-packing. The entire operation is considered to be at steady state. The reactants CO and H_2 enter the reactor to the pre-packing zone and exit the reactor from the post-packing region, which is inert and non-catalytic. It is assumed that the reaction only takes place in the catalytic bed over 15% Co/Al_2O_3 catalyst particles of identical sphericity ($\emptyset = 1$). Additional assumptions and more details are stated in each part of the model development section.

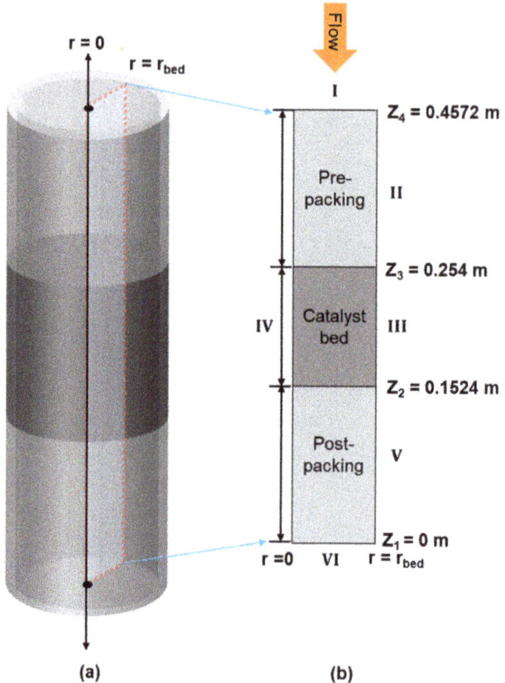

Figure 1. (**a**) Three-dimensional model and (**b**) 2D-axisymmetric cut section of the Fischer Tropsch (FT) reactor bed model cylindrical geometry.

2.1. Momentum Transport Expressions

To model fluid flow in porous media, a built-in module in COMSOL® Multiphysics called "Brinkman Equation," was adopted Equation (1). The Brinkman physics module is used to compute the fluid velocity and pressure field in single-phase flow in porous media in the laminar flow regime. This mathematical model extends Darcy's law to account for dissipation in kinetic energy due to shear stress, similar to the Naiver stokes equation [36]. This physics module comprises two main terms; the Forchheimer drag term and the convective term. The convective terms take into account the effect of inertial and viscous forces on the fluid flow through the porous medium. The Forchheimer drag term accounts for the inertial drag effects that occur in fast flows (Reynold number (*Re*) greater than unity) [37]. Considering slow flow regimes where *Re* is less than unity, the Forchheimer drag contributions are neglected. The 2D single-phase fluid flow through the PBR is described in terms of the velocity (*u*) and pressure fields (*p*), which are computed via solving the momentum equation and continuity equation (Equation (1) and Equation (2), respectively) simultaneously along with boundary conditions (Equation (3)) pertaining to (i) radial symmetry, (ii) no-slip condition at the wall, (iii) inlet mass flow, (iv) outlet pressure. The changes in volumetric gas flow rates of the reacting species during the FT reaction results in variations in the fluid density; therefore, a compressible flow formulation of the continuity equation is used. When a compressible flow is modelled using the Brinkman equation, the Mach number must be below 0.3. This condition is fulfilled in this case study.

$$\frac{1}{\epsilon_{bed}}\rho_f(u \cdot \nabla)u\frac{1}{\epsilon_{bed}} = \nabla \cdot \left[-pI + \varphi_f \frac{1}{\epsilon_{bed}}\left(\nabla u + (\nabla u)^T\right) - \frac{2}{3}\varphi_f \frac{1}{\epsilon_{bed}}(\nabla \cdot u)I \right] - \left(\frac{\varphi_f}{\kappa_{bed}} + \beta_f |u| \right)u \tag{1}$$

$$\nabla(\rho_f \mathbf{u}) = 0 \tag{2}$$

$$u = 0 \, @ \, r = r_{bed}, q_m = q_{m,o} \, @ \, z = z_4, \, p = p_e \, @z = z_1 \tag{3}$$

where, φ_f is the dynamic viscosity of the fluid, ϵ_{bed} is the porosity, ρ_f is the density of the fluid, κ_{bed} is the permeability of the porous media, β_f is the Forchheimer drag coefficient, r is the radius of the reactor bed, q_m is the mass flow rate and z is the height of the reactor bed (cartesian coordinate). A constant porosity of 0.626 was chosen for the MFECC bed, while 0.36 was chosen for the PBR per the catalyst specifications reported by Sheng et al. [29,30].

The permeability of the porous medium was calculated using the modified Ergun equation [38]:

$$\frac{1}{\kappa_{bed}} = \frac{150(1 - \epsilon_{bed})^2}{d_p^2 \epsilon_{bed}^2} + \frac{1.75 \rho_f u (1 - \epsilon_{bed})}{d_p \varphi_f \epsilon_{bed}^3} \tag{4}$$

where d_p is the particle diameter.

2.2. Mass Transport Expressions

Mass conservation equations for the pseudo homogenous reaction (assuming catalyst effectiveness as unity) is defined for each component of the reaction mixture of the following FT reactions:

$$n \, CO + (2n + 1) \, H_2 \, \rightarrow \, C_nH_{(2n+2)} + n \, H_2O \text{ (Synthesis of paraffin)} \tag{5}$$

$$n \, CO + 2n \, H_2 \, \rightarrow \, C_nH_{2n} + n \, H_2O \; \text{ (Synthesis of olefins)} \tag{6}$$

Components considered in the system are N_2, CO, H_2, H_2O, CH_4, CH_o and CH_p. N_2 was set as the mass constraint component since it is a non-reacting species. The hydrocarbon components CH_o and CH_p are the summation of olefins and paraffin products, respectively. This was done for C_1, C_2, C_3, ... , C_{15} components. The C_{15} to C_{22} hydrocarbons are lumped into one component represented by a paraffinic compound $C_{19}H_{40}$, while the higher weight hydrocarbons C_{22+} are represented by the paraffinic component $C_{22}H_{40}$. It is important to note that this study considers a cobalt-based catalyst where the rate of water-gas shift reaction is assumed to be negligible. Therefore, selectivity calculations of CO_2 have not been considered in this modeling study.

The local mass balance for species i (N_2, CO, H_2, H_2O, CH_4, CH_o, and CH_p) was described by Equation (7) using a built-in physics module "Transport of concentrated species." This physics accounts for mass transport through convection and diffusion in the axial and radial directions. The equation provided in COMSOL® Multiphysics for transport mechanism is as follows:

$$\nabla j_i + \rho_f(u \cdot \nabla) w_i = r_i \tag{7}$$

where w_i is the mass fraction of species i, j_i is the mass flux relative to the mass averaged velocity of species i and r_i is the reaction rate representing production or consumption of species i.

The diffusion model selected in this case was the Maxwell Stefan diffusion model, where the relative mass flux vector is calculated using Equation (8):

$$j_i = - \rho_f w_i \sum D_{ik} d_k \tag{8}$$

where D_{ik} represents the binary diffusion coefficient and d_k is the diffusional driving force acting on species k defined as follows:

$$d_k = \nabla x_k + \frac{1}{p}[(x_k - w_k)\nabla p] \tag{9}$$

where x_k is the mole fraction is species k.

Equations (7)–(9) are solved using appropriate boundary conditions corresponding to (i) axial symmetry, (ii) no-flux at the wall and (iii) inlet composition:

$$\frac{\partial w_i}{\partial r} = 0 \ @ \ r = r_{\text{bed}}, \ w_i = w_{i,o} \ @ \ z = z_4 \tag{10}$$

The binary diffusivities D_{ik} in the Maxwell-Stefan diffusion model are estimated using Fickian diffusivities with an empirical correlation proposed by Fuller et al. [39]:

$$D_{ik} = \frac{\sqrt{\frac{1}{M_i} + \frac{1}{M_k}}}{p\left(v_{c,i}^{\frac{1}{3}} + v_{c,k}^{\frac{1}{3}}\right)^2} \times 10^{-7} \tag{11}$$

where M_n represents the mean molar mass and v_c represents the molar volume for species i and k. The molar volume of the representative paraffin and olefin components CH_o and CH_p are calculated based on the ASF product distribution. A correlation based on the molar weight average sum using the molar volumes of the individual hydrocarbon species is used. The molar volumes of H_2O, CO, H_2, N_2 and CH_4 used in this model are: (12.7, 18.9, 7.07, 17.9 and 37.9) $\frac{cm^3}{mol}$, respectively.

2.3. Heat Transport Expressions

Energy balance within the 2D reactor domain was considered to account for the transport of heat through convection, conduction and thermal dispersion. Radiative heat transport was neglected in this case. Balance equations were solved using the simplified "Heat transfer in porous media physics" Equation (12):

$$\rho_f C_p \boldsymbol{u} \cdot \nabla T - \nabla(k_{\text{eff}} \nabla T) = (-\Delta H_{\text{rxn}}) R_{\text{CO}} \tag{12}$$

where T is the temperature inside the reactor bed, C_p is the heat capacity of the fluid mixture, k_{eff} is the effective thermal conductivity of the reactor bed, R_{CO} is the rate of consumption of carbon monoxide and ΔH_{rxn} is the heat of the reaction per mole of CO consumed.

The reaction enthalpy $(-\Delta H_{\text{rxn}})$ is an important parameter that determines the amount of heat released during the FT reaction. Previous modeling studies reported values of $(-\Delta H_{\text{rxn}})$ ranging from 150 kJ/mol to 165 kJ/mol to represent the FT reaction enthalpy [9,21,30,40,41]. Its value mainly depends on the hydrocarbon product selectivity. The $-\Delta H_{\text{rxn}}$. value used in this study was 152 kJ/mol.

The effective thermal conductivity of the bed was calculated using a volume-based average model to account for both the solid matrix and the fluid properties:

$$k_{\text{eff}} = \epsilon_{\text{bed}} k_s + (1 - \epsilon_{\text{bed}}) k_f \tag{13}$$

where k_s is the thermal conductivity of the catalyst bed and k_f is the thermal conductivity of the fluid mixture.

Equation (12) was solved along with boundary conditions corresponding to (i) radial symmetry, (ii) external cooling (heat transfer between the reactor and a constant temperature cooling medium), (iii) inlet temperature and (iv) open outflow:

$$k_{er} \frac{\partial T}{\partial r} = U_{\text{overall}} (T_c - T) @ \ r = r_{\text{bed}}, T = T_o @ \ z = z_3, \ \frac{\partial T}{\partial z} = 0 \ @ \ z = 0 \tag{14}$$

The overall heat transfer coefficient (U_{overall}) represents the overall heat transmittance from the reactor bed to the vicinity of the wall. The latter is defined using the following correlation suggested by Mamonov et al. [42]:

$$U_{\text{overall}} = \left(\frac{d_t}{8k_{er}} + \frac{1}{h_{\text{wall}}} + \frac{d_w}{k_w} + \frac{1}{h_{w,\text{ext}}}\right) \tag{15}$$

where d_t is the inner tube diameter, k_{er} is the effective radial heat coefficient of the catalyst bed, $h_{w,int}$ is the radial heat transfer coefficient near the wall, d_w is the wall thickness, k_w is the thermal conductivity of the wall and $h_{w,ext}$ is the heat transfer coefficient from the tube wall to the cooling liquid. Values for k_w and $h_{w,ext}$ were taken from Mamonov et al. [42].

(1) Radial Heat Transfer Coefficient at the Wall

The effective radial heat transfer coefficient at the wall h_{wall} is one of the main parameters that determine the rate of heat transfer in PBR. This parameter quantifies the increase in heat transfer resistance near the wall of the reactor bed. Several correlations to properly estimate the h_{wall} value has been proposed in the literature [43–51]. Specchia and Baldi proposed a two-parameter correlation that has shown to satisfactorily predict the h_{wall} value in PBR with different particle geometries [21]. This was used in the present work:

$$h_{wall} = h_{wall,o} + h_{wall,g} \tag{16}$$

$$h_{wall,o} = \frac{k_f}{d_p}\left(2\epsilon_{bed} + \frac{1 - \epsilon_{bed}}{\frac{k_f}{k_s} \times \gamma_w + \varphi_w}\right) \tag{17}$$

$$h_{wall,g} = \frac{k_f}{d_p} \times 0.0835 \times Re^{0.91} \tag{18}$$

where γ_w and φ_w are dimensionless parameters, Re represents the Reynold numbers and $h_{wall,o}$ represents the stagnant/conductive contribution while $h_{wall, g}$ represents the convective contribution.

The dimensionless parameters γ_w and φ_w in Equation (17) are dependent on the geometry of the contact surface between the particle and the wall. For spherical particles, the parameters are defined as $(\gamma_w = \frac{1}{3}, \varphi_w = 0.0024 \times \left(\frac{d_t}{d_p}\right)^{1.58})$ [47].

(2) Effective Radial Thermal Conductivity

The effective radial thermal conductivity is one of the main parameters that determine the rate of heat transfer in PBR; the parameter effecting heat transfer in PBR is the radial effective thermal conductivity k_{rad}. A two-parameter correlation that adequately predicts the effective radial heat transfer coefficient in PBR was taken from Specchia and Baldi [46]:

$$k_{rad} = k_{rad,o} + k_{rad,g} \tag{19}$$

$$k_{rad,o} = \left(\epsilon_{bed} + \frac{\beta(1 - \epsilon_{bed})}{\varphi + \frac{k_f}{k_s} \times \gamma}\right)k_f \tag{20}$$

$$k_{rad,g} = \frac{Re_{pa}Pr}{8.65\left(1 + 19.4 \times \frac{d_p}{d_t}^2\right)}k_f \tag{21}$$

where Pr represents the Prandtl number, β, γ and φ represent the ratios between characteristic lengths and the particle diameter (the particles are assumed to be spheres) and $k_{rad,o}$ represents the static/conduction contribution while $k_{rad,g}$ represents the convective contribution.

Kunii and Smith reported that for spherical particles $\beta = 1$ for almost all packed beds, $\gamma = \frac{2}{3}$ and $\varphi = 0.22(\epsilon_{bed})^2$ based on fitting of experimental data for $k_{rad,o}$ [52].

2.4. Kinetics

The rate of CO disappearance is calculated using the Yates and Satterfield (YS) kinetic model, which has been commonly used in previous modeling studies [21,25,53–57]:

$$-r_{CO}{}^{YS} = \frac{k\,p_{H_2}p_{CO}}{(1 + a\,p_{CO})^2} \tag{22}$$

$$k = A_k\,exp\left(-\frac{E_k}{RT}\right) \tag{23}$$

$$a = A_a\,exp\left(-\frac{E_a}{RT}\right) \tag{24}$$

where p_{CO}, and p_{H_2} are the partial pressures of CO and H_2, k and a are the kinetic rate constants, A_k and A_a are the pre-exponential factors and E_k and E_a are the activation energies for CO consumption.

The product selectivity for CH_4 and C_{2+} hydrocarbons is calculated using a kinetic model by Ma et al. [58] and a detailed kinetic model of Todic et al. [59], respectively. The Ma kinetic model for CH_4 formation was proven to provide a good prediction of CH_4 selectivity [55], while the detailed kinetic model developed based on the carbide mechanism showed a good prediction of the hydrocarbon product distribution [59].

The rate of formation of CH_4 by the following expression [58]:

$$r_{CH_4}{}^{Ma} = \frac{k_M p_{CO}{}^{a_M} p_{H_2}{}^{b_M}}{1 + m_M \frac{p_{H_2O}}{p_{H_2}}} \tag{25}$$

$$k_M = A_M\,exp\left(-\frac{E_M}{RT}\right) \tag{26}$$

where k_M is the rate constant, m_M is the water effect coefficient, a_M is the reaction order of CO, b_M is the reaction order of H_2, A_M is the pre-exponential factor and E_M is the activation energy for CH_4 formation.

The rate of formation of the C_{2+} hydrocarbons from the detailed kinetic model by Todic et al. [59] performance as follows:

$$r_{C_2H_4}{}^{Prod} = k_{6E,0}\,e^{2c}\,\sqrt{K_7 p_{H_2}}\,\alpha_1\alpha_2[S] \tag{27}$$

$$r_{C_nH_{2n+2}}{}^{Prod} = k_5 K_7{}^{0.5} p_{H_2}{}^{1.5}\alpha_1\alpha_2 \prod_{i=3}^{n} \alpha_i[S]\ n \geq 2 \tag{28}$$

$$r_{C_nH_{2n}}{}^{Prod} = k_{6,0}\,e^{cn}\,\sqrt{K_7 p_{H_2}}\,\alpha_1\alpha_2 \prod_{i=3}^{n} \alpha_i[S]\ n \geq 3 \tag{29}$$

where the ks represent the kinetic rate constants, Ks represent the equilibrium constants, α_n are the chain growth probabilities and $[S]$ is the fraction of vacant sites.

The chain growth probabilities dependent on the carbon number are calculated using the following expressions:

$$\alpha_1 = \frac{k_1 p_{co}}{k_{5m} p_{H_2} + k_1 p_{co}} \tag{30}$$

$$\alpha_2 = \frac{k_1 p_{co}}{k_5 p_{H_2} + k_{6,0}\,e^{2c} + k_1 p_{co}} \tag{31}$$

$$\alpha_n = \frac{k_1 p_{co}}{k_5 p_{H_2} + k_{6,0}\,e^{cn} + k_1 p_{co}}\ n > 2 \tag{32}$$

The fraction of vacant sites is calculated as follows:

$$[S] = 1 / \left\{ 1 + \sqrt{K_7 p_{H_2}} + \sqrt{K_7 p_{H_2}} \left(1 + \frac{1}{K_4} + \frac{1}{P_{H_2} K_3 K_4} + \frac{1}{K_2 K_3 K_4} \frac{p_{H_2O}}{p_{H_2}} \right) \left(\alpha_1 + \alpha_1 \alpha_2 + \alpha_1 \alpha_2 \sum_{i=3}^{n} \prod_{j=3}^{i} \alpha_j \right) \right\} \tag{33}$$

The kinetic parameters in the detailed kinetic model, YS model, and the Ma kinetic model were estimated by Todic et al. [59] and Stamenic et al. [55] using experimental data with 0.48%Re 25%Co/Al$_2$O$_3$ catalyst.

In this work, a hybrid kinetic model adopted by Stamenic et al. [55] and Bukur et al. [56] was used. In their works, they coupled the YS model, the Ma model and a detailed kinetic model based on the CO insertion mechanism. In this case study, the reaction rates for CO, H$_2$, H$_2$O, CH$_p$ (n-paraffin) and CH$_o$ (1-olefins) was defined by coupling the YS kinetics for CO consumption, Ma et al. kinetics for CH$_4$ formation and the detailed kinetic model of Todic et al. for C$_{2+}$ hydrocarbon formation based on the carbide mechanism. A normalization procedure was followed to obtain atomic balances for C, O and H. The procedure was done described below.

The rate of consumption of CO excluding methane from the detailed kinetic model (rate of C$_{2+}$ formation from the model by Todic et al.) can be calculated based on the reaction stoichiometry as:

$$(-r_{CO})_{C_{2+}}^{\text{Prod}} = \sum_{n=2}^{n} n \left(r_{C_n H_{2n+2}}^{\text{Prod}} + r_{C_n H_{2n}}^{\text{Prod}} \right) \tag{34}$$

The rate of CO consumption excluding methane from the YS model (rate of C$_{2+}$ formation from YS model) is calculated as:

$$(-r_{CO})_{C_{2+}}^{\text{YS}} = \left(-r_{CO}^{\text{YS}} \right) - \left(r_{CH_4}^{\text{Ma}} \right) \tag{35}$$

The normalized rates of formation of C$_{2+}$ hydrocarbons are obtained as follows:

$$r_{C_n H_{2n+2}} = r_{C_n H_{2n+2}}^{\text{Prod}} \times \frac{(-r_{CO})_{C_{2+}}^{\text{Prod}}}{(-r_{CO})_{C_{2+}}^{\text{YS}}} \quad n > 2 \tag{36}$$

$$r_{C_n H_{2n}} = r_{C_n H_{2n}}^{\text{Prod}} \times \frac{(-r_{CO})_{C_{2+}}^{\text{Prod}}}{(-r_{CO})_{C_{2+}}^{\text{YS}}} \quad n \geq 2 \tag{37}$$

From the stoichiometry, the H$_2$ formation rate is calculated using the individual product formation rates of the hydrocarbon species as:

$$- r_{H_2} = 3 r_{CH_4}^{\text{Ma}} + \sum_{n=2}^{n} \left[(2n+1) r_{C_n H_{2n+2}} + 2n r_{C_n H_{2n}} \right] \tag{38}$$

The rate of H$_2$O formation is equal to the rate of CO consumption based on the reaction stoichiometry:

$$r_{H_2O} = - r_{CO}^{\text{YS}} \tag{39}$$

3. Results

3.1. Comparison of Model Predictions with Experimental Data

The developed model was validated with experimental data reported by Sheng et al. [29,30] to test its robustness under different experimental conditions. Sheng et al. reported two experiments to compare between the PBR and MFECC beds under GP conditions [30]. The experiment was conducted in a stainless-steel tubular reactor with a wall thickness of 2 mm and 15 mm ID. The total height of the reactor bed was 457.2 mm compromising of a 203.2 mm pre-packing zone, a 101.6 mm catalyst bed and

152.4 mm post-packing zone. The MFECC bed consisted of $\varphi = 37.4\%$ (30% (15% Co/Al$_2$O$_3$) and 7.4% copper fibers) and $\epsilon_{\text{bed}} = 62.6\%$. The PBR was diluted to the same catalyst density as the MFECC bed with fresh alumina of different particle sizes. The overall bed space in the PBR comprised $\varphi = 65\%$ (30% (15%Co/Al$_2$O$_3$) and 34% fresh alumina) and $\epsilon_{\text{bed}} = 36\%$. The average particle size of the Catalyst in the reactor bed was 0.149 mm to 0.177 mm.

The experimental results were obtained by varying the inlet temperature over a range from 498.15 K to 528.15 K at 2 MPa pressure and syngas molar ratio (H$_2$/CO) of 2:1 at a constant Gas Space Velocity (GHSV) of 5000 h^{-1}. In all simulation runs, the parameters (inlet temperature, pressure, GHSV and H$_2$/CO ratio) were kept identical to those used in the experimental study by Sheng et al. [29,30]. A particle diameter of 0.162 mm was used in the simulation to represent the average particle size of 0.149 mm to 0.177 mm used in the experiments, as reported by Sheng et al. [30].

Four sets of results were used to validate the model with the experimental study, including CO conversion, maximum temperature deviation from the centerline to the reactor wall ($T_{\max} - T_{\text{wall}}$), CH$_4$ selectivity and C$_{5+}$ selectivity. The simulation results for the MFECC bed demonstrated a clear agreement with the experimental predictions for CO conversion, ($T_{\max} - T_{\text{wall}}$), CH$_4$ selectivity and C$_{5+}$ selectivity, as shown in Table 1. For the PBR case, the CO conversions from the modeling results consistently match with the experimentally-obtained PBR results under all conditions, as shown in Table 2. Moreover, the ($T_{\max} - T_{\text{wall}}$), CH$_4$ selectivity and C$_{5+}$ selectivity also closely match with the experimental results from 498.15 to K 518.15 K. However, at 528.15 K, the deviation between the modeling and the experimental predictions becomes higher. This deviation in model predictions from experimental data could be attributed to the sensitivity of the kinetic parameters that are generated using a ruthenium promoted Catalyst of different loading reported by Stamenic et al. [55] and Todic et al. [55,59]. The lower CH$_4$ selectivity predicted from the model at a high operating temperature (528.15 K) can be due to the fact that at high-temperature methane formation could follow multiple reaction routes on FT sites [60,61]; (1) termination of the chain growth, (2) through intermediates participating in chain propagation and (3) due to hydrogenation of surface carbon. The latter methane formation pathway does not follow the polymerization/chain growth route for FT synthesis. However, the kinetic model used in this modeling study to predict the methane selectivity has not considered the secondary pathway for methane formation, thus the methane selectivity shows greater deviation from the predicted values relative to the experimental data. Additionally, the experiments were conducted using a ruthenium catalyst which has lower methane selectivity compared to conventional cobalt-based catalysts. This also results in a deviation in C$_{5+}$ selectivity compared to the experimentally predicted values. Further analysis of the validation results reveals that the deviation of the reactor wall temperature from the centerline temperature for the PBR is higher and increases at a faster rate than the MFECC bed under all temperature conditions (498.15 K to 528.15 K). However, it can be noted that when the inlet temperature reaches 528.15 K, a drastic increase of the centerline temperature inside the PBR occurs. This rapid ignition of the PBR temperature leads to the formation of a hotspot and a rapid decrease in the catalyst activity. Such an effect is not observed in the MFECC bed due to the high thermal conductivity of the copper fibers, which aids in eliminating the formation of hotspots even at a high operating temperature (528.15 K).

A close comparison between the PBR and MFECC in terms of CO conversion indicates that the MFECC bed provides lower conversions compared to the PBR case for the same reactor temperature. However, the rapid increase in reactor temperature in the PBR, leading to hot spot formation, shifts the product selectivity towards methane and light hydrocarbon products. This effect can be observed in Tables 1 and 2, where the CH$_4$ selectivity is higher while the C$_{5+}$ selectivity is lower for the PBR compared to the MFECC bed under all temperature conditions. Thus, the high conversion levels achieved in the PBR reactors go mostly toward the formation of methane. The results discussed above demonstrate that the developed model is valid and applicable to quantitatively compare the performance of the PBR and MFECC bed in terms of thermal profiles and hydrocarbon product selectivity.

Table 1. $T_{max} - T_{wall}$, CO conversion, CH_4 selectivity, C_{5+} selectivity from the model, and experimental results for the Microfibrous-Entrapped-Cobalt-Catalyst (MFECC) reactor concerning wall temperature at 2 MPa pressure for H_2: CO ratio of 2:1 and 5000 h^{-1} Gas Space Velocity (GHSV).

T_{wall} (K)	$T_{max} - T_{wall}$ (K) Experimental	$T_{max} - T_{wall}$ (K) Model	CO Conversion Experimental	CO Conversion Model	CH_4 Selectivity Experimental	CH_4 Selectivity Model	C_{5+} Selectivity Experimental	C_{5+} Selectivity Model
498.15	0	1.65	17	24.85	13	10.64	82.1	83.23
508.15	2.5	2.59	38.25	38.13	17	13.64	80.1	80.42
518.15	4.8	4.05	51.67	54.24	19	17.49	73.8	76.81
528.15	9.5	6.16	78.75	72	21	22.48	70.2	72.1

Table 2. $T_{max} - T_{wall}$, CO conversion, CH_4 selectivity, C_{5+} selectivity from the model, and experimental results for the Packed Bed Reactor (PBR) reactor concerning the wall temperature at 2 MPa pressure for H_2: CO ratio of 2:1 and 5000 h^{-1} GHSV.

T_{wall} (K)	$T_{max} - T_{wall}$ (K) Experimental	$T_{max} - T_{wall}$ (K) Model	CO Conversion Experimental	CO Conversion Model	CH_4 Selectivity Experimental	CH_4 Selectivity Model	C_{5+} Selectivity Experimental	C_{5+} Selectivity Model
498.15	5.6	6.35	35.71	49.33	15	11.66	79.7	82.28
508.15	9	11.75	53.13	69.08	19	16.3	73.4	75.3
518.15	14.1	21.85	86.81	89.74	30	25.2	50.6	69.55
528.15	69.8	48.4	99.32	92.27	83	43.2	12.4	52.68

3.2. Comparison of Thermal Profiles

The validated 2D model for the PBR and MFECC bed was used to compare the thermal profiles of the two reactor beds using different inlet conditions; inlet temperature and GHSV. First, a side by side comparison of the thermal profiles of the PBR and MFECC bed was done to study the radial and axial temperatures of both reactor beds at 528.15 K, 2 MPa pressure, H_2/CO ratio of 2:1 and a constant GHSV 5000 h^{-1}. Figure 2 shows the temperature profile predicted by the 2D reactor model for both PBR and MFECC bed.

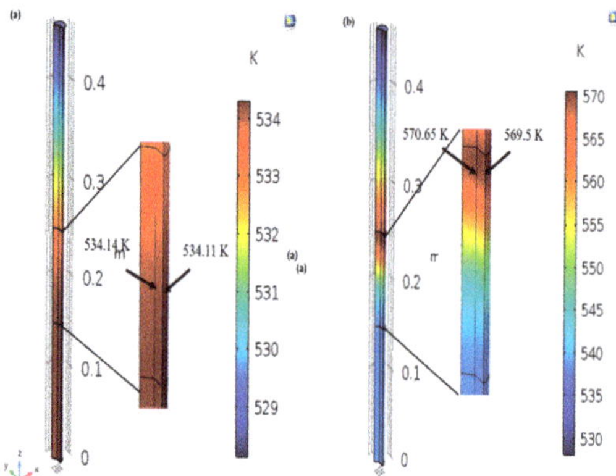

Figure 2. (**a**) Hot spot in MFECC bed and (**b**) Hotspot in PBR under GP conditions for 15 mm ID at 528.15 K, 2 MPa pressure, H_2/CO 2:1 and 5000 h^{-1} GHSV.

As can be seen from Figure 2b, a hotspot is predicted in the PBR at the centerline of the reactor and close to the reactor inlet. Giovanni et al. [40] reported a similar finding for a 2D pseudo homogenous model of a milli-scale fixed FT bed reactor using a Co-based catalyst. The axial temperature deviation for the PB reactor as predicted from the model is around 43 K, and the radial temperature gradient is around 1.15 K. As discussed previously, the temperature deviation predicted from the model for the PBR case at 528.15 K is underpredicted. Therefore, the axial temperature gradient from the experiments conducted by Sheng et al. [30] was even higher—around 70 K. The occurrence of the maximum temperature at the reactor inlet is due to the high partial pressure of the reactants at that location, which results in higher reaction rates. Therefore, heat generation is significantly higher. Moreover, under typical FT conditions, the inlet of the reactor is a region where the liquid is absent since PBR typically operate under trickle bed behavior (the liquid produced during FT trickles down the bed). The latter has negative implications on the heat transfer process inside the PBR reactor. It is worth noting that the PB reactor shows a temperature increase at the pre-packing zone, indicating that the ability of the external cooling on the reactor bed to remove excess heat is extremely insufficient, which leads to a temperature rise at the pre-packing zone and close to the inlet due to thermal diffusion. In the lower part of the reactor, the temperature decreases steadily due to the lower reaction rates in that zone, reducing the amount of heat generated during the reaction. Moreover, the effect of liquid formation at the lower part of the reactor is more prominent (trickle bed behavior), which positively affects the rate of heat transfer within the reactor bed.

The MFECC bed provided better temperature control, and a uniform temperature profile was maintained, as can be seen in Figure 2a. The maximum axial temperature rise in the MFECC bed was only 10 K, and the radial temperature gradient was 0.013 K. This reduction in hot spot formation in the MFECC bed is solely the result of the high thermal conductivity of the MFECC material. Sheng et al. experimentally determined the thermal parameters of the MFECC bed and PBR (effective radial thermal conductivity and wall heat transfer coefficient) [29]. The study reported that the radial effective thermal conductivity of MFECC was 56 times higher than that of alumina PBR in a stagnant gas, while the inside wall heat transfer coefficient was 10 times higher.

As mentioned previously, the hydrocarbon product distribution in the FT reaction strongly depends on the temperature inside the reactor bed. The hotspot formed in the PBR at 528.15 K resulted in around 100% CO conversion as per experimental results, where most of the conversion goes toward methane formation, as it is favorable at high-temperature conditions (shifting the product selectivity toward lower weight hydrocarbons). Based on the modeling results shown in Table 2, the values of CH_4 and C_{5+} selectivity in the PBR at 528.15 K were 43.2% and 52.68%, respectively. However, as mentioned previously, these values were unpredicted by the model at 528.15 K, and a higher CH_4 selectively, and thus, lower C_{5+} selectivity are expected at such a temperature condition. The experimental values of CH_4 and C_{5+} selectivity in the PBR at 528.15 K were 83% and 12.44%, respectively. For the MFECC, the uniform temperature distribution resulted in higher selectivity toward higher weight hydrocarbon products and lowered CH_4 selectivity. As can be seen from Table 1, the model predictions for CH_4 and C_{5+} selectivity in the MFECC bed at 528.15 K was 22.48% and 72.2%, respectively. The latter findings imply that the MFECC bed provides safe operation under high operational temperatures to achieve high conversions per tube pass in conventional Multi tubular fixed bed reactors/PBR without the risk of selectivity loss.

3.3. Effect of Varying the GHSV

The impact of varying the inlet gas flow rate/GHSV on the heat generation and removal for the MFECC bed and PBR was investigated. The simulations were carried out by varying the GHSV while keeping the other process parameters constant (518.15 K, 2 MPa pressure, H_2/CO 2:1). Figure 3 shows the effect of varying the GHSV on the reactor thermal behavior in terms of maximum temperature rise at the centerline of the reactor. For the PBR, increasing the GHSV from 5000 h^{-1} to 10,000 h^{-1} results in less efficient heat removal, leading the centerline temperature to increase from 566 K to 628 K. This resulted in hotspot formation and temperature runaways. This is due to the poor thermal conductivity of the PBR reactor. For the MFECC bed, the maximum temperature at the centerline of the reactor remains almost constant under all GHSV conditions (5000 h^{-1} to 10,000 h^{-1}) at 534.5 K. Therefore, operating at higher velocities induces very small changes in the thermal behavior of the MFECC bed. The observations in the thermal behavior of both the PBR and MFECC discussed above are also supported by a modeling study conducted by Sheng et al., who conducted a microscale heat transfer comparison between a PBR and an MFECC bed in a stagnant gas and flowing nitrogen gas conditions [62]. They reported that 97.2% of the total heat flux transferred within the MFECC bed was found to be transported by the continuous metal fibers. This demonstrates that the continuous metal fibers were the primary conduction path for the heat transfer inside the MFECC bed. Therefore, it is expected that changing the GHSV would not have a significant effect on the thermal profile inside the MFECC bed. Moreover, they reported the temperature distribution inside the PBR and MFECC bed at two different gas velocities (200 m/s and 500 m/s), and they found the temperature of the flowing nitrogen gas at higher gas velocity decreased significantly in PBR, while in the MFECC bed it did not change much. This finding indicates that increasing the gas velocity inside the MFECC provides stable temperature profiles and efficient heat transfer between the solid/fluid interfaces. On the other hand, increasing the gas velocity in the PBR reactor provides inadequate heat transfer rates between the solid and the fluid interface, which could be the main reason in the formation of local hotspots on catalyst particles inside PBR reactors. This indicates that operating at high gas velocities would have a detrimental effect on heat removal/management inside the PBR reactor.

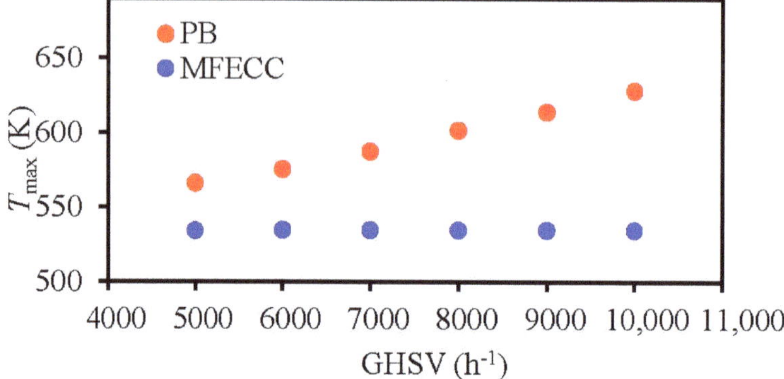

Figure 3. $T_{max} - T_{wall}$ versus GHSV for PBR and MFECC bed at 528.15 K, 2 MPa pressure, H_2: CO ratio 2:1.

Similarly, Figure 4 shows the effect of varying the GHSV on CO conversion. It can be noted that the CO conversion for the PBR case decreases from 93.45% to 83% as the GHSV increases from 5000 h^{-1} to 10,000 h^{-1}. This finding was expected since lower residence times are associated with higher gas velocities. Although the shortening of the residence time results in lower CO conversions, the total amount of syngas converted into hydrocarbons is observed to be higher for higher gas velocities. Therefore, more heat generation per mole of CO consumed is expected at higher GHSV values, which also explains the ascending trend of the centerline temperature with increasing GHSV. A similar trend is observed in the case of the MFECC bed, where the CO conversion decreases from 71.95% to 50.6% as the GHSV increases from 5000 h^{-1} to 10,000 h^{-1}. However, a steeper decrease in CO conversions with increasing GHSV was noted for the MFECC bed. This is mainly due to the high-temperature rise/deviation in the PBR reactor, which was much more prominent than in the MFECC as the GHSV was increased. This effect contributes to higher CO conversions, thus resulting in a slower decreasing rate of CO conversion in PBR. It should be noted that although increasing the GHSV might have negative implications from heat management and CO conversion standpoint, the total hydrocarbons productivity per catalyst mass is higher as the GHSV increases. The MFECC bed has shown to provide near isothermal operation, leading to improved selectivity control even when the gas velocity is increased. This raises the opportunity to achieve higher hydrocarbon productivity per catalyst mass, thus increasing the catalyst utilization.

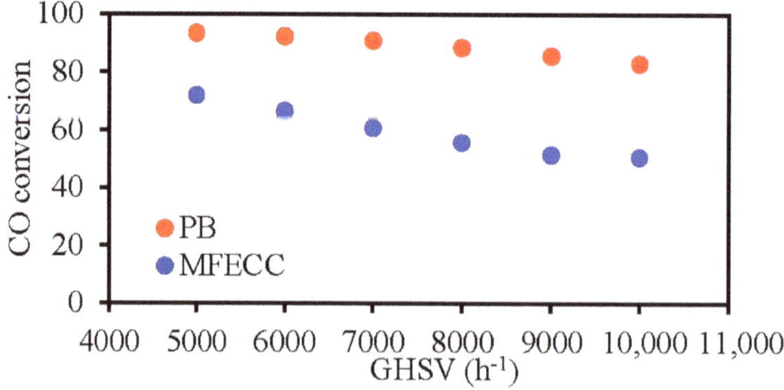

Figure 4. CO conversion versus GHSV for PBR and MFECC bed at 528.15 K, 2 MPa pressure, H_2: CO ratio 2:1.

The performance of the PBR and MFECC bed at different GHSV values was evaluated in terms of the hydrocarbon product selectivity. As mentioned previously, the CO conversions were higher inside the PBR compared to the MFECC under all GHSV conditions. Thus, the total mass productivity of the hydrocarbons (total amount of hydrocarbons produced) is expected to be higher inside the PBR for the same inlet gas velocity, due to the higher CO conversions associated with it. However, on the other hand, the selectivity toward C_{5+} hydrocarbons was significantly lower in PBR than the MFECC as shown in Figure 5. Most importantly, Figure 6 indicates that the selectivity of the most undesired product methane was significantly higher in the PBR reactor because of the relatively high-temperature gradients inside the reactor bed. The CH_4 and C_{5+} selectivity for the PBR was 42.2% and 54.4%, respectively, at 5000 h^{-1} GHSV. However, the CH_4 selectivity in the PBR reactor goes up to 100% at 10,000 h^{-1} GHSV, resulting in 0% selectivity of the C_{5+} hydrocarbons. Therefore, the total productivity of the desired hydrocarbons per catalyst mass is higher when using an MFECC bed. As mentioned previously in the introduction section, the single-pass conversion in Multi tubular/PBR is kept at 50% or lower to avoid temperature runaways. The results discussed above prove that the potential of the MFECC bed in minimizing this drawback associated with PBR. The high thermal conductivity of the MFECC bed provided efficient temperature control within the reactor bed, which offers a better opportunity to minimize the selectivity of methane while maximizing the selectivity toward C_{5+} products even at high CO conversions. This is one of the main requirements in industrial applications of FT reactors.

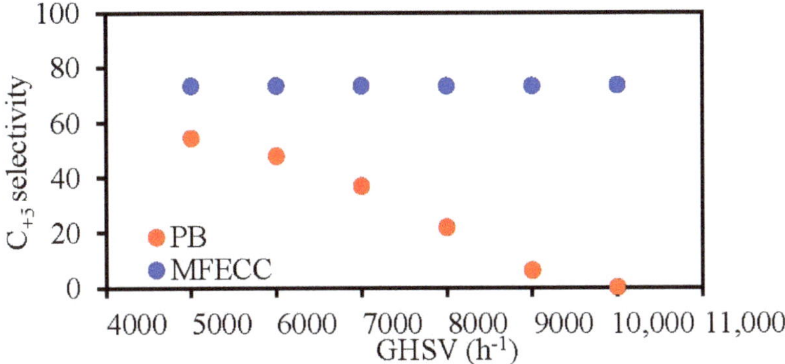

Figure 5. C_{5+} selectivity conversion versus GHSV for PBR and MFECC bed at 528.15 K, 2 MPa pressure, H_2: CO ratio 2:1.

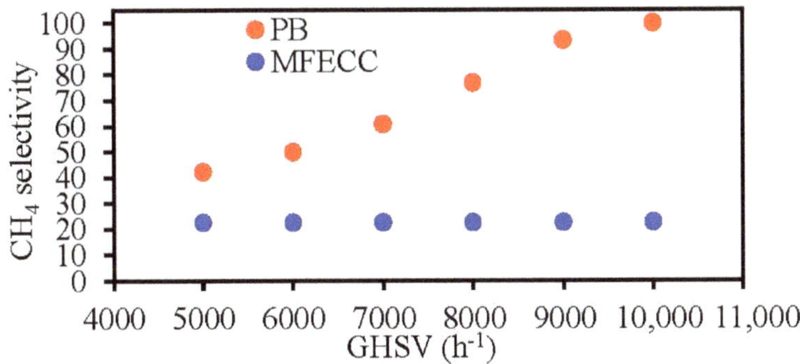

Figure 6. CH_4 selectivity versus GHSV for PBR and MFECC bed at 528.15 K, 2 MPa pressure, H_2: CO ratio 2:1.

3.4. Effect of Reactor Tube Size (Scaling-Up)

As mentioned previously, FT is a highly exothermic reaction; thus, efficient heat removal is one of the main considerations while designing commercial-scale FT reactors [7]. Therefore, the range of tube diameters used in industrial-sized PBR reactors is 20 mm to 50 mm. Larger tube sizes provide poor heat management and are more likely to suffer from the local hotspot formation, which results in the formation of methane while lowering the selectivity of the desired C_{5+} hydrocarbon products. For industrial applications, it is important to set up the FT process in a way that would provide high production of higher weight hydrocarbons (C_{5+}), with low methane selectivity and good temperature control within the reactor bed. The results discussed in the previous sections indicate that the latter can be achieved upon utilization of the novel MFECC structures of high thermal conductivity that allows uniformity in temperature distribution across reactor bed at larger tube diameters. Several simulations were performed to study the effect of scaling up the reactor tube diameter up to 102 mm on the thermal behavior of both the PBR and MFECC bed. A comparison between the base case model (15 mm ID) and the scaled-up model (102 mm ID) at 2 MPa pressure and H_2/CO ratio of (2:1) was made at different inlet temperatures 498.15 K to 528.15 K. Since changing the tube diameter affects the gas velocity/GHSV (45-fold reduction in GHSV going from 15 mm up to 102 mm, which has a prominent effect on the mass and heat transport properties), the inlet gas flow rate of the was adjusted to maintain a constant GHSV value of 5000 h^{-1} to provide a fair comparison between the base case and the scaled-up case. Figure 7 shows the maximum temperature rise inside the PBR at various inlet temperatures. If all the process parameters are kept constant, increasing the tube diameter from 15 mm to 102 mm while maintaining constant GHSV (5000 h^{-1}) results in extreme temperature runaway at all inlet temperature conditions, where the hotspot was beyond 200 K. This drastic change in the maximum temperature rise is around 40% to 50% higher than the base case scenario at 15 mm. The extremely high-temperature gradients are mainly due to high radial heat transport resistances at larger tube sizes. As a result of the high temperatures inside the PBR, the CO conversion at 102 mm was beyond 90% at all inlet temperatures, as shown in Figure 8. The reason behind this is that using larger tube diameters accompanied by the use of higher inlet gas flow rates enables higher CO conversions, thus higher hydrocarbon productivity. However, the selectivity toward methane associated with such high-temperature gradients goes up to 100% at all temperature conditions, consequently leading to 0% selectivity toward the desired C_{5+} hydrocarbons, as demonstrated in Figure 9.

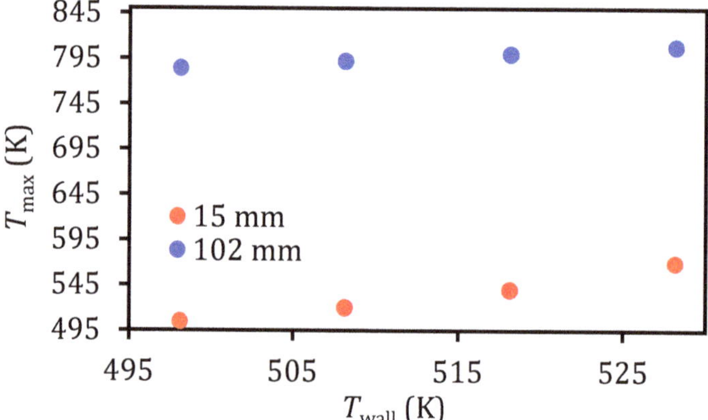

Figure 7. Maximum temperature in PBR 102 mm ID with the base case of 15 mm ID at 5000 GHSV, H_2/CO 2:1, 2 MPa pressure.

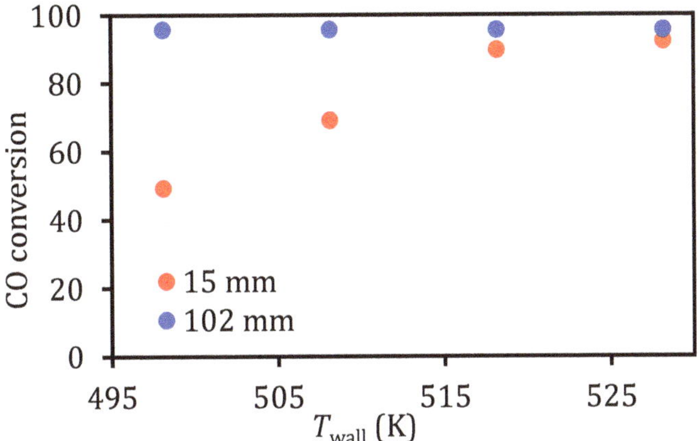

Figure 8. CO conversion in PBR 102 mm ID with the base case of 15 mm ID at 5000 GHSV, H_2/CO 2:1, 2 MPa pressure.

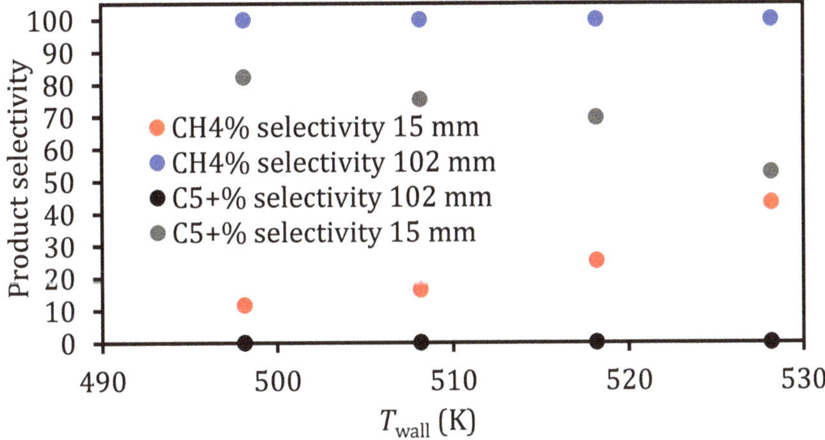

Figure 9. CH_4 and C_{5+} selectivity in PBR 102 mm with the base case of 15 mm ID at 5000 GHSV, H_2/CO 2:1, 2 MPa pressure.

On the other hand, the MFECC bed provided better temperature control relative to the PBR under the same operating conditions when the tube diameter is scaled up to 102 mm. The maximum temperature deviation from the base case (15 mm, 498.15 K to 528.15 K, GHSV 5000 h^{-1}, H_2/CO 2:1, P_{tot} 2 MPa) was below 4 K at all inlet temperature conditions, as shown in Figure 10. This is due to the high radial effective thermal conductivity of the MFECC bed, which was able to facilitate heat removal even at higher radial resistances in larger tubes. Moreover, when the tube diameter is increased from 15 mm to 102 mm while keeping a constant GHSV, the CO conversion goes up by more than 9% for all the simulated temperature conditions, as shown in Figure 11. The latter indicates a noticeable increase in hydrocarbon productivity per tube is achieved for larger tube diameters. An interesting observation is that for 102 mm diameter, the CH_4 selectivity only increases by less than 1.4%, while the C_{5+} decreases by less than 1.3% compared to the base case scenario, as shown in Figure 12. The higher CO conversion, accompanied by selectivity control at 102 mm diameter achieved using the MFECC bed, indicates that higher C_{5+} productivity is obtained relative to the base case.

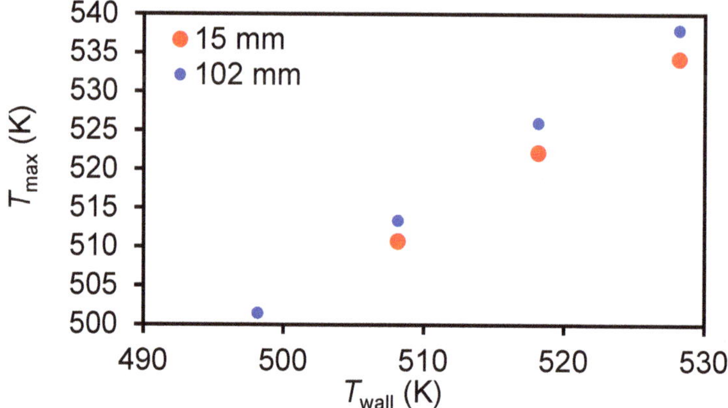

Figure 10. Maximum temperature rise in MFECC 102 mm with the base case of 15 mm ID at 5000 GHSV, H_2/CO 2:1, 2 MPa pressure.

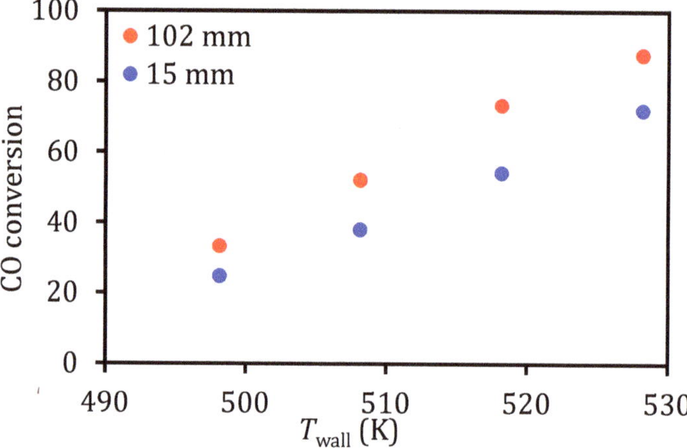

Figure 11. CO conversion in MFECC 102 mm ID with the base case of 15 mm ID at 5000 GHSV, H_2/CO 2:1, 2 MPa pressure.

It is worth noting that, as previously stated that at 528.15 K (only for the case of the PB reactor), the deviation between the modeling and the experimental predictions becomes high. The difference between the maximum temperature (centerline temperature) and the wall temperature is underestimated, thus the model under-predicts the methane selectivity relative to the experimental data. Consequently, the selectivity of C_{5+} products at 528.15 K is over-predicted by the model when compared to experimental data. In the case of scaling up the tube diameter up to 102 mm, for the MFECC case based on the aforementioned discussion, the maximum temperature deviation from the base case (15 mm, 498.15 K to 528.15 K, GHSV 5000 h^{-1}, H_2/CO 2:1, $p_{tot} = 2$ MPa) was below 4 K at all inlet temperature conditions. Therefore, extremely large temperatures are not observed in this case, hence confirming the reliability of the results for the MFECC bed. In the case of PBR, large temperature deviations are indeed observed (as previously mentioned). However, the selectivity toward methane associated with such high-temperature gradients goes up to 100% at all temperature conditions, consequently leading to 0% selectivity toward the desired C_{5+} hydrocarbons. Although the difference between the maximum temperature (centerline temperature) and the wall temperature is underestimated is this case (methane selectivity should be higher than predicted by the model) achieving a selectivity beyond 100% is physically not possible. Similarly, for the C_{5+} selectivity, the model over-predicts the C_{5+} selectivity (C_{5+} selectivity should be lower than predicted by the model) and achieving a selectivity below 0% is also physically not possible. The point to be highlighted here is that the hydrocarbon selectivity estimations are very reliable; however, higher temperature gradients are expected when such conditions are practically employed (this will not affect the product selectivity based on the above discussion).

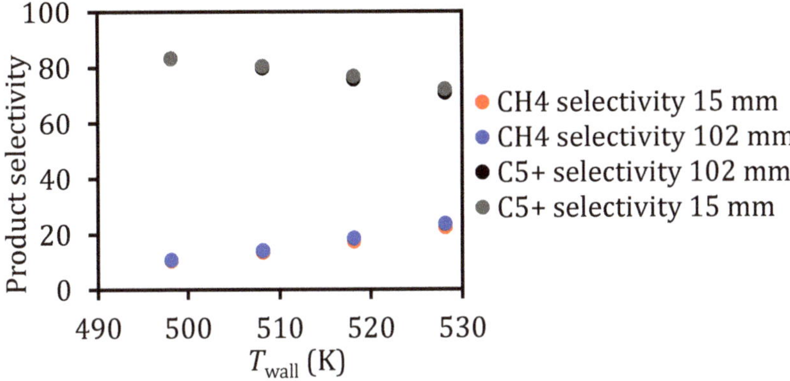

Figure 12. CH_4 and C_{5+} selectivity in MFECC 102 mm ID with the base case of 15 mm ID at 5000 GHSV, H_2/CO 2:1, 2 MPa pressure.

4. Conclusions

A 2D CFD model of an FT reactor was developed in COMSOL® Multiphysics v5.3a for two systems; MFECC bed and conventional PBR under GP-FT conditions. The possibility of scaling-up the reactor diameter tube >15 mm (up to 102 mm ID) was studied as an initial step towards the process intensification of the FT technology. The model results for both the MFECC bed and PBR provided a good match with the experimental predictions for CO conversion, maximum temperature deviation from the wall temperature, CH_4 selectivity, and C_{5+} selectivity. The influence of process parameters (inlet temperature, GHSV, tube diameter) on the hydrocarbon product selectivity, CO conversion, and hotspot formation was investigated. Results indicated that increasing the GHSV from 5000 h^{-1} to 10,000 h^{-1} in the PBR had a detrimental effect on the thermal performance of the reactor bed. This resulted in shifting the product selectivity toward undesired methane (CH_4 selectivity of 100%). Comparing with the MFECC bed simulated under equivalent conditions shows a very small temperature deviation (<7 K) and constant hydrocarbon product selectivity (~22% CH_4 selectivity and ~73% C_{5+} selectivity). Therefore, the near isothermal operation in the MFECC bed facilitates higher C_{5+} hydrocarbon productivities per tube pass at higher space velocities without the risk of selectivity loss. A comparison between the base case model (15 mm ID) and the scaled-up model (102 mm ID) revealed that the high radial effective thermal conductivity of the MFECC bed was able to provide efficient heat removal (temperature gradient < 10 K) even at higher radial distances in larger tubes. In contrast, the PBR suffered from extreme temperature runaways (>200 K), leading to poor product selectivity (100% CH_4 selectivity and 0% C_{5+} selectivity). The ability of the MFECC bed to provide near isothermal operation at 102 mm ID provides a 16-fold reduction in the total number of tubes required to achieve a targeted capacity compared to a conventional 0.025 m ID reactor bed. This study demonstrates the possibility of implementing the MFECC reactor bed in future commercial FT reactors to achieve high production rates by utilizing larger reactor tube diameters.

Author Contributions: Conceptualization, B.A.W. and N.O.E.; methodology, M.S.C.; software, A.E.A. and M.S.C.; validation, A.E.A. and M.S.C.; formal analysis, A.E.A. and M.S.C. investigation, A.E.A. and M.S.C.; resources, B.A.W. and N.O.E.; data curation, A.E.A. and M.S.C.; writing—original draft preparation, A.E.A. and M.S.C.;

writing—review and editing, B.A.W. and N.O.E.; visualization, A.E.A. and M.S.C.; supervision, N.O.E.; project administration, N.O.E.; funding acquisition, N.O.E. All authors have read and agreed to the published version of the manuscript.

Funding: This research was funded by the Qatar National Research Fund (a member of the Qatar Foundation), grant number [NPRP 7-843-2-312] and [NPRP X-100-2-024].

Acknowledgments: This paper was made possible by an NPRP award [NPRP 7-843-2-312] and [NPRP X-100-2-024] from the Qatar National Research Fund (a member of the Qatar Foundation). The statements made herein are solely the responsibility of the authors.

Conflicts of Interest: The authors declare no conflict of interest.

References

1. Challiwala, M.; Ghouri, M.; Linke, P.; El-Halwagi, M.; Elbashir, N. A combined thermo-kinetic analysis of various methane reforming technologies: Comparison with dry reforming. *J. CO2 Util.* **2017**, *17*, 99111. [CrossRef]

2. Alsuhaibani, A.S.; Afzal, S.; Challiwala, M.; Elbashir, N.O.; El-Halwagi, M.M. The impact of the development of catalyst and reaction system of the methanol synthesis stage on the overall profitability of the entire plant: A techno-economic study. *Catal. Today* **2019**. [CrossRef]

3. Abusrafa, A.E.; Challiwala, M.S.; Choudhury, H.A.; Wilhite, B.A.; Elbashir, N.O. Experimental verification of 2-Dimensional computational fluid dynamics modeling of supercritical fluids Fischer Tropsch reactor bed. *Catal. Today* **2020**, *343*, 165–175. [CrossRef]

4. Elbashir, N.O.; Chatla, A.; Spivey, J.J.; Lemonidou, A. Reaction Engineering and Catalysis Issue in Honor of Professor Dragomir Bukur: Introduction and Review. *Aristotle Univ. Thessalon.* **2020**, *343*, 1–7. [CrossRef]

5. Challiwala, M.S.; Wilhite, B.; Ghouri, M.M.; Elbashir, N. *2-D Modeling of Fischer Tropsch Packed Bed Reactor: First Step Towards Scale-Up*; AIChE: Minneapolis, MN, USA, 2017.

6. Alsuhaibani, A.S.; Afzal, S.; Challiwala, M.; Elbashir, N.O.; El-Halwagi, M.M. Process Systems Engineering and Catalysis: A Collaborative Approach for the Development of Chemical Processes. In *Computer Aided Chemical Engineering*; Muñoz, S.G., Laird, C.D., Realff, M.J., Eds.; Elsevier: Amsterdam, The Netherlands, 2019; pp. 409–414.

7. Dry, M.E. Practical and theoretical aspects of the catalytic Fischer–Tropsch process. *Appl. Catal. A Gen.* **1996**, *138*, 319–344. [CrossRef]

8. Park, N.; Kim, J.-R.; Yoo, Y.; Lee, J.; Park, M.-J. Modeling of a pilot-scale fixed-bed reactor for iron-based Fischer–Tropsch synthesis: Two-Dimensional approach for optimal tube diameter. *Fuel* **2014**, *122*, 229–235. [CrossRef]

9. Challiwala, M.S.; Wilhite, B.A.; Ghouri, M.M.; Elbashir, N.O. Multidimensional modeling of a microfibrous entrapped cobalt catalyst Fischer–Tropsch reactor bed. *AIChE J.* **2017**, *64*, 1723–1731. [CrossRef]

10. Elbashir, N.O.; Challiwala, M.S.; Sengupta, D.; El-Halwagi, M.M. System and Method for Carbon and Syngas Production, World Intellectual Property and Organization. Production, System and Method for Carbon and Syngas. WO Patent 2018187213A1, 11 October 2020.

11. Elbashir, N.; Challiwala, M.S.; Ghouri, M.M.; Linke, P.; El-Halwagi, M. *Modeling Development of a Combined Methane Fixed Bed Reactor Reformer*; Hamad bin Khalifa University Press (HBKU Press): Doha, Qatar, 2016; p. EESP2384.

12. Chatla, A.; Ghouri, M.M.; el Hassan, O.W.; Mohamed, N.; Prakash, A.V.; Elbashir, N.O. An experimental and first principles DFT investigation on the effect of Cu addition to Ni/Al2O3 catalyst for the dry reforming of methane. *Appl. Catal. A Gen.* **2020**, *602*, 117699. [CrossRef]

13. Challiwala, M.S.; Sengupta, D.; El-Halwagi, M.; Elbashir, N. *A Process Integration Approach for a Sustainable Gtl Process Using Tri-Reforming of Methane*; AIChE: Minneapolis, MN, USA, 2017.

14. Challiwala, M.S.; Ghouri, M.M.; Linke, P.; Elbashir, N.O. Kinetic and thermodynamic modelling of methane reforming technologies: Comparison of conventional technologies with dry reforming. In Proceedings of the Conference: Qatar Foundation Annual Research Conference (ARC), Doha, Qatar, 8–13 November 2015.

15. Challiwala, M.S.; Afzal, S.; Choudhury, H.A.; Sengupa, D.; El-Halwagi, M.M.; Elbashir, N.O. Alternative via Pathways Dry Reforming for CO of 2 Utilization Methane. *Adv. Carbon Manag. Technol.* **2020**, *1*, 253.

16. Omran, A.; Yoon, S.H.; Khan, M.; Ghouri, M.; Chatla, A.; Elbashir, N. Mechanistic Insights for Dry Reforming of Methane on Cu/Ni Bimetallic Catalysts: DFT-Assisted Microkinetic Analysis for Coke Resistance. *Catalysts* **2020**, *10*, 1043. [CrossRef]
17. Dry, M.E. The fischer-tropsch process: 1950–2000. *Catal. Today* **2002**, *71*, 227–241. [CrossRef]
18. Filip, L.; Zámostný, P.; Rauch, R. Mathematical model of Fischer-Tropsch synthesis using variable alpha-Parameter to predict product distribution. *Fuel* **2019**, *243*, 603–609. [CrossRef]
19. Espinoza, R.; Steynberg, A.; Jager, B.; Vosloo, A. Low temperature Fischer-Tropsch synthesis from a Sasol perspective. *Appl. Catal. A Gen.* **1999**, *186*, 13–26. [CrossRef]
20. Wender, I. Reactions of synthesis gas. *Fuel Process. Technol.* **1996**, *48*, 189–297. [CrossRef]
21. Todic, B.; Mandic, M.; Nikacevic, N.; Bukur, D.B. Effects of process and design parameters on heat management in fixed bed Fischer-Tropsch synthesis reactor. *Korean J. Chem. Eng.* **2018**, *35*, 875–889. [CrossRef]
22. Steynberg, A.; Espinoza, R.; Jager, B.; Vosloo, A. High temperature Fischer–Tropsch synthesis in commercial practice. *Appl. Catal. A Gen.* **1999**, *186*, 41–54. [CrossRef]
23. Yang, J.H.; Kim, H.-J.; Chun, D.H.; Lee, H.-T.; Hong, J.-C.; Jung, H.; Yang, J.-I. Mass transfer limitations on fixed—Bed reactor for Fischer-Tropsch synthesis. *Fuel Process. Technol.* **2010**, *91*, 285–289. [CrossRef]
24. Fratalocchi, L.; Visconti, C.G.; Groppi, G.; Lietti, L.; Tronconi, E. Intensifying heat transfer in Fischer-Tropsch tubular reactors through the adoption of conductive packed foams. *Chem. Eng. J.* **2018**, *349*, 829–837. [CrossRef]
25. Mandić, M.; Todić, B.; Živanić, L.; Nikačević, N.; Bukur, D.B. Effects of Catalyst Activity, Particle Size and Shape, and Process Conditions on Catalyst Effectiveness and Methane Selectivity for Fischer-Tropsch Reaction: A Modeling Study. *Ind. Eng. Chem. Res.* **2017**, *56*, 2733–2745. [CrossRef]
26. Sie, S.; Krishna, R. Fundamentals and selection of advanced Fischer–Tropsch reactors. *Appl. Catal. A Gen.* **1999**, *186*, 55–70. [CrossRef]
27. Botes, F.; Niemantsverdriet, J.; Van De Loosdrecht, J. A comparison of cobalt and iron based slurry phase Fischer–Tropsch synthesis. *Catal. Today* **2013**, *215*, 112–120. [CrossRef]
28. Lu, X. Fischer-Tropsch Synthesis: Towards Understanding. 2011. Available online: http://hdl.handle.net/10539/11175 (accessed on 27 September 2020).
29. Sheng, M.; Yang, H.; Cahela, D.R.; Tatarchuk, B. Novel catalyst structures with enhanced heat transfer characteristics. *J. Catal.* **2011**, *281*, 254–262. [CrossRef]
30. Sheng, M.; Yang, H.; Cahela, D.R.; Yantz, W.R., Jr.; Gonzalez, C.F.; Tatarchuk, B.J. High conductivity catalyst structures for applications in exothermic reactions. *Appl. Catal. A Gen.* **2012**, *445*, 143–152. [CrossRef]
31. Sheng, M.; Cahela, D.R.; Yang, H.; Gonzalez, C.F.; Yantz, W.R.; Harris, D.K.; Tatarchuk, B. Effective thermal conductivity and junction factor for sintered microfibrous materials. *Int. J. Heat Mass Transf.* **2013**, *56*, 10–19. [CrossRef]
32. Kalluri, R.R.; Cahela, D.R.; Tatarchuk, B. Microfibrous entrapped small particle adsorbents for high efficiency heterogeneous contacting. *Sep. Purif. Technol.* **2008**, *62*, 304–316. [CrossRef]
33. Yang, H.; Cahela, D.R.; Tatarchuk, B. A study of kinetic effects due to using microfibrous entrapped zinc oxide sorbents for hydrogen sulfide removal. *Chem. Eng. Sci.* **2008**, *63*, 2707–2716. [CrossRef]
34. Zhu, W.H.; Flanzer, M.E.; Tatarchuk, B. Nickel–zinc accordion-fold batteries with microfibrous electrodes using a papermaking process. *J. Power Sources* **2002**, *112*, 353–366. [CrossRef]
35. Greiner, A.; Wendorff, J.H. Electrospinning: A fascinating method for the preparation of ultrathin fibers. *Angew. Chem. Int. Ed.* **2007**, *46*, 5670–5703. [CrossRef]
36. Le Bars, M.; Worster, M.G. Interfacial conditions between a pure fluid and a porous medium: Implications for binary alloy solidification. *J. Fluid Mech.* **2006**, *550*, 149–173. [CrossRef]
37. Nield, D. The limitations of the Brinkman-Forchheimer equation in modeling flow in a saturated porous medium and at an interface. *Int. J. Heat Fluid Flow* **1991**, *12*, 269–272. [CrossRef]
38. Hicks, R.E. Pressure Drop in Packed Beds of Spheres. *Ind. Eng. Chem. Fundam.* **1970**, *9*, 500–502. [CrossRef]
39. Fuller, E.N.; Schettler, P.D.; Giddings, J.C. New method for prediction of binary gas-phase diffusion coefficients. *Ind. Eng. Chem.* **1966**, *58*, 18–27. [CrossRef]
40. Chabot, G.; Guilet, R.; Cognet, P.; Gourdon, C. A mathematical modeling of catalytic milli-fixed bed reactor for Fischer–Tropsch synthesis: Influence of tube diameter on Fischer Tropsch selectivity and thermal behavior. *Chem. Eng. Sci.* **2015**, *127*, 72–83. [CrossRef]

41. Ghouri, M.; Afzal, S.; Hussain, R.; Blank, J.; Bukur, D.B.; Elbashir, N. Multi-scale modeling of fixed-bed Fischer Tropsch reactor. *Comput. Chem. Eng.* **2016**, *91*, 38–48. [CrossRef]

42. Mamonov, N.A.; Kustov, L.; Alkhimov, S.A.; Mikhailov, M.N. One-dimensional heterogeneous model of a Fischer-Tropsch synthesis reactor with a fixed catalyst bed in the isothermal granules approximation. *Catal. Ind.* **2013**, *5*, 223–231. [CrossRef]

43. Bunnell, D.G.; Irvin, H.B.; Olson, R.W.; Smith, J.M. Effective Thermal Conductivities in Gas-Solid Systems. *Ind. Eng. Chem.* **1949**, *41*, 1977–1981. [CrossRef]

44. Campbell, J.; Huntington, R. Part II, Heat Transfer and Temperature Gradients. *Pet. Refin.* **1952**, *31*, 123–131.

45. De Wasch, A.; Froment, G. Heat transfer in packed beds. *Chem. Eng. Sci.* **1972**, *27*, 567–576. [CrossRef]

46. Specchia, V.; Baldi, G.; Sicardi, S. Heat transfer in packed bed reactors with one phase flow. *Chem. Eng. Commun.* **1980**, *4*, 361–380. [CrossRef]

47. Bauer, R. Effective radial thermal conductivity of packings in gas flow. *Int. Chem. Eng.* **1978**, *18*, 181–204.

48. Dixon, A.G. Wall and particle-shape effects on heat transfer in packed beds. *Chem. Eng. Commun.* **1988**, *71*, 217–237. [CrossRef]

49. Specchia, V.; Sicardi, S. Modified correlation for the conductive contribution of thermal conductivity in packed bed reactors. *Chem. Eng. Commun.* **1980**, *6*, 131–139. [CrossRef]

50. Plautz, D.A.; Johnstone, H.F. Heat and mass transfer in packed beds. *AIChE J.* **1955**, *1*, 193–199. [CrossRef]

51. Quinton, J.; Storrow, J.A. Heat transfer to air flowing through packed tubes. *Chem. Eng. Sci.* **1956**, *5*, 245–257. [CrossRef]

52. Kunii, D.; Smith, J.M. Heat transfer characteristics of porous rocks. *AIChE J.* **1960**, *6*, 71–78. [CrossRef]

53. Yates, I.C.; Satterfield, C.N. Intrinsic kinetics of the Fischer-Tropsch synthesis on a cobalt catalyst. *Energy Fuels* **1991**, *5*, 168–173. [CrossRef]

54. Vervloet, D.; Kapteijn, F.; Nijenhuis, J.; Van Ommen, J.R. Fischer-Tropsch reaction–diffusion in a cobalt catalyst particle: Aspects of activity and selectivity for a variable chain growth probability. *Catal. Sci. Technol.* **2012**, *2*, 1221–1233. [CrossRef]

55. Stamenic, M.; Dikić, V.; Mandić, M.; Todić, B.; Bukur, D.B.; Nikačević, N.M. Multiscale and Multiphase Model of Fixed Bed Reactors for Fischer-Tropsch Synthesis: Intensification Possibilities Study. *Ind. Eng. Chem. Res.* **2017**, *56*, 9964–9979. [CrossRef]

56. Bukur, D.B.; Mandić, M.; Todic, B.; Nikačević, N.M. Pore diffusion effects on catalyst effectiveness and selectivity of cobalt based Fischer-Tropsch catalyst. *Catal. Today* **2020**, *343*, 146–155. [CrossRef]

57. Maretto, C.; Krishna, R. Design and optimisation of a multi-stage bubble column slurry reactor for Fischer–Tropsch synthesis. *Catal. Today* **2001**, *66*, 241–248. [CrossRef]

58. Ma, W.; Jacobs, G.; Das, T.K.; Masuku, C.M.; Kang, J.; Pendyala, V.R.R.; Crocker, M.; Klettlinger, J.L.S.; Yen, C.H. Fischer–Tropsch Synthesis: Kinetics and Water Effect on Methane Formation over 25%Co/γ-Al$_2$O$_3$ Catalyst. *Ind. Eng. Chem. Res.* **2014**, *53*, 2157–2166. [CrossRef]

59. Todic, B.; Bhatelia, T.J.; Froment, G.F.; Ma, W.; Jacobs, G.; Davis, B.H.; Bukur, D.B. Kinetic Model of Fischer–Tropsch Synthesis in a Slurry Reactor on Co-Re/Al2O3 Catalyst. *Ind. Eng. Chem. Res.* **2013**, *52*, 669–679. [CrossRef]

60. Chernobaev, I.; Yakubovich, M.; Tripol'skii, A.; Pavlenko, N.; Struzhko, V. Investigation of the mechanism of methane formation in the fischer-tropsch synthesis on a Co/SiO$_2$ Zr IV catalyst. *Theor. Exp. Chem.* **1997**, *33*, 38–40. [CrossRef]

61. Lee, W.; Bartholomew, C.H. Multiple reaction states in CO hydrogenation on alumina-supported cobalt catalysts. *J. Catal.* **1989**, *120*, 256–271. [CrossRef]

62. Sheng, M.; Gonzalez, C.F.; Yantz, W.R.; Cahela, D.R.; Yang, H.; Harris, D.R.; Tatarchuk, B. Micro Scale Heat Transfer Comparison between Packed Beds and Microfibrous Entrapped Catalysts. *Eng. Appl. Comput. Fluid Mech.* **2013**, *7*, 471–485. [CrossRef]

 *processes*

Article

Continuous Improvement Process in the Development of a Low-Cost Rotational Rheometer

Francisco J. Hernández-Rangel [1], María Z. Saavedra-Leos [1], Josefa Morales-Morales [1,2], Horacio Bautista-Santos [2,3], Vladimir A. Reyes-Herrera [4], José M. Rodríguez-Lelis [5] and Pedro Cruz-Alcantar [1,*]

1   Coordinación Académica Región Altiplano, Universidad Autónoma de San Luis Potosí, Carretera Cedral Km, 5+600 Ejido San José de las Trojes, Matehuala, San Luis Potosí 78700, Mexico; josue.hernandez@uaslp.mx (F.J.H.-R.); zenaida.saavedra@uaslp.mx (M.Z.S.-L.); josefa.morales@uaslp.mx (J.M.-M.)
2   Tecnológico Nacional de Mexico/ Instituto Tecnológico Superior de Chicontepec, Calle Barrio Dos Caminos No. 22. Colonia Barrio Dos Caminos, Chicontepec, Veracruz 92709, Mexico; horacio.bautista@itsta.edu.mx
3   Tecnológico Nacional de México/ Instituto Tecnológico Superior de Tantoyuca, Desviación Lindero Tametate S/N, Colonia La Morita, Tantoyuca, Veracruz 92100, Mexico
4   Instituto de Energías Renovables, Universidad Nacional Autónoma de Mexico, Priv. Xochicalco s/n, Col. Centro, Temixco, Morelos 62580, Mexico; Varh@ier.unam.mx
5   Tecnológico Nacional de Mexico/Centro Nacional de Investigación y Desarrollo Tecnológico, Interior Internado Palmira S/N, Col. Palmira, Cuernavaca 62490, Mexico; jmlelis@cenidet.edu.mx
*   Correspondence: pedro.cruz@uaslp.mx

Received: 2 July 2020; Accepted: 30 July 2020; Published: 3 August 2020

Abstract: The rheological characterization of fluids using a rheometer is an essential task in food processing, materials, healthcare or even industrial engineering; in some cases, the high cost of a rheometer and the issues related to the possibility of developing both electrorheological and magnetorheological tests in the same instrument have to be overcome. With that in mind, this study designed and constructed a low-cost rotational rheometer with the capacity to adapt to electro- and magneto-rheological tests. The design team used the method of continuous improvement through Quality Function Deployment (QFD) and risk analysis tools such as Failure Mode and Effect Analysis (FMEA) and Finite Element Analysis (FEA). These analyses were prepared in order to meet the customer's needs and engineering requirements. In addition to the above, a manufacturing control based on process sheets was used, leading to the construction of a functional rheometer with a cost of USD $1500.

Keywords: rheometer; quality function deployment; design

1. Introduction

Rheology studies stress and deformation of matter that may flow, not only including liquids but also soft solids and substances such as sludge, suspensions, and human body fluids, among others [1,2]. Rheology focuses on the study of fluids, with viscosity changing when the material deforms or is submitted to certain external forces or agents, such as electric fields, magnetic fields, temperature, pressure, light, and PH, among others. Within fluids using rheological properties for very specific applications, we have magneto- and electro-rheological fluids [1,3]. Magneto-rheological fluids are smarts fluid containing particles that orient themselves and form chains in the presence of a magnetic field. In turn, electrorheological fluids are suspensions that do not conduct electricity; these fluids are known as dielectric, and when an electric stimulus is applied to these materials, some of their mechanical properties may be controlled [1–4]. A basic instrument for the study of rheological properties is the rheometer. A rheometer is capable of measuring the viscosity and elasticity of Newtonian and non-Newtonian materials in a wide range of conditions. Some of the properties that may be measured with a rheometer are viscoelasticity, elasticity limit, thixotropy, and extensional viscosity [1,5]. Rheology has wide applications in industry, in research, and in the development of products—for example, in the evaluation of paints, creams, and plasters; in healthcare; and in aeronautical and automotive applications [6–13]. A particular application in automobiles is in rheological dampers found in suspension systems that allow the adjustment of the mechanism of operation in accordance with the vehicle's needs [13,14]. In order to make the necessary operation adjustments, modern automobiles have sensors which enable the detection of road conditions, measuring the power of rebounds experienced in the suspension. Among the advantages of using suspensions based on rheological dampers is their low energy requirement. In addition, the system is reliable, and there is dynamic control of the whole automobile [15]. This last application is part of the motivation for this work, since having a rheometer would greatly help the characterization of rheological fluids. Currently, there are rheometers of varying kinds and prices [16], but most of them have a high cost and may only be used for very specific tests, and do not allow, for example, electrorheological and magnetorheological tests to be carried out on the same rheometer. This study presents the design and development of a low-cost rotational rheometer with the capacity to couple to any necessary instruments in order to carry out electro- and magneto-rheological tests. The proposed methodology considers continuous improvement processes in the design of new products widely used in the industrial and development sector—specifically, the use of the Quality Function Deployment (QFD) tool based on customer satisfaction and cost reduction [17–29], supported with Failure Mode and Effect Analysis (FMEA) risk detection tools [30–34], Finite Element Analysis (FEA) engineering analyses, and manufacturing processes. With the methodological process used, the design and construction of a functional rheometer was possible with a cost of USD $1500 with the capacity to adapt for electro- and magnetorheological tests.

2. Methodology

The methodological process for the design and manufacturing of the rheometer is shown in the following Figure 1.

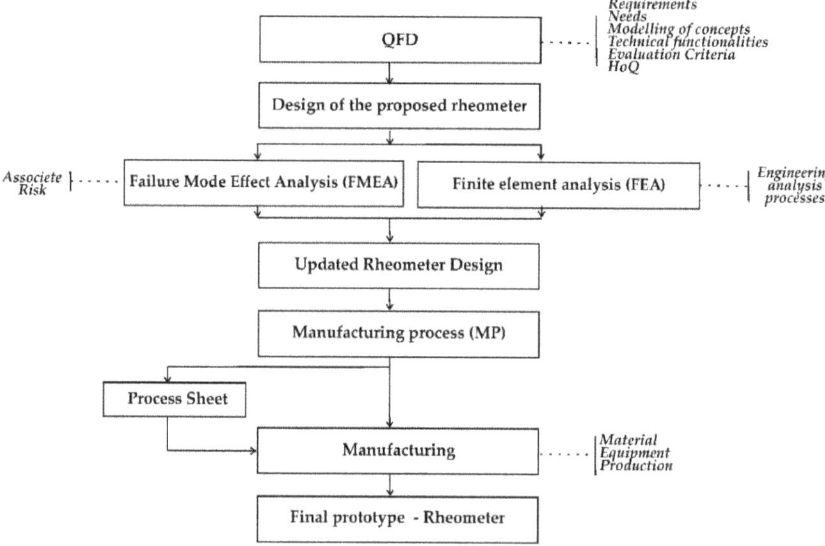

Figure 1. The methodological process.

Steps within the methodology consider the initial phase with the QFD development and the HoQ (House of Quality), where the customer requirements and the technical specifications are taken as baselines in order to design the rheometer. With the above in mind, an initial design for the rheometer is proposed to pass afterwards for a risk analysis using the FMEA technique and the analysis of finite elements (FEA). The purpose is to provide feedback on the proposed design to generate an updated final design. The last phase deals with the assurance of the manufacturing process (MP), where a tool called the process sheet was used to control activities related to the manufacturing process and finally to obtain the rheometer prototype.

2.1. Quality Function Deployment (QFD) and House of Quality (HoQ)

QFD is a method to transform users' qualitative demands into quantitative parameters, thus achieving design quality in subsystems and pieces and ultimately in the specific elements of the manufacturing process. The method was developed in 1966 in Japan and represented the impulse for the continuous improvement of products seeking optimization and rationalization in the design of its products and processes [17]. QFD is a systematic methodology with a priority of interpreting and meeting the customer's requirements during the design and manufacturing processes, promoting the interaction of the design actors (engineers, technicians, users, sellers, etc.) since the QFD is a group methodology [25,27]. The QFD approach is based on the deployment of users' expectations (the "What's") in terms of the design and parameters related to the production (the "How's") of the new product. This process is represented by a succession of double entry or matrix tables of "What/How", allowing correlations between the entries to be identified and prioritized [27]. A crucial step during QFD is the translation of the customer' needs into engineering characteristics [18].

2.1.1. Customer Requirements

Requirements are customers' declared and undeclared needs, demands, or wishes. Identifying customer requirements is considered critical to achieving effective and successful processes in product development at the design stage [35]. In some works, requirements are reported within the preliminary design stage as the formulation or decomposition of requirements [35,36]. The design and development department faces the problem of identifying the requirements in order to obtain a list of needs that will lead to the success of the design project development [37]. Customer, user, or market demands are captured in different ways: direct discussion or interviews with potential customers, surveys, focus groups, customer specifications, observation, warranty data, field reports, etc. [24]. One way to obtain simple information is through an interview, applying the method of the Five W1H (Who, What, Why, How, Where, When). The respondents can be a small number of people or even individuals among current customers, competing customers, and direct users in the same market segment [38].

2.1.2. House of Quality (HoQ)

QFD implies the construction of a matrix or a set of interconnected matrices known as "quality tables". The first matrix is called "House of Quality" (HoQ) and has two main parts: the horizontal part, containing information relevant for the customer, and the vertical part, containing the corresponding technical translation of their needs. In addition, there are the "What/Which" correlations, which allow the integration of elements related to the analysis of product competence and the identification of contradictions between different product characteristics [25]. Thus, this matrix offers the double advantage of facilitating the transition between the user world and the designer, in order to combine in one document all the data for decision making in relation to the development of the product. Commonly, the design teams have to base their estimations on their own experience, intuition, and determination [18].

Figure 2 presents the components of the HoQ, which are: customer requirements, technical parameters, relationship matrix, engineering importance, numerical values of the parameters, correlation matrix between the technical parameters, customer benchmarking, technical benchmarking, and technical importance [24].

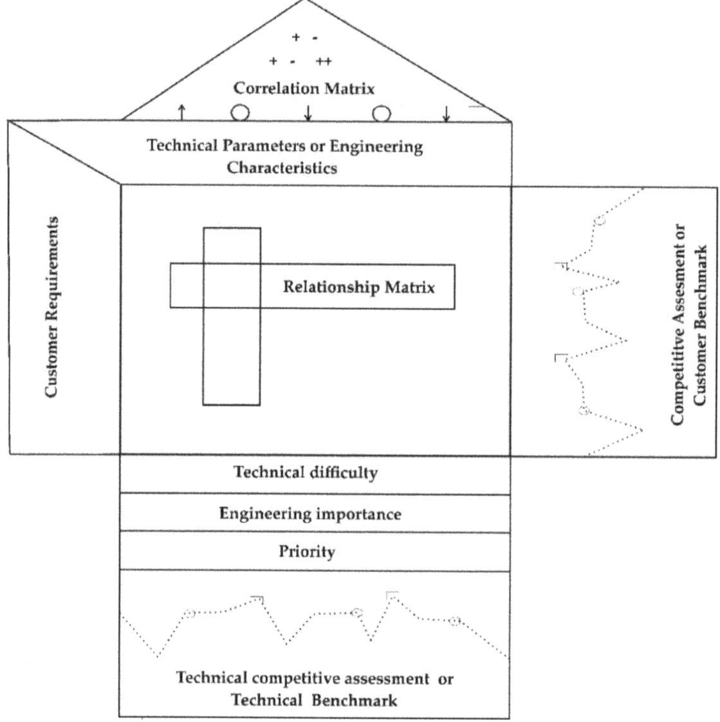

Figure 2. House of Quality (HoQ) and components.

The results expected after applying QDF through the House of Quality are the following [24]:

1. Document the environment in which the product is used (needs).
2. Classify needs in a logical order and evaluate the relative importance therof.
3. Understand customer's needs as summarized in the planning of the product, matrices, or House of Quality.
4. Identify the technical requirements directly influencing satisfaction needs in order to formulate complete technical specifications.

Commonly, the QFD is supported by several methods to improve its effectiveness, such as the TRIZ (the Russian acronym of the Theory of Inventive Problem Solving) [39,40], FMEA (Failure Mode, Effect, and Analysis), FTA (Failure Tree Analysis) [41], and diffuse theory, among others [21,28].

2.2. Failure Mode Effect Analysis (FMEA) and Finite Element Analysis (FEA)

FMEA is an analytic technique used to identify, reduce, or eliminate negative effects of the failure methods of a product, design, service, or process and its relevant causes before they are produced [22]. The technique functions by means of the identification of a concrete cause of failure mode within a system, such that the logical sequence of this condition is traced through the system up to the final effects [25]. Risk assessment is based on the occurrence, severity, and detection of a possible failure; therefore, it focuses on potential failures and not on the failures produced, as the objective is to prevent failures from occurring. The main idea is to generate a risk priority number (RPN) for each failure mode; the higher the risk number, the greater the potential risk and importance of addressing this failure mode [30].

FMEA improves the operational performance of production cycles and reduces the risk level. When using the FMEA, the risk is calculated in relation to three coefficients: the seriousness of the non-conformity (S), the non-conformity occurrence frequency (O), and the detectability of the non-conformity (D). Each coefficient is assigned a value ranging from 1 to 10, and the relative risk index is calculated (R) as shown in the following equation [30]:

$$RPN = D \times O \times S. \tag{1}$$

The values of parameters O, S, and D are taken from the tables provided by the method and are shown below in Table 1.

Table 1. D, O, and S evaluation criteria.

D Value	Detection	O value	Occurrence	S Value	Severity
1	Almost certain	1	Extremely Unlikely	1	None
2	Very High	2	Unlikely	2	Very Slight
3	High	3	Minimal	3	Minimal
4	Moderately high	4	Very Slight Probability	4	Minor
5	Moderate	5	Slight Probability	5	Moderate
6	Low	6	Low Probability	6	Significant
7	Very low	7	Medium Probability	7	Serious
8	Remote	8	Moderately High Probability	8	Very Serious
9	Very remote	9	High Probability	9	Extremely Serious
10	Absolute uncertainty	10	Extremely High Probability	10	Hazardous

A scale proposed to quantify the severity of a failure using the RPN is presented below in Table 2.

Table 2. Risk priority number (RPN) criteria.

RPN	Severity	Description
RPN < 50	Minor	This mode of failure can be detected by the company; it means a simple change in the design, unless the mode of failure is included in the standard.
50 < RPN < 100	Significant	The Priority Risk Number of this failure mode can be decreased after an adjustment to the design in the company's laboratories.
RPN > 100	Very Serious or Hazardous	This mode of failure must be reduced or eliminated through the various laboratory tests that are performed, or if necessary we must change the design concept. If it is not diminished, the product cannot be sold to the market.

FMEA is focused on predicting a possible product failure in advance in order to avoid reworking in the subsequent fabrication process, thus greatly reducing the cost of the product, shortening the cycle, and improving the manufacturing efficiency [21].

2.3. Finite Element Analysis (FEA)

Finite Element Analysis (FEA) is a numerical method for approximating the solutions to differential equations widely used in diverse problems—mainly for the modelling and simulation of engineering problems. FEA can be applied in stress analyses on structures or complicated machines, dynamic responses, or in thermal or electromagnetic analyses. FEA applications have also been extended to materials in science, biomedical engineering, geophysics, and many other fields [42,43].

2.4. Manufacturing Process (MP)

The planning process has an essential role in the manufacturing and industrial processes of home appliance and tool manufacturing and chemical processing plants. Planning a manufacturing process is transforming a portion of the design into specifications from the drawing plans in the operation instructions needed for manufacturing [44]. The planning of a manufacturing process may be defined as the systematic determination of the detailed methods with which components or parts may be economically manufactured from the initial stage to the end stage. Geometric characteristics; dimensions; tolerances; materials; and, finally, the sequence of operations, machinery, tools, and work stations are analyzed to obtain the desired final product. The first task of the planning process implies a series of steps: design data interpretation (CAD); understanding the functions, conditions, and product specifications; and carefully analyzing any assembly operations that are to be performed [44,45]. All the above may be found on a manufacturing process sheet.

Process Sheet

Process planning is a pre-manufacturing step which determines the sequence of operations or processes necessary to produce a part or a set. This is the most important step in workshops where unique products are manufactured or where the same product is manufactured infrequently. The result of this step is the process sheet, which considers the route and operations data [44,45]. A process sheet is a document listing the exact sequence of operations needed to complete the job. The route sheets are useful in manufacturing and production planning. An example of a process sheet is found in Table 3.

Table 3. Typical process sheet.

Process Capacity Sheet	Approved by:	Part No.		Application		Entered by:	Date:
		Part Name:		Number of Parts		Line:	
Step Name	Machine	Basic Time		Tool Change		Processing Capacity/Shift	Remarks
		Manual	Auto	Completion Change	Time		

3. Results

For the QDF development, we had the criteria of 12 participants, among whom were laboratory technicians, researchers, vendors, and designers who have used a rheometer at some time or have been connected to rheological tests. The process of acquiring the requirements consisted of the participation of the working group in discussions and interviews and the analysis of direct specifications for the rheometer design by a research group on intelligent fluids. In addition to the above, an extensive search of patents and scientific articles was carried out in order to explore the data reported on operation issues, components, and requirements, and, in turn, it was complemented by a market study on commercial rheometers. The criteria analyzed were the overall features, operation, aesthetics, ergonomics, measurement schemes, man-machine environments, and costs, among others. With the information found, the determination of the customer's needs and requirements was possible, as shown in Table 4. For each of the requirements considered, a quantified weight is assigned in order to prioritize them and define which ones will be of priority in the design stage. The scales to qualify the weight of requirements vary according to each author; in this case, significance values were proposed from 5, 3, and 1, where 5 is for the highest significance and 1 for the lowest significance [30].

Table 4. Customer requirements and their importance.

Customer Requirements	Importance or Weights	Customer Requirements	Importance or Weights
Easy to use	5	Minimum error margin	5
Display results on an LCD display	5	Use with any fluid—electro and magnet	3
Economic	3	Store data	3
Aesthetic	3	Automatic calculations	5
Small size	3	Easy to manufacture	5
Strong and lightweight	1	Secure	5
Portable	5	Easy maintenance	5
Detachable-interchangeable tips			3

3.1. Technical Requirements

In this section, the work team translates the customer's needs into an engineering language that is direct and easy to interpret; this is done in order that the designers may have clearly defined, in the design stage, clearly defined parameters and properties to be included in the modeling stage. Figure 3 shows the technical requirements translated into engineering language. In addition, the right side of the figure shows the relationships between the technical requirements and what will be taken as criteria for the design team's decision making.

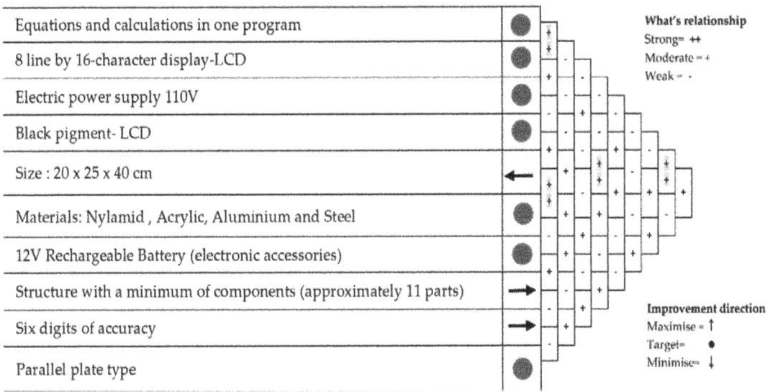

Figure 3. Technical or engineering requirements.

After the customers' requirements and technical requirements had been identified, the design team proposed the following correlations for the "What"/"How" between each of the customer's needs and each engineering characteristic, where 5 is a strong correlation, 3 is a moderate correlation, and 1 is a weak correlation. After this, estimations were compared among them. Once the significance values and the correlation values were defined, the absolute significance value is calculated (B) as the sum of the customer evaluation product multiplied by the weight corresponding to the degree of dependence [20].

$$B = \sum (I \times R). \tag{2}$$

The result is provided at the bottom of the column and shows the significance of the requirement—that is, the priority quality characteristics for a customer.

3.2. Competitive and Technical Comparison Evaluation (Customer benchmarking-Technical benchmarking)

The QFD tool through the House of Quality allows direct comparison of how its design or product ranks against the competition in fulfilling customer requirements. This analysis is very important in making decisions related to design, considering that a certain type of scale is essential to describe the excellence level of these products and our product, in order to find the route to achieve customer satisfaction. The design team, with the contribution of all its members, identifies a series of existing products to be compared—usually the proposed product and two or three of the relevant competitors [23,27]. One of the purposes of this stage is to know how the competitors' products compare to the customers' requirements compared to the product proposed by the design team. In order to carry out the competitive evaluation, four commercial rheometers were taken into consideration: HAAKE RheoStress 6000, HAAKE MARS Rheometer, ARES-G2 Rheometer, and Kinexus lab+ rheometer. Once the competitors are identified, a table is generated with the customer's requirements met by each of the competitors and the design team's proposal using a five-level scale where 1 = very low satisfaction and 5 = very high satisfaction [23]. Once the above points are obtained, the House of Quality is obtained for the design of the rheometer proposed by the design team. The House of Quality is shown in the Figure 4.

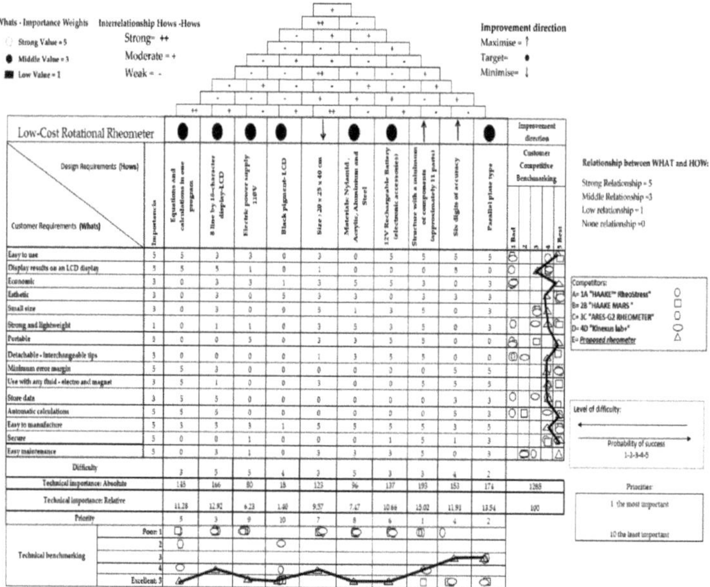

Figure 4. House of Quality (HoQ)—low cost rheometer.

In the last QFD analysis section, the importance of all the customer requirements and the contribution of each technical requirement to meeting said requirements are analyzed. This final analysis is crucial for design, as this is where the information obtained in the evaluations of the other matrices is gathered and the result of each technical requirement is evaluated and finally reordered, before proceeding to the design. From the House of Quality results, especially the results of absolute technical importance (B) and priority, it is clear that some requirements are of greater importance than others. This is closely related to the methodology, but in spite of its score, all the requirements are met and included in the design generated by the work team. From the technical importance results, the following can be concluded:

(1) The technical requirement of minimal parts obtained a high importance index of 193 and a priority order of 1. This means that the proposed design must meet the minimal necessary elements for rheometer operation; this requirement also impacts the ease of maintenance required during its lifecycle.

(2) The technical requirement of parallel plates obtained a high importance index of 174 and a priority order of 2. This technical requirement depends on how the rheological tests are carried out and the condition of the rheometer's accessories.

(3) The technical requirement for the use of an 8-line display obtained a high importance index of 166 and a priority order of 3. This technical requirement mainly affects how the rheometer user will interact with the variables that must be measured during the rheological tests.

Finally, we would obtain the remaining features, where the feature with the lowest priority is the use of dark pigments for display data, among others. The competitive and technical evaluation had the purpose of analyzing similar products which were already on the market in order to produce a rheometer with qualities not lower than those of competitors. As may be observed regarding quality in the competitive and technical evaluation section, the designing team proposal was always within a margin superior to the competition in most aspects. This shows that our product will be superior to or will keep customers much more satisfied than products existing on the market, since the proposed rheometer has a better score than the competition regarding the engineering and customers' requirements.

3.3. Preliminary Design

Once the House of Quality evaluation is complete, we proceed to the section on preliminary prototype design according to the House of Quality—QFD requirements. The design concept proposal has been established in a simple, intuitive approach based on the design team's experience, with the following stages: (i) the identification of the problems associated with satisfying customer requirements; (ii) the identification of the associated problems and the satisfaction of technical or engineering requirements; (iii) the acquisition of knowledge and foundations for the proposal of solutions; (iv) the proposal of design solutions; (v) the proposal of critical elements of operation, structure, and materials; (vi) the proposal of a measurement and control environment; (vii) the evaluation of the design proposal in terms of the functionality and the requirements to be met; and, finally, (viii) the preliminary rheometer design. The Figure 5 shows the exploded CAD model of the rheometer designed by the work team.

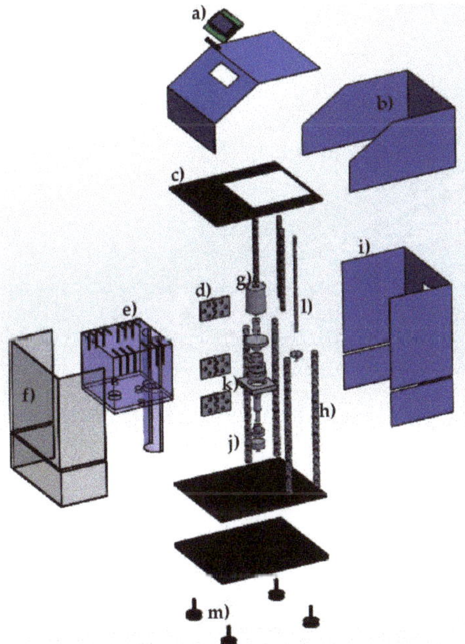

Figure 5. Design of the proposed rheometer and components. (**a**) LCD display; (**b**) sheet metal casing; (**c**) steel plates; (**d**) ball bearing hinge; (**e**) container box (the motor and electronic accessories); (**f**) acrylic protective cover; (**g**) motor; (**h**) steel rods; (**i**) galvanized sheet; (**j**) shaft, tip, and sample container assembly; (**k**) Nylamid couple, inertia flywheel, Nylamid bearing holder, and bearing assembly; (**l**) threaded rod; and (**m**) adjustable leg leveler.

Tables 5 and 6 show the complete FMEA format carried out in accordance with the procedure specified in the methodology section with the objective of analyzing the designed rheometer components, broken down into the parts considered critical. In accordance with the FMEA, each part of the rheometer is related to its possible potential failures. This will facilitate the identification and reinforcement of those areas where failures may occur, along with their correction, before creating the prototype. It will also provide its possible effects to provide design controls, as foreseen by the work team as a whole.

Table 5. Failure Mode and Effect Analysis (FMEA)—low-cost rheometer: mechanical components.

Item	Potential Failure Mode	Potential Effects	S	Potential Causes of Failure	O	Current Controls for Prevention	D	RPN	Recommended Actions
Motor	Damage to the motor armature	Component failures	10	Motor overload	8	Inspection and preventive maintenance	6	480	Proper maintenance and allow for cooling
	Bearing damage	Deterioration of the elements		Excessive load, poor lubrication					Allow cooling and lubrication
	Short circuit	Probability of damage to the other components		Contact with water, contact between wires					Protect from moisture, check connections
	Motor misalignment	Destruction of the components		Sudden movements					Avoid sudden movements
Rotor (shaft and tip)	Plate misalignment	Wrong results	5	Plate friction, overloading the motor	2	Preventive maintenance and adjustment of components	8	80	Check multi-directional leveling (bull's eye level)
	Gear wear	Do not rotate the shaft		Wear due to lack of lubricant					Preventive maintenance and lubrication
Plates	Misleveled plates	Measurement error	5	Not having a bubble leveler	2	Preventive maintenance, lubricating, and checking levels	5	50	Check multi-directional leveling
	Loosening	Test point parallelism problems		Poor installation of components					Checking component adjustment
Basis and structure	Fracture of the plates	Measurement error	5	Fracture due to misuse	2	Replace component	5	50	Replace the part

S: Severity, *O*: Ocurrency, *D*: Detectability.

Table 6. FMEA—low-cost rheometer: electrical and electronic components.

Item	Potential Failure Mode	Potential Effects	S	Potential Causes of Failure	O	Current Controls for Prevention	D	RPN	Recommended Actions
Potentiometer	Bad installation	Lack of speed control	7	Installation mistakes	2	Install components correctly, allow proper cooling	7	98	Verify correct installation
	Heating	Damaged or burnt components		Risk of short circuit					Turn off rheometer and allow cooling
LCD display	Burn	Error in the speed indicated on the LCD display	6	Short circuit	4	Protect it from falls and water contact	1	24	Only connect at the time of testing
	Break	Does not turn on the LCD display		Falls, blows, misuse					Avoid hitting the LCD display
	Short circuit	Does not turn on the LCD display		Electric shock, overheating					Protect against humidity
Battery holder	Electrical discharge	No electric power	8	No electric power	5	Use properly, protect from the humidity and clean	5	200	Check electrical connection
	Incorrect Electrical Wiring			Poorly connected wires					Check wiring and connection
	Short circuit	The other components are put at risk		overheating, contact with water					Protecting it from humidity and proper use
USB port	Installation mistakes	Disconnected USB Data is not saved	4	Bad connection of the USB ports	2	Replacing components	6	48	Verify correct installation
	Problem with the interface	Data is not saved							Remove and reinsert it
Control Panel	Installation mistakes	lack of control over functions	7	Not installing the component correctly	2	Control panel for rheometer control	5	70	Verify correct installation
	Component failure	Difficulty in pressing buttons		Poor installation, disconnected wires					Verify correct installation

S: Severity; *O*: Occurrence; *D*: Detectability.

The results from the FMEA and the relevant scores for the parameters S, O, and D were established to find the RPN of each rheometer component and its characteristic failure modes. The risk assessment was performed on both the mechanical parts and the electrical or electronic components. With the FMEA, it was discovered that the high RPN of 400 corresponds to the electric motor that transmits the torque and rotational movement to the shaft, suggesting that preventive maintenance during operation should be considered a high priority. Another element with a high RPN is the battery case, with a value of 200, which feeds electrical energy to some electronic components, and therefore should be protected from vibrations, falls, or contact with liquid. According to Table 2, these two RPN values classify the failures of these elements as serious or dangerous. With RPN values of 98, 80, and 70, the potentiometer, rotor, and panel or display failure modes are classified as significant failures. The other failure modes fall into the minor category.

3.4. Finite Element Analysis (FEA)

This section describes the Finite Element Analysis carried out on the parts that comprise the rheometer, with the objective of verifying if they meet the various established design parameters established and detecting possible improvements to the proposed design. It is worth noting that only a vibration analysis was carried out, as this is a phenomenon that greatly affects the rheometer performance through high-magnitude resonant vibrations [46]. Modal vibration analysis generates the fundamental vibration frequencies of the elements and the magnitude of their deformation.

The ANSYS Workbench software was used for the modal analysis of the rheometer components. The 3D models of the rheometer elements were built in a computer-aided design (SolidWorks) system and imported into the ANSYS Workbench preprocessing platform. The main materials used for the component analysis were structural steel and Nylamid. Steel was assigned a density property of 7850 kg/m3, a Young's modulus of 200 GPa, and a Poisson's ratio of 0.3; for Nylamid, it was assigned a density of 7850 kg/m3, a Young's modulus of 2353 Mpa, and a Poisson's ratio of 0.21. Once the materials were defined, each of the elements analyzed underwent the meshing or discretization process. The finite element SOLID185 was used for the meshing, defined by eight nodes with three degrees of freedom in each node (UX, UY, and UX) [47]. This type of finite element is suitable for modeling general 3D solid structures, allowing prismatic and tetrahedral degenerations. SOLID185 is appropriate for modal analysis, with an adequate ratio of convergence time and computational resource consumption [48]. The mesh convergence test was performed to determine the optimal mesh size using the first modal frequency as a reference. The boundary conditions and speed loads were applied specifically for each case. Modal analysis was limited to the earliest vibration modes, those closest to the rheometer operating speed.

A simulation was carried out on the full rotor (shaft, flywheel, and tip) and the coupling to the driving motor. Figure 6 shows both the meshed geometry and the shape rotor vibration that occurs at the fundamental vibration frequency of 1689.3 Hz. Vibration resonance exists when the frequency of an excitation source coincides with the fundamental frequency of a body or element—an undesirable condition which generates the greatest number of deformations and displacement in bodies. Figure 6 indicates that if there were resonance, the maximum displacements would be at the tip of the shaft where the probe is placed and would be 152.55 mm. The analysis only returns this as a point of design improvement. It is important to mention that the tip is what comes into contact with the sample container, and it would not be appropriate to have those types of vibrations in rheological tests.

Figure 7 shows the result of the analysis of the inertia flywheel, where a fundamental vibrational frequency of 10,026 Hz was obtained. The shape of resonance vibration and its magnitudes do not represent conditions that require improvement.

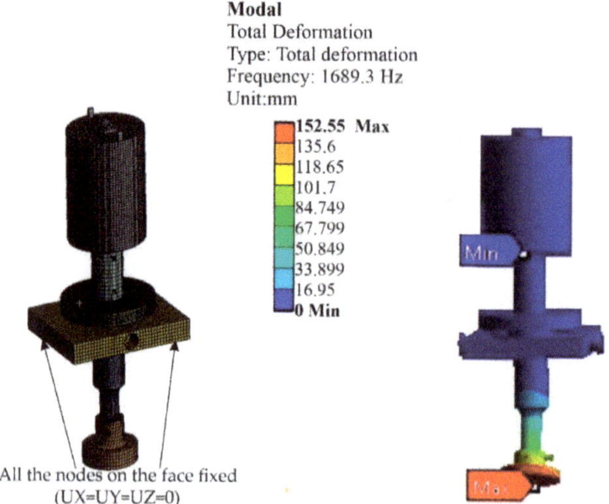

Figure 6. Total deformation of the rotor.

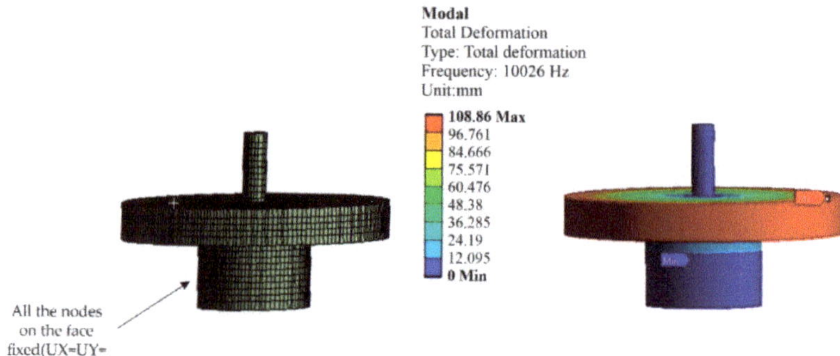

Figure 7. Total deformation of the inertia flywheel.

The tip is the part that is attached to the rotor and is what comes into contact with the fluid. Figure 8 shows the vibration analysis of the shaft and of the tip with the fundamental frequencies of 3928.2 and 28,188 Hz, respectively.

As can be seen by the shapes of the vibration, the tip of the shaft is again in a state of high vibration movements, which is consistent with the rotor analysis findings and is therefore taken into account when making design improvements. Finally, an analysis was carried out on the complete structure, as shown in Figure 9. The fundamental frequency was 331.6 Hz, and resonant displacements of 378.8 mm were found in the upper rear part of the structure, which indicates that improvements must be made to that section to avoid future performance failures in the rheometer.

With the results from the FMEA and the FEA, the following improvement actions were taken to update the proposed design:

(1) A minimal part or element analysis was performed on the rheometer design. This led to the elimination of some nonessential accessories, helping to meet the requirement of easy maintenance.
(2) Small marks were made on the plates and laminates to increase the stiffness and prevent susceptibility to excessive vibrations and loosening.

(3) The rotation shaft section was modified at the base, where the test probe is placed, to avoid excessive movements due to vibration based on the results obtained in the FMEA and FEA.

(4) The section of the upper rear part of the structure was modified to avoid movements due to vibration and possible failures caused by mechanical fatigue.

(5) The space devoted to the battery holder for the sensor power source was improved.

(6) Improvements to the electrical components were taken into account in the instrumentation section.

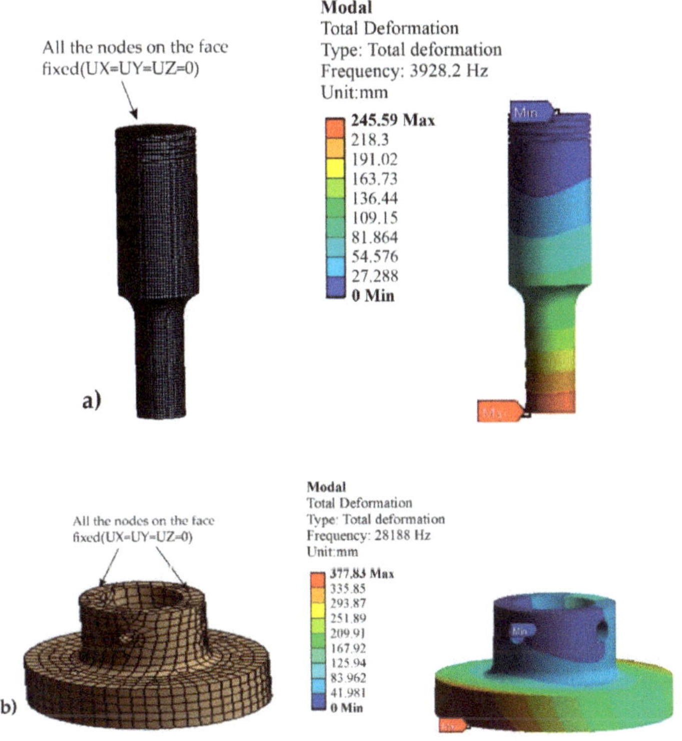

Figure 8. Total deformation—(**a**) shaft and (**b**) tip.

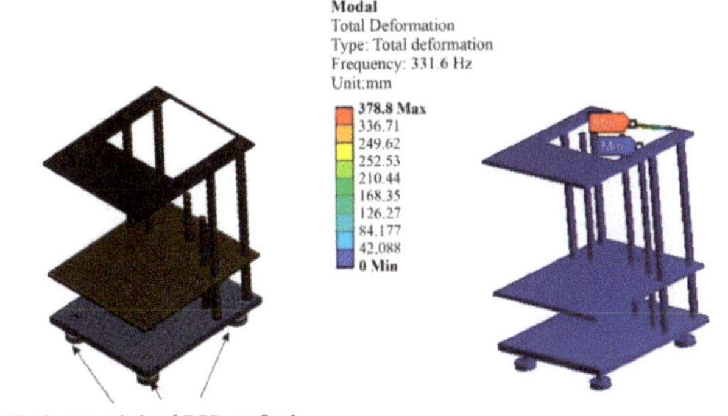

Figure 9. Total deformation—structure of steel plates and rods.

Figure 10 shows the final rheometer design drawn up in CAD as a result of the improvement points.

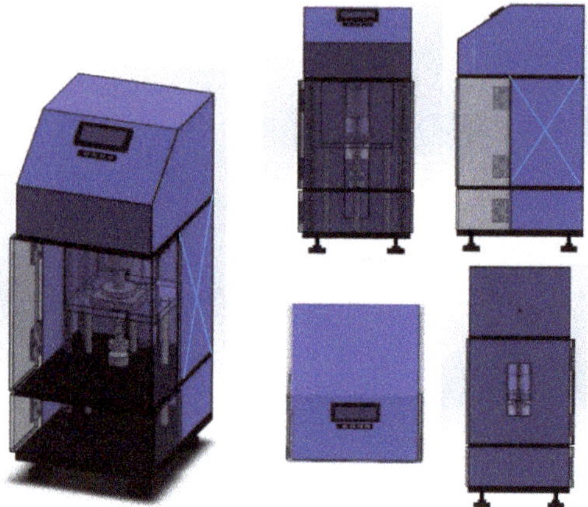

Figure 10. Final rheometer design.

3.5. Rheometer Manufacturing Process

Once the final design was finished in CAD, and based on the results from the QFD, FMEA, and FEA, the rheometer was manufactured with materials such as steel, acrylic, aluminum, laminated steel, Nylamid, and stainless steel. The manufacturing planning process and the sequence of features or processes was performed using the process sheet tool with the exact sequence of features necessary to complete the work. A process sheet was completed for each rheometer component. Table 7 shows the process sheet for the sheet metal casing of the rheometer.

Table 7. Sheet process—metal casing.

Control Manufacturing Processes (CMP)	Sheet Process "Sheet Metal Casing"		Part: 9
			Sheet: 9
Title Sheet Metal Casing	**Material** Galvanized Sheet	**Dimensions** 300 × 250 mm 160 × 80 mm	**Estimated Time** 1 h 30 min.

Op.	Description	Tools and Equipment	Working Conditions	Estimated Operation Time
1	Trace the dimensions on the cardboard sheets	Tape measure and square	Manual	20 min.
2	Using the cardboard, mark the exact measurements on the galvanized Sheet	Square and marker	Manual	10 min.
3	Cut the galvanized sheet according to the proposed measurements	Sheet metal scissors	Manual	10 min.
4	Drill holes in the galvanized sheet	Drill and bit set	Manual	15 min.
5	Bend the galvanized sheet to the proposed dimensions	Hydraulic bending press	Manual	10 min.
6	Weld the indicated flaps of the galvanized sheet	Automatic spot welder	Manual	15 min.
7	Sand the galvanized sheet	Sponge sanding pads and ultra fine sand paper	Manual	10 min.
8	Painting galvanized sheet	Air paint gun	Manual	20 min.

Comments: The dimensions indicated on the cardboard sheet are drawn first, so that there are no errors in the lines. Once the dimensions have been completed exactly, they are transferred to the sheet to avoid waste or possible errors.

Figure 11 shows some elements of the manufactured rheometer.

Figure 11. (**a**) and (**b**) Steel plates, (**c**) container box (the motor and electronic accessories), (**d**) sheet metal casing, (**e**) acrylic protective cover, (**f**) inertia flywheel, (**g**) threaded rod, (**h**) rotor (shaft ant tip), (**i**) adjustable leg leveler.

With the components manufactured, the structure assembly process is developed, as shown in Figure 12.

Figure 12. Structure assembly process.

The internal part of the rheometer is composed of the motor (12V DC—RPM: 1000RPM) and its accessories. Figure 13 shows the complete assembly, which is composed of the rotor, couplings, bearings, and inertia flywheel. This assembly is particularly important for the rheological testing process.

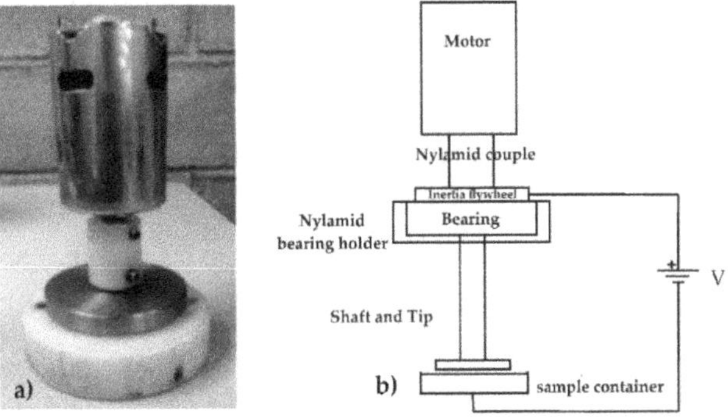

Figure 13. (**a**) Motor and inertia flywheel assembly, (**b**) electrical arrangement for electromechanical tests.

The Nylamid elements were included to isolate the motor and the electrical and electronic components from the energy used for the electro-rheological tests (Figure 13b).

3.6. Instrumentation and Implementation

To meet customer requirements regarding the environment or user interaction with the rheometer, it was decided to use the Arduino platform to manage the sensors, button panel, display, automatic software calculations, and storage of the test data. Some devices used in the instrumentation of the rheometer include the following:

(a)   Open platform Arduino UNO boards, which facilitate microcontroller programming.
(b)   A DHT11 digital temperature and humidity sensor, which uses a capacitive humidity sensor and a thermistor to measure the surrounding air and only one pin to read the data.
(c)   An FC-03 Infrared Sensor Encoder, a device responsible for converting angular movement into electrical pulses which can be interpreted by the system controller.
(d)   An Arduino button panel, control element.
(e)   An H Bridge L298N, a circuit that allows controlling the direction of the rotation and speed in DC motors.
(f)   An LCD display and an I2C Adapter for LCD displays with microcontrollers.

The separation of the parallel plates from the tip is measured from the vertical displacement bar, and the torque is estimated based on the current and intake voltage of the driving motor.

Figure 14 shows the tests carried out on the sensors.

Figure 14. Sensor and LCD display testing.

Finally, the assembly of the rheometer is completed, as shown in Figure 15.

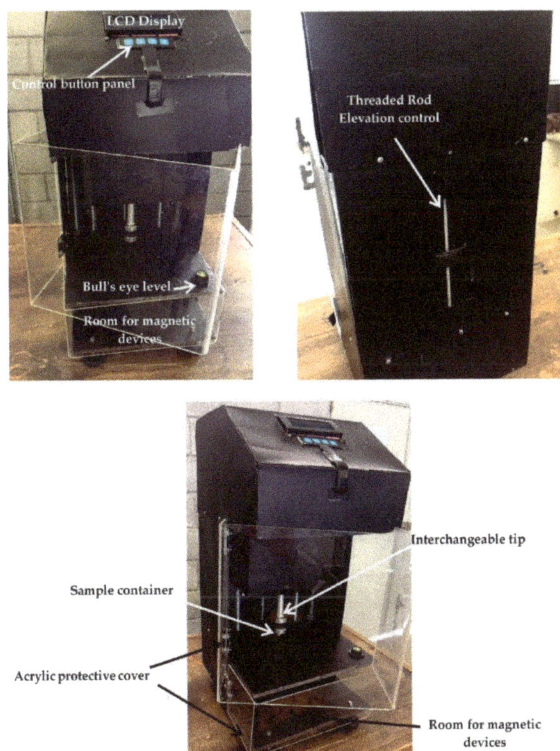

Figure 15. Manufactured rheometer—final version. Product specifications: power source 110V, sensor power source 12 V, variable speed operation from 0 to 1000 RPM, interchangeable test probes, ability to incorporate attachments to perform electro- and magneto-rheological tests.

4. Discussion

The main interest of this work was accomplished by obtaining a rotational rheometer with the required characteristics using of the methodologies set forth herein. By developing the QFD by means of the matrices in the aforementioned House of Quality, it was possible to guide this work toward the rheometer design and manufacturing stages. To manufacture the prototype, it was necessary to determine the customer, technical, or engineering requirements and establish the correlation between these. The accuracy of the customer requirements depends on the work team's environment, in addition to its knowledge and experience in the subject. In this regard, the work team had to overcome the lack of contributions from more people or customers, as rheometers are not commonly used instruments and their application is very specialized. However, with the help of research into patents and scientific articles, the customer's significant requirements were met to structure the House of Quality. Another aspect detected during the QFD elaboration process was weakness or subjectivity when performing the benchmarking, as an additional, more reliable mechanism is required to make this judgement and achieve greater reliability and determination [23]. Rheometer performance and cost are highly valued aspects on the market. The rheometer designed and built in this work has a material and supply cost of $1500 USD, not including labor—a very competitive price in comparison to commercial rheometers.

Limitations and Expected Impact

Improvements in the rheometer development process presented in this work arise from the analysis with the following limitations:

- The sample of participants considered for the requirements analysis and for the design process should be improved, involving more experts in the field of rheology, equipment vendors, users, and analysts.
- Although the methodology used in the preliminary design stage was sufficient to fulfill the objective of this work, more robust or systematic methods for the entire preliminary design process need to be adopted, as suggested by some research works in the field of design [37,49].
- A weakness found during the stage is the lack of an in-depth analysis of the technical details in the design of the proposed solution.
- It would be very convenient to update the failure mode analysis (FMEA) with field evaluations of the experimental performance in order to continue improving and adjusting the rheometer design.
- The verification stage of the experimental operation of the rheometer would give certainty for industrial production and would give feedback on the entire design process.
- An ergonomic analysis of man-machine interaction would undoubtedly strengthen the development process of the rheometer as a product.

We consider that the main impact of this work is the development of a low-cost rheometer using technology within the reach of most people, not only in the industrial sector. The implemented methodology is already well known, but it can be followed for the development of other equipment or instruments. Academically, having a rheometer like the one developed in this work will impact research in the field of rheology and will support the training of future engineering students.

5. Conclusions

The low-cost rotational rheometer developed in this study was achieved using quality control and continuous improvement tools, such as QFD-HoQ and FMEA, in addition to the use of manufacturing tools such as Manufacturing Process Management and engineering CAE tools such as the FEA. The rheometer design has the capacity to adapt in order to perform both electro- and magneto-rheological tests. During the design and construction process, the design team adhered to the customer's requirements and to the engineering requirements proposed, taking the advantage offered by the QFD. In addition, improvements in the proposed rheometer design were achieved with the FMEA technique in order to detect any possible failures and the development of an engineering analysis (FEA) to ensure the design in the structural part and the operation with which the rheometer construction was finally achieved. With the results obtained in this study, the possibility of designing rheometers was left open for electro- and magneto-rheological tests in the future.

Author Contributions: Conceptualization, F.J.H.-R., P.C.-A., and J.M.-M.; resources and materials, M.Z.S.-L., J.M.R.-L., and V.A.R.-H.; writing—original draft preparation, P.C.-A. and H.B.-S.; writing and analysis, P.C.-A. and J.M.-M.; writing—review and editing, P.C.-A. and H.B.-S. All authors have read and agreed to the published version of the manuscript.

Funding: This research received no external funding.

Acknowledgments: The authors gratefully acknowledge mechanical engineering student Elsa Guadalupe Salas Reyes for her participation in the development of this work.

Conflicts of Interest: The authors declare no conflict of interest.

References

1. Schramm, G. *A Practical Approach to Rheology and Rheometry*; Block, H., Kelly, J.P., Eds.; Gebrueder Haake: Karlsruhe, Germany, 1994.

2. Barnes, H.A.; Hutton, J.F.; Walters, K. *An Introduction to Rheology*; Rheology Series; Elsevier Science Publishers: Amsterdam, The Netherlands, 1989.

3. Halsey, T.C.; Toor, W. Structure of electrorheological fluids. *Phys. Rev. Lett.* **1990**, *65*, 2820–2823. [CrossRef] [PubMed]

4. García-Ortiz, J.H.; Sadek, S.H.; Galindo-Rosales, F.J. Influence of the Polarity of the Electric Field on Electrorheometry. *Appl. Sci.* **2019**, *9*, 5273. [CrossRef]

5. Lang, L.; Alexandrov, S.; Lyamina, E.; Dinh, V.M. The Behavior of Melts with Vanishing Viscosity in the Cone-and-Plate Rheometer. *Appl. Sci.* **2020**, *10*, 172. [CrossRef]

6. Wang, H.; Jiao, J.; Wang, Y.; Du, W. Feasibility of Using Gangue and Fly Ash as Filling Slurry Materials. *Processes* **2018**, *6*, 232. [CrossRef]

7. Wang, H.; Hassan, E.A.M.; Memon, H.; Elagib, T.H.H.; Abad AllaIdris, F. Characterization of Natural Composites Fabricated from Abutilon-Fiber-Reinforced Poly (Lactic Acid). *Processes* **2019**, *7*, 583. [CrossRef]

8. De la Cruz Martínez, A.; Delgado Portales, R.E.; Pérez Martínez, J.D.; González Ramírez, J.E.; Villalobos Lara, A.D.; Borras Enríquez, A.J.; Moscosa Santillán, M. Estimation of Ice Cream Mixture Viscosity during Batch Crystallization in a Scraped Surface Heat Exchanger. *Processes* **2020**, *8*, 167. [CrossRef]

9. Li, D.; Li, G.; Chen, Y.; Man, J.; Wu, Q.; Zhang, M.; Chen, H.; Zhang, Y. The Impact of Erythrocytes Injury on Blood Flow in Bionic Arteriole with Stenosis Segment. *Processes* **2019**, *7*, 372. [CrossRef]

10. Wang, Y.; Huang, Y.; Hao, Y. Experimental Study and Application of Rheological Properties of Coal Gangue-Fly Ash Backfill Slurry. *Processes* **2020**, *8*, 284. [CrossRef]

11. García-Ortiz, J.H.; Galindo-Rosales, F.J. Extensional Magnetorheology as a Tool for Optimizing the Formulation of Ferrofluids in Oil-Spill Clean-Up Processes. *Processes* **2020**, *8*, 597. [CrossRef]

12. Huang, W.; Wang, D.; He, P.; Long, X.; Tong, B.; Tian, J.; Yu, P. Rheological Characteristics Evaluation of Bitumen Composites Containing Rock Asphalt and Diatomite. *Appl. Sci.* **2019**, *9*, 1023. [CrossRef]

13. Cruz, P.; Martínez, F.J.; Pineda, Z.; Morales, J.; Hernández, J.F.; Reyes, M.; Figueroa, R.A. Perspectiva del uso de amortiguadores inteligentes en vehículos eléctricos. In *Tendencias Actuales de Fuentes de Energía Renovables*; UASLP: Matehuala, México, 2017; pp. 26–31. ISBN 978-607-535-017-2.

14. Pineda, R.Z.; Martínez, L.F.J.; Cruz, A.P. Amortiguadores inteligentes y reología. *Univ. Potos.* **2017**, *216*, 28–33.

15. Zapateiro, M.; Pozo, F.; Karimi, H.R.; Luo, N. Semiactive Control Methodologies for Suspension Control with Magnetorheological Dampers. *IEEE/ASME Trans. Mechatron.* **2012**, *17*, 370–380. [CrossRef]

16. da Silva, V.B.; da Costa, M.P. Influence of Processing on Rheological and Textural Characteristics of Goat and Sheep Milk Beverages and Methods of Analysis. In *Processing and Sustainability of Beverages*; Grumezescu, A.M., Holban, A.M., Eds.; Woodhead Publishing: Sawston, Cambridge, 2019; pp. 373–412. ISBN 9780128152591. [CrossRef]

17. Akao, Y. Development history of quality function deployment. In *The Customer Driven Approach to Quality Planning and Deployment*; Asian Productivity Organization: Minato, Tokyo, Japan, 1994.

18. Kuijt-Evers, L.F.M.; Morel, K.P.N.; Eikelenberg, N.L.W.; Vink, P. Application of the QFD as a design approach to ensure comfort in using hand tools: Can the design team complete the House of Quality appropriately? *Appl. Ergon.* **2009**, *40*, 519–526. [CrossRef] [PubMed]

19. Utne, I.B. Improving the environmental performance of the fishing fleet by use of Quality Function Deployment (QFD). *J. Clean. Prod.* **2009**, *17*, 724–731. [CrossRef]

20. Elena, F.; Gazizulina, A.; Eskina, E.; Ostapenko, M.; Aidarov, D. Development of QFD Methodology. In *Quality, IT and Business Operations*; Springer Proceedings in Business and Economics; Kapur, P., Kumar, U., Verma, A., Eds.; Springer: Singapore, 2018.

21. Tan, X.; Xiong, W. Improving Product Quality Based on QFD and FMEA Theory. In *2020 Prognostics and Health Management Conference (PHM-Besançon)*; IEEE: Besancon, France, 2020; pp. 274–282. [CrossRef]

22. Mukherjee, S.P. Improving Process Quality. In *Quality. India Studies in Business and Economics*; Springer: Singapore, 2019; ISBN 978-981-13-1270-0. [CrossRef]

23. Franceschini, F.; Maisano, D. A new proposal to improve the customer competitive benchmarking in QFD. *J. Qual. Eng.* **2018**, *30*, 730–761. [CrossRef]

24. Freddi, A.; Salmon, M. *Design Principles and Methodologies from Conceptualization to First Prototyping with Examples and Case Studies*; Springer International Publishing: Cham, Switzerland, 2019; ISBN 978-3-319-95341-0.

25. Almannai, B.; Greenough, R.; Kay, J. A decision support tool based on QFD and FMEA for the selection of manufacturing automation technologies. *Robot. Comput.-Integr. Manuf.* **2008**, *24*, 501–507. [CrossRef]

26. Zadry, H.R.; Rahmayanti, D.; Susanti, L.; Fatrias, D. Identification of Design Requirements for Ergonomic Long Spinal Board Using Quality Function Deployment (QFD). *Procedia Manuf.* **2015**, *3*, 4673–4680. [CrossRef]

27. Marsot, J. QFD: A methodological tool for integration of ergonomics at the design stage. *Appl. Ergon.* **2005**, *36*, 185–192. [CrossRef]

28. Mayda, M.; Choi, S. A reliability-based design framework for early stages of design process. *J. Braz. Soc. Mech. Sci. Eng.* **2017**, *39*, 2105–2120. [CrossRef]

29. Wolniak, E.R.; Sędek, A. Using QFD method for the ecological designing of products and services. *Qual. Quant.* **2009**, *43*, 695–701. [CrossRef]

30. Trafialek, J.; Kolanowski, W. Application of Failure Mode and Effect Analysis (FMEA) for audit of HACCP system. *Food Control* **2014**, *44*, 35–44. [CrossRef]

31. Liu, H.; Deng, X.; Jiang, W. Risk Evaluation in Failure Mode and Effects Analysis Using Fuzzy Measure and Fuzzy Integral. *Symmetry* **2017**, *9*, 162. [CrossRef]

32. Zheng, Y.; Johnson, R.; Larson, G. Minimizing treatment planning errors in proton therapy using failure mode and effects analysis. *Int. J. Radiat. Oncol. Biol. Phys.* **2016**, *43*, 2904–2910. [CrossRef] [PubMed]

33. Manger, R.P.; Paxton, A.B.; Pawlicki, T.; Kim, G.Y. Failure mode and effects analysis and fault tree analysis of surface image guided cranial radiosurgery. *Med. Phys.* **2015**, *42*, 2449–2461. [CrossRef] [PubMed]

34. Kim, K.M.; Yun, N.; Yun, H.J.; Dong, H.L.; Cho, H.H. Failure analysis in after shell section of gas turbine combustion liner under base-load operation. *Eng. Fail. Anal.* **2010**, *17*, 848–856. [CrossRef]

35. Fiorineschi, L.; Becattini, N.; Borgianni, Y.; Rotini, F. Testing a New Structured Tool for Supporting Requirements' Formulation and Decomposition. *Appl. Sci.* **2020**, *10*, 3259. [CrossRef]

36. Kroll, E. Design theory and conceptual design: Contrasting functional decomposition and morphology with parameter analysis. *Res. Eng. Des.* **2013**, *24*, 165–183. [CrossRef]

37. Pahl, G.; Beitz, W.; Feldhusen, J.; Grote, K.H. *Engineering Design*, 3rd ed.; Springer: London, UK, 2007. [CrossRef]

38. Jinks, T. The 5W1H Method. In *Psychological Perspectives on Reality, Consciousness and Paranormal Experience*; Palgrave Macmillan: Cham, Switzerland, 2019. [CrossRef]

39. Altshuller, G. *Creativity as an Exact Science: The Theory of the Solution of Inventive Problems*; Gordon and Breach Science Publishing: New York, NY, USA, 1984; ISBN 0-677-21230-5.

40. Yamashina, H.; Ito, T.; Kawada, H. Innovative product development process by integrating QFD and TRIZ. *Int. J. Prod. Res.* **2002**, *40*, 1031–1050. [CrossRef]

41. Whiteley, M.; Dunnett, S.; Jackson, L. Failure Mode and Effect Analysis, and Fault Tree Analysis of Polymer Electrolyte Membrane Fuel Cells. *Int. J. Hydrog. Energy* **2016**, *41*, 1187–1202. [CrossRef]

42. Hughes, T.J.R. *The Finite Element Method: Linear Static and Dynamic Finite Element Analysis*, 1st ed.; Dover Publications: Mineola, NY, USA, 2000; ISBN 978-0486411811.

43. Strang, G.; Fix, G.J. *An Analysis of the Finite Element Method*, 2nd ed.; Series in Automatic Computation; Prentice-Hall, Inc.: Englewood Clifs, NJ, USA, 1973.

44. Zhang, H.C. Manufacturing Process Planning. In *Handbook of Automation and Manufacturing System*; Dorf, R.C., Kusiak, A., Eds.; John Wiley & Sons: New York, NY, USA, 1994; pp. 587–616. ISBN 978-0-471-55218-5.

45. Swamidass, P.M. *Encyclopedia of Production and Manufacturing Management*; Springer: New York, NY, USA, 2000; p. 980. ISBN 978-0-7923-8630-8. [CrossRef]

46. Ouriev, B.N.; Uriev, N.B. Influence of vibration on structure rheological properties of a highly concentrated suspension. *Meas. Sci. Technol.* **2005**, *16*, 1691–1700.

47. *ANSYS Mechanical APDL Element Reference*; ANSYS, Inc.: Canonsburg, PA, USA, 2011.

48. *ANSYS Meshing User's Guide*; ANSYS, Inc.: Canonsburg, PA, USA, 2013.

49. Fiorineschi, L.; Papini, S.; Pugi, L.; Rindi, A.; Rotini, F. Systematic design of a new gearbox for concrete mixers. *J. Eng. Des. Technol.* **2020**. [CrossRef]

Review

Updated Principles of Sustainable Engineering

Peter Glavič

Department of Chemistry and Chemical Engineering, University of Maribor, Smetanova 17, 2000 Maribor, Slovenia; peter.glavic@um.si

Abstract: A change in human development patterns is needed, including mankind's environmental, economic, and social behavior. Engineering methods and practices have a substantial impact on the way to sustainable development. An overview of the guiding principles of sustainability, sustainable design, green engineering, and sustainable engineering is presented first. Sustainable engineering principles need to be updated to include the present state of the art in human knowledge. Therefore, the updated principles of sustainable development are presented, including traditional and more recent items: a holistic approach, sustainability hierarchies, sustainable consumption, resource scarcity, equalities within and between generations, all stakeholders' engagement, and internalizing externalities. Environmental, social, and economic impacts that respect humans' true needs and well-being are of importance to the future. The updated 12 principles include the tridimensional system's approach, precautionary and preventive approaches, and corporate reporting liability. The environmental principles comprise a circular economy with waste minimization, efficient use of resources, increased share of renewables, and sustainable production. The social pillar includes different views of equality, the engagement of stakeholders, social responsibilities, and decent work. Economic principles embrace human capital, creativity, and innovation in the development of products, processes and services, cost-benefit analysis using the Life Cycle Assessment, and the polluters must pay principle. The principles will require further development by engaging individual engineers, educators, and their associations.

Keywords: sustainability; engineering; design; principle; responsibility

Citation: Glavič, P. Updated Principles of Sustainable Engineering. *Processes* **2022**, *10*, 870. https://doi.org/10.3390/pr10050870

Academic Editor: Alfredo Iranzo

Received: 29 January 2022
Accepted: 22 April 2022
Published: 28 April 2022

Publisher's Note: MDPI stays neutral with regard to jurisdictional claims in published maps and institutional affiliations.

1. Introduction

The human population has had an exponential development, which has been very intensive in the last century and has exceeded the planetary limits in the last decades. The population has increased from 2 G (billion, 10^9) in 1927 to more than 7.9 G today and will likely reach around 9.7 G by 2050, and 10.9 G by 2100 [1]. The population growth rate has decreased from 2.2%/a (percent per year) 50 years ago to about 1.0%/a now and is projected to decline to 0.13%/a by the year 2100; the world average life expectancy has increased from 31 a in 1900 to 73.2 a now and is expected to reach 81.7 a by 2100 and 100 a by 2300 [2]. In addition, the household final expenditure per capita has doubled since 1960, from USD 3k (thousand US dollars) to USD 5.9k [3]. As the global source and sink capacities are final, this type of evolution has increased pollution and reduced resource availability to critical values.

The United Nations (UN) responded to the uncontrolled development by establishing the World Commission on Environment and Development (WCED), known better as the Brundtland Commission [4]. In 1987, the Commission issued the document Our Common Future (Brundtland Report) defining sustainable development, SD, as "development that meets the needs of the present without compromising the ability of future generations to meet their own needs." The long-term environmental strategy with 3 pillars (economic growth, environmental protection, and social equality) was proposed. The Commission called for an international meeting where more concrete initiatives and goals could be mapped out. This meeting was held in Rio de Janeiro in 1992 as a World Summit on

Sustainable Development (WSSD, also called the Earth Summit). A comprehensive plan of action, known as Agenda 21, came out of the meeting. It entailed actions to be taken globally, nationally, and locally to make life on Earth more sustainable. Earth Summits 2002 Rio+10 (2002) in Johannesburg, and Rio+20 (2012) in Rio de Janeiro continued the endeavors. The result of the third Summit was the document The Future we Want [5], which, with the 192 governments in attendance, renewed the political commitment to sustainable development and declared the commitment to the promotion of a sustainable future.

The Brundtland Report recognized that human resource development in the form of poverty reduction, gender equity, and wealth redistribution was crucial to formulating strategies for environmental conservation. The social component became necessary after the neoliberal globalization introduced by USA President Ronald Reagan and UK Prime Minister Margaret Thatcher took the free market too far—the economic pillar started to erode the social one. For decades, the USA arms industry was producing wars, causing migrations and terrorism. The second disaster is climate change, causing climate refugees.

In 2000, the UN Millennium Summit committed to achieving the eight Millennium Development Goals (MDGs) to eradicate poverty and hunger, promote gender equality, achieve universal primary education, reduce child mortality, improve maternal health, combat diseases, ensure environmental sustainability, and improve global partnership for sustainable development by 2015. At the end of the MDGs era, the Sustainable Development Goals (SDGs) with 17 "Global Goals", and 169 targets was adopted by the 193 UN member states [6]. The social pillar was extended by including reduced inequalities, peace, and justice. The economic pillar included decent work, responsible consumption and production, and climate action. In total, 230 indicators were proposed to monitor the SDGs' success.

The present situation on sustainability is critical. Our civilization has crossed four of nine "planetary boundaries": the greenhouse gases' (GHGs) concentration is causing climate change, species extinction, deforestation, and pollution from nitrogen and phosphorus [7]. The other four boundaries (ocean acidification, freshwater use, atmospheric aerosol loading, and chemical pollution with radioactive and nanomaterials) are approaching the boundary limits fast. Planetary boundaries are determining the references to environmental sustainability [8].

The European Environment Agency (EEA) in its 5-yearly State and Outlook of the European Environment Report, SOER [9] stated that decoupling environmental pressures from economic growth was incremental with only partially improved ecosystem resilience and human health. We need to accelerate progress towards decoupling in a rapidly changing global context. To achieve its 2050 vision of "living well within environmental limits", it must fundamentally transform its core societal systems: food, energy, mobility, and the built environment. Achieving such changes will require "profound changes in dominant practices, policies and thinking". Oxfam reported that eight men owned the same amount of wealth as the poorest half of the world—"A world where 1% of humanity controls as much wealth as the bottom 99% will never be stable" [10]. In total, 8% of the global workforce is "working poor"—having less than 60% of the average income [11]. In September 2017, the UN General Assembly focused on people striving for peace and a decent life for all on a sustainable planet.

Engineering has brought faster development to all countries and most people and created more equal societies with fewer existential problems. On the other hand, it has contributed to strong environmental impacts, climate change, species extinction, and increased pollution. We need technical solutions that will not cause long-term negative impacts on the environment, making human society resilient and sustainable. New approaches are needed that require (among others) a new type of engineering—a sustainable engineering approach in the design and control of systems, processes, and products. Regarding the open question: "Is technology the culprit or the saver?" [12], the right answer is: "It shall be sustainable in long term." Sustainable consumption and production goals, zero waste, a circular economy, and resource efficiency are some concepts that we are encountering in this respect. However, the environmental pillar is not enough—we must, equally as well, consider the social and economic pillars: equality, decent jobs, the engagement of all stakeholders, social responsibility, innovations, life cycle analysis, the cost–benefit approach, reporting to public, etc.

This paper builds on a review and evaluation of engineering aspects of sustainability and sustainable development, engineering design and sustainable engineering (green engineering), and sustainability principles and practices for engineers. It intends to update and refine the existing sets of sustainable engineering principles by integrating the knowledge developed in the last decade. The new set of principles shall be utilized to help fulfil the 17 SDGs and the Paris agreement more effectively and enable humanity to "Accelerate the transition to equitable, sustainable, livable, post-fossil fuels society" [13].

2. Methodology

A literature review, foreseeing the engineering developments from international organizations along with their analysis and synthesis, and personal experience are used as the methodology.

3. Engineering and Sustainable Development

Many predecessors of sustainable engineering have been developed. Some of them that still exist will be described. One of the oldest is Environmental Engineering, which existed from the beginning of our civilization. Its modern approach started in the 1970s. This is a branch of engineering that deals with the adverse environmental effects of nature and human activities on fresh water supply, water and air pollution, wastewater and waste management, energy preservation, global warming, acid rain, sanitation, and agricultural systems [14]. Courses in civil engineering as well as in chemical engineering studies exist; they are well connected with environmental science and technology.

Ecological engineering (EE), which started in the 1960s, is the "design of sustainable ecosystems that integrates human society with its natural environment for the benefits of both" [15]. It deals with restoration of "rivers, lakes, forests, grasslands, wetlands . . . , and phytoremediation sites". EE enables the design, construction, restoration, and management of ecosystems. Mitsch and Jørgensen [16] identified five classes of EE design: (1) ecosystem utilization to reduce/solve pollution problems, such as phytoremediation, (2) ecosystem imitation to resolve a source problem, e.g., forest restoration, (3) ecosystem recovery, e.g., mine land restoration, (4) ecosystem ecological modification, e.g., selective plant harvesting, (5) balanced use of ecosystems, such as sustainable agricultural systems. They identified 19 possible design principles of EE.

Cleaner production (CP, in USA Pollution prevention, PP) was initiated in the 1990s by UNEP [17] to minimize resource use, pollution, and waste in companies. It was developed by researchers, policy makers, and practitioners from many countries. UNIDO has assisted in the development of very successful National Cleaner Production Centers (NCPCs) and Programs (NCPPs) all over the world. The research field is still very active with several hundred papers every year. CP was soon enriched with the Eco-Design approach, which takes into consideration the environmental impacts of a product throughout its entire life cycle, from resources to the end-of-life scenario. Energy and material efficiencies are an important part of the eco-design, and so are renewable and/or recycled resources.

Industrial Ecology (IE) is a predecessor of circular economy—it is about shifting processes from open loop (linear) systems to the closed loop ones. IE is based on mimicry since natural systems do not know any waste. They use "industrial metabolism" (material and energy flows, design for environment or eco-design), life cycle planning, eco-industrial parks ("industrial symbiosis"), etc., to mimic natural systems. The Danish industrial park at Kalundborg [18] is the best-known example where industrial outputs from one industry serve as inputs to another one, thereby "reducing use of raw materials, pollution, and saving on waste treatment" [19]. Graedl and Allenby [20] devoted their book to IE, and green engineering principles and cases to offer practical and reasonable approaches to design decisions. Links between IE and circular economy can be found in Ghisellini et al. [21]. Saavedra et al. [22] analyzed the theoretical contribution of IE to circular economy.

All the above-mentioned engineering approaches are environmentally oriented; they are lacking the social component to be sustainable. An overview of guiding principles in engineering for sustainable development (Sustainable Engineering) is presented in the sections below. Figure 1 illustrates the hierarchy of sustainability.

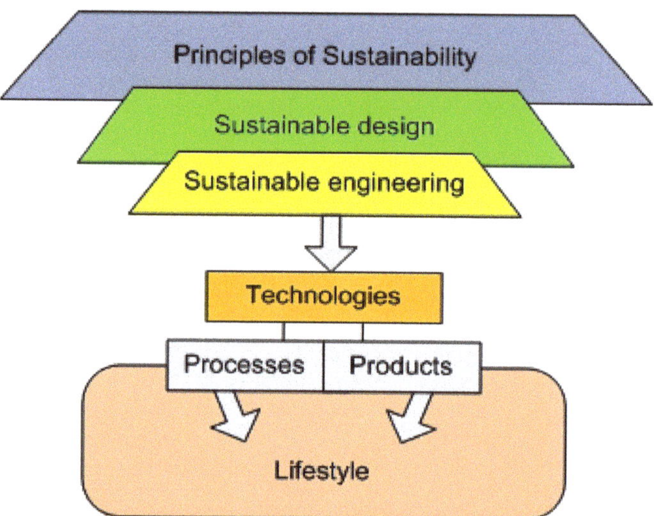

Figure 1. Hierarchy of sustainability principles, technologies, processes, and products is determining humanity lifestyles [23].

The Principles of Sustainability on the top guide the Sustainable Design, the process of thinking. The stage of Sustainable Engineering deals with the technical implementation of ideas. Sometimes it is not an easy process, and some aspects of the design may be changed or alternative solutions used. The design and engineering stages determine the Technologies, which provide the Processes and Products. The latter are sold. It is some sort of a portal through which the established principles of sustainable design and engineering affect people's lifestyle creating changes in society. Because of people's strong dependence on multiple technologies, these become the factors that can facilitate change in society and can even become tools of manipulation and initiation of global trends [24].

Engineering activities from the design and planning to operations follow the above-mentioned international activities. The content of socio-economic principles has increased with time, although the environmental ones are still predominant.

Principles of Sustainability and Sustainable Development

What is the difference between sustainability and sustainable development? Sustainability and sustainable development have somewhat different meanings. Sustainability is future oriented, while sustainable development concentrates on the methods and strategies to achieve sustainability [25]. "Sustainable development is the pathway to sustainability" [26]. Mulder et al. [12] examined the methods to be used in engineering design to achieve sustainable development. The book contains case studies and presents the results of several design projects. Table 1 presents three sets of sustainability principles, and three sets of the sustainable development ones.

Table 1. An overview of sustainability and Sustainable Development (SD) principles.

Initiative	Principles
Natural step (FSSD), sustainability principles	(1) Resource extraction, (2) Removal of chemicals, (3) Planetary equilibria, (4) Basic needs of others
Ben-Eli, core principles	(1) Resource scarcity, (2) Internalize externalities, (3) Biodiversity, (4) Equality, (5) Social life and ethics
UK Engineering Council, sustainability principles	(1) Sustainable society, (2) Professional and responsible judgement, and leadership, (3) More than legislation, (4) Resources efficiency, (5) Multiple views, (6) Risks to people and environment
Harris, basic principles of SD	(1) Inequalities and environmental damage, (2) Natural capital, (3) Population, resources, and biodiversity limits, (4) Equity, health, education, and democracy
Csaba and Nikolett, basic principles	(1) Holistic approach, (2) Solidarity, (3) Justice, (4) Resources, (5) Integration, (6) Local resources, (7) Participation, (8) Responsibility, (9) Precaution, (10) Polluter pays
University of PEI, guiding principles	(1) Professionalism, (2) Respecting diversity, (3) Collaboration, (4) Education, (5) Championship, (6) Leadership

Two years after the Brundtland report, the Natural Step framework, a non-profit NGO, was founded by Swedish scientist Karl Henrik Robért. Their approach is collectively called the "Framework for Strategic Sustainable Development"; it is a comprehensive model for planning in complex systems [27] based on four Sustainability Principles: (1) "we cannot dig stuff up from the Earth at a rate faster than it naturally returns and replenishes; (2) we cannot make chemical stuff at a rate faster than it takes nature to break it down; (3) we cannot cause destruction to the planet at a rate faster than it takes to regrow; (4) we cannot do things that cause others to not be able to fulfill their basic needs".

The Natural Step's understanding of human needs was based on the work of the Chilean economist Manfred Max-Neef [28]. He identified nine fundamental human needs that were consistent across time and cultures: subsistence, protection, affection, understanding, participation, leisure, creation, identity, and freedom. "These fundamental human needs cannot be substituted one for another and a lack of any of them represents a poverty of some kind".

In contrast to the Brundtland report, Ben-Eli [29,30] published a definition and five Core Principles of Sustainability. They were related to five fundamental domains: (1) material (material and energy flows), (2) economic (wealth formation and management), (3) life (biosphere), (4) social (human interactions), and (5) spiritual (attitudes and ethics). The resulting five Core Principles included policy and operational implications foreseeing their interplay. In short, the corresponding five Principles aimed to: (1) reduce throwing away resources, (2) internalize external costs, (3) maintain biodiversity, (4) respect equality of all humans, and (5) take care of their social life and ethics.

The UK Engineering Council has issued a set of six high-level sustainability principles and related guidance [31]: "(1) contribute to building a sustainable society, present and future; (2) apply professional and responsible judgement and take a leadership role; (3) do more than just comply with legislation and codes; (4) use resources efficiently and effectively; (5) seek multiple views to solve sustainability challenges; and (6) manage risk to minimize adverse impact to people or the environment".

Harris [32] wrote a working paper on the basic principles of sustainable development (SD): (1) rectify social inequities and environmental damage with sound economics, (2) conserve natural capital beyond market mechanisms, (3) respect population, resources, and biodiversity limits, and (4) maintain social equity, basic health, educational needs, and participatory democracy.

Csaba and Nikolett [33] defined the 10 basic principles of SD presented in Table 1.

The Environmental Advisory Council [34] of the Canadian province of Prince Edward Island (PEI) identified the guiding principles for sustainable development. They are community oriented and grouped in two chapters covering all the three pillars of SD:

1. Changing the way we work, with 12 principles addressing people and communities, needs and opportunities, problems and critical issues, policies and programs, funding and spending plans, short term decisions, awareness raising and education, etc.
2. Government shall take care of the present and future generations' interests by using 10 principles: placing people and their information at the center, including healthy and resilient environment, cost-effectively preventing environmental damage, long-term planning by using scientific knowledge, respecting risks, and uncertainties, etc.

The 22 principles were designed to build a sustainable economy, improve living standards while conserving the environment, and protect the land- and seascape. Environmental strategy and the government's environmental policy are dealt with in more detail.

The University of PEI has accepted six guiding principles on sustainability and energy management activities: (1) conduct professionally, (2) respect diversity of opinions, options, and ideas, (3) collaborate on campus to promote the development of a sustainability culture and energy efficiency, (4) education—raise awareness of sustainability fundamentals and initiatives, and energy efficiency activities, (5) work to foster change by demonstrating sustainable and energy efficient principles on campus, (6) provide guidance to the campus community to foster a sustainable and energy efficient community.

The sets of sustainability principles in Table 1 deal with the environmental, economic, and social pillars. Ben-Eli also includes the biosphere and spiritual domains. The principles are goal oriented. The UK's Engineering council deals with ethical and legislative components as well as social ones (multiple views and risk management). The UK principles are behavior oriented and in some way a step towards sustainable engineering ones.

Harris includes two or even three pillars in each basic principle. Social and environmental components are present in three basic principles each, and economic ones in two. Csaba and Nikolett include social matters in six out of ten basic principles, three from environmental ones, and only one from economics; they basically apply a management approach. Prince Edward Island's guiding principles are, in the foreground, social ones, and two of them are economic and only one is environmental—they are policy oriented.

Table 1 reveals that the most often cited principle relates to resources including their scarcity, and the need to conserve and use them efficiently; their recycling is not mentioned but natural capital should be managed sustainably and not thrown away. Biodiversity, precaution and prevention, chemicals' pollution, and land- and seascape protection are mentioned within the environmental principles. The most often cited social principles are equality and diversity of opinions, participation, and integration. Solidarity, health, education, justice, democracy, basic needs of others, social life, and ethics are mentioned, too. Internalizing externalities and the polluter pays principles are mentioned within the economic pillar. Holistic approach and professionalism are very important overarching characteristics of engineers; the latter includes professional and responsible judgement, risk management, and leadership. More than just respecting the legislation is expected from them.

4. Design for Sustainability

Design for sustainability, also called (environmentally) sustainable design, or eco-design is the way of designing products, processes, and systems that enable sustainable development. Some of its principles are summarized in Table 2.

Table 2. An overview of sustainable design principles on the time axis.

Initiative	Principles
Hannover principles	(1) Co-existence with nature, (2) Interdependence, (3) Spirit and matter, (4) Responsible design, (5) Long-term value, (6) No waste, (7) Natural energy flows, (8) Design limitations, (9) Constant improvement
Todd, eco-design principles and practices	(1) Living world matrix, (2) Laws of life, (3) Biological equity, (4) Bio-regionality, (5) Renewable energy, (6) Living systems integration, (7) Co-evolution, (8) Healing the planet, (9) Following ecology
Ryn and Covan, ecology principles	(1) Solutions from place, (2) Ecological accounting, (3) Design with nature, (4) Everyone is a designer, (5) Make nature visible
McLennan, sustainable design principles	(1) Learn from nature, (2) Respect natural resources, (3) Respect for people, (4) Respect for place, (5) Respect for future, (6) System thinking
Riel, integrated skills	(1) Product Life cycle, (2) Innovation, (2) Responsibility, (3) Networked collaboration, (4) Intercultural skills, (5) Knowledge engineering
Mattson and Wood, design principles for developing countries	(1) Local population, (2) Testing the product locally, (3) Adapting imported technology, (4) Poverty elimination, (5) Women and children, (6) Country needs, (7) Interdisciplinary teams, (8) Cooperation with governments, (9) Adaptation to world markets
Econation, design for sustainability	(1) Doing right things to support individuals, human intrinsic values, social equity and common good, local communities, health, sustainable production, and consumption, (2) Doing things right: thinking in systems, dematerialization, renewable and natural materials, biomimicry, cradle to cradle

The Hannover principles [35] contain nine principles: "(1) Insist on rights of humanity and nature to co-exist; (2) Recognize interdependence; (3) Respect relationships between spirit and matter; (4) Accept responsibility for consequences of design decisions; (5) Create safe objects of long-term value; (6) Eliminate the concept of waste; (7) Rely on natural energy flows; (8) Understand limitations of design; (9) Seek constant improvement by the sharing of knowledge".

Principles and practice of eco-design [36] are: "(1) The living world is the matrix for all designs; (2) Design should follow, not oppose, the laws of life; (3) Biological equity must determine design; (4) Design must reflect bio-regionality; (5) Use renewable energy sources; (6) Proceed by integration of living systems; (7) Design should be co-evolutionary with nature; (8) Building and design should heal the planet; (9) Design should follow a sacred ecology".

Principles of physical and social ecology [37] are: "(1) Solutions grow from place; (2) Ecological accounting informs design; (3) Design with nature; (4) Everyone is a designer; (5) Make nature visible".

McLennan [38] urged for a different design approach to abandon all negative environmental impacts. It should be purpose oriented. The six principles of sustainable design include [24]):

1. "Learn from and as natural systems (Biomimicry Principle). Nature shall serve as a model (recycling everything), as a measure (limits) and as a mentor (mimicry of natural designs).

2. Respect natural material and energy resources (Conservation Principle). Energy originates from the sun—it is abundant, every day it meets our needs for 27 years, but it must be concentrated and stored by physical, biological or chemical processes.
3. Respect for people (Human Vitality Principle) by creating healthy and friendly products, and infrastructure.
4. Respect for place (Ecosystem Principle): use local materials and local conditions.
5. Respect for the future ('Seven Generations' stewardship Principle), e.g., to consume non-renewables at a rate below replenishment, and approach zero waste.
6. System thinking (Holistic Principle)—considering impacts of design on environment (ecology), people (equity), and business (economy)".

Practical design applications vary among disciplines (process or product design, ICT, architecture, services, etc.) but some common principles have been. Sustainable design requires a holistic approach to development including the protection of natural resources and energy usage [39].

"Integrated Product Development requires understanding and predicting the whole Product Life cycle [40]. It has significant implications on the competence profiles of engineers. Integrated Design Engineers need "integrated skills" including Product Life Cycle Engineering, Innovation Driven Design, Responsible Design, Networked Collaboration, Intercultural Skills, Requirements and Knowledge Engineering. Certification rules of the ECQA (European Certification and Qualification Association) can be used to leverage these assets to a worldwide unique qualification and certification platform for Integrated Design".

Ranky [41] published 18 product design engineering principles and rules, which are very complex. They are based on a sustainable green, lean design, and assembly approach, also known as concurrent or simultaneous Green PLM (Product Lifecycle Management). The principles and rules also include Ranky's intelligent Sustainable Enterprise Engineering (iSEE:Green) concept.

Mattson and Wood [42] summarized the experiences of engineering researchers and practitioners of design for developing countries into nine guiding design principles. The social pillar is included within design as the most important one. The nine principles are: (1) design with respect to the local population and its needs, (2) test the product in the actual environment, (3) adapt imported technology to specific local and regional needs, (4) take care of urban and rural poverty elimination, (5) regard especially women and children, (6) adapt project design and management techniques to the specific needs of developing countries, (7) engage interdisciplinary teams, (8) cooperate with local and regional governments, and (9) adapt distribution to developing world markets.

Design for sustainability is an approach that puts the well-being of people and the sustainability of the environment first [43]. It is a whole system approach that considers the overall impacts of designs. As a multidisciplinary and interdisciplinary approach, it can be applied in all fields. To design and create sustainability, there are two key questions to consider:

(1) Are you doing the right things to support: individual well-being, human intrinsic values, social equity and the common good, diverse, and thriving local communities, healthy environments, and environmentally sustainable production and consumption?
(2) Are you doing things right: thinking in systems, dematerialization, renewable and natural materials, biomimicry, and cradle to cradle?

The early principles required equilibrium between human and nature rights. Later, social requirements prevailed, except in Todds' principles and practice of eco-design where environmental principles are dominant. Economic principles are regarded only in third place. McLennan pointed out three of the environmental principles (biomimicry, natural and local resources), two social ones (respect for people and future), and an overarching one (holistic or systems approach). Mattson and Wood elaborated on nine guiding design principles for developing countries, respecting local and regional population, environment, needs, social circumstances, local cooperation, management, and export. Design for sustainability is about well-being and environmental protection.

5. Sustainable Engineering

Sustainable engineering refers to the integration of social, environmental, and economic considerations into product, process, and energy system design methods. Additionally, sustainable engineering encourages the consideration of the complete product and process lifecycle during the design effort. The intent is to minimize environmental impacts across the entire lifecycle while simultaneously maximizing the benefits to social and economic stakeholders.

A review of the principles from green engineering and sustainable engineering is presented in Table 3.

Table 3. An overview of sustainable engineering principles.

Initiative	Principles
Anastas and Warner, 6 out of 12 Principles of Green Chemistry	(1) No waste production, (2) minimum risk to humans and environment, (3) minimum energy, (4) renewable resources, (5) benign end-of-live products, (6) real time analysis for pollution prevention
Anastas and Zimmerman, 12 Principles of Green Engineering	(1) Non-hazardous inputs/outputs, (2) waste prevention, (3) min. resource usage, (4) max resource/time efficiency, (5) "output pulled" resources, (6) recycle, reuse, refurbish, (7) durability not immortality, (8) no overcapacity, (9) easy recycling, (10) recycle within process, (11) extended use, (12) renewables
Sandestin, Green Engineering principles	(1) Holistic approach, (2) natural ecosystem conservation, (3) life cycle thinking, (4) safe/benign inputs, (5) min natural resources, (6) waste prevention, (7) local conditions, (8) engage all stakeholders
BASF, Eco-efficiency Analysis	(1) Customer viewpoint, (2) societal factors by using LCA, (3) economy vs. ecology, (4) social aspects
WBCSD, Eco-efficiency principles	(1) Fewer materials and (2) energy, (3) disperse toxics, (4) recyclability, (5) renewables, (6) durability, (7) greater service intensity
IPENZ principles	See Table 4
RAE principles	(1) Beyond locality and immediate future, (2) innovate and create, (3) balanced solution, (4) all stakeholders, (5) needs and wants, (6) plan/manage effectively, (7) benefit sustainability, (8) polluters pay, (9) holistic approach, (10) do right things right, (11) no cost reduction masquerade, (12) practice what you preach
SE principles	See Figures 2 and 3
Rosen, Key requirements for SE	Sustainable (1) resources and (2) processes, (3) increased efficiency, (4) reduced environmental impact when using LCA, (5) other aspects: economic affordability, equity, resource demand, safety, community, social acceptability, human needs, land use, aesthetics, lifestyles, population

Table 4. Three key principles with 15 principles of sustainability and engineering in New Zealand.

	Maintain the Viability of the Planet
1	Humans need to maintain the integrity of global and local biophysical systems.
2	Renewable resources must be managed within sustainable harvest rates and non-renewable resource depletion rates must equal the rate at which renewable substitutes take their place.
3	Technological options must favor choices that minimize the use of resources and reduce risks.
4	The material and energy intensity used in products, processes or systems needs to be reduced significantly—by 10 to 50 times—using recycling and minimization techniques.
5	Waste streams during the life cycle of products, processes or systems must be minimized to the assimilative capacity of the local and global environments.
6	Any use and production of environmentally hazardous materials must be minimized and carried out prudently if necessary.
	Providing for equity within and between generations
7	Humans, now and in the future, must have equal access to choices in life that reduce significant gaps between people in areas such as health, security, social recognition, and political influence.
8	Total consumption of resources needs to be within the environment's sustainable capacity and balanced between the affluent and those yet to fulfil their basic needs.
9	Present resource use and development must be considered over a sufficiently long timescale that future generations are not disadvantaged.
10	Those directly affected by engineering projects, products, processes, or systems must be consulted and their views incorporated into the planning and decision-making processes.
	Solving problems holistically
11	Problem solutions must be needs-based, rather than technology-driven.
12	Demand growth targets must be realistically assessed and if necessary managed, rather than simply meeting predictions.
13	A holistic, systems-based approach must be used to solve problems, rather than technology focusing on only single aspects of problems.
14	Unsustainable practices must be reduced to zero over time, and where practicable past degradation shall be addressed.
15	Problem solutions must be based on prudent risk management approaches.

5.1. Green Engineering Principles

Anastas and Warner [44] published 12 Principles of Green Chemistry—they are mainly chemicals oriented but half of them apply to engineering, too: "(1) Avoiding waste production is far better than treating or cleaning it up afterwards; (3) Feeds and products to and from a chemical process should pose as little risk to human health, and be as environmentally benign as possible; (6) Reduce energy waste and consumption by striving to operate at ambient conditions; (7) Where technically and economically possible, (chemical) feeds and raw materials should be sourced renewably; (10) Materials and chemicals should be designed, where possible, to degrade into benign and non-toxic substances at the end of their functional lives; (11) Use real time analysis for pollution prevention".

The 12 principles of green engineering [45] include: (1) non-hazardous material and energy inputs and outputs, (2) waste prevention, (3) minimum resource usage, (4) maximum resource and time efficiencies, (5) use "output pulled" energy and materials, (6) recycle, reuse, or dispose for usable purposes, (7) the design goal shall be durability, rather than immortality, (8) avoid overcapacity, (9) design for easy recycling, (10) recycle and reuse resources within your production process, (11) design for extended use, and (12) use renewables. The principles are chemical process oriented.

The nine Sandestin principles are the result of a conference entitled Green Engineering: defining the principles, which took place in May 2003 at the Sandestin Resort in Florida [46]. The Sandestin Declaration agreed upon the following draft principles: (1) holistic approach, (2) natural ecosystems conservation, (3) life cycle thinking, (4) safe and benign material and energy inputs, (5) minimum depletion of natural resources, (6) waste prevention, (7) respect local conditions, (8) improve–innovate–invent new technologies, and (9) engage all stakeholders.

At the same time, BASF developed and tested the Eco-Efficiency Analysis using over 180 industrial applications [47]. The work mostly confirmed the 12 principles listed in the Anastas and Zimmerman paper, but it was more practice oriented, and included economic and social principles. As well as the preliminary conditions in the analysis, the environmental efficiency analysis was developed using 10 principles including: (1) customer viewpoint when calculating total cost, health, safety, and risk analysis, (2) weighting societal factors by using LCA, (3) comparing the relevance of economy versus ecology, and (4) regarding social aspects (optionally).

Eco-efficiency has developed in the last two decades. As defined by the World Business Council for Sustainable Development [48], "Eco-efficiency is achieved by the delivery of competitively priced goods and services that satisfy human needs and bring quality of life, while progressively reducing ecological impacts and resource intensity throughout the life cycle to a level at least in line with the Earth's estimated carrying capacity. It is concerned with creating more value with less impact." According to Lehni [48], "the most critical features of eco-efficient companies are: (1) Producing goods and services with fewer materials. (2) Producing goods and services with less energy. (3) Dispersing lower volumes of toxic materials. (4) Improved recyclability. (5) Striving to use renewable resources. (6) Making goods that last longer, and (7) Greater service intensity of products and services".

Garcia-Serna et al. [49] presented a "broad review of disciplines and technologies concerning the design trends: The Natural Step, Biomimicry, Cradle to Cradle, Getting to Zero Waste, Resilience Engineering, Inherently Safer Design, Green Chemistry and Self-Assembly". The core of the review paper was "Green Engineering, its main definitions, scope of application, different guiding principles, a framework for design and legislative aspects". Thirteen sets of principles were presented, and a quick selection guide about using each one was given. Some of them are described in this paper; here is the list of six not yet described sets of principles: The Earth Charter Principle, The Coalition for Environmentally Responsible Economies (CERES) Principle, The Bellagio Principles, The Ahwahnee Principles, The Interface Steps of Sustainability, and Design for Environment (DfE) Key Strategies.

These green engineering principles were and still are environmentally stressed as shown by the name. Industry introduced additional economic and social components, which helped them to survive until the present. Eco-efficiency has slowed down the rate of hazardous human impacts on the environment. Sets of principles were described to help engineers design their objects easier and better.

5.2. Sustainability Principles and Practice for Engineers

The Institute of Professional Engineers New Zealand (IPENZ) started work on sustainability principles in 2003. Their time horizon was 1000 a, and they assumed that basic human needs will not change. Two years later they presented the first draft of their updated principles for engineers [50] that were further improved by numerous case studies and checklists. On this basis, they postulated three key principles (Viability of the Planet, Equity within and between generations, Solving problems holistically) with 15 principles in all [51] (Table 4).

The Royal Academy of Engineering (RAE, United Kingdom) presented 12 "Principles of Engineering for Sustainable Development" [52], which are very useful from the operational point of view. The 12 principles (Table 3) are: "(1) Look beyond your own locality and the immediate future; (2) Innovate and be creative, (3) Seek a balanced solution; (4) Seek engagement from all stakeholders; (5) Be sure you know the needs and wants; (6) Plan and manage effectively; (7) Give sustainability the benefit of any doubt; (8) If polluters must pollute, then they must pay as well; (9) Adopt a holistic, "cradle to grave" approach; (10) Do things right, having decided on the right thing to do; (11) Beware cost reductions that masquerade as value engineering, and (12) Practice what you preach".

The IPENZ principles are well developed, covering three main areas: environmental, equity, and systems approaches. They are weak on economic and societal sides, however. The RAE principles are short but very instructive for practicing engineers.

5.3. Sustainable Engineering Principles

Sustainable engineering (SE) does not harm the environment, nor does it exploit resources that belong to future generations [53]. The NAL Thesaurus [54] defines it as "the design, commercialization and use of processes and products that are feasible and economical while reducing the generation of pollution at the source and minimizing the risk to human health and the environment". It includes all the three SD pillars. Kauffman and Lee [55] edited a handbook of SE with 66 chapters and 1285 pages written by academic researchers as well as practitioners. They brought new engineering approaches to sustainable production, and innovative operational practices.

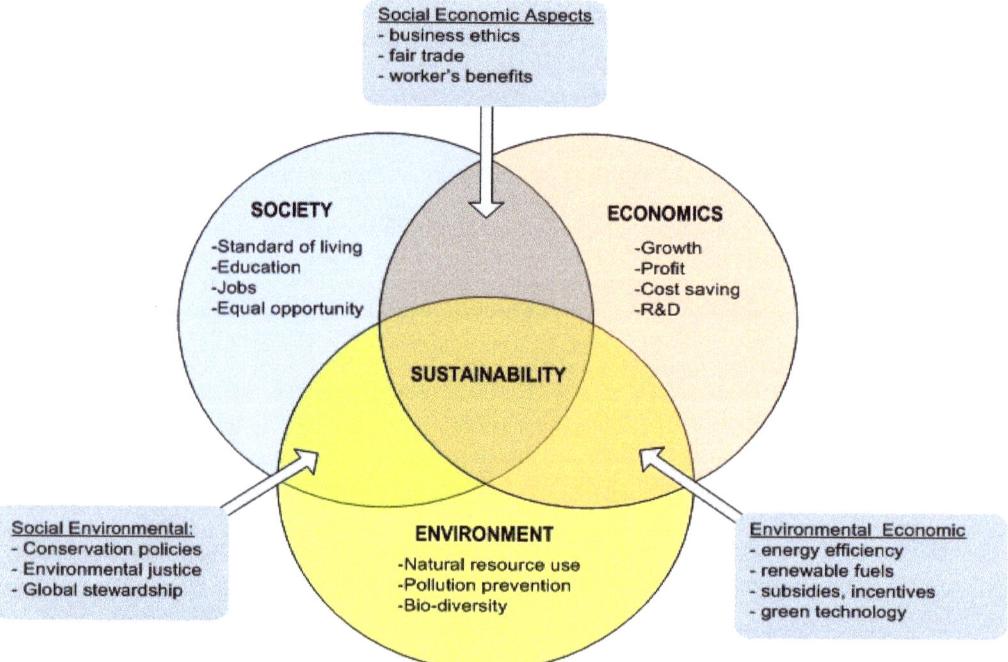

Figure 2. Interplay of the environmental, economic, and social aspects of sustainable development (Mark Fedkin [23]; adopted from the University of Michigan Sustainability Assessment [56]).

The advantages of SE in comparison with traditional green engineering are the following [23]:

- SE considers the whole system in which the product, process or services are used.
- It deals not only with technical but also with non-technical issues in synergy.
- It is designed to solve problems for the distant future, not just for the immediate period.
- As well as the local context, it considers the global one, too.

- It deals with political, ethical, and societal issues by interacting with experts in other disciplines.

Sustainable engineering principles can be shown by a triangle, having environmental, social, and economic pillars as the corners. Figure 2 presents all the tree pillars in the form of spheres, with their most important constituents listed. The environmental aspects include the use of natural resources, pollution prevention, biodiversity, and ecological health. The social aspects include standards of living, the availability of education and jobs, and equal opportunities for all members of society. The economic factors are drivers for growth, profit, reducing costs, investments into research and development, etc. Interaction of the social and economic spheres result in the formulation of combined social–economic aspects. Those are, e.g., business ethics, fair trade, and worker's benefits. The combination of economic and environmental interests facilitates increasing energy efficiency, the development of renewable fuels, green technologies, and creation of special incentives and subsidies for environmentally sound businesses. The intersection of social and environmental spheres leads to the creation of conservation and environmental protection policies, establishment of environmental justice, and global stewardship for the sustainable use of natural resources.

The diagram in Figure 3 presents a consolidated framework for sustainable engineering principles, which are in part adopted from the work of Gagnon et al. [57] and from the green engineering principles established by Sandestin Conference (Abraham and Nguyen, 2003). Gagnon et al. presented sets of principles that are most relevant to sustainable engineering. Based on general sustainable development principles and on specific engineering ones, they proposed a set of fifteen sustainable engineering principles, organized in a triangle. The principles that are closer to the extremities of the triangle are one-dimensional. Those on the sides are bi-dimensional; they have a stronger connection with the angle they are closer to. Three-dimensional principles are in the middle of the triangle; each of them being situated according to their links with the angles. These principles can serve as guidelines in a specific engineering project.

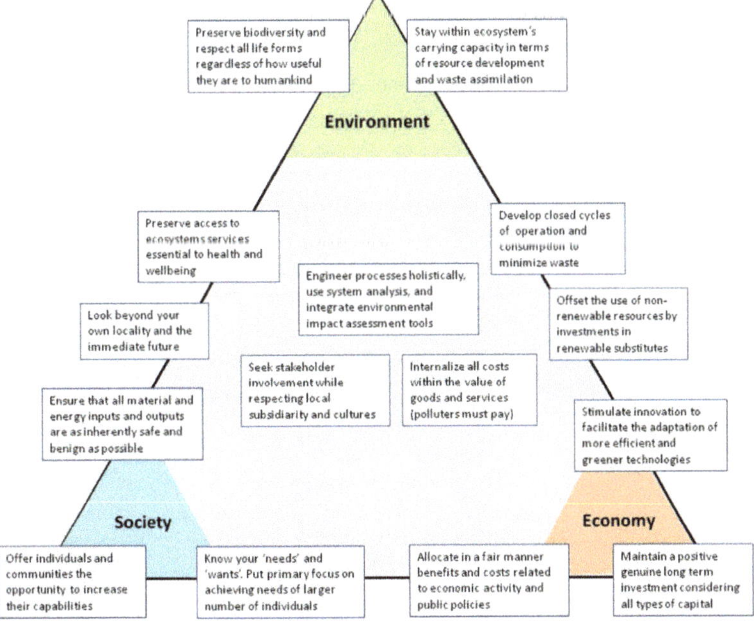

Figure 3. Classification of sustainable engineering principles versus environmental, social, and economic criteria.

Rosen [58] cited "5 key requirements for engineering sustainability: (1) sustainable resources, (2) sustainable processes, (3) increased efficiency, (4) reduced environmental impact when applying LCA, and (5) fulfilment of other aspects of sustainability: economic affordability, equity, meeting increased resource demand, safety, community involvement and social acceptability, human needs, appropriate land use, aesthetics, lifestyles, and population".

All the three approaches described above are close to the modern understanding of SE. However, they do not mention circular economy and zero waste system, which is broader and deeper than just recycling—see the Circular Economy System Diagram [59]. Moreover, human capital, social responsibility, and annual reporting are not presented as societal and 3D components, respectively. Cost-Benefit Analysis using LCA is important, too. The precautionary approach has been well known for three decades, but it is not included in the existing SE principles. Therefore, it seems reasonable to update the SE Principles.

6. Updated Sustainable Engineering Principles

Sustainable engineering (SE) principles need an update to include the present state of the art in human knowledge. Table 5 presents the missing principals in the individual SE sets studied. The updated 12 principles have been synthesized using the literature review in Sections 2–4, especially the above-described principles of sustainable development, SD [33], new trends, recent developments, and personal experience in the SD area. The SDGs have been respected, too. As well as the omissions mentioned in the last paragraph, more rapid climate change than expected, critical raw materials break out, pollution (especially plastics) spread, and acceleration of species extinction are suggested to modify the outdated principles. Increasing inequalities, extreme human wealth distribution, advantage of profits against the common good, the fraction of decent jobs reduction, etc. require changes in the behavior of engineers. They are expected to mitigate the negative consequences of human development and present a solid ground for long-term changes in engineering design. Sustainable engineering principles are an important vehicle towards sustainable development.

Environmental, social, and economic pillars are of equal importance to the future. Therefore, the updated principles (the connections with the SDG numbers are shown in parentheses) include:

A. Tridimensional principles

    1.   Holistic approach, systems thinking and management (SDG 16)
    2.   Precautionary and preventive approaches (SDG 13)
    3.   Annual sustainability reporting using GRI (SDG 4)

B. Environmental principles

    4.   Circular economy, waste minimization, sustainability hierarchies (SDG 13)
    5.   Efficient use of resources, and increased share of renewables (SDGs 6, 7)
    6.   Sustainable consumption and production (SDG 12)

C. Social principles

    7.   Equalities within and between generations (SDGs 5, 10)
    8.   Engagement of communities and all stakeholders (SDG 11)
    9.   Corporate social responsibility and decent work (SDGs 1–3, 8)

D. Economic principles

    10.   Human capital, innovations, and creativity (SDGs 4, 9)
    11.   Cost–benefit analysis using Life Cycle Assessment (SDGs 3, 13)
    12.   Internalizing externalities—polluters must pay (SDG 13)

Table 5. Comparing the updated Sustainable Engineering Principles (SEP) with the ones described in Section 4.

Updated Sustainable Engineering Principles	Green Engineering	Eco-Efficiency Analysis	Sandestin Principles	SD Triangle Figure 2	Sustainable Engineering Principles, Figure 3	Sustainable Principles and Practices IPEN2	RAE Principles of Engineering for SD
Tridimensional principles							
Holistic approach and system thinking			holistic approach	environmental stewardship, environmental justice	holistic process, use system analysis	use holistic systems-based approach	holistic approach
Reporting						prudent risk management	do things right
Environmental principles							
Sustainable consumption and production	minimum resource use, output pull, durability	environmental fingerprint	minimum resource depletion	resource use, pollution prevention, biodiversity	Biodiversity, know the needs and wants	minimum use of resources and risks, integrity	know the needs and wants
Circular economy, waste minimization, waste, and energy hierarchies	recycling, reuse, design for recycling, waste prevention		waste prevention	green technology	closed cycles	within environmental capacity, minimum waste	
Efficiency, renewable resources	maximum resource and time efficiency, renewable resources		ecosystems conservation	energy efficiency, renewable fuels		min energy and material intensity, renewable sources	
Precautionary and preventive approach	non-hazardous in/outputs, no over-capacity		safe and benign inputs	conservation policies	eco-carrying capacity, safe and benign input	minimum hazardous materials, needs-based	look beyond locality and near future
Societal principles							
Equity within, between generations				equal opportunity		equal access to choices	
Communities, all stakeholder's engagement		societal weighing	respect local cond., engage all stakeholders	living standard, jobs	stakeholder interests beyond your	consult and consider directly affected stakeholders	engage all stakeholders, balanced solutions
Social responsibility, CSR, decent work				business ethics, fair trade	health and well-being	no disadvantage for future generations	practice what you preach
Human capital, creativity, and innovations			improve, innovate, invent	education, workers' benefit, R&D	stimulate innovations in greener technologies		innovate and be creative
Economic principles							
Cost–benefit analysis when applying LCA		total costs, eco-efficiency analysis, LCA	LC thinking	growth, cost saving, profit	allocate benefits and costs	reduce unsustainable practices to zero	beware cost reduction masquerade
Internalizing externalities, polluters pay				subsidies, incentives	internalize all costs—PMP		polluters must pay

Table 5 compares the updated SE Principles with those obtained from the descriptions in Section 4. Although it is difficult to find the adequate location of each principle, each of the sets is evidently missing at least two or more of the updated principles.

7. Discussion

To accelerate the transition to equitable, sustainable, livable, post-fossil carbon societies [60], we need to do today more than yesterday and tomorrow more than today.

Scientists, engineers, policy makers, institutions, and NGOs have shown new and improved ways to a sustainable future. They need to be included in the different sets of sustainability principles. The engineering ones will include the SDGs. Although some of them are old, others are not and have to be included in the standard list of SE principles: innovation, decent work, responsible consumption, and peace present some examples of them. International and national organizations are developing new and improved approaches to sustainable development, which will be accompanied by SE activities. The pace is speeding up and, therefore, the SE Principles must be broad enough to encompass new methods and tools into everyday practice.

7.1. The Tridimensional Principles

The Holistic approach considers a global view, the Planet, instead of a local one. It is concerned with assumptions, knowledge, methods, and implications of various disciplines treating them as an integrated whole, or system. Systems thinking is about how things interact with one another; it must be differentiated from systematic (analytical, or synthetic) thinking. Environmental management using ISO 14000 family standards on LCA, and EU Environmental Management and Audit Scheme (EMAS) helps organizations (a) minimize environmental impacts of their operations (e.g., processes); (b) comply with applicable laws, regulations, and other environmentally oriented requirements; (c) apply continuous improvement approach. ISO 14000, LCA, standardizes a technique to assess environmental impacts associated with all the stages of a product's life, from raw material extraction through material processing, manufacturing, distribution, use, repair and maintenance, disposal, and recycling.

Precautionary principle requires the prevention of actions whose consequences are uncertain and/or potentially dangerous. It was first proposed in the Rio declaration. Humans will avoid or diminish activities that may lead to morally unacceptable harm to the environment or humans—to human life or health—serious and effectively irreversible actions that are inequitable to present or future generations, or do not respect the human rights of persons involved [61]. The judgement of plausibility should be grounded in scientific analysis. The principle, proposed as a guideline in environmental decision making has four central components: (1) taking preventive action in the case of uncertainty, (2) shifting the burden of proof to the proponent of an activity, (3) exploring a wide range of alternatives to potentially harmful actions, and (4) increasing public participation in decision making.

The EU definition of the precautionary principle applies where scientific evidence is insufficient, inconclusive, and preliminary scientific evaluation indicates that there are reasonable grounds of concern that the potentially dangerous effects on the environment, human, animal, or plant health may be inconsistent with the high level of protection chosen by the EU [62]. Examples of harm are environmental hazards (climate change and global warming, species extinction, effluents, acute pollution), public and human health hazards (diseases, toxic chemicals, carcinogenic substances, hazardous waste, radiation) of chemical, physical, biological, or psychosocial origins

The Global Reporting Initiative, GRI [63] is an international organization that "helps businesses, governments and other organizations understand and communicate their impacts on issues such as climate change, human rights and corruption". "A sustainability report published by a company or organization is about the economic, environmental and social impacts caused by its everyday activities including its values and governance model and demonstrates the link between its strategy and its commitment to a sustainable global economy". Sustainability reporting can help organizations to "measure, understand and communicate their economic, environmental, social, and governance performance, and then set goals, and manage change more effectively". An annual sustainability report is the "key platform for communicating sustainability performance and impacts—whether positive or negative". GRI guidelines include (a) the required general standard disclosures, and (b) specific standard disclosures with three categories: economic, environmental, and social, each with several aspects. Each aspect has several descriptions and/or indicators.

7.2. The Environmental Principles

The closed loop or circular economy (CE) model [59] has emerged from the discipline of industrial ecology (described in Section 2). It presents a transition from a linear, "take-make-waste" or "cradle to grave" production system, to a circular, or "cradle to cradle", "waste is food", biomimicry, "blue economy", and "whole system thinking". It has been used as an exemplar [20] to apply the metaphor of natural systems to the production of goods in the industrial economy. CE includes eco-design, sustainable supply, and responsible consumption. The Zero waste ideal leads to waste minimization and better utilization of resources (materials and energy). It requires separation of wastes and their collection and uses all the 12 Rs (rethink, reform, refuse, reduce, reuse, re-gift, repair, refurbish, remanufacture, repurpose, recycle, and recover) to mimic the rules in the biosphere. When using waste, it is important to use it for the highest possible level of hierarchy (Figure 4).

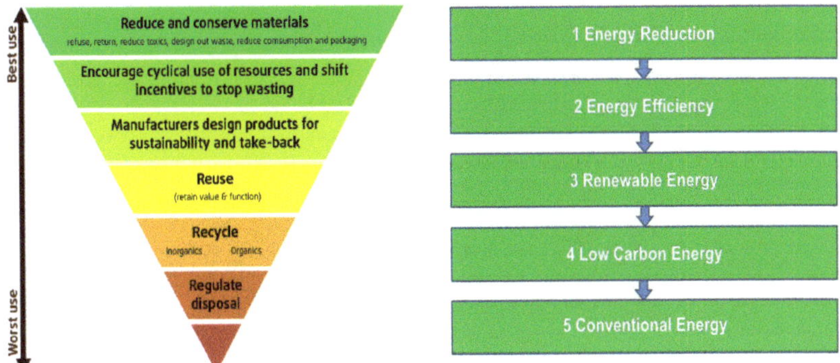

Figure 4. Two possible sustainability hierarchies.

Sustainability hierarchy considers, e.g., natural resource conservation and protection as the top priority, followed by manufacturing efficiency, product quality, judicious product selection, product repair and restoration, resource reuse, resource recycling, energy recovery, waste treatment, waste disposal, and pollution at the bottom [64]. The waste management hierarchy considers waste prevention as the highest priority, followed by waste minimization, resource reuse, and resource recycling, energy recovery, waste treatment, and waste disposal as the lowest preference. A similar approach to the energy hierarchy has energy reduction/saving (leaner) at the top, followed by energy efficiency (keener), renewable (greener) energy, low carbon, and other emissions (cleaner), and conventional (meaner) energy at the bottom.

The resources are finite, and the EU has already identified 30 critical raw materials [65]. China is the biggest producer of them. Several other countries are dominant suppliers of specific raw materials, e.g., USA (beryllium) and Brazil (niobium). A circular economy enables them to be reused but it requires a suitable eco-design. Respecting health and safety rules and including the use of materials that accumulate in the environment (e.g., greenhouse gases) are of great importance for the future climate stability. In the case of energy, cascading is used, e.g., by the closest temperature differences (pinch technology) or distance, by recycling.

Resource efficiency requires use of the Earth's limited resources in a sustainable way while minimising impacts on the environment. It allows us to create more output and to deliver greater value with less input. Resources include materials, water and energy, capital, and finance (money), human capital (staff, employees, unemployed or migrated individuals), etc. Renewables are renewable energy sources (solar, hydro, wind, biomass, biofuel, biogas, geothermal, tidal, and wave energy) as well as material ones (water, air, soil, plants, animals, and humans, agricultural, aqua-cultural, and forestry products—food, biomass, timber, paper, cotton and wool, bioplastics, and bio-chemicals).

Examples from industry, sustainable cities and communities, universities, schools, etc., demonstrate that the proposed principles can be put into practice bringing substantial improvements. Let us consider, for example, achievements of the European chemical industry in the period 1990–2019 [66]: total greenhouse gas (GHG) emissions fell by −54% (–154 Mt/a), GHG intensity (GHGs per unit of production) by −54%, and GHG emissions per energy production by −47%. While chemical production increased by +47%, specific energy consumption decreased by −47%, showing that decoupling of the two is viable.

Sustainable consumption and production (SCP) was identified as one of the three overarching objectives of, and essential requirements for, sustainable development: "Fundamental changes in the way societies produce and consume are indispensable for achieving global sustainable development with the developed countries taking the lead and with all countries benefiting from the process. Poverty eradication, changing unsustainable patterns of production and consumption, and protecting and managing the natural resource base of economic and social development are essential requirements" [67]. SCP is included in the SDGs under No 12: "SCP is about promoting resource and energy efficiency, sustainable infrastructure, and providing access to basic services, green and decent jobs, and a better quality of life for all. Its implementation helps to achieve overall development plans, reduce future economic, environmental, and social costs, strengthen economic competitiveness, and reduce poverty".

7.3. The Societal Principles

Intergenerational equity is a concept that humans "hold the natural environment of our planet in common with other species, other people, and with past, present and future generations" [68]. The present generation includes children, youth, adults, and seniors. They inherited the Earth from previous generations and must pass it on to future generations in good condition. The continued depletion of natural resources in the past diminishes their opportunities in the future. National debt is an example of intergenerational inequity, as future generations will suffer paying it off. Intra-generational equity must ensure equal opportunities and equal access to choices for all customers and employees, irrespective of gender, sexuality, race, nationality, or religion.

Stakeholder engagement means that the public has the right to participate in decision making. Engineers must respect public opinions in project outcomes [69]. Engagement of communities and all stakeholders in projects for the developing world has been explained in Section 3. It is important to achieve the triple bottom line, and it relates to both the corporate social responsibility and the GRI. The ISO 9000 standard series on quality management systems helps organizations to meet the customers' and other stakeholders' needs as well as other requirements related to a product (e.g., quality control).

Social/societal responsibility (SR) is a self-evaluation tool for companies regarding their impact on society [70]. SR is an ethical framework for individual engineers and scientists, too [71]. The word "social" refers to "having to do with human beings living together as a group in a situation in which their dealings with one another affect their common welfare" [72]. The definition of the word "societal" is more restrictive but is roughly included within the wider definition of "social"; it is defined as "of or pertaining to society". "Social/societal responsibility" regards new technologies, which can be found in responsible innovation, and it originates from technology assessment. Technology Assessment emerged in the 1970s to reduce the (negative) side effects of new technologies (motivated by the pollution caused by the growing industrial system).

Social responsibility relates to all citizens, while corporate social responsibility (CSR) refers to companies taking responsibility for their impact on society. The European Commission [73] believes that "CSR is important for the sustainability, competitiveness, and innovation of EU enterprises and the EU economy. It brings benefits for risk management, cost savings, access to capital, customer relationships, and human resource management". The Commission promotes CSR in the EU and encourages enterprises to adhere to international guidelines and principles. "CSR policy functions as a self-regulatory mechanism whereby a business monitors and ensures its active compliance with the spirit of the law, ethical standards and national or international norms" [70].

ISO 26000 voluntary standard (Guidance on social responsibility) helps to integrate social responsibility into company's values and practices; it includes seven key principles: "accountability, transparency, ethical behaviour, respect for stakeholder interests, respect for the rule of law, respect for international norms of behaviour, and respect for human rights". Every user of ISO 26000 should also consider its seven core subjects: "organizational governance, human rights, labour practices, environment, fair operation practices, consumer issues, and community involvement and development".

Decent work has been defined by the United Nations Economic and Social Council [74] as employment that "respects the fundamental rights of the human person as well as the rights of workers in terms of conditions of work safety and remuneration". It respects "the physical and mental integrity of the worker in the exercise of his/her employment". According to the International Labour Organization [75], "decent work involves opportunities for work that are productive and deliver a fair income, security in the workplace and social protection for families". It offers "better prospects for personal development and social integration, freedom for people to express their concerns, organize and participate in the decisions that affect their lives, and equality of opportunity and treatment for all women and men." Health and safety are constituent elements of workplace security.

7.4. The Economic Principles

Human capital (HC) includes social and economic components—the former was dealt with in SR and decent work, the latter is regarded here. It includes personal competences, knowledge, skills, know-how, expertise, habits, and social attributes such as creativity and mobility. Employees need permanent, long-life education, and training. HC consists of attracting talents, innovators, and experts while developing the knowledge and skills of employees by using lifelong learning. It needs good relations between the employees and relaxed relations of workers with management. HC increases productivity, added value, and a company's profit.

Innovation is a joint product of human capital and creativity. "The capacity to innovate is seen to be a function of a region's ability to attract human capital and to provide low barriers to entry for talented and creative people of all backgrounds" [76]. Research, technical development, and innovations are indispensable organizational forms of exploiting human capital. Responsible research and innovation (RRI) require high ethical standards and gender equality in research. Policymakers' responsibility for avoiding the harmful effects of innovation belongs to the social principle.

"Open innovation was defined as the use of purposive inflows and outflows of knowledge to accelerate internal innovation, and expand the markets for external use of innovation, respectively" [77]. When using open innovation, boundaries of the organisation become permeable and that allows combining the company resources with the external co-operators. Open innovation is a potential answer to tap into creativity and knowledge outside an organisation. It exploits a creative power of users, communities, and customers to co-develop new products, services, and processes. Innovation is a joint product of human capital and creativity. "The capacity to innovate is seen to be a function of a region's ability to attract human capital and to provide low barriers to entry for talented and creative people of all backgrounds" [76].

"Creativity is the capability or act of conceiving something original or unusual while innovation is the implementation of something new" [78]. Creativity brings new ideas—inventions. Invention must be distinguished from innovation, which is the realization of an invention. Nowadays, all employees should be creative, and companies must take care to stimulate their inventions and develop them into innovations.

Two of the most popular decision support tools today are Cost–Benefit Analysis (CBA) and Life Cycle Assessment (LCA). These two tools provide the decision makers with different information. The CBA objective is to maximize overall utility to society, whereas the LCA aims at minimizing harmful impacts on the environment. Since the methodologies are answering different questions, they should not be considered as competing but rather as complementary tools. The LCA and CBA are decision support tools and not decision-making tools because they provide information that normally needs to be complemented with legal, social, economic, or technical information before decisions are made.

The CBA evaluates the expenses and advantages of a product, process or system, project, policy, or program to society, e.g., implementing a given waste policy or building a treatment facility. If the net benefit is positive, the project will, as a rule, be implemented. A clear advantage of the CBA is that the result is expressed in a well-known measure, money.

The LCA considers the impacts of a product, process, or service on the environment during its whole lifetime, from raw material extraction and energy consumption to waste and emissions and to product disposal. Its result is the life cycle inventory, LCI [79]. Life cycle costing (LCC) or life cycle cost analysis (LCCA) is used to minimize the total cost of a product, service, or system during its life span. Eco-efficiency analysis considers the economic and life cycle environmental effects of a product or process; different scenarios can be conducted [47]. The LCA is based on ISO 14000 family standards

The polluter pays principle (PPP) requires the firm or consumer to pay the total social cost (both private and external), rather than just the private cost [80]. The EU member states are expected to promote this principle and "internalize the externalities". An externality is a cost (e.g., pollution) or benefit (e.g., public health) caused by an external party; it can be positive or negative. Governments can pose a tax to the producer of the externality and thereby internalize its cost. Companies can include the cost or benefit of the externality into the product price.

8. Conclusions

Climate change is following the "hysteresis" effect [81]. Since the preindustrial era, the temperature has grown slowly because of increased greenhouse gas (GHG) emissions, speeding up exponentially later to approach a new equilibrium state. If we want to lower the temperatures just to the preindustrial level, the hysteresis curve will require GHGs to be lowered to below the concentrations in the preindustrial era; by lowering them to the preindustrial level, the temperatures will not fall to the historical level. The same kind of speeding up is taking place in the societal area—since the mid nineteen seventies, the neoliberal economy has increased the inequality gap, which has speeded up in the last few decades, and during the last crisis has reached a state where 1% of the richest individuals possess the same value of wealth as the poorest half of the world's population.

Engineers have a special, crucial role in these circumstances. They must reverse the hysteresis curve of human development. They must halt the increasing GHG emissions and then lower them to slow down the climate change. They also need to design processes, products, and services to adapt to and mitigate the changing environment. By applying life cycle thinking, resource efficiency, and waste minimization, engineers are expected to develop a circular economy and prevent resource and species extinction. By using renewable sources and internalizing environmental costs, they can improve economic sustainability. Sustainable consumption and production are very much depending on the design and operation of engineers.

Engineers have an important influence on the increased quantity and quality of decent, green jobs. They can design for a better life and more equality instead of designing for higher profit. They can participate in stakeholder engagement, increased consumer awareness, and respect their true needs. Fostering a lifelong education of employees, developing their creativity and innovativeness, and enacting corporate social responsibility and professional ethics, they can bring about a just, social, and equal society.

An overview of sustainability, sustainable design, and sustainable design principles has been presented. Twelve principles of sustainable engineering have been proposed in the form of 12 updated sustainable engineering principles, and each one explained. They enable designers and engineers to improve their process, product, and service designs, and optimize resource (re)use. The principles will require constant updating, which is the primary duty of engineers themselves and their associations. This paper is one of the many steps on this path and it can stimulate other colleagues to keep developing sustainable design, engineering, and operations in their everyday work.

Environmental principles respect the ecosystem's carrying capacity and preserve the biodiversity. The Holistic approach considers a global view, the Planet, instead of a local one. Holistic approaches are concerned with the assumptions, knowledge, methods, and implications of various disciplines and treat them as an integrated whole or system. Systems thinking is about how things interact with one another; it must be differentiated from systematic (analytical or synthetic) thinking. The LCA is used to assess environmental impacts associated with all the stages of a product's life from raw material extraction through the materials' processing, manufacture, distribution, use, repair and maintenance, and disposal or recycling.

The updated societal principles speak in favor of improved equalities including the distribution of wealth. An increased number and quality of decent and green jobs is of utmost importance. True needs in the value chain and the well-being of consumers will be respected. Awareness raising and education level, justice in human development, and respect to ethics will be improved. The industry of arms and wars will be limited and controlled globally. "Sustainable consumption needs products and services that have a minimal impact on the environment enabling future generations to meet their needs. Sustainable production as the creation of goods and services using processes and systems shall be safe and healthful for workers, communities, and consumers".

Regarding the economic principles, precautionary thinking will be respected, and the polluter pays imperative will improve existing consumption patterns. Quality control and annual reporting need to include all the three sustainability pillars. Resource productivity will be increased, while processes, products, and services are optimized. Employees represent human capital, which will be supported, lifelong education enabled, and their creativity maintained at the highest possible level. Profits are important for sustainable future development of businesses, but they will not be regarded as a target by themselves.

International reports from IPCC, EEA, etc., show an increasingly worrying situation—new approaches are required. Therefore, sustainable engineering principles must be adapted constantly to these changes to keep sustainable design at the forefront of the changes for improved future sustainability.

Funding: This research received no external funding.

Data Availability Statement: No new data were created or analyzed in this study. Data sharing is not applicable to this article.

Conflicts of Interest: The author declares no conflict of interest.

References

1. United Nations, Department of Economic and Social Affairs. Population Facts, December 2019, No. 2019/6. Available online: https://www.un.org/en/development/desa/population/publications/pdf/popfacts/PopFacts_2019-6.pdf (accessed on 29 Januray 2022).
2. Roser, M.; Ortiz-Ospina, E.; Ritchie, H. Life Expectancy, Our World in Data. 2019. Available online: https://ourworldindata.org/life-expectancy (accessed on 29 January 2022).
3. The World Bank. Household Final Consumption Expenditure per Capita. 2022. Available online: https://data.worldbank.org/indicator/NE.CON.PRVT.PC.KD (accessed on 15 April 2022).
4. Brundtland Report. *Report of the World Commission on Environment and Development–Our Common Future*; United Nations: New York, NY, USA, 1987.
5. The Future We Want. In Proceedings of the United Nations Conference on Sustainable Development, Rio+20, Rio de Janeiro, Brazil, 20–22 June 2012.

6. United Nations. Final List of Proposed SDGs Indicators. UN Statistical Commission (UNSC). 2016. Available online: https://sustainabledevelopment.un.org/content/documents/11803Official-List-of-Proposed-SDG-Indicators.pdf (accessed on 15 March 2022).

7. Rockström, J.; Steffen, W.; Noone, K.; Persson, Â.; Chapin, F.S.; Lambin, E.; Lenton, T.M.; Scheffer, M.; Folke, C.; Schellnhuber, H.; et al. Planetary boundaries: Exploring the safe operating space for humanity. *Ecol. Soc.* **2009**, *14*, 32. [CrossRef]

8. Bjørn, A.; Chandrakumar, C.; Boulay, A.M.; Doka, G.; Fang, K.; Gondran, N.; Hauschild, M.Z.; Kerkhof, A.; King, H.; Margni, M.; et al. Review of life cycle-based methods for absolute environmental sustainability assessment and their applications. *Environ. Res. Lett.* **2020**, *15*, 083001. [CrossRef]

9. European Environment Agency. The European Environment—State and Outlook. SOER. 2020. Available online: https://www.eea.europa.eu/soer/2020/at-a-glance (accessed on 17 April 2022).

10. Oxfam. Ten Richest Men Double Their Fortunes in Pandemic while Incomes of 99 Percent of Humanity Fall. 2022. Available online: https://www.oxfam.org/en/press-releases/ten-richest-men-double-their-fortunes-pandemic-while-incomes-99-percent-humanity (accessed on 31 March 2022).

11. International Labour Organization, ILO. The Working Poor. 2019. Available online: https://ilo.org/wcmsp5/groups/public/---dgreports/---stat/documents/publication/wcms_696387.pdf (accessed on 15 April 2022).

12. Mulder, K. *Sustainable Development for Engineers: A Handbook and Resource Guide*; Routledge: New York, NY, USA, 2006.

13. Huisingh, D. How Can Building Upon Circular Economies Help us to Accelerate the Transition to Equitable, Sustainable, Livable, Post-Fossil Carbon Societies? Sustainable Circularity Symposium. 2019. Available online: https://static.uni-graz.at/fileadmin/projekte/circular/2019_DonaldHuisinghHowcanbuildinguponCircular_Economyhelpus.pdf (accessed on 18 April 2022).

14. Davis, M.L.; Cornwell, D.A. *Introduction to Environmental Engineering*, 4th ed.; McGraw-Hill: New York, NY, USA, 2006.

15. Mitsch, W.J. What is Ecological Engineering? *Ecol. Eng.* **2012**, *45*, 5–12. [CrossRef]

16. Mitsch, W.J.; Jorgensen, S.E. *Ecological Engineering and Ecosystem Restoration*; John Wiley & Sons: New York, NY, USA, 2003.

17. Weizsäcker, E.U.; Lardelel, J.A.; Hargroves, K.C.; Hudson, C.; Smith, M.H.; Rodrigues, M.A.E. Decoupling Technologies, Opportunities and Policy Options. UNEP IRP. 2014. Available online: https://www.resourcepanel.org/sites/default/files/documents/document/media/-decoupling_2_technologies_opportunities_and_policy_options-2014irp_decoupling_2_report-1.pdf (accessed on 5 April 2022).

18. Kalundborg. Kalundborg Symbiosis. 2019. Available online: http://www.symbiosis.dk/en/ (accessed on 7 August 2021).

19. Frosch, R.; Gallopoulos, N.E. Strategies for Manufacturing. *Sci. Am.* **1989**, *261*, 144–152. [CrossRef]

20. Graedel, T.E.; Allenby, B.R. *Industrial Ecology and Sustainable Engineering*; Pearson: London, UK, 2010.

21. Ghisellini, P.; Cialani, C.; Ulgiati, S. A review on circular economy: The expected transition to a balanced interplay of environmental and economic systems. *J. Clean. Prod.* **2016**, *114*, 11–32. [CrossRef]

22. Saavedra, Y.M.B.; Iritani, D.R.; Pavan, L.R.; Ometto, A.R. Theoretical contribution of industrial ecology to circular economy. *J. Clean. Prod.* **2018**, *170*, 1514–1522. [CrossRef]

23. Fedkin, M.V. EME 807 Course: Technologies for Sustainability Systems. Penn State University. 2017. Available online: https://www.e-education.psu.edu/eme807/node/575 (accessed on 15 March 2022).

24. Cushman-Roisin, B. Sustainable Design Principles. ENGS 44. 2017. Available online: http://engineering.dartmouth.edu/~{}d30345d/courses/engs44/designprinciples.pdf (accessed on 27 March 2022).

25. UNESCO. Sustainable Development. 2017. Available online: https://en.unesco.org/themes/education-sustainable-development/what-is-esd/sd (accessed on 26 March 2022).

26. Circular Ecology. Sustainability and Sustainable Development Guide. Available online: https://circularecology.com/introduction-to-sustainability-guide.html (accessed on 29 January 2022).

27. The Natural Step. The Framework for Strategic Sustainable Development. Available online: https://redamaltea.es/wp-content/uploads/2017/05/Framework.pdf (accessed on 18 March 2022).

28. Max-Neef, M.A.; Elizalde, A.; Hopenhayn, M. *Human Scale Development*; The Apex Press: New York, NY, USA; London, UK, 1991.

29. Ben-Eli, M. Sustainability: The Five Core Principles, A New Framework. 2009. Available online: http://www.sustainabilitylabs.org/ecosystem-restoration/wp-content/uploads/2015/08/Sustainability-The-Five-Core-Principles.pdf (accessed on 29 March 2022).

30. Ben-Eli, M.U. Sustainability: Definition and five core principles, a systems perspective. *Sustain. Sci.* **2018**, *13*, 1337–1343. [CrossRef]

31. Bogle, D.; Seaman, M. The Six Principles of Sustainability. 2010. Available online: https://www.engc.org.uk/engcdocuments/internet/website/tce%20article%20on%20sustainability%20March%202010.pdf (accessed on 15 March 2022).

32. Harris, J.M. *Basic Principles of Sustainable Development*; Tufts University: Medford, MA, USA, 2000; p. 19. Available online: http://ase.tufts.edu/gdae/publications/working_papers/Sustainable%20Development.pdf (accessed on 26 March 2022).

33. Csaba, J.; Nikolett, S. Environmental Management. 2008. Available online: http://www.tankonyvtar.hu/en/tartalom/tamop425/0032_kornyezetiranyitas_es_minosegbiztositas/ch04s02.html (accessed on 15 March 2022).

34. Environmental Advisory Council. Principles of Sustainable Development. *Prince Edward Island.* 2013. Available online: https://www.princeedwardisland.ca/sites/default/files/publications/principles_of_sustainable_development.pdf (accessed on 27 March 2022).

35. McDonough, W.; Braungart, M. *The Hannover Principles: Design for Sustainability*; William McDonough & Partners: Charlottesville, VA, USA, 1992.

36. Todd, N.J.; Todd, J. *From Eco-Cities to Living Machines: Principles of Ecological Design*; North Atlantic Books: Berkeley, CA, USA, 1994.
37. Van der Ryn, S.; Cowan, S. *Ecological Design*; Island Press: Washington, DC, USA, 1996.
38. McLennan, J.F. *The Philosophy of Sustainable Design*; Ecotone Publishing Company LLC: Kansas City, MI, USA, 2004.
39. Horsley Witten Group. Sustainable Design Principles, Boston. 2017. Available online: http://www.horsleywitten.com/profile/sustainable-design/sustainable-design-principles/ (accessed on 26 March 2022).
40. Riel, A.; Tichkiewitch, S.; Grajewski, D.; Weiss, Z.; Draghici, A.; Draghici, G.; Messnarz, T. Formation and Certification of Integrated Design Engineering Skills. In Proceedings of the International Conference on Engineering Design, Palo Alto, CA, USA, 24–27 August 2009; pp. 161–170.
41. Ranky, P.G. Sustainable Green Product Design and Manufacturing/Assembly Systems Engineering Principles and Rules with Examples. In Proceedings of the 2010 IEEE International Symposium on Sustainable Systems and Technology (ISSST), Arlington, VA, USA, 17–19 May 2010.
42. Mattson, C.A.; Wood, A.E. Nine Principles for Design for the Developing World as Derived from the Engineering Literature. *J. Mech. Des.* **2014**, *136*, 121403. [CrossRef]
43. Econation. Design for Sustainability. 2018. Available online: https://econation.co.nz/design-basics/ (accessed on 8 February 2022).
44. Anastas, P.T.; Warner, J.C. *Green Chemistry: Theory and Practice*; Oxford University Press: New York, NY, USA, 1998; p. 30.
45. Anastas, P.T.; Zimmerman, J.B. Through the 12 Principles of Green Engineering. *Environ. Sci. Technol.* **2003**, *37*, 94A–101A. [CrossRef] [PubMed]
46. Abraham, M.A.; Nguyen, N. Green engineering: Defining the principles—Results from the Sandestin conference. *Environ. Prog.* **2003**, *22*, 233–236. [CrossRef]
47. Shonnard, D.R.; Kircher, A.; Saling, P. Industrial Applications Using BASF Eco-Efficiency Analysis: Perspective on Green Engineering Principles. *Environ. Sci. Technol.* **2003**, *37*, 5340–5348. [CrossRef] [PubMed]
48. Lehni, M. *Eco-Efficiency: Creating more Value with Less Impact*; World Business Council for Sustainable Development, Conches: Geneva, Switzerland, 2000.
49. García-Serna, J.; Pérez-Barrigón, L.; Cocero, M.J. New trends for design towards sustainability in chemical engineering: Green engineering. *Chem. Eng. J.* **2007**, *133*, 7–30. [CrossRef]
50. Boyle, C.; Coates, G. Sustainability Principles and Practice for Engineers. *IEEE Technol. Soc. Mag.* **2005**, *24*, 32–39. [CrossRef]
51. Boyle, C.B.; Coates, G.K.; Macbeth, A.; Shearer, I.; Wakim, N. Sustainability and Engineering in New Zealand, Practical Guide for Engineers. In *Engineering New Zealand*; IPENZ: Wellington, New Zealand, 2017.
52. Royal Academy of Engineering, RAE. *Engineering for Sustainable Development*; Guiding Principles: London, UK, 2005; 52p. Available online: http://www.raeng.org.uk/publications/reports/engineering-for-sustainable-development (accessed on 29 March 2022).
53. UNESCO. Sustainable Engineering. 2017. Available online: http://www.unesco.org/new/en/natural-sciences/science-technology/engineering/sustainable-engineering/ (accessed on 15 March 2022).
54. NAL Agricultural Thesaurus. National Agricultural Library, United States Department of Agriculture. 2014. Available online: https://agclass.nal.usda.gov/vocabularies/nalt/concept?uri=https%3A//lod.nal.usda.gov/nalt/136603 (accessed on 17 April 2022).
55. Kauffman, J.; Lee, K.M. *Handbook of Sustainable Engineering*; Springer: Berlin/Heidelberg, Germany, 2013.
56. Rodriguez, S.I.; Roman, M.S.; Sturhahn, S.C.; Terry, E.H. *Sustainability Assessment and Reporting for the University of Michigan's Ann Arbor Campus*; Report No. CSS02-04; Center of Sustainable Systems: Ann Arbor, MI, USA, 2002.
57. Gagnon, B.; Leduc, R.; Savard, L. Sustainable Development in Engineering: A Review of Principles and Definition of a Conceptual Framework. *Environ. Eng. Sci.* **2009**, *26*, 1459–1472. [CrossRef]
58. Rosen, M.A. Engineering Sustainability: A Technical Approach to Sustainability. *Sustainability* **2012**, *4*, 2270–2292. [CrossRef]
59. MacArthur. Schools of Thought. *Ellen MacArthur Foundation.* 2017. Available online: https://www.ellenmacarthurfoundation.org/circular-economy/schools-of-thought/cradle2cradle (accessed on 30 March 2022).
60. Zhao, L.; Mao, G.; Wang, Y.; Du, H.; Zou, H.; Zuo, J.; Liu, Y.; Huisingh, D. How to achieve low/no- fossil carbon transformations: With a special focus upon mechanisms, technologies and policies. *J. Clean. Prod.* **2017**, *163*, 15–23. [CrossRef]
61. UNESCO. Precautionary Principle. *United Nations Educational, Scientific and Cultural Organization, the World Commission on the Ethics of Scientific Knowledge and Technology (COMEST).* 2005. Available online: http://unesdoc.unesco.org/images/0013/001395/139578e.pdf (accessed on 30 March 2022).
62. European Commission. *The Precautionary Principle: Decision-Making under Uncertainty*; European Commission: Brussels, Belgium, 2017; Available online: https://ec.europa.eu/environment/integration/research/newsalert/pdf/precautionary_principle_decision_making_under_uncertainty_FB18_en.pdf (accessed on 29 January 2022).
63. GRI. Global Reporting Initiative. 2017. Available online: https://www.globalreporting.org/information/sustainability-reporting/Pages/default.aspx (accessed on 29 December 2021).
64. Envirobiz Group. Improving the Waste Management Hierarchy "The Sustainability Hierarchy", Minneapolis. 2010. Available online: https://www.envirobiz.com/Sustainability_Hierarchy.pdf (accessed on 28 March 2022).
65. European Commission. Critical Raw Materials, Brussels. 2020. Available online: https://ec.europa.eu/growth/sectors/raw-materials/areas-specific-interest/critical-raw-materials_en (accessed on 29 January 2022).

66. Cefic. Facts and Figures of the European Chemical Industry. 2022. Available online: https://cefic.org/a-pillar-of-the-european-economy/facts-and-figures-of-the-european-chemical-industry/ (accessed on 29 January 2022).
67. Johannesburg Plan of Implementation of the World Summit on Sustainable Development, Chapter III/14–23. 2002. Available online: http://www.un.org/esa/sustdev/documents/WSSD_POI_PD/English/WSSD_PlanImpl.pdf (accessed on 22 March 2022).
68. Weiss, E.B. In Fairness to Future Generations and Sustainable Development. *Am. Univ. Int. Law Rev.* **1992**, *8*, 19–26.
69. Morrissey, B. The Importance of Stakeholder and Community Engagement in Engineering Projects. 2015. Available online: http://www.engineersjournal.ie/2015/04/21/importance-stakeholder-community-engagement-engineering-projects/ (accessed on 29 March 2022).
70. Rasche, A.; Morsing, M.; Moon, J. *Corporate Social Responsibility Strategy: Communication, Governance, Cambridge*; Cambridge University Press: Cambridge, UK, 2017.
71. Barnaby, W. Science, technology, and social responsibility. *Interdiscipl. Sci. Rev.* **2000**, *25*, 20–23. [CrossRef]
72. Your Dictionary. Social Definition. Available online: https://www.yourdictionary.com/social (accessed on 29 January 2022).
73. European Commission. Corporate Social Responsibility. 2017. Available online: http://ec.europa.eu/growth/industry/corporate-social-responsibility_sl (accessed on 29 March 2022).
74. Economic and Social Council, Committee on Economic, Social and Cultural Rights. The Right to Work. In *General Comment*; No. 18, II/7; United Nations: New York, NY, USA, 2005.
75. International Labour Organization, ILO. Decent Work. 2017. Available online: http://www.ilo.org/global/topics/decent-work/lang--en/index.htm (accessed on 29 March 2022).
76. Lee, S.Y.; Florida, R.; Gates, G. Innovation, Human Capital, and Creativity. *Int. Rev. Public Adm.* **2010**, *14*, 13–24. [CrossRef]
77. Chesbrough, H.W. *Open Innovation: The New Imperative for Creating and Profiting from Technology*; Harvard Business School Press: Boston, MA, USA, 2003.
78. Hunter, S. *Out Think: How Innovative Leaders Drive Exceptional Outcomes*; Wiley: Hoboken, NJ, USA, 2013.
79. Villanueva, A.; Kristensen, K.B.; Hedal, N. A Quick Guide to LCA and CBA in Waste Management. Danish Topic Centre on Waste and Resources, Copenhagen. 2006. Available online: https://cri.dk/files/dokumenter/artikler/filea951.pdf (accessed on 29 March 2022).
80. Pettinger, T. Polluter Pays Principle. Economics Helps. 2013. Available online: https://www.economicshelp.org/blog/6955/economics/polluter-pays-principle-ppp/ (accessed on 25 March 2022).
81. Glavič, P. Natural laws dominate the human society. *Chem. Eng. Trans.* **2009**, *18*, 523–530. [CrossRef]

Perspective

Process Design and Sustainable Development—A European Perspective

Peter Glavič *, Zorka Novak Pintarič and Miloš Bogataj

Faculty of Chemistry and Chemical Engineering, University of Maribor, Smetanova 17, SI-2000 Maribor, Slovenia; zorka.novak@um.si (Z.N.P.); milos.bogataj@um.si (M.B.)
* Correspondence: peter.glavic@um.si

Abstract: This paper describes the state of the art and future opportunities for process design and sustainable development. In the Introduction, the main global megatrends and the European Union's response to two of them, the European Green Deal, are presented. The organization of professionals in the field, their conferences, and their publications support the two topics. A brief analysis of the published documents in the two most popular databases shows that the environmental dimension predominates, followed by the economic one, while the social pillar of sustainable development is undervalued. The main design tools for sustainability are described. As an important practical case, the European chemical and process industries are analyzed, and their achievements in sustainable development are highlighted; in particular, their strategies are presented in more detail. The conclusions cover the most urgent future development areas of (i) process industries and carbon capture with utilization or storage; (ii) process analysis, simulation, synthesis, and optimization tools, and (iii) zero waste, circular economy, and resource efficiency. While these developments are essential, more profound changes will be needed in the coming decades, such as shifting away from growth with changes in habits, lifestyles, and business models. Lifelong education for sustainable development will play a very important role in the growth of democracy and happiness instead of consumerism and neoliberalism.

Keywords: process design; sustainable development; chemical industry; process industry; megatrends; design tools

Citation: Glavič, P.; Pintarič, Z.N.; Bogataj, M. Process Design and Sustainable Development—A European Perspective. *Processes* **2021**, *9*, 148. https://doi.org/10.3390/pr9010148

Received: 14 November 2020
Accepted: 11 January 2021
Published: 13 January 2021

1. Introduction

In the Introduction, some basic information about process design and sustainable development will be presented, separately for both of them as well as together as process design for the sustainable development era. Then, we shall proceed with the three dimensions of process design—environmental, economic, and social, and continue with process design tools for achieving them. As a case study, the results of sustainable design in chemical, biochemical, and process industries will be presented (process industries include cement, ceramics, food, glass, iron and other metals, oil and gas, plastics, pulp and paper production, waste incineration, etc.), and the article will finish with speculative future development described in the conclusions. However, before going into details, we need to stress the long-term character of process design. Process plants are designed for one or two decades, at least, but most of them operate for several decades. Moreover, process systems will be characterized as circular economy units; they will have to be maintained for longer operation, reused or refurbished for similar future processes, or have its parts and equipment recycled for another process or purpose. Therefore, we need to design for the future. Therefore, it is important to estimate the future development in the company, branch, production, and value chain, in a national and global context. One of the important perspectives is the study of megatrends in the world.

Global megatrends of future development are speculating about our fate: "Trends are an emerging pattern of change likely to impact how we live and work. Megatrends are large, social, economic, political, environmental, or technological changes that are slow to form, but once in place can influence a wide range of activities, processes, and perceptions, possibly for decades. They are the underlying forces that drive change in global markets, and our everyday lives [1]." Although megatrends are not deterministic, they can help us in planning and developing products, processes, and services for future customers. Many studies on megatrends are available, the most popular being the ones of Ernst & Young [2], the European Environment Agency [3] and Pricewaterhouse Coopers [4].

Comparing the most reliable reports on megatrends, some common beliefs can be observed. The following six megatrends and their implications are shown as a synthesis of the four megatrend reports [1–4]:

1. Climate change—(a) Air pollution with greenhouse gases (GHGs) emissions, (b) Exponential climate impacts (extreme weather events acceleration, air–land–oceans heating, polar ice caps, permafrost and glaciers melting, sea-level rise, wildfires, deforestation and deserts), (c) Loss of biodiversity and ecosystem services. Several implications will happen because of the climate change, e.g.,: (i) Decarbonization, reforestation, green buildings, carbon capture with utilization or storage, (ii) Tax on GHGs emissions, (iii) Beyond GDP (gross domestic product) metrics.

2. Resource scarcity—(a) Increased strain on the planet's resources including degraded soil, (b) Food–water–energy nexus, (c) Critical raw materials. Implications: (i) Zero waste, circular economy and increased efficiency, (ii) Shift from fossil fuels to renewable energy and bio-based raw materials, (iii) Microbiomes (bacteria, archaea, fungi, viruses, and nanoplankton), synthetic biology (intersection of biology and technology).

3. Shifting economic power—(a) Emerging economies (E7—China, India, Brazil, Mexico, Russia, Indonesia, and Turkey) as the growth markets, (b) Global demographics change (different population growth rates), (c) Techno-economic cold war. Implications: (i) Power shift from the west (G7 (Group of Seven)—Canada, France, Germany, Italy, Japan, UK, and USA) to the east (E7), (ii) Industry 4.0 (the fourth industrial revolution, use of cyber-physical systems), (iii) Consumer preferences are changing, e.g., in the food industry (organic and fresh food, online delivery).

4. Technological breakthrough—(a) The pace of change is exponential, not linear, (b) Data are the new oil, (c) Automation and robotization (many jobs will be replaced by machines/robots). Implications: (i) Digitalization—AI (Artificial Intelligence), big data, 3D printing, 5G (the 5th generation) network, IoT (Internet of Things, 26 billion "things" are connected by the internet), (ii) Increased research and innovation, (iii) Industry 5.0 (interaction of human intelligence and cognitive computing).

5. Demographic and social changes—(a) Population continues to grow, (b) More old people and fewer children, (c) Income inequality rises. Implications: (i) Healthcare spending (rise of expenses, saving for retirement), (ii) Education for sustainable development, lifelong learning, creativity, entrepreneurship, (iii) Higher taxation of high incomes and succession duties.

6. Rapid urbanization—(a) Migration to the cities (megacities), (b) Life is better in the cities. Implications: (i) Smart cities, new infrastructure, (ii) Healthcare and security (changing disease burdens and risk of pandemics, crimes and terror—surveillance, monitoring), (iii) Consumer behaviors change (resources will be shared, move from energy suppliers to mobility solutions).

Special studies exploring future trends in different areas exist. Let us mention one of them—New Energy Outlook (NEO) [5], which is forecasting the trends in the energy field, which is one of the most important sources of process industries (see next paragraph). NEO has three major parts: (1) Economic Transition Scenario (ETS), (2) NEO Climate Scenario (NCS), and (3) Implications for Policy. The Executive Summary has six chapters, each one with several scenarios: (a) Energy and emissions, (b) Power, (c) Transport (general, road, shipping, aviation, rail), (d) Buildings, (e) Industry, and (f) Climate.

Industry consumes 29 % of total final energy. Energy consumption grows at an average of 0.6 %/a (per year) and will reach 149 EJ (exajoules, 10^{18} J) by 2050. Steel and chemicals production are the two largest energy consumers in industry, which are responsible for 19 % and 18 % of final energy use in the sector in 2019. They are followed by cement, at 14 %, and aluminium processes, at 6 %. Around 12 % of all fossil fuels consumed in the industry are used as a feedstock for non-energy purposes (from petrochemicals to plastics). In 2050, the sector will account for around 34 % of emissions from fuel combustion, up from 25 % in 2019. Energy demand for steel will grow 50 %, for aluminum will grow 80 %, and for plastics will grow 100 %. High investments in energy—wind 3.3 T\$ (trillion US dollars = 10^{12} \$) and solar 2.8 T\$—are expected by 2050. Prices of renewable wind and solar energy are forecast to fall by about 50 % [6].

1.1. European Green Deal

Process design is and will be increasingly dependent on political decisions in the future. Climate change and loss of biodiversity are an existential threat to Europe and the world. Therefore, the European Commission responded to the first two of the most important risks, mentioned in the above Megatrends, by accepting the European Green Deal (EGD) [7]. It aims to "transform the Union into a modern, resource-efficient and competitive economy where:

- There are no net emissions of greenhouse gases by 2050;
- Economic growth is decoupled from resource use;
- No person and no place are left behind."

The EGD will have a deep influence on life in the European Union, both at personal and enterprise levels. It will also very deeply hit the chemical and process industries.

EU has met its GHG "emissions reduction target for 2020 and has put forward a plan to further cut emissions—at least 55 % by 2030. By 2050, Europe aims to become the world's first climate-neutral continent. Climate action is at the heart of the EGD—an ambitious package of measures ranging from severely cutting greenhouse gas emissions, to investing in cutting-edge research and innovation, to preserving Europe's natural environment. Its action plan aims to:

- Boost the efficient use of resources by moving to a clean, circular economy;
- Restore biodiversity and cut pollution."

One of the first activities is the European Commission's proposal of the European Climate Law, which is a legally binding target of net-zero greenhouse gas emissions by 2050. A system for monitoring progress and taking further action if needed is planned. "Reaching this target will require action by all sectors of the economy, including:

- Investing in environmentally friendly technologies,
- Supporting industry to innovate,
- Rolling out cleaner, cheaper, and healthier forms of private and public transport,
- Decarbonizing the energy sector,
- Ensuring that buildings are more energy-efficient,
- Working with international partners to improve global environmental standards."

EGD and a European COVID-19 response can address Europe's climate, biodiversity, pollution, economic, political and health crises, and at the same time strengthen its institutions and reignite popular support for the European project. SYSTEMIQ and The Club of Rome published a report *A System Change Compass* concentrating on the drivers and pressures that lead to these environmental challenges and on solutions and required changes to the current economic operating model [8]. The report (a) foresees radical resource decoupling and sustainability, (b) offers a system perspective, (c) starts from the human drivers for change, (d) offers a set of principles for support, and (e) takes the natural system as a starting point. To achieve this system-level change, the report addresses three fundamental barriers for the change: (1) shared policy orientations at the overall system level, (2) systemic orientation for each economic ecosystem, and (3) a shared target picture and roadmap for Europe's next industrial backbone.

The System Change Compass offers the following:

- Each of the 10 principles has three orientations giving 30 system-level political orientations for the overarching system as a checklist for policymakers;
- Eight ecosystem and three to five ecosystem orientations (directions) for Europe's industrial backbone;
- Over 50 Champion orientations (directives) that form a view of industrial priorities.

The 10 principles with their orientations are including the following redefinitions:

1. Prosperity—from economic growth to fair and social economics;
2. Natural resources—consumption and development decoupled, a shift to responsible usage;
3. Progress—from economic activities/sectors to societal needs within planetary boundaries;
4. Metrics—from GDP growth to natural capital and social indicators;
5. Competitiveness—EU based on low-carbon products, services, and digital optimization;
6. Incentives—aligned with the Green Deal ambitions and economic ecosystems;
7. Consumption—from individual identity to an individual, shared, and collective identity;
8. Finance—from subsidizing "old" industries to supporting economic ecosystems;
9. Governance—from top–down to transparent, flexible, inclusive participatory one;
10. Leadership—from traditional to system one, based on an intergenerational agreement.

The eight economic ecosystems with over 50 Champions are resulting in industrial priorities:

1. Healthy food (organic, no waste, water, urban agriculture, alternative proteins, etc.);
2. Built environment (planning, ownership, buildings repurpose and retrofit, net zero, circular);
3. Intermodal mobility (high-speed railways, green aviation and shipping, ride-sharing, etc.):
4. Consumer goods (product–service, product sharing, maintenance, and value retention);
5. Nature-based (degraded land restoration, urban greening, ecotourism, paid ecosystem services, forest, sea, marine, and land protection);
6. Energy (renewables, hydrogen, low-carbon fuels, smart metering, carbon capture, grids);
7. Circular materials (value chain systems, asset recovery, and reverse logistics, markets for secondary materials, high-value material recycling, materials-service, 3D printing, etc.);
8. Information and processing (distributed manufacturing, high-speed infrastructure, etc.).

1.2. Process Design

Process Design (PD) is the choice and sequencing of processing steps and their interconnections for desired physical and/or chemical transformation of materials [9]. The steps include several unit operations: reaction, separation, mixing, heating, cooling, pressure change, particle size reduction or enlargement, etc. Today, the design is governed by the circular economy, which requires design for repair, reuse, recovery, refurbishment, restoration, and recycling [10]. Process design is distinct from equipment design, which is closer to the design of unit operations. Process design can be the design of new facilities or it can be the modification or expansion of existing ones. The process design can be divided into three basic steps: synthesis, analysis, and optimization [11].

Design starts with process synthesis—the choice of technology and combinations of industrial units to achieve goals. First, product purities, yields, and throughput rates shall be defined. Modeling and simulation software is often used by design engineers. Simulations can identify weaknesses in a design and allow engineers to choose better alternatives. However, engineers still rely on heuristics, intuition, and experience when designing a process. Human creativity is an important element in complex designs.

Process analysis is usually made up of three steps: solving energy and material balances, sizing and costing the equipment, and evaluating the economic worth, safety, operability, etc. of the chosen flow sheet.

Optimization involves both structural and parametric optimization. Structural optimization is more difficult, and it includes equipment selection and interconnection between the units. Parameter optimization is regarding stream compositions and operating conditions such as temperature and pressure.

Several decisions have to be made during the design of each process while respecting the aforementioned objectives: i.e., constraints (capital investment), social conditions (employment, health and safety), environmental impacts (emissions, waste, resource efficiency, operating and maintenance costs), and other factors such as reliability, redundancy, flexibility, and variability in feedstock and product. Process design documentation includes the following:

- Simple block flow diagrams (BFD, rectangles and lines indicating major material or energy flows, stream compositions, and stream and equipment pressures and temperatures),
- More complex process flow diagrams (PFD) or process flowsheets with major unit operations, material and energy balances,
- Piping and instrumentation diagrams (P&ID, piping class, pipe size, valves and process control schemes), and specifications (written design requirements of all major equipment items).

Working Party of the European Federation of Chemical Engineering (EFCE) on Computer-Aided Process Engineering (CAPE) is organizing annual events—the European Symposium on Computer-Aided Process Engineering (ESCAPE) in which researchers and practitioners in the field of computer-aided process systems engineering from academia and industry come together. Process engineering focuses on the design, operation, control, optimization, and intensification of chemical, physical, and biological processes from a vast range of industries: agriculture, automotive, biotechnical, chemical, food, material development, mining, nuclear, petrochemical, pharmaceutical, and software development. The application of systematic computer-based methods to process engineering is called "process systems engineering". Papers presented at the ESCAPE events are all published in Elsevier publications, the CAPE Proceedings Series *Computer-Aided Chemical Engineering* [12].

In the United States of America (US), a nonprofit organization CACHE (Computer Aids for Chemical Engineering) organizes the Foundations of Computer Aided Process Design (FOCAPD) international conferences, focusing exclusively on the fundamentals and applications of computer-aided design for the process industries. The conference is organized every five years and brings together researchers, educators, and practitioners to identify new challenges and opportunities for process and product design. Papers from the conferences are published by the Elsevier CAPE Book series as *Proceedings of the International Conference on Foundations of Computer-Aided Process Design*.

1.3. Sustainable Development

Sustainable development (SD) must meet the needs of the present without compromising the ability of future generations to meet their own needs [13]. The Amsterdam Treaty of European Union (EU) sets out the EU "vision for a sustainable development of Europe based on balanced economic growth and price stability, a highly competitive social market economy, aiming at full employment and social progress, and a high level of protection and improvement of the quality of the environment." Transforming our World: the 2030 Agenda for Sustainable Development, including its 17 Sustainable Development Goals (SDGs) and 169 targets, was adopted in 2015 by Heads of State and Government at a special United Nations (UN) summit. The Agenda is a commitment to eradicate poverty and achieve sustainable deveopment by 2030 worldwide.

The Chemical Sector SDG Roadmap is an "initiative led by a selection of leading chemical companies and industry associations, convened by the World Business Council for Sustainable Development (WBCSD), to explore, articulate, and help realize the potential of the chemical sector to leverage its influence and innovation to contribute to the SDG agenda" [14]. Building on the Responsible Care program and other sustainability initiatives, "the European Chemical Industry Council (Cefic) and its members have developed a Sustainability Charter and agreed on a roadmap to foster innovation" [15]. They focused on resources in the "four critical areas to progress sustainable development: low-carbon economy, resource efficiency, circular economy and human protection".

The International Conference on Sustainable Development (ICSD) is organized annually by the European Center of Sustainable Development (ECSD) in collaboration with other partners; conference papers are published in the open-access European Journal of Sustainable Development, issued by the ECSD [16]. Conference proceedings are good sources of recent research and development in the area.

The American Institute of Chemical Engineers (AIChE) and the Association of Pacific Rim Universities (APRU, a network of leading universities linking the Americas, Asia, and Australasia) have organized the Conference on Engineering Sustainable Development in December 2019 [17]. They are going to organize the 2nd Engineering Sustainable Development Conference in December 2020, both conferences addressing the UN 2030 Agenda for Sustainable Development and the 17 SDGs.

The Asia Pacific Institute of Science and Engineering (APISE) is organizing International Conferences on Environmental Engineering and Sustainable Development (CEESD) annually; papers are published in the IOP (Institute of Physics) Conference Series: Earth and Environmental Science.

1.4. Process Design and Sustainable Development

Process Design and Sustainable Development (PD&SD) started with the ecodesign (ecological design, also called green design or environmentally conscious design), which considered the environmental impact of a product throughout its entire life cycle only. A typical example is green engineering design [18], which evolved from the green chemistry principles [19]. As sustainable development (SD) has also economic and social components, the additional SD principles have been integrated into engineering design [20]. Today, sustainable development is a part of engineering principles [21,22].

Crul and Diehl published a handbook on Design for Sustainability (D4S) [23]. Ceschin described the evolution of design for sustainability [24] and Acaroglu overviewed sustainable design strategies [25]. The generic conventional engineering design process is including four phases: (1) planning and problem definition, (2) conceptual analysis, (3) preliminary design, and (4) detailed design [12].

Many textbooks on chemical process design are on the market. An older one is dealing with preliminary analysis and evaluation of processes, the analysis using rigorous models, and basic concepts in process synthesis with optimization approaches [26]. Economic evaluation is dealt with, heat and power integration are described to reduce energy consumption, and safety is the only social topic mentioned. In some textbooks, sustainable development and environmentally sound design (prevent/minimize, recycle/reuse, and recovery) are also described using a few pages [27]. More recent ones are adding process intensification, steam system and cogeneration, environmental design for atmospheric emissions, water systems, and clean process technology, as well as inherent safety chapters [28].

Professional literature on PD&SD was more advanced in the past decades, as design engineers had to respect laws and regulations regarding environmental protection, labor protection, and occupational safety in the approval procedures [29]. Newer literature is including natural resource and environmental challenges, sustainable materials identification, sustainability improvements of engineering designs, evaluation of sustainable designs, and monetizing their benefits besides the legislative framework [30]. A sustainability engineering approach is also including Total Quality Management [31] and Life-Cycle Assessment (LCA) [32].

2. Process Design for Sustainability

The publication statistics search in Scopus [33] includes article titles, abstracts, and keywords. It contains an abstract and citation database with over 25 100 titles (articles, conference papers, books, etc.). Searching for the four words: process, design, sustainable, and development yielded 16 135 documents, 2 869 of them in open access. There was a constant rise in the number of publications since the year 1999 (54 documents), reaching 1 859 documents in 2019. By subject area, most of them belong to Engineering (7 060) and Environmental Science (4 090); they are followed by Energy (2 848), Social Sciences (2 777), and Computer Science (2 471). Of these, 7 487 of them are articles, 6 330 are conference papers, 1 085 are reviews, 703 are book chapters, and 294 are conference reviews. Most of the articles were published in J. Cleaner Production (456) and in Sustainability journal (290). The authors with the most publications are still coming from the EU and USA; the most frequent affiliations are located in EU and China: Delft University (170), Politecnico di Milano (123), Wageningen University & Research (114), Danmarks Tekniske Universitet (108), and Chinese Academy of Sciences (99). The most frequent keywords are sustainable development (9 604) and sustainability (2 512), followed by design (1 840), product design (1 511), and life cycle (1 514); process design is not so often mentioned.

Similar statistics in the Web of Science (WoS) Core Collection database [34] showed 8 915 documents (14 823 in WoS All Databases); a steady growth was realized in the last four years—from 772 units in 2016 to 1 234 ones in 2019. Most of them (2 557) belong to the categories of environmental science and studies, 1 528 belong to green sustainable science and technology, 808 belong to environmental engineering, and 620 belong to energy and fuels. Articles (5 610) are prevailing, followed by papers in proceedings (2 700), reviews (818), and book chapters (260). Regarding the organizations, Wageningen University Research (102), Delft University of Technology (98), Centre National de la Recherche Scientifique (89), Helmholtz Association (88), and Chinese Academy of Sciences (86) are on the top.

The WoS Core Collection base covers more than 21 419 journals, books, and conference proceedings, while the Web of Science platform includes 34 586 journals, books, proceedings, patents, and datasets. As it was impossible to review several thousand documents, the highly cited ones in the field (121 documents) were selected. Examining their titles lead to 43 documents, and by reading their abstracts, 16 articles were selected for a closer look.

2.1. Environmental Dimension

Most of the selected 16 articles deal with environmental sustainability; however, the economic dimension is included in only nine of them—mainly as a criterion for process optimization. The social dimension (health) is present in two of them. Optimal design of chemical processes and supply chains is concentrated on energy efficiency as well as waste and water management [35]. Multiple criteria decision making (MCDM) [36] and Life-Cycle Assessment (LCA) [37] are the tools most often mentioned. Various metrics are used to assess the sustainability of processes; the three most popular ones are presented here:

- United States Environmental Protection Agency's (EPA) "Gauging Reaction Effectiveness for the ENvironmental Sustainability of Chemistries with a multi-Objective Process Evaluator (GREENSCOPE [38]) tool provides scores for the selected indicators in the economic, material efficiency, environmental and energy areas having about 140 indicators in four main areas: material efficiency (26), energy (14), economics (33) and environment (66)";
- The Tool for the Reduction and Assessment of Chemical and other environmental Impacts (TRACI 2.0 [39]) "for sustainability metrics, life-cycle impact assessment, industrial ecology, and process design impact assessment for developing increasingly sustainable products, processes, facilities, companies, and communities"; it is containing human health criteria-related effects, too; and
- The mass-based green chemistry metrics, extended to the "environmental impact of waste, such as LCA, and metrics for assessing the economic viability of products" and processes [31].

Sustainability-oriented innovations (SOIs) in small and medium-sized enterprises (SMEs) are integrating ecological and social aspects into products, processes, and organizational structures [40].

Five out of the 16 articles dealt with biofuels. Purified biogas is an essential source of renewable energy that can act as a substitute for fossil fuels; anaerobic co-digestion is a pragmatic method to resolve the difficulties related to substrate properties and system optimization in single-substrate digestion processes [41]. The synthesis of important biofuels using biomass gasification, key generation pathways for their production, the conversion of syngas to transportation fuels together with process design and integration, socio-environmental impacts of biofuel generation, LCA, and ethical issues were discussed [42]. A multi-objective possibilistic programming model was used to design a second-generation biodiesel supply chain network under risk; the proposed model minimized the total costs of biodiesel supply chain from feedstock supply to customers besides minimizing the environmental impact [43]. The cultivation, harvesting, and processing of microalgae for second-generation biodiesel production, including the design of microalgae production units (photo-bioreactors and open ponds) was described [44]. A multi-objective optimization model based on a mathematical programming formulation for the optimal planning of a biorefinery was developed, considering the optimal selection of feedstock, processing technology, and a set of products [45].

Circular economy topics are the second most numerous ones within 16 articles. The first one traced the conceptualizations and origins of the Circular Economy (CE), researched its meanings, explored its antecedents in economics and ecology, and discussed how the CE was operationalized in business and policy [46]; the authors proposed a revised definition of the CE to include the social dimension. Another contribution proposed a new unified concept of Circular Integration that combined elements from Process Integration, Industrial Ecology, and Circular Economy into a multi-dimensional, multi-scale approach to minimize resource and energy consumption [47].

High-pressure technologies involving sub- and supercritical fluids offer a possibility to obtain new products with special characteristics or to design new processes that are environmentally friendly and sustainable [48]. Sustainable product–service systems offer service by lending the product to a customer—they attempt to create designs that are sustainable in terms of environmental burden and resource use whilst developing product concepts as parts of sustainable whole systems that provide a service or function to meet essential needs [49].

2.2. Economic Dimension

For most managers in industry, economic performance is the most important criterion for decisions on investing money in production and energy facilities [50]. Economic performance indicators are well known, and process and product designs are usually carried out by maximizing profits or minimizing costs [51]. Other criteria are used less frequently, e.g., the network for the conversion of waste materials into useful products has been optimized using the maximum return on investment [52].

Techno-economic evaluations of process alternatives with different criteria lead in some cases to the same best solution, as Ziyai et al. [53] showed by comparing the three biodiesel production scenarios with the criteria net present value, internal rate of return, payback period, discounted payback period, and return on investment. In general, optimization using different economic criteria leads to different optimal process solutions [54]. These processes differ not only in economic performance but also in operational efficiency and environmental impact [55]. This phenomenon is particularly evident in more precise mathematical models [56], which include sufficient trade-offs between investments on one hand and benefits on the other, such as higher conversion, higher product purity, the higher degree of heat integration between process streams. Applying the correct economic criteria can lead to more sustainable solutions; for example, the net present value criterion provides optimal process solutions that strike a balance between long-term stable cash flow generation, moderate profitability, and moderate environmental impact [57].

With the introduction of the concept of sustainable development, criteria other than economic indicators have become more important in process design, thereby promoting the reduction of negative environmental impacts and the improvement of social performance. When designing sustainable processes, the techno-economic, environmental, and social criteria of various process alternatives are evaluated and the most suitable solution is selected from among them, whereby compromises between all criteria are sought [58]. More systematic approaches use multi-objective optimization. The most common method is to generate equivalent non-dominant Pareto solutions that show a range of solutions where the improvement of one criterion leads to the deterioration of other criteria [59]. However, Pareto curves are not best suited for decision making, because the decision-maker usually must choose one alternative for realization, which requires additional multi-criteria analyses of Pareto solutions [60].

Another approach is to transform the multi-objective optimization into a single-criterion optimization by monetarizing all pillars of sustainability, which means that in addition to the economic criterion, environmental and social impacts are also expressed in monetary terms. However, this is not an easy task, as environmental and especially social impacts cannot simply be expressed in monetary terms. Environmental impacts are expressed in terms of the burdens and reliefs of the environment. They can be monetized using the eco-cost system [61], which expresses the cost of environmental pollution at the price necessary to prevent it. Greenhouse gas emissions can be monetized with a CO_2 tax. Novak Pintarič et al. [62] showed that a deviation from the economic optimum for investments in emission-reducing technologies can lead to a reduction of the tax due to lower emissions, which can compensate for economic loss to a certain extent. The point on the Pareto curve was called the "Economic–Environmental Break Even".

Sustainable process designs include various concepts to achieve sustainable solutions; examples are cleaner production [63], zero waste processes [64], zero carbon emission technologies [65], LCA environmental impact assessment in early design phases [66], eco-efficiency indicators [67], etc. Recently, the concept of the circular economy has become particularly popular and sometimes even overcomes the term sustainable development, although the terms are by no means equivalent [68]. The concept of circularity is already being used in the design and optimization of technologies and processes, such as the recovery of hydrogen from industrial waste gases [69] or the development of the novel indicator Plastic Waste Footprint to facilitate an improvement of circularity in the use of plastics [70].

Process systems engineering offers many approaches and tools for the design of process solutions in the field of circular economy and sustainable development, such as the synthesis of processes and supply chains with mathematical programming, process integration, optimization and intensification, multi-objective and multi-level optimization, optimization under uncertainty conditions, etc. [71]. The fact is that circular economy projects, especially those that solve the waste problem, are hardly economically successful based on classic economic criteria; for example, recycling of plastics is not economically viable at low fractions of recycled material [72]. However, it is important to look at these projects in a broader perspective and to include all the three dimensions of sustainable development into design and strategic decision-making.

2.3. Social Dimension

In process design, economic and social dimensions are important in addition to environmental performance. To achieve a sustainable and circular economy, it is necessary to develop all the three pillars of sustainable development as well as SDGs and take them into account in process design. While economic and environmental aspects are well established and quantified, social criteria are far less developed. The integration of social effects into process design is difficult, and little research has been conducted, although it is becoming increasingly important in both the academic and business environments [73]. The monetization of all the three pillars of sustainable development has been used to synthesize processes and supply networks with sustainability criteria such as sustainable profit [74] and sustainable net present value [75].

Each company has to take care of their customers, employees, owners or shareholders, and local community to fulfill their requirements. Companies shall respect the corporate social responsibility standard, ISO 26000 [76]; the standard is voluntary, and it is based on the following:

(a) Seven key principles: accountability, transparency, ethical behavior, respect for stakeholder interests, respect for the rule of law, respect for international norms of behavior, and respect for human rights;

(b) Seven core subjects: organizational governance, human rights, labor practices, the environment, fair operating practices, consumer issues, and community involvement and development.

What can a process designer do in this respect?

Regarding the employees, it is the most important to design a process, its equipment, and products in the way that enables safety and health protection at work (occupational health and safety, OHS) [77]. It is including the physical, mental, and social wellbeing of workers. OHS is achieved by using process monitoring and control, automation, even robots to prevent contact with dangerous substances, fires and explosions, accidents at work, release heavy burdens, etc. Digitalization and computer-aided operation of plants are used increasingly to release the workers from process malfunctions, unexpected events, or even accidents.

Similar requirements are valid for customers using products of the process industries. This is particularly important for chemicals, which can have negative effects on customers' health, safety and wellbeing. The products shall be long-lasting, without weak elements, easy to maintain, and user friendly. Product design is critical for its dismantling and recycling at the end of life. Take back or product–service systems are used increasingly to enable circular economy with the reuse of materials and energy in the waste products. Every designer shall use Life-Cycle Assessment (LCA) methodology [78] to evaluate impacts throughout the supply chain, from raw materials extraction to processing, use, and end-of-life treatment, applying the principles of the circular economy.

The local community is strongly connected with processes and operation of the company located within its boundaries. Most employees are coming from the neighborhood, and employment is enabling their families to live better. The local population is very sensitive to any radiation and emissions into the air, water, or land around the factory. Process design has to take care of their health and safety by proper process design as well as by planning sensors, monitoring, and measurement units in the surroundings of the company buildings. Often, the process of surplus heat can be used for heating public buildings or even residents' houses. Zero-waste, wastewater treatment and reuse, and hazardous waste recycling are important principles guiding process design. The selection and monitoring of indicators shall be carried out well in advance, during the process design.

Most of the companies are using the Global Reporting Initiative (GRI) to communicate their impacts on people (human rights, corruption, etc.) and the planet [79]. GRI's framework for sustainability reporting helps companies identify, collect, and report their impacts in a clear way.

2.4. Process Design Tools and Sustainability

The Process Systems Engineering (PSE) Community has fully embraced the concept of "sustainability" as one of the leading guides in process design. Although it is difficult to pinpoint the exact time when the three pillars of sustainability (i.e., economic, environmental, and social) were considered and emphasized simultaneously in the design of chemical processes, one may argue that even the works published as early as the late 1970s [80] and early 1980s [81] directly addressed at least two of the pillars of sustainable process design—economic and environmental ones. Although the incentives to develop what we now regard as a sustainable process may have been purely economic at the time, the enabling insight was the ability to view a chemical process as a system—the system that is not isolated from its environment, but a system that interacts with the environment.

Fast-forward five decades of research in the field of PSE, the approaches to designing sustainable chemical processes rely heavily on computer-aided tools. These tools enable simulation, analysis, optimization, and synthesis of chemical processes at various spatial and time scales. They range from computer-aided molecular design [82], simulations of transport phenomena (heat and mass transfer in single or multiphase flows) [83], simulations of single-unit operations [84] and whole processes [85] to the synthesis and optimization of processes [86,87] and complete supply networks [88]. The widely accepted approach to assess the sustainability of a given process design is the Life-Cycle Sustainability Assessment (LCSA), which is commonly performed to compare different process design alternatives [89] after the feasible designs have been identified. On the flip side, if a composite sustainability criterion, for example, the sustainability profit [90], is incorporated directly into the process synthesis and optimization phase as an objective function, the most sustainable designs can be obtained directly without the need of a posteriori LCSA assessment.

The PSE computational tools enable a practical way to analyze the performance of a wide range of product–process engineering problems as well as to identify the possibilities for improvement However, some software packages come together with a high license price, and although the price can be justified with the benefits gained, it very often remains an obstacle, especially for small engineering companies. However, in the last few years, the open-source initiatives have begun to offer freely available alternatives to the paid versions (Table 1). Provided that quality matches those of their paid counterparts, a greater adaptation of these tools in the industry can be expected.

Table 1. Licensed and free computational tools, used for process simulation, synthesis/optimization, and sustainability assessment.

Software	Description	Web Site	License
ANSYS Fluent [91]	Computational Fluid Dynamics	www.ansys.com	Licensed
OpenFoam [92]	Computational Fluid Dynamics	www.openfoam.com	Free
Aspen Plus [93]	Process Simulation	www.aspentech.com	Licensed
DWSIM [94]	Process Simulation	dwsim.inforside.com.br	Free
GAMS [95]	Mathematical programming and optimization	www.gams.com	Licensed
Pyomo [96]	Mathematical programming and optimization	www.pyomo.org	Free
GABY [97]	LCA and sustainability assessment	www.gabi-software.com	Licensed
OpenLCA [98]	LCA and sustainability assessment	www.openlca.org	Free

Identifying what could generally be considered as mitigating solutions to complex problems is necessary, although such solutions may not be sufficient to achieve the goals of sustainable development in the long term. A real breakthrough will be achieved by identifying innovative restorative solutions. In this context, the Process Systems Engineering (PSE) should develop tools that simultaneously address the whole (bio)-chemical supply network. To harvest the synergistic effects among the constituents of the supply network to a greater extent, the traditional (bio)-chemical supply network should be expanded to include additional elements (e.g., nano-robots, molecular machines, labs-on-chips, or micro-processes) and linked to other supply networks (energy, agriculture, food, etc.) to form circular and sustainable system-wide supply networks.

Despite many achievements and contributions of the PSE community that undeniably contributed to the development of the modern biochemical and chemical industry, there are no professional tools and hardly any academic ones that are specialized in providing innovative solutions to these complex problems.

A noteworthy initiative to develop an advanced computer platform to support innovative conceptual design and process intensification is the IDEAS PSE Framework [99]. The platform addresses the capability gap between state-of-the-art simulation packages and algebraic modeling languages (AMLs) by integrating an extensible, equation-oriented process model library within the open-source Pyomo AML, which addresses challenges in formulating, manipulating, and solving large, complex, structured optimization problems.

The second initiative is MIPSYN-GLOBAL [100]. It is being built on the foundations of its predecessor MIPSYN [101], making use of knowledge and experience gained in the decades of research in the field of PSE. The development of MIPSYN-Global encompasses all the four basic PSE tasks: (i) development of advanced synthesis concepts, algorithms, and strategies; (ii) modeling; (iii) development of synthesizer tools; and (iv) development of different applications.

3. Case Study

The European Union (EU) is the second-largest chemicals producer in the world—with 565 M€ (million euros), it is behind China (1 198 M€) but before NAFTA (North American Free Trade Agreement—USA and Canada, 530 M€); EU a positive trade balance [102]. About 96 % of all manufactured goods rely on chemistry. The chemical industry is the fourth-largest producer after automotive, food, and machinery/equipment ones; with the 16 % added value, it is the leading sector in the EU. A total of 29 000 small, medium, and large companies are offering 1.2 million jobs, which is 12 % of EU manufacturing employment. Labor productivity in chemicals is 77 % higher than the manufacturing average, and salaries are 50 % higher. It is also the largest investor in EU manufacturing. The chemical industry is spending 10 G€/a (billion euros per year) for research and innovation.

3.1. European Chemical Industry Council

The European Chemical Industry Council (Cefic) is the European association for the chemical industry. Cefic developed the Sustainable Development Vision in 2012. It was based on the Responsible Care program—a global, voluntary initiative developed autonomously by the chemical industry. It was initiated by the Canadian Chemical Producers' Association—CCPA in 1985, and it is now adopted by almost 90 % of the global chemical industry. It aimed to improve health, safety, and environmental performance. Cefic's Sustainable Development program started in 2016; it aims at the transition toward a safe, resource-efficient, circular, and low-carbon society. It is organized around the four sustainability focus areas of the Cefic Charter: Create Low-Carbon Economy, Conserve Resource Efficiency, Connect Circular Economy, and Care for People and Planet [103]:

- Enabling the transition to a low carbon economy by:
  - Promoting innovation and stimulation of breakthrough technologies development in energy-efficient chemicals processes,
  - Offering market solutions consistent with low-carbon requirements,
  - Fostering the development and use of sustainable and renewable raw materials,
  - Fostering the use of sustainable and renewable energy and raw materials with a focus on cost and accessibility,
  - Innovating for chemical energy storage, and
  - Developing fuels and building blocks built on CO_2;
- Driving resource efficiency across global value chains and their operations by:
  - Designing sustainable solutions needing fewer resources over the entire life cycle and allowing easy reuse and recycling,
  - Maximizing material recovery and reuse,
- Promoting the adoption of circular economy principles to prevent waste, achieve low-carbon economy, and enhance resource efficiency;
- Preventing harm to humans and the environment throughout the entire life cycle by:
  - Mitigating risks, including assessment of substitutes,
  - Promoting the uptake of safe substances, materials, and solutions,
  - Minimizing negative environmental impacts on biodiversity and ecosystems,
  - Facilitating reuse, recycling, and recovery with steady information flows on products.

In the period 1991–2017, chemical production rose by 84 % while energy consumption was reduced by 16 % and energy intensity was reduced by 54 % (–40 % in the whole industry) [102]. Fuel and energy consumption was reduced by 24 % in the same period. In the period 1990–2017, greenhouse gas (GHG) emissions have been reduced by 58 % or 190 Mt/a, from 330 Mt/a down to 160 Mt/a of CO_2 equivalent. GHG emissions per energy consumption have been reduced by 48 %, and GHG intensity per production was reduced by 76 %. In the period 2007–2017, acidifying emission intensity fell by 40 %, nitrogen emission intensity fell by 48 %, and non-methane volatile organic compounds intensity fell by 48 %. These results are typical cases of decoupling economic activity from resource and environmental impacts.

Cefic supported the Green Deal and Europe's ambition to become climate neutral by 2050. In May 2020, the eight-point vision for Europe in 2050 was adopted:

1. "The world has become more prosperous and more complex, with a volatile geopolitical environment that brings more economic and political integration within most regions, but more fragmentation between them.
2. Europe has developed its own different but competitive place in the global economy.
3. The European economy has gone circular, recycling all sorts of molecules into new raw materials. The issue of plastic waste in the environment has been tackled.
4. Climate change continues to transform our planet. European society is close to achieving net-zero greenhouse gas emissions while keeping all Europeans citizens and regions on board.
5. Europeans have set the protection of human health and the environment at the center of an uncompromising political agenda.
6. European industry has become more integrated and collaborative in an EU-wide network of power, fuels, steel, chemicals, and waste recycling sectors.
7. Digitalization has completely changed the way people work, communicate, innovate, produce, and consume and brought unprecedented transparency to value chains.
8. The United Nations SDGs are at the core of European business models and have opened business opportunities as market shares increase for those who provide solutions to these challenges."

Cefic has welcomed the European Commission proposal for the European Climate Law, turning the climate neutrality objective into legislation and aiming to achieve progress on the global adaptation goal. However, besides "what" the EU aims to achieve, the "how" is also important, as it will allow the EU to turn this ambition into reality. Cefic puts forward several proposals aiming to clarify, complement, or adjust certain provisions by ensuring:

- A sound and detailed definition of climate-neutrality providing a signal for long-term investments;
- A level-playing field for industry across the EU through union-wide emission reduction mechanisms (i.e., the EU Emissions Trading System, ETS);
- That all sectors of the economy contribute to the climate-neutrality objective through fair burden-sharing;
- Progress on the enabling framework for the transformation of the EU economy, in line with the trajectory for achieving climate-neutrality.

3.2. Chemicals Strategy for Sustainability

Cefic calls for a sustainability strategy that recognizes the essential role of chemicals to deliver climate ambitions and integrates multiple facets of chemicals management including safety, circularity, resource efficiency, environmental footprint, science, and innovation. The following should be the key components of the strategy:

1. Consolidating and promoting the solid foundation Europe has already built, primarily REACH regulation (Registration, Evaluation, Authorization, and Restriction of Chemicals) by its improvement, better implementation, and enforcement;

2. Adopting a proportionate and robust approach for managing to emerge, scientifically complex issues;
3. Enabling the development of truly sustainable and competitive European solutions to deliver the Green Deal.

Cefic had welcomed the EU approach to adopt the new Industrial Strategy, basing it on the European industrial ecosystems; actors agreed that the Recovery Plan should be organized around these ecosystems.

SusChem is the European Technology Platform for Sustainable Chemistry. It is a forum that brings together industry, academia, policymakers, and the civil society. An important part of SusChem is a network of national platforms (NTPs). "SusChem's mission is to initiate and inspire European chemical and biochemical innovation to respond effectively to societal challenges by providing sustainable solutions". SusChem recognizes "three overarching and interconnected challenge areas [104]:

1. Circular economy and resource efficiency—transforming Europe into a more Circular Economy. (a) Materials design for durability and/or recyclability, (b) Safe by design for chemicals and materials (accounting for circularity, (c) Advanced processes for alternative carbon feedstock valorization (waste, biomass, CO/CO_2), (d) Resource efficiency optimization of processes, (e) Advanced materials and processes for sustainable water management, (f) Advanced materials and processes for the recovery and reuse of critical raw materials and/or their sustainable replacement, (g) Industrial symbiosis, (h) Alternative business models, (i) Digital technologies to increase value chain collaboration, (j) informing the consumer and businesses on reuse and recyclability;
2. Low-carbon economy—mitigating climate change with Europe becoming carbon neutral: (a) Advanced materials for the sustainable production of renewable electricity, (b) Advanced materials and technologies for renewable energy storage, (c) Advanced materials for energy efficiency in transport and buildings, (d) Electrification of chemical processes and use of renewable energy sources, (e) Increased energy efficiency of process technologies, enabled by digital technologies, (f) Energy-efficient water treatment, (g) Industrial symbiosis via the better valorization of energy streams, (h) Alternative business models;
3. Protecting environmental and human health—safe by design for materials and chemicals (functionality approach, methodologies, data, and tools): (a) Improve the safety of operations through process design, control, and optimization, (b) Zero liquid discharge processes, (c) Zero waste discharge processes, (d) Technologies for reducing GHGs emissions, (e) Technologies for reducing industrial emissions, (f) Sustainable sourcing of raw materials, (g) Increasing transparency of products within value chains through digital technologies, (h) Alternative food technologies, (i) Novel therapeutics and personalized medicine, (j) Sustainable agriculture, forestry, and soil health-related technologies, (k) Biocompatible materials for health applications."

The new SusChem's Strategic Innovation and Research Agenda, SIRA, has five chapters:

1. "Introduction with an overview where to find the challenge areas;
2. Advanced materials: composites and cellular materials (lightweight, insulation properties), 3D printable materials, bio-based chemicals and materials, additives, biocompatible and smart materials, materials for electronics, membranes, materials for energy storage (batteries), coating materials and aerogels;
3. Advanced processes (for energy transition and circular economy): new reactor design concepts and equipment, modular production, separation process technologies, new reactor and process design utilizing non-conventional energy forms (plasma, ultrasound, microwave), electrochemical, electrocatalytic, and photo-electrocatalytic processes, power-to-heat (heat pumps, electrical heating technologies), hydrogen production with low-carbon footprint, power-to-chemicals (syngas, methanol, fuel, methane, ammonia), catalysis, industrial biotechnology, waste valorization, advanced water management;

4. Enabling digital technologies: laboratory 4.0 (digital R&D), process analytical technologies (PAT), cognitive plants (real-time process simulation, monitoring, control and optimization, advanced (big) data analytics and artificial intelligence, predictive maintenance, digital support of operators and human–process interfaces, data sharing platforms and data security, coordination and management of connected processes at different levels, and distributed-ledger technologies.
5. Horizontal topics: sustainability assessment innovation, safe by design approach for chemicals and materials, building on education and skills capacity in Europe."

3.3. Process Industry

SPIRE (Sustainable Process Industry through Resource and Energy Efficiency) is the "European contractual public–private partnership (cPPP) involving the cement, ceramics, chemicals, engineering, minerals, non-ferrous metals, steel, and water sectors under the Horizon 2020 program. It has been successfully developing breakthrough and key enabling technologies and sharing best practices along all stages of existing value chains to enable a competitive, energy and resource-efficient process industry in Europe. SPIRE's new Vision 2050: "Towards the next generation of European Process Industries—Enhancing our cross-sectoral approach in research and innovation" foresees an integrated and digital European Process Industry, delivering new technologies and business models that address climate change and enable a fully circular society in Europe with enhanced competitiveness and impact for jobs and growth" [105]. They are contributing 6.3 million jobs in the EU. The SPIRE community has initiated 77 innovative projects with a total estimated private investment of 3 G€ (billion euros) in the last five years. Their turnover increased by an estimated 25 %—double the EU average.

SPIRE's Vision is that "the future of Europe lies in a strongly enhanced cooperation across industries—including SMEs—and across borders to become physically and digitally interconnected. Innovative "industrial ecology" business models will be developed to foster the redesign of the European industrial network. Four "technology drivers" will help the Process Industries achieve their SPIRE ambitions." Two transversal topics—industrial symbiosis and digitalization—will support and accelerate the transformations:

1. "Electrification of industrial processes as a pathway towards carbon neutrality: adaptation of industrial processes to the switch towards renewable electricity (e.g., electrochemistry, electric furnaces or kilns, plasma, or microwave technologies).
2. Energy mix and use of hydrogen as an energy carrier and feedstock: renewable electricity, low-carbon fuels, bio-based fuels, waste-derived fuels.
3. Capture and use of CO_2 from industrial exhaust gases (capture, collection, intermediate storage, pre-treatment, feeding and processing technologies, intelligent carbon management)."
4. Resource efficiency and flexibility; full re-use, recycling or recovery of waste as alternative resources: collection, sorting, transportation, pre-treatment and feeding technologies; all possible resource streams to be considered and explored (notably plastic waste, metallurgical slags, non-ferrous metals, construction and demolition waste, etc.); zero water discharge, maximal recovery of sensible heat from wastewater, the substitution of chemical solvents by water (e.g., in bio-based processes); full traceability of value chains as a crucial instrument to deploy circular business models and customers' growing demand for product-related information.
5. Industrial symbiosis technologies including industrial–urban symbiosis models.
6. Digitalization of process industries has a tremendous potential to dramatically accelerate change in resource management, process control, and in the design and the deployment of disruptive new business models.

The research and innovation efforts of Process Industries under the SPIRE 2050 Vision ultimately want to enhance and—wherever possible—enlarge the underlying value to society generated by their businesses while (a) achieving overall carbon neutrality, (b) moving toward zero-waste-to-landfill, and (c) enhancing the global competitiveness of their sectors.

4. Conclusions

The above results show that the most urgent future development areas of the process industry are climate change with GHGs emissions and ecosystems (the terrestrial one is affected by drought, wildfires, floods, glacier melting or species extinction; marine through temperature rise, ocean acidification, and sea-level rise), energy with renewable sources and efficiency, (critical) raw materials and other resources, water resources and recycling, zero waste and circular economy and resource efficiency, supply chain integration, process design and optimization, process integration and intensification, industrial ecology and life cycle thinking, industrial–urban symbiosis, product design for circularity, digitalization, sustainable transport, green jobs, health and safety, hazardous materials and waste, customer satisfaction, education, and lifelong learning.

The chemical and process industry associations (Cefic, SusChem) and their projects (SPIRE, SIRA) have added great value; therefore, they should be practiced in other continents, too. Companies and professional associations must respect international agreements and conventions (SDGs, Paris Agreement, EGD), declarations, and recommendations.

There is no doubt that existing and innovative future technologies for the efficient management of GHGs, water, energy, and raw materials will play a crucial role in transforming current chemical and bio-chemical processes into more sustainable ones. Due to their high cost, some of these technologies would need to be co-funded by governments, while others could be implemented as long-term investments at the corporate level. For example, Norway has recently announced to fund a first large-scale carbon capture and storage project "Longship" [106] (1.5 billion €). The cement and waste-to-energy plants involved in the project plan to reduce their CO_2 emissions by 50 % by capturing and storing CO_2 in an underwater reservoir in the North Sea. On the other hand, a Dutch brewery [107] has recently implemented an innovative green fuel alternative that comes in the form of metal powders [108]. Iron powder is considered as a high-density energy storage medium. It burns at high temperatures to form iron oxide, which can be reduced back to iron by electrolysis using renewable energy sources (e.g., photovoltaics) in a carbon-free cycle.

The critical step toward more sustainable processes, regardless of the novel technologies available, is the necessary shift in mindset from chasing short-term financial gains to pursuing long-term, sustainable financial, environmental, and social benefits. This step is required not only at the governance and corporate level, but also at the level of each individual.

Based on past experience, it is safe to say that advanced computational PSE tools will play an important role in the development of future chemical and biochemical processes. Today, process analysis, simulation, and synthesis/optimization tools are generally used in a sandbox mode—i.e., either in isolation from each other or in a sequential/iterative procedure. To identify truly innovative, mitigating, and perhaps even restorative solutions, these tools would need to be linked into a system that simultaneously enables detailed multi-scale modeling [109], process intensification (i.e., reduction of energy and resource requirements, waste production and equipment size, out-of-the-box process solutions/schemes) [110], and LCSA analysis.

Projects to develop sustainable processes are very demanding, both in terms of knowledge and the financial investment required to implement them. Circular economy projects often do not provide much added value. This is particularly problematic when it is cheaper to manufacture products from virgin materials than to process waste materials into secondary raw materials. The development and implementation of sustainable processes are highly interdisciplinary and involve laboratory research, pilot plant trials, process set-up, and commissioning. This is followed by the manufacturing and marketing of products and efficient waste management, which includes the reuse, recycling, and processing of waste into value-added products, fuels, secondary raw materials, or energy recovery. There is no doubt that engineers are already developing efficient computer-aided tools for developing sustainable technologies, including key enabling technologies and sustainable processes, supply chains, and networks that promote greater efficiency, waste reduction, closed loops, and eco-design. However, this will certainly not be enough to transform society from a linear to a circular economy. We believe there is still a long way to go, as changes will be needed in many areas of society, i.e., at the level of business, education, finance, politics, legislation, and society as a whole.

At the enterprise level, efforts should focus on building and optimizing value chains in which stakeholders are linked through raw material extraction, product manufacturing, transportation, collection, sorting, and processing of waste into secondary raw materials, functional materials, and energy. The aim should be to promote such industrial projects that balance economic efficiency, environmental impact, and social wellbeing. It is necessary to promote the growth of bio-based products and to seek market niches for such products.

In the field of education, young people must be encouraged to study science, technology, engineering, and mathematics (STEM), as these areas are crucial for the development of sustainable technologies and processes and the circular economy. Curricula need to be strengthened with attractive contents for young people and practical examples of green chemistry, cleaner production, eco-design, recycling, key enabling technologies, etc.

Experts in the social sciences such as psychology and sociology must also be involved in the development of sustainable processes and products, as people need to change many deep-rooted habits and understand the impact of these changes on society and the environment in order to accept them as their own. The transition from a linear to a circular society must include the reduction of inequalities in society, more equality, justice, solidarity, participation and inclusion of citizens. Art must also be involved, because products made from secondary raw materials, for example, must also be aesthetically designed if people are to accept them.

Developing sustainable technologies and implementing sustainable projects can require large financial investments, so the role of financial institutions and policymakers is also important. They must create the conditions for funding to be available for environmentally beneficial projects in the field of renewable energy, secondary raw materials, functional materials, key enabling technologies, etc. It is necessary to increase investment in education, research, innovation, and development.

The transition to a circular economy and a sustainable society can be promoted to some extent by political agreements and legal norms that impose restrictions on countries and companies in terms of emissions, proportions of recycled materials, the use of renewable resources and secondary raw materials, etc. However, in the long term, changes in existing political and wider social systems are needed, moving toward greater participation and balance, with the long-term sustainable progress of society and the protection of the environment taking precedence over the partial interests of individuals.

Author Contributions: Conceptualization, P.G.; methodology, P.G., Z.N.P. and M.B., investigation, P.G., Z.N.P. and M.B.; writing—original draft preparation, P.G., Z.N.P. and M.B.; writing—review and editing, P.G., Z.N.P. and M.B.; supervision, P.G. All authors have read and agreed to the published version of the manuscript.

Funding: This research was co-funded by the Slovenian Research Agency (Research Program P2-0032 and Project J7-1816).

Institutional Review Board Statement: Not applicable.

Informed Consent Statement: Not applicable.

Data Availability Statement: Not applicable.

Conflicts of Interest: The authors declare no conflict of interest.

References

1. Fisk, P. Megatrends 2020–2030 . . . What They Mean for You and Your Business, and How to Seize the New Opportunities for Innovation and Growth. Available online: https://www.thegeniusworks.com/2019/12/mega-trends-with-mega-impacts-embracing-the-forces-of-change-to-seize-the-best-future-opportunities/ (accessed on 8 November 2020).

2.   Ernst & Young. Are You Reframing Your Future or Is the Future Reframing You? EYQ 3rd Ed. 2020. Available online: https://assets.ey.com/content/dam/ey-sites/ey-com/en_gl/topics/megatrends/ey-megatrends-2020-report.pdf (accessed on 8 November 2020).

3.   European Environment Agency, EEA. European Environment—State and Outlook 2015: Assessment of Global Megatrends. Available online: File:///C:/Users/glavic/AppData/Local/Temp/Assessment%20of%20global%20megatrends-1.pdf (accessed on 8 November 2020).

4.   PricewaterhouseCoopers. PwC: 5 MegaTrends Affecting Your Business in 2019, Parts 1 and 2. 2017. Available online: https://brandminds.live/pwc-5-megatrends-affecting-your-business-in-2019-2-of-2/ (accessed on 8 November 2020).

5.   Moore, J.; Henbest, S. New Energy Outlook 2020, Executive Summary. BloombergNEF (New Energy Finance). Available online: https://assets.bbhub.io/professional/sites/24/928908_NEO2020-Executive-Summary.pdf (accessed on 9 November 2020).

6.   New Energy Oultook 2017, Bloomberg New Energy Finance. Available online: https://data.bloomberglp.com/bnef/sites/14/2017/06/NEO-2017_CSIS_2017-06-20.pdf (accessed on 9 November 2020).

7.   European Commission. European Green Deal. Available online: https://ec.europa.eu/clima/policies/eu-climate-action_en (accessed on 6 September 2020).

8.   Ballweg, M.; Bukow, C.; Delasalle, F.; Dixson-Declève, S.; Kloss, B.; Lewren, I.; Metzner, J.; Okatz, J.; Petit, M.; Pollich, K.; et al. A System Change Compass–Implementing the European Green Deal in a Time of Recovery. 2020. Available online: https://clubofrome.org/wp-content/uploads/2020/10/System-Change-Compass-Full-report-FINAL.pdf (accessed on 7 November 2020).

9.   TPE Design Engineering. Available online: https://www.tpede.co.za/old-copies/process-design (accessed on 15 October 2020).

10.   Ellen MacArthur Foundation, Towards the Circular Economy, Volume 1–2. 2013. Available online: https://www.ellenmacarthurfoundation.org/assets/downloads/publications/Ellen-MacArthur-Foundation-Towards-the-Circular-Economy-vol.1.pdf (accessed on 15 October 2020).

11.   Montastruc, L.; Belletante, S.; Pagot, A.; Negny, S.; Raynal, L. From conceptual design to process design optimization: A review on flowsheet synthesis. *Oil Gas Sci. Technol. Rev.* **2019**, *74*, 80. [CrossRef]

12.   Computer Aided Process Engineering. Book Series. Available online: https://www.sciencedirect.com/bookseries/computer-aided-chemical-engineering/vol/46/suppl/C (accessed on 15 October 2020).

13.   European Commission, DG Environment. Available online: https://ec.europa.eu/environment/sustainable-development/index_en.htm (accessed on 15 October 2020).

14.   World Business Council for Sustainable Development (WBCSD). Chemical Sector SDG Roadmap. 2018. Available online: http://docs.wbcsd.org/2018/07/Chemical_Sector_SDG_Roadmap.pdf (accessed on 15 October 2020).

15.   The European Chemical Industry Council (Cefic). A Solution Provider for Sustainability. 2019. Available online: https://cefic.org/our-industry/a-solution-provider-for-sustainability/ (accessed on 15 October 2020).

16.   European Center of Sustainable Development (ECSD). Available online: https://ecsdev.org/ (accessed on 16 October 2020).

17.   Conference on Engineering Sustainable Development 2019 co-Hosted by AIChE-APRU. Available online: https://apru.org/event/2019-international-conference-on-technical-and-engineering-challenges-of-addressing-sustainable-development/ (accessed on 16 October 2020).

18.   Anastas, P.T.; Zimmerman, J.B. Design through the Twelve Principles of Green Engineering. *Environ. Sci. Technol.* **2003**, *37*, 94A–101A. [CrossRef]

19.   Anastas, P.T.; Warner, J. *Green Chemistry, Theory and Practice*; Oxford University Press: London, UK, 1998.

20.   Gagnon, B.; Leduc, R.; Savard, L. Sustainable development in engineering: A review of principles and definition of a conceptual framework. *Environ. Eng. Sci.* **2009**, *26*, 1459–1472. [CrossRef]

21.   *The Royal Society of Engineering*; Guiding Principles; Engineering for Sustainable Development: London, UK, 2005.

22.   American Chemical Society, Green Chemistry Institute. Design Principles for Sustainable and Green Chemistry and Engineering: Washington. Available online: https://www.acs.org/content/acs/en/greenchemistry/principles.html (accessed on 16 October 2020).

23.   Crul, M.R.M.; Diehl, J.C. *Design for sustainability: A Step by Step Approach*; UNEP: Paris, France, 2009; Available online: https://wedocs.unep.org/handle/20.500.11822/8742 (accessed on 16 October 2020).

24.   Ceschin, F. Evolution of design for sustainability: From product design to design for system innovations and transitions. *Des. Stud.* **2016**, *47*, 118–163. [CrossRef]

25.   Acaroglu, L. Quick Guide to Sustainable Design Strategies. Available online: https://medium.com/disruptive-design/quick-guide-to-sustainable-design-strategies-641765a86fb8 (accessed on 17 October 2020).

26.   Biegler, L.T.; Grossman, I.E.; Westerberg, A.W. *Systematic Methods of Chemical Process Design*; Prentice Hall: Upper Saddle River, NJ, USA, 1999.

27.   Koolen, J.L.A. *Design of Simple and Robust Process Plants*; Wiley-VCH: Weinheim, Germany, 2001.

28.   Smith, R. *Chemical Process: Design and Integration*, 2nd ed.; Wiley: Hoboken, NJ, USA, 2014.

29.   Bernecker, G. *Plannung und Bau Verfahrenstechnischer Anlagen*; Springer: Berlin, Germany, 2001.

30.   Allen, D.; Shonnard, D.R. *Sustainable Engineering: Concepts, Design and Case Studies*; Prentice Hall: Upper Saddle River, NJ, USA, 2012.

31.   Perl, J. Sustainability Engineering: A Design Guide for the Chemical Process Industry; Springer: Berlin/Heidelberg, Germany. Available online: https://www.springer.com/gp/book/9783319324937 (accessed on 9 November 2020).

32. Luu, L.Q.; Halog, A. Life Cycle Sustainability Assessment; A Holistic Evaluation f Social, Economic, and Environmental Impacts. In *Sustainability in the Design, Synthesis and Analysis of Chemical Engineering Processes*; Ruiz-Mercado, G., Cabezas, H., Eds.; Elsevier Science Direct: Amsterdam, The Netherlands, 2016.
33. Scopus, Document Search. Available online: https://www.scopus.com/search/form.uri?display=basic (accessed on 31 October 2020).
34. Web of Science, Core Collection Results. Available online: http://apps.webofknowledge.com/Search.do?product=WOS&SID=F4 AauXxtQuY92K5WeIQ&search_mode=GeneralSearch&prID=9b5b11c6-5dfb-4e80-b5eb-3143c6c4f66b (accessed on 30 September 2020).
35. Nikolopoulou, A.; Ierapetritou, M.A. Optimal design of sustainable chemical processes and supply chains: A review. *Comput. Chem. Eng.* **2012**, *44*, 94–103. [CrossRef]
36. Kumar, A.; Sah, B.; Singh, A.R.; Deng, Y.; He, X.; Kumar, P.; Bansal, R.C. A review of multi criteria decision making towards sustainable renewable energy development. *Renew. Sustain. Energy Rev.* **2017**, *69*, 596–609. [CrossRef]
37. Corominas, L.; Foley, J.; Guest, J.S.; Hospido, A.; Larsen, H.F.; Morera, S.; Shaw, A. Life cycle assessment applied to wastewater treatment: State of the art. *Water Res.* **2013**, *47*, 5480–5492. [CrossRef]
38. Li, S.; Mirlekar, G.; Ruiz-Mercado, G.J.; Lima, F.V. Development of Chemical Process Design and Control for Sustainability. *Processes* **2016**, *4*, 23. [CrossRef]
39. Bare, J. TRACI 2.0: The Tool for the Reduction and Assessment of Chemical and Other Environmental Impacts 2.0. *Clean Technol. Environ.* **2011**, *13*, 687–696. [CrossRef]
40. Klewitz, J.; Hansen, E.G. Sustainability-Oriented Innovation of SMEs: A Systematic Review. *J. Clean. Prod.* **2014**, *65*, 57–75. [CrossRef]
41. Siddique, M.N.I.; Wahid, Z.A. Achievements and perspectives of anaerobic co-digestion: A review. *J. Clean. Prod.* **2018**, *194*, 359–371. [CrossRef]
42. Singh, S.V.; Ming, Z.; Fennell, P.S.; Shah, N.; Anthony, E.J. Progress in biofuel production from gasification. *Prog. Energy Combust.* **2017**, *61*, 189–248.
43. Babazadeh, R.; Razmi, J.; Pishvaee, M.S.; Rabbani, M. A sustainable second-generation biodiesel supply chain network design problem under risk. *Omega* **2017**, *66*, 258–277. [CrossRef]
44. Mata, T.M.; Martins, A.A.; Caetano, N.S. Microalgae for biodiesel production and other applications: A review. *Renew. Sustain. Energy Rev.* **2010**, *14*, 217–232. [CrossRef]
45. Santibañez-Aguilar, J.E.; González-Campos, J.B.; Ponce-Ortega, J.M.; Serna-González, M.; El-Halwagi, M.M. Optimal Planning of a Biomass Conversion System Considering Economic and Environmental Aspects. *Ind. Eng. Chem. Res.* **2011**, *50*, 8558–8570. [CrossRef]
46. Murray, A.; Skene, K.; Haynes, K. The Circular Economy: An Interdisciplinary Exploration of the Concept and Application in a Global Context. *J. Bus. Ethics* **2017**, *140*, 369–380. [CrossRef]
47. Walmsley, T.G.; Ong, B.H.Y.; Klemeš, J.J.; Tan, R.R.; Varbanov, P.S. Circular Integration of processes, industries, and economies. *Renew. Sustain. Energy Rev.* **2019**, *107*, 507–515. [CrossRef]
48. Knez, Ž.; Markočič, E.; Leitgeb, M.; Primožič, M.; Knez Hrnčič, M.; Škerget, M. Industrial applications of supercritical fluids: A review. *Energy* **2014**, *77*, 235–243. [CrossRef]
49. Roy, R. Sustainable product-service systems. *Futures* **2000**, *32*, 289–299. [CrossRef]
50. Galli, B.J. How to Effectively Use Economic Decision-Making Tools in Project Environments and Project Life Cycle. *IEEE Trans. Eng. Manag.* **2020**, *67*, 932–940. [CrossRef]
51. Khalid, M.; Aguilera, R.P.; Savkin, A.V.; Agelidis, V.G. On maximizing profit of wind-battery supported power station based on wind power and energy price forecasting. *Appl. Energy* **2018**, *211*, 764–773. [CrossRef]
52. Nicoletti, J.; Ning, C.; You, F. Optimizing return on investment in biomass conversion networks under uncertainty using data-driven adaptive robust optimization. In *Computer Aided Chemical Engineering, Proceedings of the 29th European Symposium on Computer Aided Process Engineering, Eindhoven, The Netherlands, 16–19 June 2019*; Kiss, A.A., Zondervan, E., Lakerveld, R., Özkan, L., Eds.; Elsevier: Amsterdam, The Netherland, 2019; pp. 67–72.
53. Ziyai, M.R.; Mehrpooya, M.; Aghbashlo, M.; Omid, M.; Alsagri, A.S.; Tabatabaei, M. Techno-economic comparison of three biodiesel production scenarios enhanced by glycerol supercritical water reforming process. *Int. J. Hydrog. Energy* **2019**, *44*, 17845–17862. [CrossRef]
54. Novak Pintarič, Z.; Kravanja, Z. Selection of the economic objective function for the optimization of process flow sheets. *Ind. Eng. Chem. Res.* **2006**, *45*, 4222–4232. [CrossRef]
55. Cisternas, L.A.; Lucay, F.; Gálvez, E.D. Effect of the objective function in the design of concentration plants. *Miner. Eng.* **2014**, *63*, 16–24. [CrossRef]
56. Kasaš, M.; Kravanja, Z.; Novak Pintarič, Z. Suitable modeling for process flow sheet optimization using the correct economic criterion. *Ind. Eng. Chem. Res.* **2011**, *50*, 3356–3370. [CrossRef]
57. Kasaš, M.; Kravanja, Z.; Novak Pintarič, Z. Achieving Profitably, Operationally, and Environmentally Compromise Flow-Sheet Designs by a Single-Criterion Optimization. *AIChE J.* **2012**, *58*, 2131–2141. [CrossRef]
58. Argoti, A.; Orjuela, A.; Narváez, P.C. Challenges and opportunities in assessing sustainability during chemical process design. *Curr. Opin. Chem. Eng.* **2019**, *26*, 96–103. [CrossRef]
59. Lee, Y.S.; Graham, E.J.; Galindo, A.; Jackson, G.; Adjiman, C.S. A comparative study of multi-objective optimization methodologies for molecular and process design. *Comput. Chem. Eng.* **2020**, *136*, 106802. [CrossRef]

60. Wang, Z.; Parhi, S.S.; Rangiah, G.P.; Jana, A.K. Analysis of Weighting and Selection Methods for Pareto-Optimal Solutions of Multiobjective Optimization in Chemical Engineering Applications. *Ind. Eng. Chem. Res.* **2020**, *59*, 14850–14867. [CrossRef]

61. TU Delft. Data on Eco-Costs 2017. Available online: www.ecocostsvalue.com/EVR/model/theory/subject/5-data.html (accessed on 26 October 2020).

62. Novak Pintarič, Z.; Varbanov, P.S.; Klemeš, J.J.; Kravanja, Z. Multi-Objective Multi-Period Synthesis of Energy Efficient Processes under Variable Environmental Taxes. *Energy* **2019**, *189*, 116182. [CrossRef]

63. Fan, Y.V.; Chin, H.H.; Klemeš, J.J.; Varbanov, P.S.; Liu, X. Optimisation and process design tools for cleaner production. *J. Clean. Prod.* **2020**, *247*, 119181. [CrossRef]

64. López-Delgado, A.; Robla, J.I.; Padilla, I.; López-Andrés, S.; Romero, M. Zero-waste process for the transformation of a hazardous aluminum waste into a raw material to obtain zeolites. *J. Clean. Prod.* **2020**, *255*, 120178. [CrossRef]

65. Salkuyeh, Y.K.; Adams, T.A., II. Integrated petroleum coke and natural gas polygeneration process with zero carbon emissions. *Energy* **2015**, *91*, 479–490. [CrossRef]

66. Karka, P.; Papadokonstantakis, S.; Kokossis, A. Environmental impact assessment of biomass process chains at early design stages using decision trees. *Int. J. Life Cycle Ass.* **2019**, *24*, 1675–1700. [CrossRef]

67. Mangili, P.V.; Prata, D.M. Preliminary design of sustainable industrial process alternatives based on eco-efficiency approaches: The maleic anhydride case study. *Chem. Eng. Sci.* **2020**, *212*, 115313. [CrossRef]

68. Geissdoerfer, M.; Savaget, P.; Bocken, N.M.P.; Hultink, E.J. The Circular Economy-A new sustainability paradigm? *J. Clean. Prod.* **2017**, *143*, 757–768. [CrossRef]

69. Yáñez, M.; Ortiz, A.; Brunaud, B.; Grossmann, I.E.; Ortiz, I. The use of optimization tools for the Hydrogen Circular Economy. In *Computer Aided Chemical Engineering, Proceedings of the 29th European Symposium on Computer Aided Process Engineering, Eindhoven, The Netherlands, 16–19 June 2019*; Kiss, A.A., Zondervan, E., Lakerveld, R., Özkan, L., Eds.; Elsevier: Amsterdam, The Netherland, 2019; pp. 1777–1782.

70. Klemeš, J.J.; Fan, Y.V.; Jiang, P. Plastics: Friends or foes? The circularity and plastic waste footprint. *Energy Sources Part A* **2020**, 1–17. [CrossRef]

71. Avraamidou, S.; Baratsas, S.G.; Tian, Y.; Pistikopoulos, E.N. Circular Economy-A challenge and an opportunity for Process Systems Engineering. *Comput. Chem. Eng.* **2020**, *133*, 106629. [CrossRef]

72. Genc, A.; Zeydan, O.; Sarac, S. Cost analysis of plastic solid waste recycling in an urban district in Turkey. *Waste Manag. Res.* **2019**, *37*, 906–913. [CrossRef]

73. Bubicz, M.E.; Barbosa-Póvoa, A.P.F.D.; Carvalho, A. Incorporating social aspects in sustainable supply chains: Trends and future directions. *J. Clean. Prod.* **2019**, *237*, 117500. [CrossRef]

74. Zore, Ž.; Čuček, L.; Kravanja, Z. Synthesis of sustainable production systems using an upgraded concept of sustainability profit and circularity. *J. Clean. Prod.* **2018**, *201*, 1138–1154. [CrossRef]

75. Zore, Ž.; Čuček, L.; Širovnik, D.; Novak Pintarič, Z.; Kravanja, Z. Maximizing the sustainability net present value of renewable energy supply networks. *Chem. Eng. Res. Des.* **2018**, *131*, 245–265. [CrossRef]

76. ISO 26000 Standard. *Guidance on Social Responsibility*; International Organization for Standardization: Geneva, Switzerland, 2010.

77. ISO 45 Standardards. *Rgye Circular Economy s Extraction to Processing, Use and End of Life Tretment Recycling its in 001, Occupational Health and Safety*; Cember 12, 2020; International Organization for Standardization: Geneva, Switzerland, 2018.

78. ISO 14040 and 14044. *Environmental Management–Life Cycle Assessment*; International Organization for Standardization: Geneva, Switzerland, 2006.

79. Global Reporting Initiative, GRI Standards. Available online: https://www.globalreporting.org/ (accessed on 12 December 2020).

80. Umeda, T.; Itoh, J.; Shikoro, K. Heat Exchanger Synthesis. *Chem. Eng. Prog.* **1978**, *74*, 70–76.

81. Linnhoff, B.; Hindmarsh, E. The pinch design method for heat exchanger networks. *Chem. Eng. Sci.* **1983**, *38*, 745–763. [CrossRef]

82. Austin, N.D.; Sahinidis, N.V.; Trahan, D.W. Computer-aided molecular design: An introduction and review of tools, applications, and solution techniques. *Chem. Eng. Res. Des.* **2016**, *116*, 2–26. [CrossRef]

83. Strniša, F.; Tatiparthi, V.S.; Djinović, P.; Pintar, A.; Plazl, I. Ni-containing CeO_2 rods for dry reforming of methane: Activity tests and a multiscale lattice Boltzmann model analysis in two model Geometries. *Chem. Eng. J.* **2020**, 127498. [CrossRef]

84. Rossetti, I. Reactor Design, Modelling and Process Intensification for Ammonia Synthesis. In *Sustainable Ammonia Production; Inamuddin*; Boddula, R., Asiri, A.M., Eds.; Green Energy and Technology; Springer: Berlin/Heidelberg, Germany, 2020; pp. 17–48.

85. Dimian, A.C.; Bildea, C.S.; Kiss, A.A. Methanol. In *Applications in Design and Simulation of Sustainable Chemical Processes*; Dimian, A.C., Bildea, C.S., Kiss, A.A., Eds.; Elsevier: Amsterdam, The Netherlands, 2019; pp. 101–145.

86. Tula, A.K.; Eden, M.R.; Gani, R. Process synthesis, design and analysis using a process-group contribution method. *Comput. Chem. Eng.* **2015**, *81*, 245–259. [CrossRef]

87. Zhang, X.; Song, Z.; Zhou, T. Rigorous design of reaction-separation processes using disjunctive programming models. *Comput. Chem. Eng.* **2018**, *111*, 16–26. [CrossRef]

88. Potrč, S.; Čuček, L.; Zore, Ž.; Kravanja, Z. Synthesis of Large-Scale Supply Networks for Complete Long-term Transition from Fossil to Renewable-based Production of Energy and Bioproducts. *Chem. Eng. Trans.* **2020**, *81*, 1039–1044.

89. Xu, D.; Lv, L.; Ren, J.; Shen, W.; Wei, S.; Dong, L. Life Cycle Sustainability Assessment of Chemical Processes: A Vector-Based Three-Dimensional Algorithm Coupled with AHP. *Ind. Eng. Chem. Res.* **2017**, *56*, 11216–11227. [CrossRef]

90. Zore, Ž.; Čuček, L.; Kravanja, Z. Syntheses of sustainable supply networks with a new composite criterion–Sustainability profit. *Comput. Chem. Eng.* **2017**, *102*, 139–155. [CrossRef]
91. ANSYS. Ansys: Engineering Simulation & 3D Design Software. Available online: www.ansys.com (accessed on 15 October 2020).
92. OpenFOAM. The Open Source CFD Toolbox. Available online: www.openfoam.com (accessed on 15 October 2020).
93. AspenPlus. The Leading Process Simulation Software in the Chemical Industry. Available online: www.aspentech.com (accessed on 15 October 2020).
94. DWSIM. DWSIM-Chemical Process Simulator. Available online: dwsim.inforside.com.br (accessed on 15 October 2020).
95. GAMS. The General Algebraic Modelling System. Available online: www.gams.com (accessed on 15 October 2020).
96. Pyomo. Available online: www.pyomo.org (accessed on 15 October 2020).
97. GABY. Gaby-Software. Available online: www.gabi-software.com (accessed on 15 October 2020).
98. OpenLCA. The Life Cycle and Sustainability Software. Available online: www.openlca.org (accessed on 15 October 2020).
99. IDEAS PSE Framework. Available online: https://idaes.org/ (accessed on 15 October 2020).
100. MIPSYN-GLOBAL. Available online: http://mipsyn-global.fkkt.um.si/ (accessed on 15 October 2020).
101. Kravanja, Z. Challenges in sustainable integrated process synthesis and the capabilities of an MINLP process synthesizer MipSyn. *Comput. Chem. Eng.* **2010**, *34*, 1831–1848. [CrossRef]
102. Cefic, Facts and Figures. 2020. Available online: https://cefic.org/app/uploads/2019/01/The-European-Chemical-Industry-Facts-And-Figures-2020.pdf (accessed on 4 October 2020).
103. Cefic sustainability Charter. Available online: https://cefic.org/app/uploads/2019/01/Cefic-Sustainability-Charter-TeamingUp-For-A-SustainableEurope.pdf (accessed on 4 November 2020).
104. SusChem, European Technology Platform for Sustainable Chemistry. Strategic Innovation and Research Agenda, SIRA. Available online: https://cefic.org/app/uploads/2020/02/SusChem-SIRA-2020.pdf (accessed on 6 September 2020).
105. Sustainable Process Industry through Resource and Energy Efficiency, SPIRE 2050 Vision. 2018. Available online: https://cefic.org/app/uploads/2019/02/SPIRE-vision-2050.pdf (accessed on 6 November 2020).
106. Hjuske, A.K. The Government Launches 'Longship' for Carbon Capture and Storage in Norway. Available online: https://bit.ly/3ewvVUI (accessed on 18 October 2020).
107. Blain, L. World First: Dutch Brewery Burns Iron as a Clean, Recyclable Fuel. Available online: https://bit.ly/2JEZeJb (accessed on 5 November 2020).
108. Bergthorson, J.M. Recyclable metal fuels for clean and compact zero-carbon power. *Prog. Energy Combust.* **2018**, *68*, 169–196. [CrossRef]
109. Floudas, C.A.; Niziolek, A.M.; Onel, O.; Matthews, L.R. Multi-scale systems engineering for energy and the environment: Challenges and opportunities. *AICHE J.* **2016**, *62*, 602–623. [CrossRef]
110. Bielenberg, J.; Palou-Rivera, I. The RAPID Manufacturing Institute–Reenergizing US efforts in process intensification and modular chemical processing. *Chem. Eng. Process.* **2019**, *138*, 49–54. [CrossRef]

MDPI

St. Alban-Anlage 66

4052 Basel

Switzerland

www.mdpi.com

Processes Editorial Office

E-mail: processes@mdpi.com

www.mdpi.com/journal/processes